Books in this series are devoted exclusively to problems - challenging, difficult, but accessible problems. They are intended to help at all levels - in college, in graduate school, and in the profession. Arthur Engels "Problem-Solving Strategies" is good for elementary students and Richard Guys "Unsolved Problems in Number Theory" is the classical advanced prototype. The series also features a number of successful titles that prepare students for problem-solving competitions.

More information about this series at https://link.springer.com/bookseries/714

Problem Books in Mathematics

Series Editor

Peter Winkler
Department of Mathematics
Dartmouth College
Hanover, NH
USA

Răzvan Gelca • Ionuţ Onişor • Carlos Yuzo Shine

Geometric Transformations

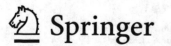 Springer

Răzvan Gelca
Mathematics and Statistics
Texas Tech University
Lubbock
TX, USA

Ionuţ Onişor
Colegiul National de Informatica Tudor
Vianu
Bucharest, Romania

Carlos Yuzo Shine
Colégio Etapa
São Paulo
São Paulo, Brazil

ISSN 0941-3502   ISSN 2197-8506   (electronic)
Problem Books in Mathematics
ISBN 978-3-030-97848-8   ISBN 978-3-030-89117-6   (eBook)
https://doi.org/10.1007/978-3-030-89117-6

Mathematics Subject Classification: 97G50, 97U40, 54H15, 51M04

This Springer imprint is published by the registered company Springer Nature Switzerland AG
The registered company address is: Gewerbestrasse 11, 6330 Cham, Switzerland

In memory of Rodica Gelca

Preface

There is a true geometry which is not [...] intended to be merely an illustrative form of more abstract investigation. Its problem is to grasp the full reality of the figures of space, and to interpret—and this is the mathematical side of the question—the relations holding for them as evident results of the axioms of space-perception.

This thought was penned by Felix Klein in his notes on the *Erlangen Program*. The *Erlangen Program* proposed a view of geometry that shifts the focus from geometric objects to groups of transformations that act upon them. The accent is placed on understanding the transformations and the properties of space they preserve. Our book does embrace this perspective, but is more of a hybrid between the old Euclidean geometry and Klein's ideas. We let transformations act on the entire configuration or just part of it, then exercise our intuition on the result, or we recognize a geometric transformation hidden inside the configuration itself.

We have selected mostly Olympiad problems, because in mathematical Olympiads the geometry of lines and circles is still alive. Elementary geometry is a valuable instrument for building space-perception and offers probably the best introduction to the concept of a group: a set of transformations that contains the compositions and inverses of all of its elements. And familiarity with groups of geometric transformations is nowadays a must, since they play a central role in physics, both classical and quantum, and with it in geometry. With this book, we want to shape the reader's mind into thinking about *geometry in motion* as opposed to the *static* view of Euclid's *Elements*.

As for the method of proof, we use both the synthetic and the analytic, where appropriate, though we put the accent on the first. To motivate this, we refer to another quote from the *Erlangen Program*:

The distinction between modern synthesis and modern analytical geometry must no longer be regarded as essential, inasmuch as both subject-matter and methods of reasoning have gradually taken a similar form in both. [...] Although the synthetic method has more to do with space-perception and thereby imparts a rare charm to its first simple developments, the realm of space-perception is nevertheless not closed to the analytic method, and the formulae of analytic geometry can be looked upon as a precise and perspicuous statement of geometrical relations.

It is the "rare charm" of the synthetic method that we aim to reveal with most of the problems; it is also unimaginable to practice geometry without a good spatial intuition, and that is why we favor the synthetic method. However, the coordinate-based approach is easier to generalize to other realms of geometry and to relate to other areas of mathematics. Thus, in many problems we have included analytic and synthetic solutions side by side, so that the reader can see how numbers and figures interact. And many a time we integrate the analytic and the synthetic in the same argument.

Besides building good geometric intuition, this text opens a window towards other parts of mathematics, giving thus a mild introduction to groups of transformations and illustrating how symmetry groups appear in combinatorics and number theory. There is a description of circular transformations; they are useful at elementary level in inversive geometry, and are an essential tool in contemporary research on hyperbolic geometry in two and three dimensions. It is important to point out that some groups of transformations present in the book, and their three-dimensional counterparts, are the simplest examples of Lie groups, and have proved essential in modeling classical and quantum physics.

There is a vast body of mathematics related to geometric transformations, too voluminous to be enclosed in the confines of a single book. We had to be selective, so we have decided to reduce our scope to just the Euclidean plane, and there, to those transformations that can be modeled with complex coordinates. All transformations appearing in this book can be placed within the framework of complex affine transformations and complex linear fractional transformations, and their conjugates. There is no discussion of two-dimensional real projective geometry (such as conics) or of two-dimensional real affine geometry.

To teach the tools and tricks of geometric transformations, we apply the following structure to each of the first three chapters ("Isometries," "Homotheties and Spiral Similarities," and "Inversions"). We begin with a discussion of theoretical results, followed by a few theoretical questions. Next, several applications of the methods are explained in detail, including some classical theorems in Euclidean geometry. This is followed by what you, the reader, await with excitement: problems to solve. The problems are listed in some increasing order of difficulty, but to keep the element of surprise and to stimulate ingenuity, there is no grouping based on common ideas. If the challenge is too big, hints for all problems can be found in the middle of the book. Additionally, to help with the learning, all problems have detailed solutions at the end of the book, often multiple solutions, some of which have been discovered by experienced problem solvers. Even if you are successful in solving a problem, and it would be good if you could explore and find more approaches, you should always read the solutions from the end of the book. Not only because they might teach you new tricks, but also because they contain commentaries about the method, and for problems whose authors and sources are known, these are mentioned there. The last chapter, "A Synthesis," shorter but more challenging than the others, puts all transformations from this book on common ground, and contains problems that can be tackled with diverse techniques.

Geometric Transformations was carefully crafted by three experienced mathematical Olympiad coaches on three continents. Răzvan Gelca has trained the United States International Mathematical Olympiad Team for many years, and has also served as deputy leader of this team, as well as leader of the team that represented the United States at the Romanian Master of Mathematics. He has helped organize the USA Mathematical Olympiad. Ionuţ Onişor has coached Romanian students for the International Mathematical Olympiad and has taken Romanian teams to international mathematics competitions. Carlos Yuzo Shine has been the head coach of the Brazilian International Mathematical Olympiad Team as well as the academic chair of the Brazilian Mathematical Olympiad, and has taken several times the Brazilian team to the International Mathematical Olympiad. The authors have benefited from the support, encouragement, and advice of their parents and of their colleagues, as well as of Georgiana Onişor and Titu Andreescu, for which they are deeply grateful. They have authored books on mathematical Olympiads in the past, and they have now come together to write a book on geometry. As for the need of such a book, Johannes Kepler once said: *At ubi materia, ibi Geometria.*[1]

Lubbock, TX, USA   Răzvan Gelca

Bucharest, Romania   Ionuţ Onişor

São Paulo, Brazil   Carlos Yuzo Shine

---

[1] Where there is matter, there is geometry.

Contents

Part I Problems

1 Isometries 3
  1.1 Theoretical Results About Isometries 3
      1.1.1 Definition and Basic Properties 4
      1.1.2 Translations, Rotations, Reflections...................... 8
      1.1.3 Isometries as Composition of Reflections 16
      1.1.4 Compositions of Isometries 21
      1.1.5 Discrete Groups of Isometries............................. 29
      1.1.6 Theoretical Questions About Isometries.................... 31
  1.2 Isometries in Euclidean Geometry Problems......................... 32
      1.2.1 Some Constructions and Classical Results in
            Euclidean Geometry That Use Isometries 32
      1.2.2 Examples of Problems Solved Using Isometries 46
      1.2.3 Problems in Euclidean Geometry to be Solved
            Using Isometries 52
  1.3 Isometries Throughout Mathematics 57
      1.3.1 Geometry with Combinatorial Flavor 57
      1.3.2 Combinatorics of Sets 62
      1.3.3 Number Theory 64
      1.3.4 Functions 67

2 Homotheties and Spiral Similarities 71
  2.1 A Theoretical Introduction to Homotheties 71
      2.1.1 Definition and Properties 71
      2.1.2 Groups Generated by Homotheties......................... 75
      2.1.3 Problems About Properties of Homotheties 78
  2.2 Problems in Euclidean Geometry That Use Homothety 78
      2.2.1 Theorems in Euclidean Geometry Proved Using
            Homothety.. 78

          2.2.2   Examples of Problems Solved Using Homothety 88

          2.2.3   Problems in Euclidean Geometry to be Solved

                    Using Homothety 93

   2.3   Homothety in Combinatorial Geometry; Scaling 97

   2.4   A Theoretical Study of Spiral Similarities........................... 100

          2.4.1   The Definition and Properties of Spiral Similarities 100

          2.4.2   The Center of a Spiral Similarity: The Generic Case 102

          2.4.3   The Center of a Spiral Similarity: The Case

                    $A' = B$, Symmedians Revisited 105

          2.4.4   Spiral Similarities and Miquel's Theorem 107

          2.4.5   Compositions of Spiral Similarities 109

          2.4.6   Groups Generated by Spiral Similarities..................... 110

          2.4.7   Theoretical Questions About Spiral Similarities 113

   2.5   Spiral Similarity in Euclidean Geometry Problems 113

          2.5.1   Similar Figures and the Circle of Similitude 113

          2.5.2   Examples of Problems Solved Using Spiral Similarities... 120

          2.5.3   Problems in Euclidean Geometry to be Solved

                    Using Spiral Similarities................................... 125

3   **Inversions** 129

   3.1   Theoretical Results About Inversion................................ 129

          3.1.1   The Definition of Inversion and Some of Its Properties 129

          3.1.2   Inverses of Lines and Circles 133

          3.1.3   Möbius Transformations..................................... 138

          3.1.4   Möbius Transformations Versus Isometries,

                    Spiral Similarities, and Inversions; Inversion and

                    Circular Transformations 141

          3.1.5   Linear Fractional Transformations of the Real Line 145

          3.1.6   The Invariance of Angles 149

          3.1.7   Inversion with Negative Ratio 152

          3.1.8   Circles Orthogonal to the Circle of Inversion.............. 153

          3.1.9   The Limiting Points of Two Circles........................ 155

          3.1.10   Problems with Theoretical Flavor About

                    Properties of Inversion and Möbius Transformations 158

   3.2   Inversion in Euclidean Geometry Problems 159

          3.2.1   Applications of Inversion to Proving Classical Results 159

          3.2.2   Examples of Problems Solved Using Inversion 178

          3.2.3   Problems in Euclidean Geometry to be Solved

                    with Inversion (or with Möbius Transformations).......... 198

4   **A Synthesis** 207

   4.1   Bringing Together All Transformations 207

          4.1.1   Some Examples 208

          4.1.2   Some Problems.. 230

4.2 A Story of Complete Quadrilaterals 234
    4.2.1 Miquel's Theorem and $\sqrt{bc}$ Inversion 234
    4.2.2 Some Classical Results 235
    4.2.3 Problems About Complete Quadrilaterals 244

Part II   Hints

5   Isometries 249

6   Homotheties and Spiral Similarities 255

7   Inversions 261

8   A Synthesis 265

Part III   Solutions

9   Isometries 271

10   Homotheties and Spiral Similarities 369

11   Inversions 437

12   A Synthesis 513

Index 577

Part I
Problems

Chapter 1
Isometries

1.1 Theoretical Results About Isometries

Our discussion begins with those transformations that lie at the heart of the notion of equality in geometry. We tend to identify two geometric figures and call them *equal* (or *congruent* when we are very rigorous) if we can place one on top of the other to coincide exactly. This means that we take one figure and move it across the plane without breaking or deforming it until it overlaps the other figure. The motions that we perform, to which we add the reflection into a mirror, are called *isometries* (from the Greek words *isos* meaning "equal" and *metron* meaning "measure"). Figure 1.1 illustrates several instances of planar figures being mapped into one another by isometries.

We have promised in the introduction to draw a parallel between the synthetic and the analytic method. Because we work in the plane, we have two options for coordinates: real or complex. There is logic in opting for the latter. Complex numbers have been introduced for solving polynomial equations, as such they come endowed with a multiplication, and this multiplication is somehow related to the richness of Euclidean geometry. For us, the plane is identified with the set $\mathbb{C}$ of complex numbers, and we convene to pass from synthetic to analytic by replacing uppercase letters by lowercase letters: the point P in the Euclidean plane becomes complex number $p \in \mathbb{C}$ (said differently p is the complex coordinate of P). Figure 1.2 shows an instance of how points in the plane acquire complex coordinates. We will use the notation $\Re z$ and $\Im z$ for the real and the imaginary parts of z and $|z|$ and $\arg z$ for the absolute value and argument of z.

© The Author(s), under exclusive license to Springer Nature Switzerland AG 2022
R. Gelca et al., *Geometric Transformations*, Problem Books in Mathematics,
https://doi.org/10.1007/978-3-030-89117-6_1

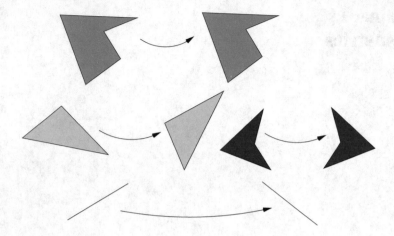

Fig. 1.1 How figures are transformed under isometries

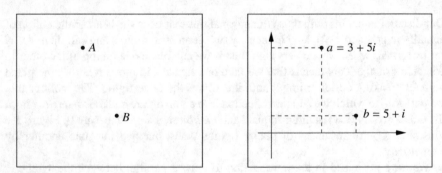

Fig. 1.2 Passing from synthetic to analytic geometry

1.1.1   Definition and Basic Properties

It is desirable to reach a good level of understanding of the isometries before starting to solve problems. For that reason we begin with a theoretical discussion.

Definition An *isometry* of the plane is a transformation that preserves distances between points, meaning that if A maps to A' and B maps to B', then the segments AB and $A'B'$ have equal lengths.

In complex coordinates, the distance between the points A and B is $|a - b|$. So, by switching to analytic geometry, we have the following definition.

Definition An *isometry* in $\mathbb{C}$ is a function $f : \mathbb{C} \to \mathbb{C}$ such that

$$|f(z) - f(w)| = |z - w| \text{ for all } z, w \in \mathbb{C}.$$

As we will see below, isometries preserve not just distances, but every property that can be phrased in the language of Euclidean geometry. They lie behind the notion of congruence, being the most natural transformations of Euclidean geometry. We now embark on the task of understanding their structure and properties. The first steps are easier in coordinates, as the functional equation $|f(z) - f(w)| = |z - w|$ can be solved explicitly, though with some effort, and yields a surprisingly simple solution.

Theorem 1.1 *The function f is an isometry of $\mathbb{C}$ if and only if $f(z) = rz + s$ or $f(z) = r\bar{z} + s$, where r, s are complex constants with $|r| = 1$.*

Proof If $f(z) = rz + s$ then

$$|f(z) - f(w)| = |r(z - w)| = |r||z - w| = |z - w|;$$

while if $f(z) = r\bar{z} + a$, then

$$|f(z) - f(w)| = |r(\bar{z} - \bar{w})| = |r||\bar{z} - \bar{w}| = |z - w|,$$

so in each case f is an isometry.

Now suppose that f is an isometry. Let $f(0) = s$. Since $1 = |1 - 0| = |f(1) - f(0)| \neq 0$, if we set $r = f(1) - f(0)$, then $r \neq 0$. We can define thus g such that $f(z) = r \cdot g(z) + s$, namely, $g(z) = \frac{f(z)-s}{r}$. Notice that

$$|g(z) - g(w)| = \left| \frac{f(z) - f(w)}{r} \right| = \frac{|f(z) - f(w)|}{|r|} = |z - w|,$$

so g is an isometry, too. We have $g(0) = 0$ and $g(1) = 1$.

The equality $|g(z) - g(0)| = |z - 0|$ is equivalent to $|g(z)| = |z|$, while $|g(z) - g(1)| = |z - 1|$ is equivalent to $|g(z) - 1| = |z - 1|$. Using the fact that $|z|^2 = z \cdot \bar{z}$, we can transform these two conditions into

$$g(z) \cdot \overline{g(z)} = z \cdot \bar{z} \text{ and } (g(z) - 1) \cdot (\overline{g(z)} - 1) = (z - 1)(\bar{z} - 1).$$

Subtracting the first relation from the second, we can transform this into

$$g(z) \cdot \overline{g(z)} = z \cdot \bar{z} \text{ and } g(z) + \overline{g(z)} = z + \bar{z}.$$

We now deduce that $\{z, \bar{z}\}$ and $\{g(z), \overline{g(z)}\}$ are the solution sets to the same quadratic equation (recall Viète's relations for a quadratic equation), so the two pairs must be equal. Hence $\{z, \bar{z}\} = \{g(z), \overline{g(z)}\}$, that is, for every $z \in \mathbb{C}$, either $g(z) = z$ or $g(z) = \bar{z}$.

It remains to prove that there do not exist nonreal numbers z, w such that $g(z) = z$ and $g(w) = \bar{w}$ simultaneously (we exclude z or w real because $r = \bar{r}$ for every real r). Were this to happen, then

$$|z - w| = |g(z) - g(w)| = |z - \overline{w}|,$$

which means that z is at the same distance from w and $\overline{w}$; alas, z lies on the perpendicular bisector of w and $\overline{w}$. But this perpendicular bisector is the real axis, so z must be a real number, which is ruled out by our assumption. So either $g(z) = z$ for all z or $g(z) = \overline{z}$ for all z. From here we deduce that $f(z) = rz + s$ or $f(z) = r\overline{z} + s$, and the theorem is proved.   □

A simple functional equation with an elegant algebraic solution. But in order to decrypt the geometric meaning of the solution, we need more work. In the next section, we will understand geometrically the multiplication by a number of absolute value 1, the addition of a complex number, and the taking of the conjugate, and this will shed light on the geometric interpretation of Theorem 1.1. This will be a recurring theme: to discover the geometric meaning hidden inside algebraic manipulations.

For the moment let us notice that, as an immediate consequence, we obtain the following result.

Theorem 1.2 *An isometry maps a segment to a segment equal to it, a line to a line, a circle to a circle of the same radius, and a triangle to a triangle congruent to it. An isometry also preserves angles.*

Proof We only discuss the case $f(z) = rz + s$ for some $r, s \in \mathbb{C}$ with $|r| = 1$; the case $f(z) = r\overline{z} + s$ is left to the reader. In this case, for every $a, b \in \mathbb{C}$ and $t \in \mathbb{R}$, the image of $ta + (1 - t)b$ is

$$rta + r(1 - t)b + s = t(ra + s) + (1 - t)(rb + s).$$

For $t \in [0, 1]$, this means that the segment with endpoints a, b is mapped to the segment with endpoints $ra + s, rb + s$, and note that $|ra + s - rb - s| = |r||a - b| = |a - b|$. By allowing t to roam freely in $\mathbb{R}$, we deduce that the line passing through a and b is mapped to the line passing through $ra + s$ and $rb + s$.

A complex number z satisfies the equation $|z - a| = R$ if and only if it satisfies the equation $|(rz + s) - (ra + s)| = R$ (here we use the fact that $|r| = 1$). So z belongs to the circle of center a and radius R if and only if $f(z) = rz + s$ belongs to the circle whose center is $f(a) = ra + s$ and radius is R. In other words, the image through an isometry of a circle is a circle of the same radius, and whose center is the image of the center of the circle.

Finally, let ABC be a triangle, and let A', B', C' be the images of ABC through the isometry. Because the isometry maps AB to $A'B'$, BC to $B'C'$, and AC to $A'C'$, the triangle $A'B'C'$ is the image of the triangle ABC and is congruent to it. Therefore the angles of ABC and the angles of $A'B'C'$ are equal. Since every angle can be placed in a triangle, isometries preserve angles.   □

By decomposing polygons into triangles, we obtain that an isometry maps a polygon to a polygon congruent to it. It is not hard to see that all special points, segments, lines, or circles (e.g., orthocenter, centroid, circumcenter, medians, angle bisectors, altitudes, incircle) of a triangle are mapped to the same special points, segments, lines, or circles of the image.

As a by-product of Theorem 1.1, we obtain that isometries are bijections from the plane to itself. Indeed, the inverse of $f(z) = rz + s$ is

$$f^{-1}(w) = \frac{1}{r}w - \frac{s}{r},$$

and the inverse of $f(z) = r\overline{z} + s$ is

$$f^{-1}(w) = \frac{1}{\overline{r}}\overline{w} - \frac{\overline{s}}{\overline{r}}.$$

We conclude that every isometry has an inverse, and the inverse is an isometry as well; it is also immediate to see that the composition of two isometries is an isometry, as distances are preserved (see Fig. 1.3). These properties characterize one of the most fundamental concepts in mathematics, which we will now introduce because it will allow us to formulate certain results from this book in a more concise language and because a new concept always gives rise to new ideas.

Definition A *group of transformations* of a given set is a set G of bijective maps from that set to itself with the property that the composition of any two elements of G is in G and the inverse of every element of G is in G.

Isometries form a group of transformations of the plane.

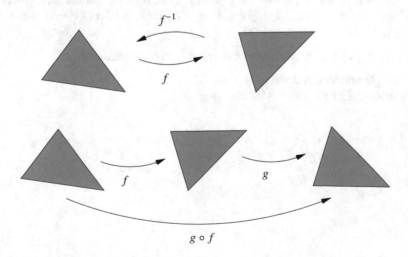

Fig. 1.3 Inverses and compositions of isometries

1.1.2   Translations, Rotations, Reflections

In this section we introduce some particular classes of isometries.

Definition Given a vector $\vec{v}$, the *translation* τ by the vector $\vec{v}$ is the transformation of the plane that maps a point A to the point $A' = \tau(A)$ such that $\overrightarrow{AA'} = \vec{v}$.

A translation is depicted in Fig. 1.4. If we write the translation vector $\vec{v} = (a, b)$ in complex coordinates as $v = a + bi$, then the translation is given by the equation

$$\tau(z) = z + v.$$

This clarifies the geometric meaning of the transformation defined by adding a complex number to the variable: it is the translation by the vector described by that number. Translations are therefore particular cases of isometries.

You can immediately check that

- all line segments of the form XX' are parallel and congruent;
- a single point A and its image $A' = \tau(A)$ uniquely determine τ, since $\vec{v} = \overrightarrow{AA'}$ is the only parameter of the translation;
- the line segments AB and $A'B'$ are parallel and of equal lengths, consequently translation maps a line to a line parallel to it;
- if AB is not parallel to $\vec{v}$ and if $\vec{v} \neq 0$, then $AA'B'B$ is a parallelogram.

Definition Given a point O and an angle α (measured counterclockwise), the *rotation* about O by angle α is the transformation ρ of the plane that maps O to itself and any other point A to a point $A' = \rho(A)$ such that $\angle AOA' = \alpha$ and A and A' are at the same distance from O.

Figure 1.5 shows an example of a rotation. You can check that

- the angle between AB and $A'B'$ is α;
- the triangles OAB and $OA'B'$ are congruent;

Fig. 1.4 A translation τ

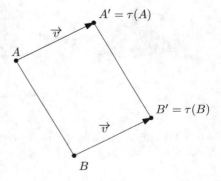

Fig. 1.5 A rotation ρ. The center is O and the angle is $\alpha = 130°$

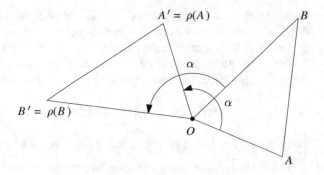

- the triangles OXX' are all similar to an isosceles triangle with angles α, $90° - \alpha/2$ and $90° - \alpha/2$;
- knowing the center of rotation O, a single point X, and its image X' is enough to determine the rotation; in fact, $\alpha = \angle XOX'$;
- when O is not given, two points A and B and their images A' and B' are required in order to find the rotation; the center O lies on both the perpendicular bisectors of AA' and BB' (since $OA = OA'$ and $OB = OB'$), so it lies at their intersection, and if the two perpendicular bisectors coincide, then O is the intersection of AB and $A'B'$;
- if the rotation is not trivial, the center of rotation is its only fixed point.

Because the triangles OAB and $OA'B'$ are congruent, the segments AB and $A'B'$ are equal. Consequently, rotations preserve distances; they are isometries.

When writing rotations in coordinates, we see the reason for using complex numbers. To understand rotations, it is better to work with complex numbers in trigonometric form

$$z = k(\cos \theta + i \sin \theta).$$

This is because of the identity

$$\arg(zw) = \arg(z) + \arg(w),$$

which follows from trigonometry

$$k_1(\cos \alpha + i \sin \alpha)k_2(\cos \beta + i \sin \beta)$$
$$= k_1 k_2[(\cos \alpha \cos \beta - \sin \alpha \sin \beta) + i(\cos \alpha \sin \beta + \sin \alpha \cos \beta)]$$
$$= k_1 k_2[\cos(\alpha + \beta) + i \sin(\alpha + \beta)].$$

Theorem 1.3 *The rotation by α about z_0 is then given by the formula*

$$\rho(z) = r(z - z_0) + z_0, \text{ where } r = \cos \alpha + i \sin \alpha.$$

Proof There is nothing to prove if $\alpha = 0$. If $\alpha \neq 0$, then the point z_0 is the unique fixed point of this transformation. The length of the segment joining z_0 and z is $|z - z_0|$, and the length of the segment joining z_0 and $\rho(z)$ is $|r(z - z_0)| = |z - z_0|$. Also the segment joining z_0 and z forms with the x-axis, an angle whose argument is $\arg(z - z_0)$, while the segment joining z_0 and z forms with the x-axis, the angle

$$\arg r(z - z_0) = \arg r + \arg(z - z_0) = \alpha + \arg(z - z_0).$$

Thus the angle between the two segments is α, and the theorem is proved.   □

The formula for rotation can also be written as $\rho(z) = rz + z_0(1 - r)$. On the other hand, if $f : \mathbb{C} \to \mathbb{C}$, $f(z) = rz + s$ with $|r| = 1$ and $r \neq 1$, then f defines a rotation; we can find the angle of rotation $\alpha = \arg(r)$ and the center $z_0 = s/(1-r)$. The imaginary number i, which has been introduced for solving the equation $x^2 + 1 = 0$, stands for the 90° rotation, and so it is related to the concept of orthogonality in geometry.

There is an elegant way of looking at rotations which uses some elements of real analysis. It is based on Euler's number

$$e = 1 + \frac{1}{1!} + \frac{1}{2!} + \frac{1}{3!} + \frac{1}{4!} + \cdots = 2.71828\ldots$$

and the associated power series

$$e^x = 1 + \frac{x}{1!} + \frac{x^2}{2!} + \frac{x^3}{3!} + \frac{x^4}{4!} + \cdots.$$

This series converges to (i.e., approximates) a real number for all real numbers x. But more is true. This definition can be extended to imaginary numbers, and we have

$$e^{\alpha i} = 1 + \frac{\alpha}{1!}i - \frac{\alpha^2}{2!} - \frac{\alpha^3}{3!}i + \frac{\alpha^4}{4!} + \frac{\alpha^5}{5!}i - \cdots$$

Comparing this with two other series expansions

$$\cos\alpha = 1 - \frac{\alpha^2}{2!} + \frac{\alpha^4}{4!} - \frac{\alpha^6}{6!} + \cdots,$$

$$\sin\alpha = \frac{\alpha}{1!} - \frac{\alpha^3}{3!} + \frac{\alpha^5}{5!} - \frac{\alpha^7}{7!} + \cdots,$$

where α is *necessarily* measured in radians, we deduce Euler's formula

$$e^{\alpha i} = \cos\alpha + i\sin\alpha.$$

The reader uncomfortable with series expansions can take this as the definition of $e^{\alpha i}$. The trigonometric identities

$$\cos(\alpha + \beta) = \cos \alpha \cos \beta - \sin \alpha \sin \beta, \quad \sin(\alpha + \beta) = \sin \alpha \cos \beta + \cos \alpha \sin \beta$$

correspond to the multiplicative properties of the exponential function

$$e^{(\alpha+\beta)i} = e^{\alpha i} e^{\beta i}, \quad \left(e^{\alpha i}\right)^n = e^{n\alpha i}.$$

A complex number is written in trigonometric form as

$$z = re^{i\alpha}, \text{ where } r = |z| \text{ and } \alpha = \arg z.$$

With this notation, we can write the counterclockwise rotation about the origin by the angle α (measured in radians) as

$$z \mapsto z' = e^{\alpha i} z.$$

Then the counterclockwise rotation about a point w by angle α is defined by the equation

$$\frac{z' - w}{z - w} = e^{\alpha i},$$

which gives $z' = e^{\alpha i} z - e^{\alpha i} w + w$.

To summarize, $e^{i\alpha}$ can be used as a short-hand writing for $\cos \alpha + i \sin \alpha$, and trigonometry dictates that it has the good multiplicative properties that powers have. But you have to be careful and use radians, so the 60° rotation about the origin is $z \mapsto e^{\pi i/3} z$.

In the context of rotations, it is appropriate to open a parenthesis and discuss directed angles modulo π (or modulo 180° if we use the old-fashioned Babylonian measurements). Given two lines ℓ and ℓ', the angle of the rotation that maps ℓ into ℓ' is ambiguous; it can take two different values even in the interval $[0, 2\pi)$. There are two ways of resolving this ambiguity. The first is to orient the lines, for example, by choosing two points on each. As such, the angle $\angle(AB, CD) \in [0, 2\pi)$ is the angle of the rotation that maps AB to CD such that the image of $\overrightarrow{AB}$ points in the same direction as $\overrightarrow{CD}$. This is the convention that we adopt when we do not say anything explicitly.

The second approach is to notice that two angles of rotation differ by multiples of π, so the ambiguity disappears when working modulo π (see Fig. 1.6). Now the angle $\angle(\ell, \ell')$ is defined unambiguously. Note that $\angle(\ell', \ell) = -\angle(\ell, \ell')$.

Whenever we declare to be working with *directed angles modulo* π, when we write $\angle ABC$, we mean the angle $\angle(AB, BC)$ between the lines AB and BC, which is the angle of the rotation that takes AB to BC, reduced modulo π.

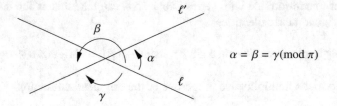

Fig. 1.6 Definition of directed angles

If a, b, c are the complex coordinates of A, B, C, then the directed angle $\angle ABC$ modulo π is

$$\arg \frac{c - b}{a - b} \quad (\text{mod } \pi).$$

The points A, B, C are collinear if and only if for some other point D we have $\angle DAB = \angle DAC$, or in complex numbers, if and only if

$$\arg \frac{b - a}{d - a} = \arg \frac{c - a}{d - a} \quad (\text{mod } \pi).$$

In a triangle ABC,

$$\angle BAC + \angle CBA + \angle ACB = \angle(AB, AC) + \angle(BC, AB) + \angle(AC, BC) = 0.$$

Also, if A, B, C are not collinear, then a point D is on the circumcircle of ABC if and only if, as directed angles modulo π, $\angle DAC = \angle DBC$. In complex numbers, the condition that four points lie on a circle is

$$\arg \frac{c - a}{d - a} - \arg \frac{c - b}{d - b} = \arg \left(\frac{c - a}{d - a} : \frac{c - b}{d - b} \right) = 0 \quad (\text{mod } \pi),$$

that is,

$$\frac{c - a}{d - a} : \frac{c - b}{d - b} \in \mathbb{R}.$$

And if A, B, C are on a line, then this is also the condition for D to lie on the same line. The quantity

$$\frac{c - a}{d - a} : \frac{c - b}{d - b} = \frac{a - c}{a - d} : \frac{b - c}{b - d},$$

which we denote by (a, b, c, d) and call *cross-ratio*, seems to be important, if only for checking that points are concyclic or collinear. Note that $(a, b, c, d) =$

$(c, d, a, b) = (d, c, b, a) = (a, b, d, c)^{-1}$. The cross-ratio depends only on the distances between the points and the angles they form, so it does not depend on the system of coordinates. We can therefore talk about the *cross-ratio of four points in the plane*.

Before moving on, let us state what was explained above as a theorem.

Theorem 1.4 (Condition for Concyclicity) *Four points A, B, C, D lie on a circle or on a line if and only if the cross-ratio of their complex coordinates is a real number.*

Now let us return to our discussion of isometries.

Definition Given a point O in the plane, the *reflection* over the point O is the transformation σ of the plane that maps O to itself and any other point A to a point A' such that O is the midpoint of AA'.

A reflection over a point z_0 in the complex plane is therefore a function σ : $\mathbb{C} \to \mathbb{C}$, such that $\frac{\sigma(z)+z}{2} = z_0$. This clearly implies that $\sigma(z) = -z + 2z_0$. This transformation does not bring anything new as it is the rotation about the point z_0 by $180°$.

Definition Given a line ℓ in the plane, the *reflection* σ over the line ℓ is the transformation of the plane that maps a point A on ℓ to itself, and a point A that does not belong to ℓ to a point $A' = \sigma(A)$ such that AA' is perpendicular to ℓ, A and A' are in opposite half-plane determined by ℓ, and the distances from A and A' to ℓ are equal.

The reader can visually notice (on Fig. 1.7) that

- all segments XX' have ℓ as perpendicular bisector;
- a single point X that does not lie on ℓ and its image X' determine a reflection over a line ℓ; ℓ is the perpendicular bisector of XX';
- the lines AB and $A'B'$ are either parallel to ℓ or meet at a point on ℓ;
- every reflection σ is an *involution*, that is, $\sigma \circ \sigma$ is the identity. This implies that lines AB' and $A'B$ are either parallel to ℓ or meet at a point on ℓ.

It is somewhat complicated to write reflections over lines in complex coordinates, because one can write the equation of a line in many ways. We nevertheless have the following result:

Theorem 1.5 *Let f be an isometry. Then f is a reflection over a line if and only if*

$$f(z) = r\bar{z} \text{ or } f(z) = -\frac{s}{\bar{s}}\bar{z} + s$$

where $r, s \in \mathbb{C}$, $|r| = 1$ and $s \neq 0$.

Proof The simplest reflection is over the real axis, $\sigma(z) = \bar{z}$.

Next, let σ be the reflection over some line ℓ given by the parametric equation $z = z_0 + vt$, $t \in \mathbb{R}$ being the parameter and z_0 and v being constants with $|v| = 1$.

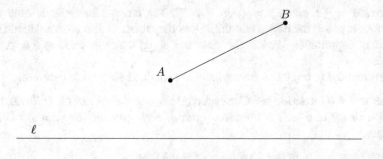

Fig. 1.7 A reflection σ over a line ℓ

We transform ℓ into the real axis by means of translations and rotations: first we translate by $-z_0$, so z and $\sigma(z)$ become $z - z_0$ and $\sigma(z) - z_0$, respectively; the line becomes ℓ' given by $z = t \cdot v$, $t \in \mathbb{R}$. Then we rotate by $-\arg(v)$: ℓ' becomes the real axis, $z - z_0$ becomes $\frac{z-z_0}{v}$, and $\sigma(z) - z_0$ becomes $\frac{\sigma(z)-z_0}{v}$. In other words, the reflection over the real axis maps $\frac{z-z_0}{v}$ to $\frac{\sigma(z)-z_0}{v}$. Therefore

$$\frac{\sigma(z) - z_0}{v} = \overline{\left(\frac{z - z_0}{v}\right)}$$

which yields, using the fact that $v/\overline{v} = v^2$, the following formula for the reflection:

$$\sigma(z) = v^2\overline{z} + z_0 - v^2\overline{z_0}. \tag{1.1}$$

So $\sigma(z)$ must be of the form $\sigma(z) = r\overline{z}+s$, $|r| = 1$. If we identify the coefficients of $\sigma(z)$, we find $r = v^2$ and $z_0 - v^2\overline{z_0} = s$, that is, $s = z_0 - r\overline{z_0}$. Since z_0 can be *any* point of ℓ, the line must have the equation $z - r\overline{z} = s$. Conjugation gives $\overline{z} - z/r = \overline{s}$ which is equivalent to $z - r\overline{z} = -r\overline{s}$, so $s = -r\overline{s}$. This happens if either $s = 0$ or $r = -s/\overline{s}$. The case $s = 0$ gives $\sigma(z) = v^2\overline{z} = r\overline{z}$, and the other case gives $\sigma(z) = -\frac{s}{\overline{s}}\overline{z} + s$.

Notice that we can recover the equation of the line ℓ from the formula for σ: it is $z - r\overline{z} = s$, which is equivalent to $\overline{s}z + s\overline{z} = |s|^2$, if $s \neq 0$, and $z = r\overline{z}$ if $s = 0$. Notice also that ℓ passes through $s/2$ and that v is any square root of $-\frac{s}{\overline{s}}$, that is, $v = \pm i\omega$, $\omega = \cos\arg(s) + i\sin\arg(s)$. $\qquad\square$

As a consequence of the proof, we obtain the formula for the reflection over a line determined by two points.

Proposition 1.6 *Let a, b be two distinct complex numbers. Then the reflection of a point z over the line determined by a and b is*

$$\sigma(z) = \frac{a-b}{\overline{a}-\overline{b}}\overline{z} + \frac{\overline{a}b - a\overline{b}}{\overline{a}-\overline{b}}.$$

Proof The parametric equation of the line is $z = a + (b-a)t$. We have seen in the proof of the previous result that

$$\frac{\sigma(z)-a}{b-a} = \overline{\left(\frac{z-a}{b-a}\right)}.$$

Hence the formula.   □

Once reflections have been introduced, it is appropriate to talk about the orientation of polygons, a concept that plays a major role in this book. A (nonskew) polygon is *oriented counterclockwise* if, when reading the names of its vertices in order, the interior is on the left, and it is *oriented clockwise* if, when reading the names of its vertices in order, the interior is on the right. In Fig. 1.8, the triangle ABC is oriented counterclockwise, and the triangle $A'B'C'$ is oriented clockwise. It is not hard to see that rotations and translations preserve orientation, while reflection changes orientation (e.g., in Fig. 1.8, the two triangles are images of each other through the reflection over the vertical dotted line).

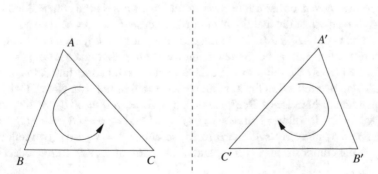

Fig. 1.8 Orientation of triangles

1.1.3 Isometries as Composition of Reflections

It looks like the aforementioned transformations are only particular cases of isometries of the plane and that there might exist more general types of isometries. But Theorem 1.1 can now be read in the dialect of synthetic geometry, and it tells that every isometry is either a rotation centered at the origin, a translation, the reflection over the x-axis, a composition of a translation and a rotation centered at the origin, or the composition of the reflection over the x-axis with a translation and/or a rotation centered at the origin. We can do better than that, for example, we can write all isometries of the plane in terms of reflections alone, as the next result shows.

Theorem 1.7 *Every isometry of the plane is the composition of two or three reflections over lines.*

Proof Let f be an isometry of the plane. Then f transforms any ray into a ray, by preserving the order of the points, and every half-plane into a half-plane.

Let A and B be two points and let h be a half-plane bounded by the line AB. With the standard notation, let $A' = f(A)$, $B' = f(B)$, and $h' = f(h)$. We distinguish the following cases:

Case 1. The rays $|AB$ and $|A'B'$ coincide. Then $A' = A$ and $B' = B$, so the entire line AB is fixed by f. Then either the entire plane is fixed by f, in which case we can write f as the square of any reflection, or f is the reflection over AB, in which case f is the cube of this reflection, as well.

Case 2. The rays $|AB$ and $|A'B'$ are mapped into each other by a reflection σ over a line. Compose f with this reflection to reduce the problem to Case 1. In this situation $f \circ \sigma$ is either a reflection in which case f is the composition of this reflection and σ or the identity map in which case $f = \sigma = \sigma \circ \sigma \circ \sigma$.

Case 3. The rays $|AB$ and $|A'B'$ do not coincide, nor do they map one to the other by a reflection. Let σ_1 be the reflection over a line that takes A to $f(A) = A'$, and let $B_1 = \sigma_1(B)$. Let also σ_2 be the reflection, over some line ℓ, that takes B_1 to B'. The two reflection lines are drawn as dotted in Fig. 1.9. Notice that ℓ is the perpendicular bisector of $B_1 B'$, and since $A'B_1 = AB = A'B'$, A' lies on ℓ, so $\sigma_2(A') = A'$. It follows that $\sigma_2 \circ \sigma_1(A) = A'$ and $\sigma_2 \circ \sigma_1(B) = \sigma_2(B_1) = B'$. Then $\sigma_2 \circ \sigma_1$ takes A to A' and B to B', so either $f = \sigma_2 \circ \sigma_1$ (two reflections) or f is the composition of $\sigma_2 \circ \sigma_1$ and a reflection over $A'B'$ (three reflections). $\qquad\square$

The theorem shows that reflections generate the group of isometries. Let us observe that the composition of an even number of reflections preserves the orientation of polygons, while the composition of an odd number of reflections changes orientation meaning that what was originally counterclockwise, is now clockwise. The isometries obtained as compositions of an even number of reflections form therefore a group, the group of orientation preserving isometries. As a corollary of Theorem 1.7 we obtain:

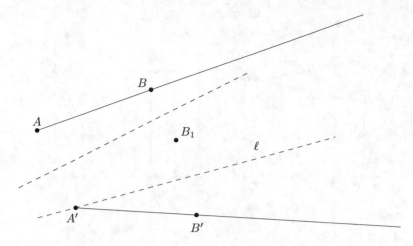

Fig. 1.9 Case 3

Theorem 1.8 *Every orientation-preserving isometry is the composition of two reflections over lines.*

But we have a better result:

Theorem 1.9 *Every orientation-preserving isometry is either a translation or a rotation.*

Proof 1 Let f be an orientation-preserving isometry of the plane. First notice that we only need to find a translation or a rotation that takes two points A and B to $A' = f(A)$ and $B' = f(B)$, respectively. Indeed, if P is a point not belonging to the line AB, then $f(P)$ is determined by A', B' and the orientation of ABP. If P is on the line AB, then the image $P' = f(P)$ of a point P on line AB is uniquely determined by the conditions $P' \in A'B'$, $PA = P'A'$, and $PB = P'B'$.

Having in mind the technique of finding the center of a rotation, consider the perpendicular bisectors a of AA' and b of BB'. We distinguish the following cases:

Case 1. (Fig. 1.10). The lines a and b have a non-empty intersection. Let O be the intersection point of a and b. Then $OA = OA'$, $OB = OB'$, and, since $AB = A'B'$, the triangles OAB and $OA'B'$ are congruent. If these two triangles have the same orientation, we have the following equalities of angles between lines:

$$\angle(OA, OA') = \angle(OA, OB) + \angle(OB, OA')$$
$$= \angle(OA', OB') + \angle(OB, OA') = \angle(OB, OB').$$

From here we deduce that f is a rotation about O by the angle $\alpha = \angle(OA, OA')$. If the two triangles have inverse orientations, then there is a reflection that takes AB to $A'B'$, and so we must have $a = b$. If AB and $A'B'$ are both parallel to

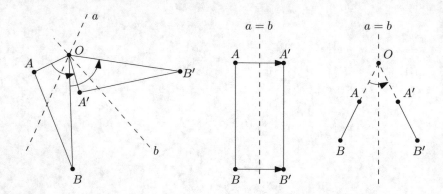

Fig. 1.10 Case 1

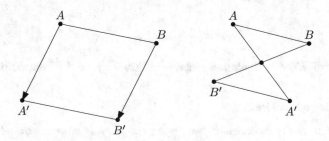

Fig. 1.11 Case 2

$a = b$, there is a translation that takes A to A' and B to B'; otherwise, $AA'B'B$ is an isosceles trapezoid with non-parallel sides AB and $A'B'$ and a rotation with center on the intersection of lines AB and $A'B'$ does the job.

Case 2. (Fig. 1.11) The lines a and b do not intersect. Then a and b are parallel and distinct. In this case AA' and BB' are also parallel. If A, A', B, B' lie on the same line and A and B are in the same order as A' and B', then $\overrightarrow{AA'} = \overrightarrow{BB'} = \overrightarrow{v}$ and f is a translation by $\overrightarrow{v}$; if A and B are in reversed order as compared to A' and B', then the midpoints of AA' and BB' coincide, but then $a \cap b \neq \emptyset$, which is ruled out by the hypothesis.

On the other hand, if $A, A', B,$ and B' do not lie on the same line, they determine a trapezoid. If the trapezoid is not a parallelogram, then AB and $A'B'$ are either its diagonals or the non-parallel sides. Either way, since $AB = A'B'$, the trapezoid is isosceles and then $a = b$, a contradiction. So the trapezoid is a parallelogram, meaning that either $\overrightarrow{AA'} = \overrightarrow{BB'} = \overrightarrow{v}$ and f is a translation by $\overrightarrow{v}$, or $\overrightarrow{AA'} = -\overrightarrow{BB'}$, and f is a reflection over the midpoint of AA' and BB', which is a rotation by $180°$.

□

Proof 2 There is a slick proof of the theorem using complex coordinates that the reader might have noticed. Since conjugation reverses orientation, an orientation-preserving isometry in $\mathbb{C}$ can only be of the form $f(z) = rz + s$, $|r| = 1$. If $r = 1$ we have a translation by s; if $r \neq 1$, we have a rotation by $\arg(z)$ about $s/(1-r)$.

$\square$

Unfortunately, we cannot describe an orientation-reversing isometry as a single reflection, but we still have the following result:

Theorem 1.10 *Every orientation-reversing isometry is the composition of a unique reflection over a line and a unique translation parallel to the line of reflection.*

Proof 1 Let f be an orientation-reversing isometry, and let $A' = f(A)$ and $B' = f(B)$ be the images of two distinct points A and B under f.

First we find a reflection σ over a line ℓ and a translation τ by a vector $\vec{v}$ parallel to ℓ such that $f = \tau \circ \sigma$. Again, it suffices to show that $\tau \circ \sigma(A) = A'$ and $\tau \circ \sigma(B) = B'$. Let M and N be the midpoints of AA' and BB', respectively (if A' coincides with A, we take $A = A'$ as the midpoint).

If $M \neq N$ (see Fig. 1.12), then ℓ is the line through M and N. Notice that $d(A, \ell) = d(A', \ell)$ and $d(B, \ell) = d(B', \ell)$, so the reflection σ over ℓ takes A to A_1 and B to B_1 such that $A'A_1$ and $B'B_1$ are both parallel to ℓ. Also, A and A' are on opposite sides of ℓ, and the same is true about B and B'. Hence $\angle(AB, \ell) = \angle(\ell, A'B')$, and thus, because of the reflection, lines A_1B_1 and $A'B'$ are parallel. Since $A_1B_1 = A'B'$, we find that $A_1B_1B'A'$ is a parallelogram, and thus there is a translation τ by the vector $\vec{v} = \overrightarrow{A_1A'}$ which is parallel to ℓ. We conclude that $f = \tau \circ \sigma$.

If $M = N$ (see Fig. 1.13), then $ABA'B'$ is a parallelogram. In this case, we take ℓ to be the line perpendicular to both AB and $A'B'$ and passing through M. Then σ takes A to $A_1 \in AB$ and B to $B_1 \in AB$, and $d(A_1, \ell) = d(A', \ell)$ and $d(B_1, \ell) = d(B', \ell)$, with A_1 and A' on the same side of ℓ, and B_1 and B' on the

Fig. 1.12 The case $M \neq N$

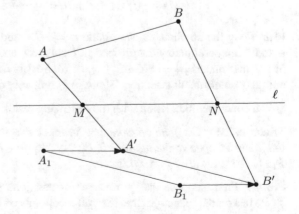

Fig. 1.13 The case $M = N$

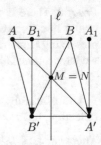

same side of ℓ. Therefore, $A_1 A'$ and $B_1 B'$ are both parallel to ℓ, and we can define τ to be the translation by $\overrightarrow{A_1 A'}$.

Now we prove that the composition is unique. Let $f = \tau' \circ \sigma'$ be a composition of a reflection σ' over a line ℓ' and a translation τ' parallel to ℓ'. We have $d(A', \ell') = d(A, \ell')$ and $d(B, \ell') = d(B', \ell')$, which means that both midpoints M of AA' and N of BB' lie on ℓ'. If $M \neq N$, we have $\ell = \ell'$, that is, $\sigma' = \sigma$. Now $\tau' = f \circ \sigma^{-1} = \tau$, and we are done. If $M = N$, then again $ABA'B'$ is a parallelogram. Since τ' is a translation, the line through $A_1' = \sigma'(A)$ and $B_1' = \sigma'(B)$ is parallel to $A'B'$, so $A_1' B_1'$ is parallel to AB. However, a reflection maps a line to another parallel line if and only if the lines are perpendicular or parallel to the line of reflection. The latter case makes σ' map AB to $A'B'$, but with the points A' and B' in reversed order, as A_1' and B_1', so we cannot translate A_1' to A' and B_1' to B' simultaneously; the former case leads to $\sigma = \sigma'$ and, as before, $\tau' = \tau$.   □

Proof 2 The complex number proof is again shorter, but it obscures completely the geometric intuition. An orientation-reversing isometry f in $\mathbb{C}$ is of the form $f(z) = r\bar{z} + s$. Our goal is to find a complex t and a real number u such that f is a composition of $g(z) = -(t/\bar{t})\bar{z} + t$ and $h(z) = z + i\omega u$, where $\omega = \cos \arg(t) + i \sin \arg(t)$. We compute

$$h(g(z)) = -(t/\bar{t})\bar{z} + t + i\omega u = -\omega^2 \bar{z} + |t|\omega + i\omega u.$$

Identifying the coefficients, we obtain $r = -\omega^2$ and $s = |t|\omega + i\omega u$. From here u and t can be uniquely determined in terms of r and s: pick ω as the square root of $-r$ that makes $|t| = \Re(s\omega^{-1}) \geq 0$; then $|t|$ is defined that way, and thus t is uniquely determined; also, $u = \Im(s\omega^{-1})$ is uniquely determined.   □

We conclude this section with the following result:

Theorem 1.11 *Let f be an isometry. Suppose that the points $P_1, P_2, \ldots, P_n$ on a line ℓ are mapped to the points $P_1', P_2', \ldots, P_n'$ on a line ℓ'. Then the midpoints of $P_1 P_1', P_2 P_2', \ldots, P_n P_n'$ are collinear.*

Proof 1 First, suppose that f reverses orientation. Since the midpoint of P and $f(P)$ lies on the reflection line, then all midpoints of $P_i P_i'$ lie on the reflection line.

If f preserves orientation, compose f with the reflection σ over ℓ, that is, consider $f \circ \sigma$. Since $\sigma(\ell) = \ell$, $f \circ \sigma$ maps P_i to P_i' and reverses orientation, and we apply the previous result to $f \circ \sigma$ instead. □

Proof 2 This can also be seen with complex numbers: If $f(z) = r\bar{z} + s$ reverses orientation, the midpoint of z_i and $f(z_i)$ is $\frac{z_i + f(z_i)}{2} = \frac{z_i + r\bar{z_i} + s}{2}$. Let ω be a square root of r; then

$$\frac{z_i + f(z_i)}{2} = \frac{z_i + \omega^2 \bar{z_i} + s}{2} = \omega \frac{\bar{\omega} z_i + \omega \bar{z_i}}{2} + \frac{s}{2} = \frac{s}{2} + \omega \Re(\bar{\omega} z_i),$$

and hence all midpoints lie on the line $z = \frac{s}{2} + \omega t, t \in \mathbb{R}$.

If $f(z) = rz + s$ preserves orientation, then $g(z) = \frac{z + f(z)}{2}$ is an affine function[1] of z that takes ℓ to the set of midpoints of points from ℓ. Every affine function g maps a line to another line. So the midpoints are on a line. □

Remark This result is also true if we consider points Q_i such that $\overrightarrow{P_i Q_i} = t \overrightarrow{P_i P_i'}$, t being a fixed real number. But this is a job for another transformation: homothety (stay tuned!).

At this moment we understand well enough the structure of isometries themselves. The next section is devoted to understanding the outcome of composing isometries.

1.1.4 Compositions of Isometries

It is clear that a composition of several isometries is an isometry itself, since it preserves distances. In the previous section, we have learned that an isometry is a rotation or a translation (when it preserves orientation) or the composition of a unique reflection and a unique translation by a vector parallel to the line of reflection (when it reverses orientation). We have also seen that if we interpret points as complex numbers, the isometries are the functions defined by the formulas $f(z) = rz + s$ or $f(z) = r\bar{z} + s, |r| = 1$.

Our next goal is to specify the outcome of all possible compositions of two isometries of the plane. Let us first deal with the composition of two orientation-preserving isometries. We already know that this composition preserves orientation, and therefore it is either a translation or a rotation. Let us be more specific. We begin with a result that was mentioned before.

Theorem 1.12 *The composition of two translations τ_1 and τ_2 by $\vec{u}$ and $\vec{v}$ is a translation by $\vec{u} + \vec{v}$.*

---

[1] Meaning a function of the form $z \mapsto az + b$.

Proof Let A be any point in the plane, $A_1 = \tau_1(A)$ and $A' = \tau_2 \circ \tau_1(A) = \tau_2(A_1)$. Thus $\overrightarrow{AA'} = \overrightarrow{AA_1} + \overrightarrow{A_1A'} = \overrightarrow{u} + \overrightarrow{v}$. Notice that this also proves that translations commute, that is, $\tau_2 \circ \tau_1 = \tau_1 \circ \tau_2$.   □

To summarize, translations form a commutative group: the compositions of the translations by vectors $\overrightarrow{v}$ and $\overrightarrow{w}$ are the translation of vector $\overrightarrow{v + w}$, and the inverse of the translation of vector $\overrightarrow{v}$ is the translation of vector $-\overrightarrow{v}$. The elements of this group are parametrized by the complex numbers.

Theorem 1.13 *Let ρ be the rotation about point O by angle α, and let τ be the translation by $\overrightarrow{v}$. Then*

(i) *the composition $\rho \circ \tau$ is a rotation by α about the unique point O_1 such that $O_1O = O_1O'$ and $\angle O'O_1O = \alpha$, where $O' = \tau^{-1}(O)$. The point O_1 lies at the intersection of the perpendicular bisectors of OO' and OO'', where $O'' = \rho \circ \tau(O)$;*

(ii) *the composition $\tau \circ \rho$ is a rotation by α about the unique point O_2 such that $O_2O = O_2O_3$ and $\angle OO_2O_3 = \alpha$, where $O_3 = \tau(O)$. The point O_2 is the intersection of the perpendicular bisector of OO_3 and the line O_3O'';*

(iii) *if O_1 and O_2 are defined as above, then $\overrightarrow{O_1O_2} = \overrightarrow{v}$.*

Proof 1 The argument can be followed on Fig. 1.14. First we prove that both compositions are rotations. To this end, let $A \neq B$ be points in the plane, and let A' and B' be the images of A and B, respectively, under either composition. Because the angle between a segment and its translation is zero, and because $\angle(AB, A'B') = \alpha$, neither of the two compositions can be a translation, both having to be rotations by the angle α.

The rest is straightforward. We have

$$\rho \circ \tau(O') = \rho \circ \tau \circ \tau^{-1}(O) = \sigma(O) = O,$$

Fig. 1.14 Composition of a rotation and a translation and vice versa

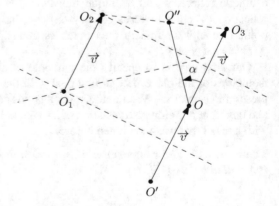

because every rotation fixes its center. It follows that O_1 must be such that $\angle O'O_1O = \alpha$ and $O_1O = O_1O'$. In particular, O_1 must lie on the perpendicular bisector of OO'. Because $O'' = \rho \circ \tau(O)$, O_1 must also lie on the perpendicular bisector of OO'', so (i) is proved.

The proof of (ii) is similar: $\tau \circ \rho(O) = \tau(O) = O_3$, so $O_2O = O_2O_3$ and $\angle OO_2O_3 = \alpha$. Also, $\angle O''O_3O = 90° - \alpha/2 = \angle O_2O_3O$, so the points O_2, O_3, and O'' are collinear.

For (iii), $\tau(O_1) = \tau(\rho \circ \tau(O_1)) = \tau \circ \sigma(\tau(O_1))$, so $\tau(O_1)$ is a fixed point of $\rho \circ \tau$. However, the only fixed point of a rotation is its center, therefore $\tau(O_1) = O_2$. And this is equivalent to $\overrightarrow{O_1O_2} = \overrightarrow{v}$.   □

Proof 2 In complex numbers, let $\rho(z) = rz + s$ and $\tau(z) = z + v$ be a rotation and a translation, respectively. Then $\rho \circ \tau(z) = r(z + v) + s = rz + rv + s$ is a rotation with center $(rv + s)/(1 - r) = rv/(1 - r) + s/(1 - r)$, and $\tau \circ \rho(z) = rz + s + v$ is a rotation with center $(s + v)/(1 - r) = s/(1 - r) + v/(1 - r)$. One can decrypt from these numbers the geometric properties listed in the statement.   □

The composition of two rotations about the same center O is a rotation by the sum of their angles (or the identity if this sum is zero). What if the centers are different?

Theorem 1.14 *Let $O_1 \neq O_2$ be points in the plane. The composition $\rho_2 \circ \rho_1$ of two rotations ρ_1 about O_1 by α and ρ_2 about O_2 by β is*

(i) A rotation by $\alpha + \beta$ about the point O that is the intersection of the perpendicular bisectors of O_1O_1', $O_1' = \rho_2(O_1)$, and O_2O_2', $O_2' = \rho_1^{-1}(O_2)$, if $\alpha + \beta$ is not a multiple of $360°$;
(ii) A translation by $\overrightarrow{v} = \overrightarrow{O_1O_1''}$, $O_1'' = \sigma_2(O_1)$ if $\alpha + \beta$ is a multiple of $360°$.

Proof 1 By Theorem 1.9, the resulting composition is either a rotation or a translation. Let A and B be distinct points in the plane, and define $A_1 = \rho_1(A)$, $B_1 = \rho_1(B)$, $A' = \rho_2 \circ \rho_1(A) = \rho_2(A_1)$ and $B' = \rho_2 \circ \rho_1(B) = \rho_2(B_1)$. Then $\angle(AB, A'B') = \angle(AB, A_1B_1) + \angle(A_1B_1, A'B') = \alpha + \beta$. So if $\alpha + \beta$ is not a multiple of $360°$, we have a rotation by $\alpha + \beta$ (as all figures are rotated by this angle), and if $\alpha + \beta$ is a multiple of $360°$, then $\overrightarrow{AB} = \overrightarrow{A'B'}$, and we have a translation. Now we only need to find the parameters of the composition map.

If $\alpha + \beta$ is not a multiple of $360°$, then we take advantage of the fact that the rotation fixes its center, and compute $\rho_2 \circ \rho_1(O_1) = \rho_2(O_1) = O_1'$ and $\rho_2 \circ \rho_1(O_2') = \rho_2 \circ \rho_1(\rho_1^{-1}(O_2)) = \rho_2(O_2) = O_2$. So the center of rotation O lies on the perpendicular bisectors of O_1O_1' and O_2O_2'. Notice also that since $O_1 \neq O_2$, $O_1' = \rho_2(O_1) \neq O_1$ and $O_2' = \rho^{-1}(O_2) \neq O_2$, the perpendicular bisectors are well defined.

If $\alpha + \beta$ is a multiple of $360°$, and we are in the presence of a translation, then we only need to find the image of one point. We choose O_1, and then $\sigma_2 \circ \sigma_1(O_1) = \sigma_2(O_1) = O_1''$. So the translation vector is $\overrightarrow{v} = \overrightarrow{O_1O_1''}$.   □

Proof 2 For the analytical proof, if $\rho_1(z) = r_1 z + s_1$ and $\rho_2(z) = r_2 z + s_2$ are rotations, $\rho_2 \circ \rho_1(z) = r_1 r_2 z + r_2 s_1 + s_2$ is a translation by $r_2 s_1 + s_2 = \rho_2(s_1)$ if $r_1 r_2 = 1$ and a rotation by $\arg(r_1 r_2)$ about $(r_2 s_1 + s_2)/(1 - r_1 r_2)$ otherwise. $\square$

We conclude that rotations do not form a group. The rotations about a fixed point O *do* form a group (here you have to include the rotation of angle zero, which is the identity map). This is the famous (multiplicative) group $U(1) = \{e^{\alpha i} \mid \alpha \in \mathbb{R}\}$ that plays an important role in electromagnetism (because $e^{i\alpha}$ is written in terms of sine and cosine, which are used for representing waves).

Now we turn to orientation-reversing isometries. Let us study first the composition of two reflections.

Theorem 1.15 *The composition $\sigma_2 \circ \sigma_1$ of two reflections σ_1, over line ℓ_1, and σ_2, over line ℓ_2, is*

 (i) *the identity if $\ell_1 = \ell_2$;*
 (ii) *a translation by a vector orthogonal to both ℓ_1 and ℓ_2, directed from ℓ_1 to ℓ_2, and length $2d(\ell_1, \ell_2)$ if ℓ_1 and ℓ_2 are parallel and distinct;*
 (iii) *a rotation with center the intersection of ℓ_1 and ℓ_2 and angle $2\angle(\ell_1, \ell_2)$ if ℓ_1 and ℓ_2 are not parallel.*

Proof 1 Part (i) is immediate. For (ii), arguing on Fig. 1.15, we introduce a *signed distance* from a point P to ℓ_1 as being positive if P and ℓ_2 are in different half-planes with respect to ℓ_1 and negative otherwise. Then we let $d = d(\ell_1, \ell_2)$ be the signed distance from any point of ℓ_2 to ℓ_1. We have $d(P, \sigma_1(P)) = 2d(P, \ell_1)$ and

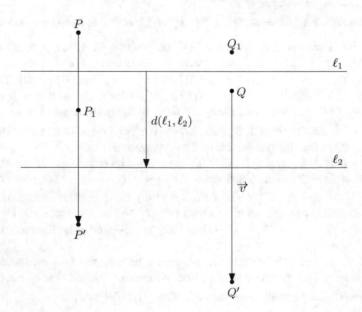

Fig. 1.15 Composition of two reflections over parallel lines

Fig. 1.16 Composition of two reflections over non-parallel lines

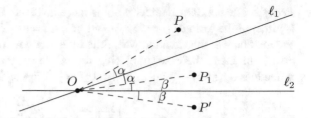

$d(\sigma_1(P), \sigma_2 \circ \sigma_1(P)) = 2(d - d(P, \ell_1))$. Therefore $d(P, \sigma_2 \circ \sigma_1(P)) = 2d(P, \ell_1) + 2(d - d(P, \ell_1)) = 2d$, so if we let $P' = \sigma_2 \circ \sigma_1(P)$ then the vector $\overrightarrow{PP'}$ is a constant.

If ℓ_1 and ℓ_2 are not parallel, they meet a unique point O (Fig. 1.16). Let P be any point in the plane, $P_1 = \sigma_1(P)$ and $P' = \sigma_2 \circ \sigma_1(P) = \sigma_2(P_1)$. We know that ℓ_1 is the perpendicular bisector of PP_1, so $OP = OP_1$. Also, ℓ_2 is the perpendicular bisector of $P_1 P'$, so $OP_1 = OP'$. Therefore, $OP = OP'$, and moreover, $\angle POP' = \angle(OP, OP_1) + \angle(OP_1, OP') = 2\angle(\ell_1, OP_1) + 2\angle(OP_1, \ell_2) = 2\angle(\ell_1, \ell_2)$, which proves (iii). □

Proof 2 For the complex number proof, if ℓ_1 and ℓ_2 are two lines that meet at z_0, we can express their equations as $z = z_0 + tv_1$ and $z = z_0 + tv_2$. So the reflections over ℓ_1 and ℓ_2 are, respectively, $\sigma_1(z) = v_1^2 \overline{z} + z_0 - v_1^2 \overline{z_0}$ and $\sigma_2(z) = v_2^2 \overline{z} + z_0 - v_2^2 \overline{z_0}$. Then

$$\sigma_1 \circ \sigma_2(z) = v_1^2 \overline{\sigma_2(z)} + z_0 - v_1^2 \overline{z_0} = v_1^2 (\overline{v_2}^2 z + \overline{z_0} - \overline{v_2}^2 z_0) + z_0 - v_1^2 \overline{z_0}$$

$$= v_1^2 \overline{v_2}^2 z - v_1^2 \overline{v_2}^2 z_0 + z_0 = z_0 + v_1^2 \overline{v_2}^2 (z - z_0),$$

which is a rotation of angle $\arg(v_1^2 \overline{v_2}^2) = 2\angle(\ell_1, \ell_2)$ about z_0.

If ℓ_1 and ℓ_2 are parallel, then we can suppose that ℓ_1 and ℓ_2 have equations $\Im(z) = k$ and $\Im(z) = -k$. For $z = a + bi$, we have $\sigma_1(z) = a + (2k - b)i$ and $\sigma_2 \circ f_1(z) = a + (-2k - 2k + b)i = z - 4ki$, which is a translation by $-4ki$. □

Theorem 1.16 *Two reflections σ_1 and σ_2 commute (i.e., $\sigma_1 \circ \sigma_2 = \sigma_2 \circ \sigma_1$) if and only if $\sigma_1 = \sigma_2$ or the lines of reflection are perpendicular.*

Proof 1 If $\sigma_1 \circ \sigma_2 = \sigma_2 \circ \sigma_1$, then

$$(\sigma_1 \circ \sigma_2) \circ (\sigma_1 \circ \sigma_2) = \sigma_1 \circ \sigma_2 \circ \sigma_2 \circ \sigma_1 = \sigma_1 \circ 1 \circ \sigma_1 = \sigma_1 \circ \sigma_1 = 1,$$

so applying $\sigma_1 \circ \sigma_2$ twice yields the identity transformation. However, $\sigma_1 \circ \sigma_2$ is a rotation, a translation, or the identity map (by Theorem 1.9). It cannot be a translation because a translation by $\overrightarrow{v}$ applied twice is a translation by $2\overrightarrow{v}$. If it is a rotation by α, then applying it twice yields a rotation by 2α. This is the identity only if $\alpha = 180°$, which means that the angle between the reflection lines is $180°/2 = 90°$. And if it is the identity map, then $\sigma_2 = \sigma_1^{-1} = \sigma_1$. □

Proof 2 The complex number proof is more involved. By Theorem 1.5, a reflection is either of the form $\sigma(z) = r\overline{z}$, $|r| = 1$ or $\sigma(z) = -\frac{s}{\overline{s}}\overline{z} + s$ for some $s \in \mathbb{C}$, $s \neq 0$. We look only at the situation where the reflections are of the latter form, which can always be made the case by placing the origin off the lines. The composition of reflections $\sigma_1(z) = -\frac{s_1}{\overline{s_1}}\overline{z} + s_1$ and $\sigma_2(z) = -\frac{s_2}{\overline{s_2}}\overline{z} + s_2$ is

$$\sigma_1 \circ \sigma_2(z) = -\frac{s_1}{\overline{s_1}}\overline{\sigma_2(z)} + s_1 = \frac{s_1\overline{s_2}}{\overline{s_1}s_2}z - \frac{s_1\overline{s_2}}{\overline{s_1}} + s_1.$$

The composition $\sigma_2 \circ \sigma_1$ is obtained by interchanging the indices, so σ_1 and σ_2 commute if and only if

$$\frac{s_2\overline{s_1}}{\overline{s_2}s_1} = \frac{s_1\overline{s_2}}{\overline{s_1}s_2} \quad \text{and} \quad -\frac{s_2\overline{s_1}}{\overline{s_2}} + s_2 = -\frac{s_1\overline{s_2}}{\overline{s_1}} + s_1$$

and this is equivalent to

$$\frac{s_2}{\overline{s_2}} = \pm\frac{s_1}{\overline{s_1}} \quad \text{and} \quad -\frac{s_2\overline{s_1}}{\overline{s_2}} + s_2 = -\frac{s_1\overline{s_2}}{\overline{s_1}} + s_1.$$

If $s_2/\overline{s_2} = s_1/\overline{s_1}$, then

$$-\frac{s_1\overline{s_1}}{\overline{s_1}} + s_2 = -\frac{s_2\overline{s_2}}{\overline{s_2}} + s_1$$

is equivalent to $-s_1 + s_2 = -s_2 + s_1$, and this of course is equivalent to $s_1 = s_2$. The latter is equivalent to $\sigma_1 = \sigma_2$.

If $s_2/\overline{s_2} = -s_1/\overline{s_1}$, the second equation is an identity:

$$\frac{s_1\overline{s_1}}{\overline{s_1}} + s_2 = \frac{s_2\overline{s_2}}{\overline{s_2}} + s_1.$$

So $s_2/\overline{s_2} = -s_1/\overline{s_1}$ is equivalent to $2\arg(s_2) = \pm\pi + 2\arg(s_1)$, that is, $\arg(s_2) = \pm\pi/2 + \arg(s_1)$. This last relation means the lines of reflection are orthogonal. $\square$

Let us study what happens when we compose a reflection with a rotation and then with a translation.

Theorem 1.17 *Let ρ be the rotation about the point O by the angle α, and let σ be the reflection over the line ℓ. Then*

(i) $\rho \circ \sigma$ is the composition of a reflection and a translation. The reflection line r passes through the midpoint O_1 of the segment determined by O and $O' = \sigma(O)$ such that $\angle(\ell, r) = \alpha/2$, and the translation vector is $\vec{v} = \overrightarrow{O_1 O_2}$, O_2 being the intersection of r and $\rho(\ell)$;

(ii) $\sigma \circ \rho$ is the composition of a reflection and a translation. The reflection line s passes through the midpoint O_1 of the segment determined by O and $O' =

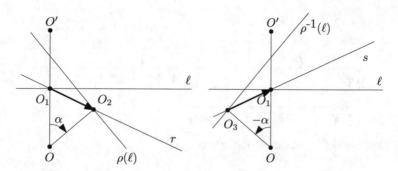

Fig. 1.17 Composition of a reflection and a rotation

$\sigma(O)$ such that $\angle(\ell, s) = -\alpha/2$, and the translation vector is $\vec{v} = \overrightarrow{O_3 O_1}$, O_3 being the intersection of s and $\rho^{-1}(\ell)$.

Proof Because both $\rho \circ \sigma$ and $\sigma \circ \rho$ reverse orientation, both are compositions of a reflection and a translation (Theorem 1.10). We need to find the reflection line and the translation vector in each case. The reasoning can be followed on Fig. 1.17.

For $\rho \circ \sigma$, let $O' = \sigma(O)$. Then

$$\rho \circ \sigma(O') = \rho \circ \sigma \circ \sigma(O) = \rho(O) = O,$$

because O is fixed by ρ. So, because in Theorem 1.10 the translation vector is parallel to the reflection line, the reflection line passes through the midpoint O_1 of OO'. Since O_1 belongs to both ℓ and the reflection line r, the translation vector is $\overrightarrow{O_1 O_2}$, where $O_2 = \rho \circ \sigma(O_1) = \rho(O_1)$. We want to better understand O_2. Because O_1 lies on ℓ, $\rho(O_1)$ lies on $\rho(\ell)$, and thus $O_2 = \rho(O_1)$ is the intersection point of r and $\rho(\ell)$ and $\vec{v} = \overrightarrow{O_1 O_2}$. Finally, if A is any other point of ℓ, then the ray $|O_1 A$ is rotated by α, so the reflection line is the angle bisector of $\angle(|O_1 A, \rho(|O_1 A))$, and it makes an angle of $\angle \alpha/2$ with ℓ. Since $\angle(\ell, r) = \alpha/2 \neq \alpha = \angle(\ell, \sigma(\ell))$, r and $\rho(\ell)$ meet at a single point, so O_2 is unique.

Now we examine $\sigma \circ \rho$. Notice that $\sigma \circ \rho(O) = \sigma(O) = O'$, so again O_1 belongs to the reflection line s. Instead of looking for $\sigma \circ \rho(O_1)$, we determine

$$(\sigma \circ \rho)^{-1}(O_1) = \rho^{-1} \circ \sigma^{-1}(O_1) = \rho^{-1} \circ \sigma(O_1) = \rho^{-1}(O_1),$$

to be able to take advantage of the fact that $\sigma(O_1) = O_1$. But $O_1 \in \ell$, so $\rho^{-1}(O_1) \in \rho^{-1}(\ell)$, and O_3 is the intersection point of s and $\rho^{-1}(\ell)$. By a similar argument as for the proof of (i), we show that $\angle(\ell, s) = -\alpha/2$. $\square$

With Theorem 1.10 in mind, let us find out what happens if we compose a reflection with a translation that is *not* parallel to the reflection line.

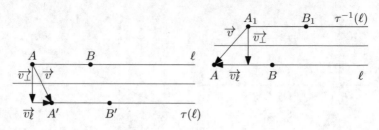

Fig. 1.18 Composition of a reflection and a translation

Theorem 1.18 *Let τ be the translation by vector $\vec{v}$, and let σ be the reflection over the line ℓ, with ℓ and $\vec{v}$ not parallel. Let $\vec{v} = \vec{v_\ell} + \vec{v_\perp}$, with $\vec{v_\ell}$ parallel to ℓ and $\vec{v_\perp}$ orthogonal to ℓ.*

(i) The composition $\tau \circ \sigma$ is a composition of a reflection over line ℓ_1 obtained by translating ℓ by $\frac{1}{2}\vec{v_\perp}$ and a translation by $\vec{v_\ell}$.

(ii) The composition $\sigma \circ \tau$ is a composition of a reflection over line ℓ_2 obtained by translating ℓ by $-\frac{1}{2}\vec{v_\perp}$ and a translation by $\vec{v_\ell}$.

Proof Because both transformations $\tau \circ \sigma$ and $\sigma \circ \tau$ reverse orientation, both are compositions of a reflection and a translation (Theorem 1.10). We need to determine the reflection line and the translation vector for each case. Let A and B be two points on ℓ (see Fig. 1.18).

We have $\tau \circ \sigma(|AB) = \tau(|AB)$, so the line ℓ is mapped to a line parallel to it. The reflection line of $\tau \circ \sigma$ is then the line that is halfway between ℓ and $\tau(\ell)$, which is obtained by translating ℓ by $\frac{1}{2}\vec{v_\perp}$. The translation vector is then the projection of $\overrightarrow{AA'}$, where $A' = \tau \circ \sigma(A)$, onto ℓ. However, since A lies on ℓ, $\tau \circ \sigma(A) = \tau(A)$, and $\overrightarrow{AA'} = \vec{v}$, so the translation vector is $\vec{v_\ell}$.

For (ii), consider $\tau^{-1}(|AB)$, so $\sigma \circ \tau(\tau^{-1}(|AB)) = \sigma(|AB) = |AB$. Thus the reflection line of $\sigma \circ \tau$ is halfway between ℓ and $\tau^{-1}(\ell)$ and is obtained by translating ℓ by $-\frac{1}{2}\vec{v_\perp}$. The translation vector is again the projection of $\overrightarrow{A_1 A_2}$ onto ℓ, where $A_1 = \tau^{-1}(A)$ and $A_2 = \sigma \circ \tau(\tau^{-1}(A)) = \sigma(A) = A$. This means that $\overrightarrow{A_1 A_2} = \vec{v}$, and so the translation vector is $\vec{v_\ell}$.   □

Notice that this theorem also proves that a translation and a reflection commute if and only if the translation vector is parallel to the reflection line. With this at hand, it becomes easier to compose an orientation-reversing isometry with some other isometry. For instance, the composition of a rotation and an orientation-reversing isometry is another orientation-reversing isometry equivalent to composing a rotation, translation, and reflection, in order. This is the same as composing a rotation (obtained by composing the rotation and the translation) and a reflection. We leave the remaining cases to the reader.

In complex numbers, it is often easier to work directly with orientation-reversing isometries as functions of the form $z \mapsto r\bar{z} + s$ and compose them with other isometries. For instance, if $f(z) = r_1\overline{z} + s_1$ and $g(z) = r_2\overline{z} + s_2$ are two orientation-reversing isometries, then $f \circ g(z) = r_1\overline{g(z)} + s_1 = r_1\overline{r_2}z + r_1\overline{s_2} + s_1$, which is a translation by $v = r_1\overline{s_2} + s_1$ if $r_1\overline{r_2} = 1$ (which means $r_1 = r_2$), and a rotation by $\arg(r_1\overline{r_2})$ about $(r_1\overline{s_2} + s_1)/(1 - r_1\overline{r_2})$ otherwise.

1.1.5 Discrete Groups of Isometries

There are some groups of geometric transformations of importance in combinatorial geometry. We will examine those that come in handy when solving the combinatorial problems listed later in this chapter.

First, there are the groups of translations that keep invariant the tilings of the plane by equilateral triangles, squares, or hexagons that are shown in Fig. 1.19. We prove now that each of these groups is isomorphic to the group $\mathbb{Z} \times \mathbb{Z}$ endowed with the composition law $(m, n) + (m', n') = (m + m', n + n')$.

Theorem 1.19 *Assume that the plane is tiled with equilateral triangles such that one of these triangles has vertices of complex coordinates $0, 1, e^{\frac{\pi i}{3}}$. Then the group of translations that maps this tiling to itself is isomorphic to $\mathbb{Z} \times \mathbb{Z}$ with the isomorphism defined by $\tau_1 \mapsto (1, 0)$ and $\tau_2 \mapsto (0, 1)$ where τ_1 is the translation by the vector $\overrightarrow{v_1}$ defined by the complex number 1 and $\overrightarrow{v_2}$ is the vector defined by the complex number $e^{\frac{\pi i}{3}}$.*

Proof An arbitrary translation τ that maps the tiling to itself is determined by $\tau(T)$, where T is the triangle with vertices $0, 1, e^{\frac{\pi i}{3}}$ mentioned in the statement. We want to write

$$\tau = \tau_1^m \tau_2^n.$$

The integer number n is uniquely determined by counting how many rows we go up or down from the level of T to that of $\tau(T)$. The integer number m is uniquely determined by counting how many steps we go left or right from $\tau_2^n(T)$ to $\tau(T)$. The isomorphism is $\tau_1^m \tau_2^n \mapsto (m, n)$, and the theorem is proved. $\qquad \square$

Similarly one can prove the following two results:

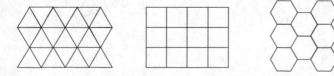

Fig. 1.19 The tessellations of the plane by regular polygons

Theorem 1.20 *The group of translations that keeps invariant the tiling of the plane by squares whose vertices are at the points of integer coordinates is isomorphic to* $\mathbb{Z} \times \mathbb{Z}$ *with the isomorphism defined by* $\tau_1 \mapsto (1,0)$ *and* $\tau_2 \mapsto (0,1)$ *where the translation vectors of* τ_1 *and* τ_2 *are* $\vec{v_1} = (1,0)$ *and* $\vec{v_2} = (0,1)$.

Theorem 1.21 *Assume that the plane is tiled with regular hexagons such that one of these hexagons has three consecutive vertices of complex coordinates* $e^{-\frac{\pi i}{3}}, 0, e^{\frac{\pi i}{3}}$. *Then the group of translations that maps this tiling to itself is isomorphic to* $\mathbb{Z} \times \mathbb{Z}$ *with the isomorphism defined by* $\tau_1 \mapsto (1,0)$ *and* $\tau_2 \mapsto (0,1)$ *where* τ_1 *is the translation by the vector* $\vec{v_1}$ *defined by the complex number* $\sqrt{3}e^{\frac{\pi i}{6}}$ *and* $\vec{v_2}$ *is the vector defined by the complex number* $\sqrt{3}e^{-\frac{\pi i}{6}}$.

A finite group that is omnipresent in applications is the one comprising the rotations that map a regular polygon with n vertices into itself. It is not hard to see (to boost your intuition use Fig. 1.20) that this group is

$$G = \{1, \rho, \rho^2, \ldots, \rho^{n-1}\},$$

where ρ is the rotation by $\frac{2\pi}{n}$ about the center of the polygon. This group is *cyclic*, meaning that is has an element with the property that any other element is some power of this element.

The last group that we will need is the group of symmetries of a nonsquare rectangle (Fig. 1.21). This is a four-element group known as the Klein 4-group, defined by

$$K = \{a, b, c, e \mid a^2 = b^2 = c^2 = e, ab = ba = c, bc = cb = a, ac = ca = b\}.$$

Here of course e is the identity element, namely, the identity transformation of the rectangle, a is the reflection over a vertical axis, b is the reflection over a horizontal axis, and c is the reflection over the center of the rectangle.

Fig. 1.20 The group of
rotations of a regular polygon

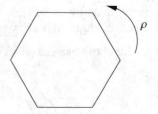

Fig. 1.21 The Klein 4-group

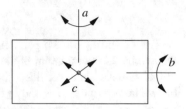

1.1.6 Theoretical Questions About Isometries

Our narrative continues with a number of questions of theoretical nature. Their aim is to test and deepen your understanding of the structure of the group of isometries, and to endow you with some useful lemmas. They will be followed in the next section by more inviting problems about applications of isometries to Euclidean geometry.

1 The triangles ABC and $A'B'C'$ are mapped into each other by the reflections over two distinct lines. Prove that both triangles are equilateral.

2 Does there exist a set of points in the plane that has exactly two centers of symmetry, meaning that it is mapped into itself by the reflections over exactly two points?

3 Show that if f is an isometry such that $f \circ f = 1$, then f is either the identity map, the reflection over a point, or the reflection over a line.

4 Let $O_1, O_2, \ldots, O_{2n}$ be points in the plane, and let σ_k be the reflection over O_k for all $k = 1, 2, \ldots, 2n$. Prove that $\sigma_1 \circ \sigma_2 \circ \cdots \circ \sigma_{2n-1} \circ \sigma_{2n}$ is the identity map if and only if $\overrightarrow{O_1O_2} + \overrightarrow{O_3O_4} + \cdots + \overrightarrow{O_{2n-1}O_{2n}} = \overrightarrow{0}$.

5 Given the reflections $\sigma_1, \sigma_2, \sigma_3, \sigma_4, \sigma_5$ over five points in the plane, prove that

$$\sigma_1 \circ \sigma_2 \circ \sigma_3 \circ \sigma_4 \circ \sigma_5 = \sigma_5 \circ \sigma_4 \circ \sigma_3 \circ \sigma_2 \circ \sigma_1.$$

6 Let a, b, c be three distinct lines in the plane, and let $\sigma_a, \sigma_b, \sigma_c$ be the reflections over these lines. Prove that

$$\sigma_a \circ \sigma_b \circ \sigma_c = \sigma_c \circ \sigma_b \circ \sigma_a$$

if and only if the lines a, b, c either have a common point or are parallel to one another.

7 Show that the composition of the reflections over three lines that intersect pairwise at three distinct points has no fixed points.

8 Let a, b, c, d be four distinct lines, and let $\sigma_a, \sigma_b, \sigma_c, \sigma_d$ be the reflections over these lines, respectively. How are the lines arranged in the plane if

$$\sigma_a \circ \sigma_b \circ \sigma_c \circ \sigma_d = \sigma_b \circ \sigma_a \circ \sigma_d \circ \sigma_c?$$

9 Let ℓ_1 and ℓ_2 be two parallel lines, and let ℓ be a line perpendicular to them. Denote by M the midpoint of the segment determined by the intersection points of ℓ_1 and ℓ_2 with ℓ, and consider $A, B \in \ell$ such that M is the midpoint of AB. Let a, b be two lines parallel to ℓ_1 and ℓ_2 that pass through A and B, respectively. Denote by $\sigma_{\ell_1}, \sigma_{\ell_2}, \sigma_a, \sigma_b, \sigma_A$, and σ_B the reflections over the lines ℓ_1, ℓ_2, a, b and the points A, B, respectively. Prove that

$$\sigma_{\ell_2} = \sigma_b \circ \sigma_{\ell_1} \circ \sigma_a = \sigma_B \circ \sigma_{\ell_1} \circ \sigma_A.$$

10 Show that all axes of symmetry of a polygon intersect at one point. (An axis of symmetry is a line such that the reflection over that line maps the polygon to itself.)

11 Using a straight edge and a compass, construct a pentagon knowing the midpoints of its sides.

12 Two cars travel at equal speeds on two roads that are nonparallel straight lines. Show that there is a point from which the two cars are, at any moment, at equal distances.

1.2 Isometries in Euclidean Geometry Problems

Isometries are ubiquitous in Euclidean geometry. Many a time they show up explicitly, such as when noticing that the orthocenter lies at the intersection of the three reflections of the circumcircle over the sides, or that the rotation of a segment by 60° around one of its endpoints gives rise to an equilateral triangle.

It is hence time to show how isometries come into play in the old-fashioned, but beautiful realm of Euclidean geometry.

1.2.1 Some Constructions and Classical Results in Euclidean Geometry That Use Isometries

We introduce below two objects from the geometry of a triangle that are constructed using reflections over the angle bisectors of a triangle.

I. The Symmedians of a Triangle
Given two lines that pass through the same vertex of a triangle (i.e., two cevians of the same vertex), these lines are called *isogonal* if they reflect into each other over the bisector of the angle at that vertex.

Fig. 1.22 The definition of
the symmedian

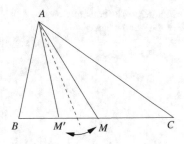

Definition A symmedian of a triangle is the reflection of a median of the triangle
over the corresponding angle bisector.

So the median and symmedian of a given vertex provide a particular example
of isogonal lines. The definition of the symmedian is illustrated in Fig. 1.22, where
M is the midpoint of the side BC and the line AM' is the reflection of the line
of support of the median AM over the angle bisector of $\angle A$. A triangle has three
symmedians, one for each median. The next result shows how to construct them.

Theorem 1.22 (Construction of Symmedians) *Let ABC be a triangle. Then the
symmedian from A passes trough the point where the tangents to the circumcircle at
B and C intersect.*

We will give two proofs of this result:

Proof 1 Let T be the intersection of the tangents to the circumcircle at B and C, let
M be the midpoint of BC, and let α, β, γ denote the measures of the three angles
of the triangle. Examining Fig. 1.23, we notice that $BM = MC$ and $BT = TC$.
We can put these equalities to work using the Law of Sines in the triangles ABM,
ACM, ABT, and ACT

$$\frac{AM}{\sin \beta} = \frac{BM}{\sin \angle BAM}, \quad \frac{AM}{\sin \gamma} = \frac{CM}{\sin \angle CAM},$$

$$\frac{AT}{\sin(\alpha + \beta)} = \frac{BT}{\sin \angle BAT}, \quad \frac{AT}{\sin(\alpha + \gamma)} = \frac{CT}{\sin \angle CAT},$$

where for the last two equalities we used the fact that $\angle ABT = \alpha + \beta$ and $\angle ACT =
\alpha + \gamma$.

Using the abovementioned equalities of segments, we obtain

$$\frac{\sin \angle BAM}{\sin \angle CAM} = \frac{\sin \angle CAT}{\sin \angle BAT}.$$

And we have $\angle BAM + \angle CAM = \angle CAT + \angle BAT = \alpha$, so by denoting the
measures of $\angle CAM$ and $\angle BAT$ by x and y, respectively, we obtain

Fig. 1.23 The construction
of the symmedian

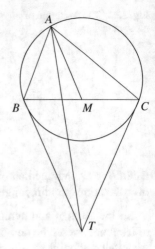

$$\frac{\sin(\alpha - x)}{\sin x} = \frac{\sin(\alpha - y)}{\sin y}.$$

Using the subtraction formula for sine, this further yields

$$\sin\alpha \cot x - \cos\alpha = \sin\alpha \cot y - \cos\alpha.$$

But the cotangent function is one-to-one on $(0, \pi)$, so this equality implies $x = y$, that is, $\angle CAM = \angle BAT$, showing that AT is symmedian.   □

Proof 2 Draw a line PQ through T ($P \in AB$, $Q \in AC$) that is antiparallel to BC, namely, such that $\angle ACB = \angle APQ$ and (consequently) $\angle ABC = \angle AQP$ (see Fig. 1.24). We have

$$\angle TBP = 180° - \angle ABC - \angle CBT = 180° - \angle ABC - \angle BAC$$

$$= \angle ACB = \angle BPT$$

and similarly $\angle TCQ = \angle TQC$. It follows that the triangles TPB and TQC are isosceles, and hence $TB = TP$ and $TC = TQ$. But $TB = TC$, so T is the midpoint of PQ. And the triangle APQ reflects over the angle bisector of $\angle BAC$ to a triangle $AP'Q'$ whose side $P'Q'$ is parallel to BC (the antiparallel becomes parallel). In that case the image T' of T under this reflection is on the line AM. Consequently the line AT is the reflection of AM; it is the symmedian.   □

   This construction of the symmedian leads naturally to the concept of a harmonic quadrilateral. We say that four points A, B, D, C that appear in this order on a circle form a *harmonic quadrilateral* if

$$\frac{AB}{AD} : \frac{CB}{CD} = 1 \iff AB \cdot CD = AD \cdot BC.$$

Fig. 1.24 Proof of the characterization of symmedians

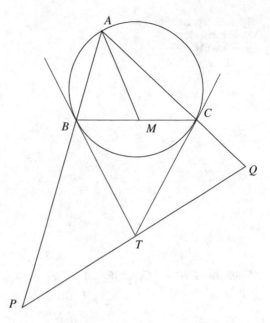

In complex coordinates, a quadrilateral A, B, C, D is harmonic if and only if

$$\frac{a-b}{a-d} : \frac{c-b}{c-d} = -1.$$

This is the condition that the cross-ratio (a, c, b, d) is equal to -1 (and a little algebra shows that this is equivalent to $(a, b, c, d) = -1$).

Proposition 1.23 *Given a triangle, the point where the symmedian intersects the circumcircle forms with the vertices a harmonic quadrilateral. Conversely, given a harmonic quadrilateral, in the triangle formed by three vertices, the diagonal that is not side of the triangle is a symmedian.*

Proof We argue on Fig. 1.25. Let ABC be the triangle, and let D be the intersection of the symmedian from A with the circumcircle. For the direct implication, note that from the similarity of triangles TBA and TDB ($\angle DTB = \angle BTA$, $\angle DBT = \angle BAT = \frac{1}{2} \overset{\frown}{BD}$), we obtain

$$\frac{BA}{BD} = \frac{TB}{TD},$$

and from the similarity of TCA and TDC, we obtain

$$\frac{CA}{CD} = \frac{TC}{TD}.$$

Fig. 1.25 Symmedians give
rise to harmonic
quadrilaterals

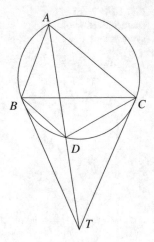

Given that $TB = TC$, we obtain

$$\frac{BA}{BD} = \frac{CA}{CD},$$

so $ABDC$ is harmonic.

For the converse, note that the condition for $ABDC$ to be harmonic can be rewritten as

$$\frac{AB}{AC} = \frac{DB}{DC}$$

so D is the second intersection point of the Apollonian circle determined by the points A and B and the ratio $k = AB/AC$ with the circumcircle ABC. Since this intersection point is unique, it must be the point where the symmedian intersects the circumcircle. We recall that the Apollonian circle determined by the points A and B and the ratio $k > 0$ is the locus of the points P with the property that $PA/PB = k$ (the proof that this locus is a circle makes the object of Problem 165).   □

We will visit harmonic quadrilaterals once more in the chapter about inversion.

II. The Isogonal Conjugate of a Point
The definition of the isogonal conjugate of a point is based on the following result.

Theorem 1.24 (Isogonal Conjugate of a Point) *Given a triangle, the reflections over the angle bisectors of three concurrent cevians are three cevians that are also concurrent.*

In other words, if three cevians of a triangle intersect, then the reflections of these cevians over the respective bisectors also intersect. If P and Q are the intersections of the two groups of cevians, then P and Q are called *isogonal conjugates*, and each point is said to be the isogonal conjugate of the other. As an example, the orthocenter is the isogonal conjugate of the circumcenter (prove it!). The isogonal conjugate of the centroid is called the *Lemoine point*. We will revisit this point in Sect. 2.5.1. Again, we give two different proofs.

Proof 1 Let us change slightly the phrasing of the theorem: In the interior of the triangle ABC, there are the points P and Q such that the segments AP and AQ form equal angles with the angle bisector of $\angle BAC$ and the segments BP and BQ form equal angles with the angle bisector of $\angle ABC$. Prove that CP and CQ form equal angles with the angle bisector of $\angle BCA$.

   To prove this, reflect P over AB, AC, and BC to obtain the points P_C, P_B, and P_A, respectively, as shown in Fig. 1.26. Then, the fact that AP and AQ form equal angles with the angle bisector of $\angle BAC$ implies that AQ is the angle bisector of $\angle P_B A P_C$. Indeed, working with oriented angles, if R is a point on the angle bisector of $\angle BAC$, then

$$\angle QAP_B = \angle PAP_B - 2\angle PAR = \angle RAP_B - \angle PAR$$

$$= \angle RAB + \angle RAB + \angle PAR - \angle PAR = 2\angle RAB$$

$$= \angle BAC$$

and

$$\angle P_C AQ = \angle P_C AP + 2\angle PAR = \angle P_C AR + \angle PAR$$

$$= \angle CAR + \angle CAR - \angle PAR + \angle PAR = 2\angle CAR$$

$$= \angle BAC.$$

So AQ bisects $\angle P_B A P_C$. Because $AP_B = AP_C$, the triangle $AP_B P_C$ is isosceles, so the angle bisector AQ is also the perpendicular bisector of the segment $P_B P_C$. Hence Q is on the perpendicular bisector of $P_B P_C$.

Fig. 1.26 Reflection of P over the sides

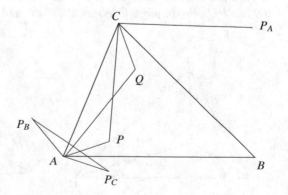

Repeating the argument, we deduce that the conditions from the statement imply that Q is on the perpendicular bisectors of the segments $P_A P_C$ as well. But then Q is the circumcenter of triangle $P_A P_B P_C$, and so Q must be on the perpendicular bisector of $P_A P_B$, which is equivalent to the fact that CP and CQ form equal angles with the angle bisector of $\angle C$.   □

Proof 2 Let the cevian from A form the angles α_1 and α_2 with AB and AC, respectively; the cevian from B form the angles β_1 and β_2 with BC and BA, respectively; and cevian from C form the angles γ_1 and γ_2 with CA and CB, respectively. By the trigonometric version of Ceva's Theorem, the fact that the three cevians intersect implies that

$$\frac{\sin\alpha_1}{\sin\alpha_2} \cdot \frac{\sin\beta_1}{\sin\beta_2} \cdot \frac{\sin\gamma_1}{\sin\gamma_2} = 1.$$

If we reflect the cevians over the corresponding bisectors, then the first of them forms the angles α_2 and α_1 with AB and AC, respectively; the second forms the angles β_2 and β_1 with BC and BA, respectively; and the third forms the angles γ_2 and γ_1 with CA and CB, respectively (Fig. 1.27). And, again by Ceva's Theorem written in trigonometric form, the condition that the three cevians intersect is

$$\frac{\sin\alpha_2}{\sin\alpha_1} \cdot \frac{\sin\beta_2}{\sin\beta_1} \cdot \frac{\sin\gamma_2}{\sin\gamma_1} = 1,$$

and this is just the above relation flipped over. Hence the conclusion.   □

We point out that a proof of Ceva's Theorem can be found in the next chapter (Problem 116). It is important to know that the construction of the isogonal conjugate also works when the triangle ABC is degenerate, with the vertex A at infinity. In that case the triangle consists of the segment BC and the two parallel rays $|BX$ and $|CY$ (Fig. 1.28).

Then the isogonal conjugate of a point P is the intersection of the reflections of PB and PC over the angle bisectors of $\angle XBC$ and $\angle YCB$ and the reflection of the

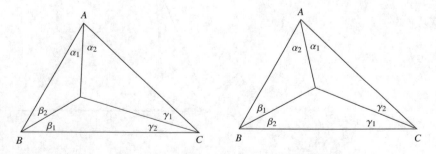

Fig. 1.27 Trigonometric proof of the existence of the isogonal conjugate

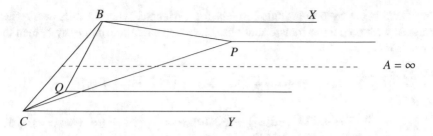

Fig. 1.28 The existence of the isogonal conjugate in the case where the triangle is degenerate

line through P that is parallel to BX over the line parallel to BC that runs through the midpoint of BC.

The existence of Q can be shown by adapting the first proof of Theorem 1.24, or by using a limiting argument with $A \to \infty$, where we notice that the angle bisector of $\angle BAC$ becomes the line that is at equal distance from BX and CY, as by the Bisector Theorem, the angle bisector of $\angle BAC$ cuts BC in the ratio AB/AC, and this ratio tends to 1 when A tends to infinity.

As a direct corollary of the first proof of Theorem 1.24, we obtain the following two characterizations of the isogonal conjugate of a point.

Proposition 1.25

(i) *Let P be a point in the plane of the triangle ABC, and let P_A, P_B, P_C be reflections of P over the lines BC, CA, and AB, respectively. Then the isogonal conjugate of P with respect to the triangle ABC is the circumcenter of $P_A P_B P_C$.*

(ii) *Let P be a point in the plane of the triangle ABC, and let L, M, N be the projections of P onto the lines BC, CA, and AB, respectively. Then the perpendiculars from A, B, and C onto the lines MN, LN, and LM, respectively, are concurrent at the isogonal conjugate of P with respect to the triangle ABC.*

Note that (ii) follows from the fact that MN, LN, and LM are parallel to $P_B P_C$, $P_A P_C$, and $P_A P_B$, respectively.

III. Area in Complex Coordinates
Here is a slick proof, via translations and rotations, of the "shoelace" formula for the area, which we phrase in complex coordinates, in the spirit of this book.

Theorem 1.26 (Gauss) *Let $a_1, a_2, \ldots, a_n$ be the coordinates of the vertices of a non-self-intersecting polygon that is oriented counterclockwise. Then the area of the polygon is equal to*

$$\frac{1}{2}\Im(\overline{a_1}a_2 + \overline{a_2}a_3 + \cdots + \overline{a_{n-1}}a_n + \overline{a_n}a_1).$$

Proof We begin by proving this formula for a triangle, where it falls apart when attacked with geometric transformations. If we translate the triangle by z, then the formula reads

$$\frac{1}{2}\Im[\overline{(a_1 + z)}(a_2 + z) + \overline{(a_2 + z)}(a_3 + z) + \overline{(a_3 + z)}(a_1 + z)]$$

$$= \frac{1}{2}\Im(\overline{a_1}a_2 + \overline{a_2}a_3 + \overline{a_3}a_1) + \frac{1}{2}\Im[(a_1 + a_2 + a_3)\overline{z} + \overline{(a_1 + a_2 + a_3)}z]$$

$$+ \frac{3}{2}\Im|z|^2.$$

And this is equal to the area of the nontranslated triangle

$$\frac{1}{2}\Im(\overline{a_1}a_2 + \overline{a_2}a_3 + \overline{a_3}a_1)$$

because the second and the third terms are real numbers. It follows that we can translate the triangle so that $a_3 = 0$, without changing the value of either the area or of the formula from the statement. It suffices to prove that, in this case, the area is equal to $\frac{1}{2}\Im\overline{a_1}a_2$. Rotate the triangle so that a_1 becomes positive and real. The rotation does not change the area, nor does it change the formula because

$$\overline{e^{i\alpha}a_1}e^{i\alpha}a_2 = e^{i(\alpha-\alpha)}\overline{a_1}a_2 = \overline{a_1}a_2.$$

And in this case the area is indeed the imaginary part of $\frac{1}{2}\Im\overline{a_1}a_2$, because this is $\frac{1}{2} \times$ base$\times$height (Fig. 1.29). Note that because the triangle is oriented counterclockwise, the imaginary part of a_2 is positive (when it is oriented clockwise, the imaginary part of a_2 is negative, and consequently the formula needs a minus sign in front of it).

We have thus proved the formula for the area of a triangle. For a general polygon, start by dissecting the polygon into triangles (Fig. 1.30). This can be done easily if the polygon is convex; if the polygon is not convex, the existence of the dissection can be proved by induction on the number of vertices once we show that interior diagonals always exist. To show the existence of an interior diagonal, pick a vertex,

Fig. 1.29 Proof of the formula for the area: the simplest case

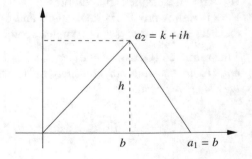

Fig. 1.30 Proof of the
formula for the area of a
polygon

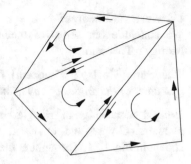

and from it look inside a polygon. If you see another vertex, join it to the chosen
one by an interior diagonal. If you do not see another vertex, then the two vertices
adjacent to the chosen vertex can be joined by an interior diagonal.

Once the polygon has been decomposed into triangles, add the area formulas for
all these triangles. Note that every internal diagonal joining some a_j and a_k belongs
to two triangles, and in the sum it gives rise to a term of the form $\frac{1}{2}\Im \overline{a_j}a_k$ and a term
of the form $\frac{1}{2}\Im \overline{a_k}a_j$. But $\overline{a_j}a_k + \overline{a_k}a_j$ is a real number, so its imaginary part is zero.
Hence the contribution of internal diagonals is zero, and the formula is proved.   □

Remark An easy check shows that

$$\Im(\overline{a_1}a_2 + \overline{a_2}a_3 + \overline{a_3}a_1) = - \begin{vmatrix} 1 & 1 & 1 \\ a_1 & a_2 & a_3 \\ \overline{a_1} & \overline{a_2} & \overline{a_3} \end{vmatrix},$$

so we can also write the area in terms of determinants (in fact it is the behavior of
determinants under geometric transformations that was used in the solution). This
formula can be used to check the collinearity of three points a_1, a_2, a_3, as the points
are collinear if and only if the triangle they form has area zero, that is, if and only if
$\Im(\overline{a_1}a_2 + \overline{a_2}a_3 + \overline{a_3}a_1) = 0$. We can thus write the equation of the line through two
points a_1, a_2 as

$$\Im(\overline{a_1}a_2 + \overline{a_2}z + \overline{z}a_1) = 0 \text{ or } \begin{vmatrix} 1 & 1 & 1 \\ a_1 & a_2 & z \\ \overline{a_1} & \overline{a_2} & \overline{z} \end{vmatrix} = 0.$$

For completeness, we recall the formula for a 3×3 determinant

$$\begin{vmatrix} a_{11} & a_{12} & a_{13} \\ a_{21} & a_{22} & a_{23} \\ a_{31} & a_{32} & a_{33} \end{vmatrix} = a_{11}(a_{22}a_{33} - a_{23}a_{32}) + a_{12}(a_{23}a_{31} - a_{21}a_{33})$$

$$+ a_{13}(a_{21}a_{32} - a_{31}a_{22}).$$

IV. Morley's Theorem

We conclude our series of theoretical examples with Alain Connes' proof of Morley's Theorem.

Theorem (Morley's Theorem) *The three points of intersection of the adjacent trisectors of the angles of any triangle form an equilateral triangle.*

Proof Let the original triangle be ABC, whose angles $\angle A$, $\angle B$, $\angle C$ are oriented counterclockwise, and let $A'B'C'$ be the triangle determined by the trisectors, as shown in Fig. 1.31. Throughout the solution, we measure angles in radians.

The first observation that Connes made was that the intersections of consecutive trisectors are the fixed points of pairwise products of rotations ρ_A, ρ_B, ρ_C, around the vertices A, B, C of the triangle, of angles equal to two thirds of the corresponding angles of the triangle. More precisely:

- A' is the unique fixed point of $\rho_B \circ \rho_C$,
- B' is the unique fixed point of $\rho_C \circ \rho_A$,
- C' is the unique fixed point of $\rho_A \circ \rho_B$.

To see this, note, with the aid of Fig. 1.32, that $\rho_C(A')$ is the reflection A'' of A' over BC and $\rho_B(A'') = A'$, so $\rho_B \circ \rho_C(A') = A'$.

His second observation is that

$$\rho_A^3 \circ \rho_B^3 \circ \rho_C^3 = 1.$$

Fig. 1.31 Morley's Theorem

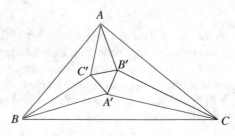

Fig. 1.32 Alain Connes' first observation

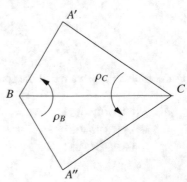

Indeed, ρ_A^3 is the rotation about A by $2\angle A$, which is the composition of the reflection σ_{AB} over AB followed by the reflection σ_{AC} over AC, and similarly for the other two rotations. Then

$$\rho_A^3 \circ \rho_B^3 \circ \rho_C^3 = \sigma_{AC} \circ \sigma_{AB} \circ \sigma_{AB} \circ \sigma_{BC} \circ \sigma_{BC} \circ \sigma_{AC} = 1,$$

because $\sigma_{AB}^2 = \sigma_{BC}^2 = \sigma_{AC}^2 = 1$.

Another way to see this is to first observe that the transformation $\rho_A^3 \circ \rho_B^3 \circ \rho_C^3$ rotates figures by $2\angle A + 2\angle B + 2\angle C = 2\pi$. This means that it does *not* rotate figures, and therefore it is either the identity map or a translation. Also

$$\rho_A^3 \circ \rho_B^3 \circ \rho_C^3(C) = \rho_A^3 \circ \rho_B^3(C) = \rho_A^3(\sigma_{AB}(C))$$

$$= \sigma_{AB}(\sigma_{AB}(C)) = C.$$

Since the transformation has a fixed point, it cannot be a translation; hence, it is the identity map.

Note also that $\rho_A \circ \rho_B \circ \rho_C$ is an isometry which rotates figures by $\frac{2}{3} \times \pi = \frac{2\pi}{3}$, so it is a *rotation* by $\frac{2\pi}{3}$, and Connes first tried to see if this is the rotation that maps $A'B'C'$ to itself. Unfortunately, it is not.

This is the moment to switch to complex numbers. Let α, β, and γ be the complex coordinates of A', B', and C', respectively. We have the following characterization of equilateral triangles, which is of interest in itself.

Lemma 1.27 *The points α, β, and γ are the vertices of an equilateral triangle oriented counterclockwise if and only if*

$$\alpha + e^{\frac{2\pi i}{3}}\beta + e^{\frac{4\pi i}{3}}\gamma = 0.$$

Proof The condition that γ rotates to α around β by $e^{\frac{\pi i}{3}}$ is that

$$\frac{\alpha - \beta}{\gamma - \beta} = e^{\frac{\pi i}{3}}.$$

Rewrite this as

$$\alpha + (e^{\frac{\pi i}{3}} - 1)\beta - e^{\frac{\pi i}{3}}\gamma = 0.$$

Using

$$e^{\frac{\pi i}{3}} - 1 = -\frac{1}{2} + \frac{\sqrt{3}}{2}i = e^{\frac{2\pi i}{3}} \quad \text{and} \quad -e^{\frac{\pi i}{3}} = -\frac{1}{2} - \frac{\sqrt{3}}{2}i = e^{\frac{4\pi i}{3}},$$

we obtain the equivalent condition

$$\alpha + e^{\frac{2\pi i}{3}} \beta + e^{\frac{4\pi i}{3}} \gamma = 0,$$

and the lemma is proved. □

Returning to the proof of Morley's Theorem, using Theorem 1.1, we write

$$\rho_A(z) = a_1 z + b_1, \quad \rho_B(z) = a_2 z + b_2, \quad \rho_C(z) = a_3 z + b_3,$$

where

$$a_1 = e^{\frac{2i\angle A}{3}}, \quad a_2 = e^{\frac{2i\angle B}{3}}, \quad a_3 = e^{\frac{2i\angle C}{3}}.$$

Note that because $\angle A + \angle B + \angle C = \pi$

$$a_1 a_2 a_3 = e^{\frac{2\pi i}{3}}.$$

Now Morley's Theorem follows by applying Connes' Theorem below to ρ_A, ρ_B, ρ_C and then using Lemma 1.27. □

Theorem (Connes' Theorem) *Consider the transformations*

$$f_j : \mathbb{C} \to \mathbb{C}, \quad f_j(z) = a_j z + b_j, \quad j = 1, 2, 3.$$

Assume that $a_1 a_2$, $a_1 a_3$, $a_2 a_3$, and $a_1 a_2 a_3$ are all different from both 0 and 1. The following are equivalent:

(i) $f_1^3 \circ f_2^3 \circ f_3^3 = 1$,
(ii) $\omega^3 = 1$ and $\alpha + \omega \beta + \omega^2 \gamma = 0$ where $\omega = a_1 a_2 a_3$ and α, β, γ are the (unique) fixed points of $f_2 \circ f_3$, $f_3 \circ f_1$, and $f_1 \circ f_2$, respectively.

Proof Condition (i) reads

$$a_1^3 a_2^3 a_3^3 z + (a_1^2 + a_1 + 1)b_1 + a_1^3 (a_2^2 + a_2 + 1)b_2 + (a_1 a_2)^3 (a_3^2 + a_3 + 1)b_3 = z,$$

for all z. Identifying coefficients we obtain $\omega^3 = a_1^3 a_2^3 a_3^3 = 1$ and

$$(a_1^2 + a_1 + 1)b_1 + a_1^3 (a_2^2 + a_2 + 1)b_2 + (a_1 a_2)^3 (a_3^2 + a_3 + 1)b_3 = 0. \quad (1.2)$$

All that remains to show is that this equality is equivalent to

$$\alpha + \omega \beta + \omega^2 \gamma = 0.$$

For this we have to compute the coordinates of α, β, γ. Note that

$$f_1 \circ f_2(z) = a_1 a_2 z + a_1 b_2 + b_1,$$

$$f_2 \circ f_3(z) = a_2 a_3 z + a_2 b_3 + b_2,$$

$$f_3 \circ f_1(z) = a_3 a_1 z + a_3 b_1 + b_3,$$

where we point out that each of these is a rotation so it has a unique fixed point. We compute the fixed points to be

$$\alpha = \text{fix}(f_2 \circ f_3) = \frac{a_2 b_3 + b_2}{1 - a_2 a_3} = \frac{a_1 a_2 b_3 + a_1 b_2}{a_1 - \omega},$$

$$\beta = \text{fix}(f_3 \circ f_1) = \frac{a_3 b_1 + b_3}{1 - a_3 a_1} = \frac{a_2 a_3 b_1 + a_2 b_3}{a_2 - \omega},$$

$$\gamma = \text{fix}(f_1 \circ f_2) = \frac{a_1 b_2 + b_1}{1 - a_1 a_2} = \frac{a_3 a_1 b_2 + a_3 b_1}{a_3 - \omega}.$$

The reader should notice that we transformed the denominators from quadratic expressions to linear expressions with the goal of simplifying the equation $\alpha + \omega\beta + \omega^2\gamma = 0$ as much as possible. Now substitute the values of α, β, and γ into this equation and then multiply by the common denominator to obtain the equivalent equation

$$(a_1 a_2 b_3 + a_1 b_2)(a_2 - \omega)(a_3 - \omega) + \omega(a_2 a_3 b_1 + a_2 b_1)(a_1 - \omega)(a_3 - \omega)$$

$$+ \omega^2 (a_1 a_3 b_2 + b_1 a_3)(a_1 - \omega)(a_2 - \omega) = 0.$$

This looks discouraging, but after multiplying out

$$a_2 b_3 \omega + b_2 \omega - a_1 a_2^2 b_3 \omega - b_3 \omega^2 - a_1 a_2 b_2 \omega - a_1 a_3 b_2 \omega + a_1 a_2 b_3 \omega^2 + a_1 b_2 \omega^2$$

$$+ b_1 a_3 \omega^2 + b_3 \omega^2 - a_2 a_3^2 b_1 \omega^2 - b_1 - a_2 a_3 b_3 \omega^2 - a_2 a_1 b_3 \omega^2 + a_2 a_3 b_1 + a_2 b_3$$

$$+ a_1 b_2 + b_1 - a_1^2 a_3 b_2 - b_2 \omega - a_3 a_1 b_1 - a_2 a_3 b_1 + a_1 a_3 b_2 \omega + a_3 b_1 \omega$$

we notice that quite a few terms cancel out. We are left with

$$a_2 b_3 (1 + \omega) + a_1 b_2 (1 + \omega^2) a_3 b_1 (\omega + \omega^2) - a_1 a_2^2 b_3 \omega - a_1 a_2 b_2 \omega - a_2 a_3^2 b_1 \omega^2$$

$$- a_2 a_3 b_3 \omega^2 - a_1^2 a_3 b_2 - a_1 a_3 b_1 = 0.$$

This can be further transformed using the equality $1 + \omega + \omega^2 = 0$ into the equivalent identity

$$a_2 b_3 \omega^2 + a_1 b_2 \omega + a_3 b_1 + a_1 a_2^2 b_3 \omega + a_1 a_2 b_2 \omega - a_2 a_3^2 b_1 \omega^2 - a_2 a_3 b_3 \omega^2$$

$$- a_1^2 a_3 b_2 - a_1 a_3 b_1 = 0.$$

In order to turn this into (1.2), we group the terms with respect to b_1, b_2, b_3

$$(a_3 + a_1a_3 + a_2a_3^2\omega^2)b_1 + (a_1\omega + a_1a_2\omega + a_1^2a_3)b_2$$
$$+(a_2\omega^2 + a_2a_3\omega^2 + a_1a_2\omega)b_3 = 0. \tag{1.3}$$

Comparing the coefficient of b_1 in (1.2) and (1.3), we notice that

$$a_3 + a_1a_3 + a_2a_3\omega^2 = a_3 + a_1a_3 + a_1^2a_2^3a_3^4 = a_3\left(1 + a_1 + \frac{1}{a_1}\right)$$

$$= \frac{a_3}{a_1}(a_1^2 + a_1 + 1).$$

And now it is straightforward to check that (1.3) becomes (1.2) after multiplying by a_1/a_3. The theorem is proved.   □

1.2.2   Examples of Problems Solved Using Isometries

Let us now present a few actual mathematics competition problems. The first has appeared in the All-Russian Mathematical Olympiad in 2002.

Problem 1.1 Let O be the circumcenter of the triangle ABC. Choose M and N on the sides AB and AC, respectively, such that $\angle MON = \angle A$. Prove that the perimeter of the triangle MAN is greater than or equal to the length of the side BC.

Solution 1 The aim is to rearrange the sides of the triangle MAN so that they form a polygonal line connecting B to C. To this end consider the rotations ρ_1 and ρ_2 about O, the first mapping A to B, and the second mapping A to C, and let $M' = \rho_1(M)$ and $N' = \rho_2(N)$ (Fig. 1.33).

Fig. 1.33 A transformation of MAN into a polygonal line

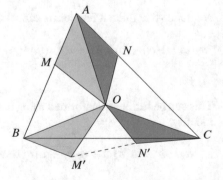

The triangle OBM' is the image of the triangle OAM through ρ_1, so $BM' = AM$, and the triangle OCN' is the image of the triangle OAN through ρ_2, so $CN' = AN$. But

$$\angle MOM' = \angle AOB = 2\angle C \text{ and } \angle NON' = \angle AOC = 2\angle B.$$

Hence

$$\angle M'ON' = 360° - \angle MON - \angle MOM' - \angle NON'$$

$$= 360° - \angle A - 2\angle C - 2\angle B = \angle A.$$

So the triangles MON and $M'ON'$ are congruent, having two pairs of sides and the angle between them equal. It follows that $M'N' = MN$. As the shortest path between two points is the segment joining them, we can write

$$AM + MN + NA = BM' + M'N' + N'C' \geq BC.$$

This is the desired inequality.   □

Solution 2 Reflect A over the lines OM and ON to the points B' and C', respectively. Then $AM = B'M$ and $AN = C'N$. Therefore

$$AM + MN + NA = B'M + MN + NC' \geq B'C',$$

where for the last inequality we have used again the fact that the shortest path between two points is the segment joining them (Fig. 1.34).

On the other hand

$$\angle B'OC' = \angle B'OM + \angle MON + \angle NOC' = \angle AOM + \angle MON + \angle AON$$

$$= 2\angle MON = 2\angle A = \angle BOC.$$

Fig. 1.34 Another transformation of MAN into a polygonal line

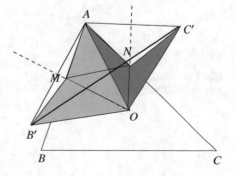

Also, $B'O = AO = C'O$, and since $AO = BO = CO$, we obtain that $BO = B'O$ and $CO = C'O$. Consequently the triangles BOC and $B'OC'$ are congruent, and therefore $BC = B'C'$. The conclusion follows.   □

This is a standard method for proving metric equalities or inequalities, which can be applied in some of the problems below: map by isometries the segments to turn them into the sides of a triangle (or polygon), and then use metric relations in a triangle (or polygon).

If we rotate a point A about O by $60°$ to a point A', then the triangle AOA' is equilateral, being isosceles and having a $60°$ angle (see Fig. 1.35). So equilateral triangles and $60°$ rotations go hand in hand, and this is the subject of the next problem, which was communicated to us by Onofre Campos.

Problem 1.2 Let $ABCD$ be a convex quadrilateral such that there exists a point M in the plane satisfying $AM = MB$, $CM = MD$, and $\angle AMB = \angle CMD = 120°$. Prove that there exists a point N in the plane such that the triangles AND and CNB are both equilateral.

Solution There is a $60°$ rotation hidden in this problem. To bring it to light, note that the point M is the center of a $120°$ rotation that takes A to B and C to D (this is where we use the fact that $ABCD$ is convex, so that A, B, C, and D are properly cyclically ordered). This rotation takes therefore the segment AC to the segment BD. So the two diagonals of the quadrilateral are equal and make an angle of $120°$. But then the diagonals also make an acute angle of $180° - 120° = 60°$ (see Fig. 1.36).

If now we focus on the equal segments AC and DB, and consider the orientation preserving isometry that maps A to D and C to B, then by Theorem 1.9, this transformation is a rotation by $60°$, which is the angle formed by AC and DB.

Fig. 1.35 A $60°$ rotation yields an equilateral triangle

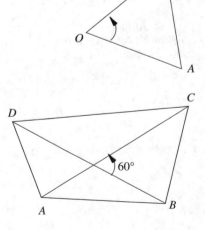

Fig. 1.36 Equal diagonals forming a $60°$ angle

Fig. 1.37 The translation
$P \mapsto P'$

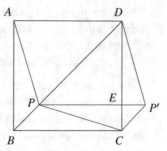

We can therefore choose N to be the center of this $60°$ rotation, as on the one hand $NA = ND$ and $\angle AND = 60°$ and on the other hand $NC = NB$ and $\angle BNC = 60°$, and so the triangles AND and CNB are equilateral, as desired. □

Problem 1.3 Let $ABCD$ be a square and let P be a point inside it. Prove that $\angle APB + \angle CPD = 180°$ if and only if P lies on at least one of the diagonals AC and BD.

Solution When the problem involves parallelograms (a square is a very special parallelogram!) and angles that sum up to nice angles, like $180°$ or $90°$, it might be useful to perform a translation. And indeed, this is what we do here. We translate P by $\overrightarrow{AD}$ to obtain P', as shown in Fig. 1.37. Then $\angle CP'D = \angle BPA$, and $\angle APB + \angle CPD = 180°$ is equivalent to $\angle CP'D + \angle CPD = 180°$. Since P is inside the square, $CP'DP$ is convex, and so $\angle APB + \angle CPD = 180°$ if and only if $CP'DP$ is cyclic.

Now let us assume that $CP'DP$ is cyclic. Note that $PP' = AD = CD$, so PP' and CD are equal chords in the same circle. The two arcs they respectively subtend are either equal or add up to the whole circle. So we either have $\angle CPD = \angle PCP'$ or $\angle CPD + \angle PCP' = 180°$. In the first case, we would have $\angle CPD + \angle PDP' = 180°$. Thus in both cases $CPDP'$ is a trapezoid, with either $PD \parallel CP'$ or $PC \parallel DP'$. But, because of the translation, $CP' \parallel BP$ and $DP' \parallel AP$, so either $PD \parallel BP$ or $PC \parallel AP$, which means that either P is on BD or it is on AC.

Conversely, if P is on say BD, then by symmetry, $\angle CPD = \angle APD$, and the latter is the supplement of $\angle APB$. □

So in this problem you noticed that the sum of two angles is $180°$, and you know that this leads to cyclic quadrilaterals. However, the angles were facing the opposite ways, but the square $ABCD$, which is a parallelogram, yields at least two translations, and each translation transports one angle to make it face the other! Of course, there are other possible transformations. Try reflecting across the diagonal BD, for example.

We conclude our discussion with a more difficult problem, where reflection over a line is used.

Problem 1.4 Let ABC be a triangle and let D be a point on the segment BC different from B and C. The circumcircle of ABD meets the segment AC again at an interior point E. The circumcircle of ACD meets the segment AB again at an interior point F. Let A' be the reflection of A over the line BC. The lines $A'C$ and DE meet at P, and the lines $A'B$ and DF meet at Q. Prove that the lines AD, BP and CQ are concurrent, or are all parallel.

This problem was given at the Romanian Master of Mathematics in 2016. We present the three official solutions, all of which rely on the reflection σ over the line BC. This reflection has been introduced already in the statement of the problem, so it is natural to expect that it will play an important role in the solution.

Solution 1 To see why we should expect σ to be useful, we start with the observation that

$$\angle BDF = \angle BAC = \frac{1}{2} \, \overset{\frown}{CDF} \text{ and } \angle CDE = \angle BAC = \frac{1}{2} \, \overset{\frown}{BDE},$$

from where we deduce that $\angle BDF = \angle CDE$, and so the lines DE and DF are images of each other under σ. On Fig. 1.38, we have marked a few more elements that arise from the action of σ such as the point $P' = \sigma(P)$, where the lines $AC = \sigma(A'C)$ and $DF = \sigma(DE)$ meet, the point $Q' = \sigma(Q)$ where the lines $AB = \sigma(A'B)$, and $DE = \sigma(DF)$ meet. We have also marked the point M where the line AD intersects for the second time the circumcircle of ABC together with $M' = \sigma(M)$. To solve the problem, we will check that the lines $DM' = \sigma(AD)$, $BP' = \sigma(BP)$ and $CQ' = \sigma(CQ)$ are concurrent.

A carefully drawn figure should disclose additional information about M', namely, that it lies at the intersection of the circumcircles of BDF and CDE. And indeed, using the cyclic quadrilaterals $ABMC$ and $AFDC$, we can write

$$\angle(BM', M'D) = -\angle(BM, MD) = \angle(BM, MA) = \angle(BC, CA) = \angle(BF, FD),$$

which shows that M' lies on the circumcircle of BDF. A similar argument shows that M' lies on the circumcircle of CDE. We deduce that the line DM' is the radical axis of the circumcircles of BDF and CDE.

On the other hand, P' lies on both lines AC and DF, so it must be the radical center of the circumcircles of ABC, ADC, and BDF, and hence the line BP' is the radical axis of the circumcircles of BDF and ABC. Similarly, the line CQ' is the radical axis of the circumcircles of CDE and ABC. We conclude that the lines DM', BP', and CQ' are concurrent at the radical center of the circumcircles of ABC, BDF and CDE, or are all parallel. Consequently, the lines $AD = DM$, BP, and CQ are also concurrent or all parallel, as desired. $\square$

The other two solutions are based on Desargues' Theorem (Problem 115) and Pappus' Theorem (proved in Sect. 2.2.1), respectively. The solutions were found by the member of the problem committee Ilya Igorevich Bogdanov.

Solution 2 You can follow this solution on Fig. 1.39. We start again with the reflection σ and conclude as above that the lines AC and DF meet at $P' = \sigma(P)$

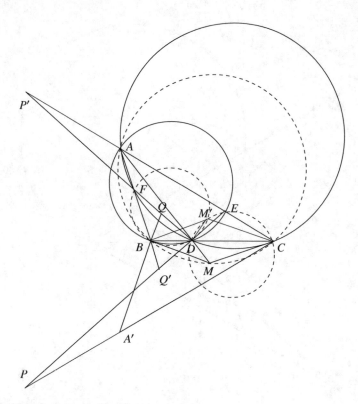

Fig. 1.38 First solution to Problem 1.4

and the lines AB and DE meet at $Q' = \sigma(Q)$. From here we deduce that the lines PQ and $P'Q' = \sigma(PQ)$ meet at some point R on the line BC (possibly at infinity). Since the pairs of lines $(CA; QD)$, $(AB; DP)$, $(BC; PQ)$ meet at three collinear points, namely, P', Q', and R, respectively, the triangles ABC and DPQ are perspective, i.e., the lines AD, BP, CQ are concurrent (or parallel), by Desargues' Theorem.   □

Solution 3 Arguing on Fig. 1.40, we let the lines BE and CF meet at X. Using inscribed angles in the cyclic quadrilaterals $BDEA$ and $CDFA$, we have $\angle XBD = \angle EAD = \angle XFD$, so the quadrilateral $BFXD$ is cyclic. Similarly, the quadrilateral $CEXD$ is cyclic. Then X is the point M' from the first solution, and consequently $X' = \sigma(X) = M$, showing that X' is on the line AD.

Let $E' = \sigma(E)$ and $F' = \sigma(F)$. The points Q, B, F' are collinear, and the points P, C, E' are also collinear, so we can apply Pappus' Theorem (Theorem 2.14 from Sect. 2.2.1) to infer that $X' = BE' \cap CF'$, $D = F'P \cap QE'$, and $BP \cap CQ$ are collinear points. It follows that the lines BP, CQ, DX' are concurrent. But the lines DX' and AD coincide, and we are done.

□

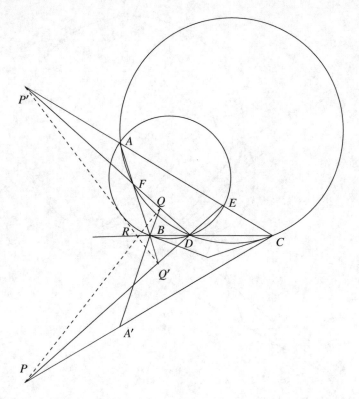

Fig. 1.39 Second solution to Problem 1.4

1.2.3   Problems in Euclidean Geometry to be Solved Using Isometries

Below is list of geometry problems for you to solve. Many of these problems can be solved by other methods, of course, but we insist that you look at them from the point of view of geometric transformations and find solutions that use isometries. We remind you that all problems have hints in the middle of the book and solutions at the end of the book.

13 Inside a square $A_1A_2A_3A_4$ consider a point P. Take the perpendiculars from A_2 to A_1P, from A_3 to A_2P, from A_4 to A_3P, and from A_1 to A_4P. Show that these four perpendiculars intersect at one point.

14 Let $ABCD$ be a quadrilateral, and let M and N be the midpoints of the sides AD and BC, respectively. Prove that if

$$MN = \frac{AB + CD}{2}$$

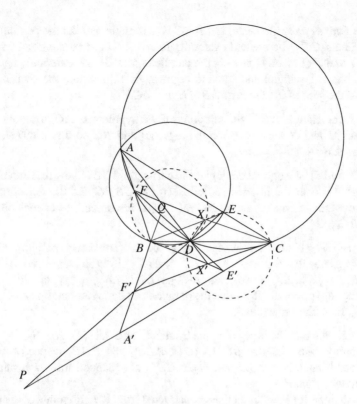

Fig. 1.40   Third solution to Problem 1.4

then $ABCD$ is a trapezoid.

15 On the sides AB and AC of a triangle ABC, construct in the exterior the squares $ABDE$ and $ACFG$. Let the point I be such that $AGIE$ is a parallelogram.

(a) Prove that EG is perpendicular to the median from A and is twice its length.
(b) Prove that AI is perpendicular to BC.
(c) Prove that DC and BF intersect on the altitude from A.
(d) Let O be the circumcenter of the triangle AEG. Prove that the line AO is a symmedian in the triangle ABC.

16 Consider a square inscribed in a parallelogram in such a way that each vertex of the square is on another side of the parallelogram. Show that the perpendiculars from the vertices of the parallelogram onto the sides of the square form another square.

17 Do there exist noncongruent rectangles $ABCD$ and $MNPQ$ that do not have the same center, such that $AM = BN = CP = DQ$?

18 *(The Pappus Area Theorem)* Let ABC be a triangle, and let the parallelograms $ABMN$ and $ACPQ$ be erected externally to ABC. Let R be the point where the lines MN and PQ meet. Construct the parallelogram $BCST$ externally having the sides CS and BT parallel and equal to segment RA. Show that the area of $BCST$ is equal to the sum of the areas of $ABMN$ and $ACPQ$.

19 In a right triangle ABC, the midpoint of the hypotenuse AC is denoted by O. The points M and N are chosen on the legs AB and BC so that $\angle MON = 90°$. Prove that $AM^2 + CN^2 = MN^2$.

20 Let P and Q be isogonal conjugates in a triangle ABC. Show that the reflection of the line AP over the internal angle bisector of $\angle BPC$ and the reflection of the line AQ over the internal angle bisector of $\angle BQC$ are reflections of each other over the line BC.

21 Let $n > 1$ and let $A_1A_2 \ldots A_{2n}$, be a (nonskew) polygon with $2n$ sides. Translate the vertices $A_1, A_3, \ldots, A_{2n-1}$ by the same vector to the points $A_1', A_3', \ldots, A_{2n-1}'$ so that $A_1'A_2A_3'A_4 \ldots, A_{2n-1}'A_{2n}$ is still a (non-selfintersecting) polygon. Show that the polygons $A_1A_2 \ldots A_{2n}$ and $A_1'A_2A_3'A_4 \ldots, A_{2n-1}'A_{2n}$ have the same area.

22 (a) Let $ABCD$ be a cyclic quadrilateral, and let H_a, H_b, H_c, H_d be the orthocenters of the triangles BCD, ACD, ABD, and ABC, respectively. Prove that the quadrilaterals $ABCD$ and $H_aH_bH_cH_d$ are mapped into each other by a reflection over a point.
(b) Show that the four perpendiculars constructed from the midpoints of the sides of $ABCD$ onto the opposite sides meet at one point.

23 Let Γ be a semicircle of diameter AB. The point C lies on the diameter AB, and the points D and E lie on the arc $\overset{\frown}{AB}$, with E between B and D. The tangents to Γ at D and E meet at F. Suppose that $\angle ACD = \angle ECB$. Prove that $\angle EFD = \angle ACD + \angle ECB$.

24 Show that if one can inscribe three equal squares in a triangle, then the triangle is equilateral.

25 Let ABC be a triangle and let ℓ be a line. Let also A_0, B_0, C_0 be the projections onto ℓ of the vertices A, B, C, respectively. Prove that the perpendiculars from A_0, B_0, C_0 to the lines BC, AC, AB intersect at one point.

26 On the sides AB, BC, CD, DA of an arbitrary convex quadrilateral $ABCD$ construct in the exterior four squares, and let M_1, M_2, M_3, M_4 be their respective centers. Prove that the segments M_1M_3 and M_2M_4 are perpendicular and have equal lengths.

27 Let ABC be an equilateral triangle. A line parallel to AC intersects the sides AB and BC at M and P, respectively. Let D be the center of the equilateral triangle BMP, and let E be the midpoint of the segment AP. Find the angles of the triangle DEC.

28 Let M and N be the midpoints of the sides CD and DE, respectively, of a regular hexagon $ABCDEF$ of center O. Let L be the intersection point of the lines AM and BN. Prove that

(a) the triangle ABL and the quadrilateral $DMLN$ have equal areas;
(b) $\angle ALO = \angle OLN = 60°$;
(c) $\angle OLD = 90°$.

29 Let M be a point inside the convex quadrilateral $ABCD$ such that $ABMD$ is a parallelogram. Prove that if $\angle CBM = \angle CDM$ then $\angle ACD = \angle BCM$.

30 Let ABC be an acute triangle. Let also AD, BE, and CF be the altitudes (with D on BC, E on AC, and F on AB), and let M be the midpoint of the side BC. The circumcircle of the triangle AEF meets the line AM at A and X. The line AM meets the line CF at Y. Let AD and BX meet at the point Z. Prove that the lines YZ and BC are parallel.

31 *(The Butterfly Theorem)* Let A be the midpoint of a chord c of a circle. Choose B and C on c such that $AB = AC$. Let MN and PQ be two arbitrary chords through B and C, respectively. Let R and S be the respective intersections of PM and NQ with c. Prove that $AR = AS$.

32 Let ABC and BCD be two equilateral triangles that share one side. An arbitrary line through D intersects AB at F and AC at E. Find the angle between the lines BE and CF.

33 On the sides of an arbitrary triangle ABC, construct the equilateral triangles BCA_1, ACB_1, and ABC_1 such that A_1 and A are separated by the line BC, B_1, and B are separated by the line AC, but C_1 and C are on the same side of the line AB. Let M be the center of the triangle ABC_1. Prove that the triangle MA_1B_1 is isosceles with $\angle A_1MB_1 = 120°$.

34 Given a convex n-gon $\mathcal{P}$ with no parallel sides, and a point O interior to it, prove that there are no more than n lines through O that bisect the area of $\mathcal{P}$.

35 Let $ABCD$ be a convex quadrilateral with $\angle BAD = \angle BCD = 120°$. Construct in the exterior equilateral triangles ABM, BCN, CDP, and DAQ. Prove that CQ is parallel to AN and CM is parallel to AP.

36 Let $ABCDE$ be a convex pentagon such that $BC \parallel AE$, $AB = BC + AE$, and $\angle ABC = \angle CDE$. Let M be the midpoint of CE, and let O be the circumcenter of the triangle BCD. Given that $\angle DMO = 90°$, prove that $2\angle BDA = \angle CDE$.

37 Let M, N, P be points on the sides BC, CA, AB of a triangle ABC, respectively. Let M_1 be reflection of M with respect to the midpoint of BC, let N_1 be the reflection of N with respect to the midpoint of AC, and let P_1 be the reflection of P with respect to the midpoint of AB. Prove that if the perpendiculars to the respective sides of the triangle ABC erected at the points M, N, and P meet at one point, then the perpendiculars to the sides erected at M_1, N_1, P_1 also meet at one point.

38 A convex hexagon $AC_1BA_1CB_1$ satisfies the relations $AB_1 = AC_1$, $BC_1 = BA_1$, $CA_1 = CB_1$, and $\angle A + \angle B + \angle C = \angle A_1 + \angle B_1 + \angle C_1$. Prove that the area of the triangle ABC is half the area of the hexagon.

39 Six points are chosen on the sides of an equilateral triangle ABC as follows: A_1, A_2 on BC, B_1, B_2 on CA, and C_1, C_2 on AB, in such a way that they are the vertices of a convex hexagon $A_1A_2B_1B_2C_1C_2$ with equal side lengths. Prove that the lines A_1B_2, B_1C_2, and C_1A_2 are concurrent.

40 Let ABC be an isosceles triangle with $AB = AC$, and let I be its incenter. A line r passing through I meets the sides AB and AC at the points D and E, respectively. Let F and G be points on the side BC such that $BF = CE$ and $CG = BD$. Show that the angle $\angle FIG$ does not depend on r.

41 Let A, B, C be fixed points in the plane. A man starts from a certain point P_0 and walks directly to A. At A, he turns his direction by $60°$ to the left and walks to P_1 such that $P_0A = AP_1$. He repeats the same action starting at P_1 and using the vertex B to arrive at P_2. After he does the same action 1986 times successively around the points A, B, C, A, B, C, ..., he returns to the starting point (this means that $P_{1986} = P_0$). Prove that the triangle ABC is equilateral and oriented counterclockwise.

42 Let C_1 and C_2 be two congruent circles centered at O_1 and O_2, which intersect at A and B. Take a point P on the arc $\overset{\frown}{AB}$ of C_2 that lies inside C_1. The line AP meets again C_1 at C, the line CB meets C_2 again at D, and the angle bisector of $\angle CAD$ intersects C_1 and C_2 at E and L, respectively. Let F be the reflection of D with respect to the midpoint of the segment PE. Prove that there exists a point X in the plane satisfying $\angle XFL = \angle XDC = 30°$ and $CX = O_1O_2$.

43 *(Fagnano's Theorem)* Prove among all triangles inscribed in a given acute triangle, the one determined by the feet of the altitudes has minimal perimeter.

44 Let $ABCD$ be an isosceles trapezoid with $AB \| CD$. Let E be the midpoint of AC. Denote by ω and Ω the circumcircles of the triangles ABE and CDE, respectively. Let P be the crossing point of the tangent to ω at A with the tangent to Ω at D. Prove that PE is tangent to Ω.

45 *(The Erdős-Mordell Theorem)* Let ABC be a triangle and let P be a point in its interior. Denote by d_a, d_b, and d_c the respective distances from P to BC, AC, and AB. Show that

$$PA + PB + PC \geq 2(d_a + d_b + d_c).$$

46 You are given a rusty compass whose opening became stuck at a fixed width. You are also given three arbitrary points A, B, C in the plane. Using the rusty compass, and no other tool, construct a fourth point D such that the lines AB and CD are perpendicular.

47 A convex polygonal surface P lies on a flat wooden table. You are allowed to drive some nails into the table. The nails must not go through the polygonal surface P, but they may touch its boundary. We say that a set of nails *blocks* P if the nails make it impossible to move P without lifting it off the table. What is the minimum number of nails that suffices to block any convex polygon P?

1.3 Isometries Throughout Mathematics

At this moment, with isometries in mind, we take a turn toward other fields of mathematics.

1.3.1 Geometry with Combinatorial Flavor

Let us make a first stop in combinatorial geometry, where we commence with a problem from a 2005 Romanian Team Selection Test for the International Mathematical Olympiad.

Problem 1.5 Prove that if any two vertices of a polygon are at distance at most 1 from each other, then the area of the polygon is less than $\frac{\sqrt{3}}{2}$.

Solution The solution that we present parallels the one found by the student Alin Purcaru during the test. It goes like this: Let us call the polygon K. The condition from the hypothesis implies that every band of width 1 determined by two parallel lines can be translated to a position that contains K inside the union of the interior and the boundary of the band.

Now consider three such bands that make angles of $\frac{2\pi}{3}$ with each other and that contain K in their intersection. These bands form an equiangular hexagon with distances between opposite sides equal to 1. We claim that

(i) the area of the hexagon is less than or equal to $\frac{\sqrt{3}}{2}$;
(ii) the diagonals joining opposite vertices are equal to $\frac{2}{\sqrt{3}}$.

Both properties hold for the regular hexagon with side equal to $\frac{1}{\sqrt{3}}$, in which the distance between opposite sides is 1. To prove the two claims in general, note that the geometric figure determined by the intersection of two of the bands is always the same (a parallelogram with an angle of $60°$), so it is the location of the third band that determines the particular figure.

For (i) it suffices to prove that the area decreases when we modify the regular hexagon by translating just one of the bands in the direction perpendicular to the lines that bound it. Examining Fig. 1.41, we see that in the process of translation, we lose one trapezoid and we gain another. Both trapezoids have equal altitudes and

Fig. 1.41 Proof of the bound
for the area

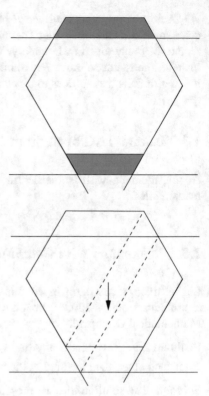

Fig. 1.42 Proof of the
invariance of the length of the
diagonals

equal angles, but the bases of the one that we lose are larger than the bases of the
one that we gain. So the area decreases, as desired.

The proof of (ii) becomes transparent when we glimpse at Fig. 1.42. When
translating one of the bands, the diagonals themselves are being translated (watch
the parallelograms determined by the diagonals!).

From (i) we deduce that K lies inside a figure of area $\frac{\sqrt{3}}{2}$, so the area of K cannot
exceed this number. But (ii) shows that K cannot be the entire hexagon (for then it
would contain two points at distance $\frac{2}{\sqrt{3}} > 1$), so the inequality is strict. $\square$

The second problem that we present was published by Yu.P. Lysov in the journal
Kvant (Quantum). It employs rotations.

Problem 1.6 Let S be the union of n closed connected subsets of the unit circle.
It is known that for every rotation ρ about the center of the circle, the sets S and
$\rho(S)$ overlap. What is the smallest value that the sum of the measures of the sets can
have?

A closed connected subset of the unit circle is either a point, a closed arc, or the
whole circle. The measure of a point is simply 0, while the measure of an arc is its
length.

Solution Let σ be the sum of the measures of the n connected sets, which sets we label in counterclockwise order. Rotate the circle by the angle ϕ, and denote by A_{jk} those values of ϕ for which the jth set of the original circle overlaps with the kth set of the rotated one. It is clear that A_{jk} is a set whose measure is the sum of the measures of the jth and kth sets (A_{jk} is itself either a point, an arc, or the whole circle). Moreover, the union of the sets A_{jk} should contain all angles, since for every rotation some jth and kth arcs overlap. Thus, for the sum of the measures of the sets A_{jk}, we have the inequality

$$\sum_{j,k} m(A_{jk}) \geq 2\pi.$$

But

$$\sum_{j,k} m(A_{jk}) = 2n\sigma,$$

as each connected set is counted $2n$ times, once when it pairs with the rotates of each of the n sets and once when it is itself a rotate and pairs with each of the original n sets. Hence

$$2n\sigma \geq 2\pi,$$

so $\sigma \geq \frac{\pi}{n}$.

The value $\sigma = \frac{\pi}{n}$ is attained when one set is the arc of length $\frac{\pi}{n}$ centered at 1, and each of the other $n - 1$ sets consists of one of the points $e^{\frac{2\pi i k}{n}}$, $k = 1, 2, \ldots, n - 1$. This example is shown in Fig. 1.43. And if you are not completely convinced that the sum of the measures of the sets A_{jk} equals $2n\sigma$, you can compute it explicitly for this example. $\quad\square$

Here are more problems that you should solve with isometries in mind.

48 Given a triangular cake and a box in the shape of its mirror image, show that the cake can be cut into three slices so that the slices now fit inside the box. (The cake has icing, so we are not allowed to flip it over and place it upside down inside the box.)

Fig. 1.43 Example with minimal sum of measures for $n = 6$

49 Two equal squares overlap to determine an octagon. The sides of the first square are colored red, while the sides of the second square are colored blue. Prove that the sum of the lengths of the red sides of the octagon is equal to the sum of the lengths of its blue sides.

50 Let n, p be natural numbers such that $6 \leq n$ and $3 \leq p \leq n - p$. The vertices of a regular n-gon are colored either red or black: p vertices red and $n - p$ vertices black. Show that there are two congruent polygons, each having at least $\lfloor p/2 \rfloor$ vertices, one with all vertices red and the other with all vertices black.

51 (a) Inside the unit square there is a set of points that is the union of finitely many polygonal surfaces and has the property that the distance between any two points of the set is never equal to 0.001. Prove that the area of the set is less than 0.34.
(b) Show that the area of this set is actually less than 0.287.

52 Every point of integer coordinates in the plane is colored either red or blue. Prove that there is an infinite set of points of the same color that has a center of symmetry.

53 Given a finite number of disks of radius 1 in the plane, denote by S the area of their union. Prove that one can choose finitely many of these disks that are pairwise disjoint and have the sum of their areas greater than $\frac{2}{9}S$.

54 Several chords are constructed in a circle of radius 1. Prove that if every diameter intersects at most k chords, then the sum of the lengths of the chords is less than $k\pi$.

55 Let $n \geq 2$ be an integer. In some group of $2n$ people, each person has at most $n - 1$ enemies. Show that the people can be seated at a round table so that no person sits next to an enemy.

56 On a cylindrical surface of radius r, unbounded in both directions, consider n points and a surface S of area strictly less than 1. Prove that by rotating around the axis of the cylinder and then translating in the direction of the axis by at most $\frac{n}{4\pi r}$ units, one can transform S into a surface that does not contain any of the n points.

57 Let $n \geq 3$ be a positive integer; and let us consider n guests that sit at a round table. A move consists of two neighbors exchanging seats. What is the smallest number of moves one has to perform so that after performing these moves the guests sit in reverse order.

58 Let $S_1, S_2, \ldots, S_n$ be a collection of equilateral triangles with one side parallel to the x axis and all pointing up. The triangles may overlap. For each S_i, let T_i be its medial triangle (formed by the midpoints of its sides). Finally, let S be the union of all S_i triangles, and let T be the union of all T_i triangles. Prove that

$$\text{area}(S) \leq 4\,\text{area}(T).$$

59 *(The Chord Theorem)*

(a) Let Γ be a polygonal line in the plane with endpoints A and B. Prove that for each positive integer n there exists a segment parallel to AB with endpoints in Γ having length $\frac{1}{n}AB$.

(b) Show that for every number $\alpha \in (0, 1)$ that is not of the form $\frac{1}{n}$, there always exists a polygonal line Γ with endpoints A and B having the property that there is no segment parallel to AB and with endpoints in Γ whose length is αAB.

60 A flea jumps between two intersecting lines in the plane, a and b, on a path $A_1 B_1 A_2 B_2 \ldots$ such that:

   (i) $A_i \in a$ and $B_i \in b$ for all i;
   (ii) $A_1 B_1 = B_1 A_2 = A_2 B_2 = B_2 A_3 = \cdots$;
   (iii) for all j, $A_j = A_{j+1}$ if and only if $A_j B_j \perp a$, and similarly $B_j = B_{j+1}$ if and only if $B_j A_{j+1} \perp b$.

Prove that the flea's path is periodic if and only if the lines a and b make a rational angle (when measured in degrees).

   And some problems about chessboards.

61 On an $n \times n$ square lattice, two players mark alternatively one (length 1) segment of the lattice. The game is won by the player who is the first to create some closed polygonal line (formed by segments marked by both). Does any of the two players have a winning strategy?

62 On an arbitrarily large chessboard, consider a generalized knight which can jump p squares in one direction and q in the other. Show that such a knight can only return to the initial position after an *even* number of jumps.

63 A rectangular bar of chocolate is divided by longitudinal and transversal hollows into 50 small equal square portions. Two people play the following game: The first breaks the bar along some hollow into two rectangular parts. Then the players take turns breaking one of the resulting pieces along a hollow. The player who is the first to break off a piece with no hollows (a) loses, (b) wins. Which of the two players can guarantee winning?

64 Can one place the letters a, b, c in the boxes of a 100×100 chessboard so that every 3×4 rectangle of the lattice contains three as, four bs, and five cs?

65 (a) Consider the tessellation of the plane by equal squares. For what $n \geq 2$ it is possible to color the squares by n colors such that the centers of the squares of the same color form a square lattice of the plane and all these lattices have equal squares and parallel sides?

(b) Consider a tessellation of the plane by equal hexagons. For what $n \geq 2$ it is possible to color the hexagons by n colors so that the centers of the hexagons of the same color form an equilateral triangular lattice and these lattices have equal triangles and parallel sides?

1.3.2   Combinatorics of Sets

More combinatorics problems for you to solve follow after this example from the
Bulgarian Mathematical Olympiad.

Problem 1.7   There are 2000 white balls in a box. There are also unlimited supplies
of white, green, and red balls, initially outside the box. At each turn, we can replace
two balls in the box with one or two balls as follows: two whites with a green; two
reds with a green; two greens with a white and a red; a white and a green with a red;
and a green and a red with a white.

(a)   After finitely many of the above operations there are three balls left in the box.
      Prove that at least one of them is green.
(b)   Is it possible that after finitely many operations only one ball is left in the box?

Solution   We consider the group of rotations of a square (mentioned in Sect. 1.1.5),
consisting of the rotations $1, \rho, \rho^2, \rho^3$ by 0°, 90°, 180°, and 270°. This group is
exhibited in Fig. 1.44.

   Assign the rotation ρ to each white ball, ρ^2 to each green ball, and ρ^3 to each
red ball. A quick check shows that the given operations preserve the product of
the rotations assigned to all balls in the box. This product is initially $\rho^{2000} = 1$.
If three balls were left in the box, none of them green, then the composition of
their associated transformations would be either ρ or ρ^3, a contradiction. Hence, if
three balls remain, at least one is green, proving the claim in part (a). Furthermore,
because no ball has the assigned rotation equal to 1, the box must contain at least two
balls at any time. This shows that the answer to the question in part (b) is negative.

$\square$

Fig. 1.44   The group of
rotations of a square

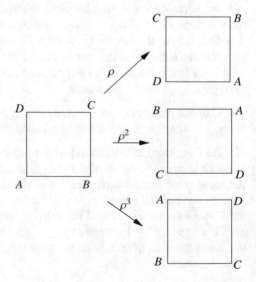

Associating algebraic structures to combinatorial configurations and the construction of mathematical objects that are invariant under transformations is a fundamental idea in mathematics.

66 There are n trees arranged in a circle, and initially on each tree there is a bird. At every moment, two birds fly in opposite directions to the neighboring trees. For which n it is possible for all birds to meet on one tree?

67 Let A and B be two sets of real numbers. Define

$$A + B = \{a + b \mid a \in A \text{ and } b \in B\}.$$

Prove that $|A + B| \geq |A| + |B| - 1$, and find all cases where equality occurs (here and below $|A|$ denotes the number of elements of the set A).

68 Let a, b be fixed positive integers, and let A and B be finite sets so that:

(i) A and B are disjoint;
(ii) if an integer i is in either A or B, then $i + a$ is in A or $i - b$ is in B.

Prove that $a|A| = b|B|$.

69 Let $a_1, a_2, \ldots, a_n$ and $b_1, b_2, \ldots, b_n$ be real numbers such that $a_i < b_i$ for all $i = 1, 2, \ldots, n$ and $b_1 + b_2 + \cdots + b_n < 1 + a_1 + a_2 + \cdots + a_n$. Prove that there is a real number c such that for all $i = 1, 2, \ldots, n$ and for every integer k

$$(a_i + k + c)(b_i + k + c) > 0.$$

70 Let $A = \{1, 2, \ldots, n\}, n > 1$. A permutation $\sigma : A \to A$ is called an involution if $\sigma \circ \sigma = 1_A$. Find the minimal number k such that every permutation is the composition of k involutions.

71 Two subsets of the positive integers are called congruent if one is obtained from the other by a translation. Is it possible to partition the set of positive integers into infinitely many subsets that are all infinite and congruent to one another?

72 Let $\leftarrow$ denote the left arrow[2] key on a standard keyboard. If one opens a text editor and types the keys "$ab \leftarrow cd \leftarrow\leftarrow e \leftarrow\leftarrow f$," the result is "*faecdb*." We say that a string B is *reachable* from a string A if it is possible to insert some amount of $\leftarrow$'s into A, such that typing the resulting characters produces B. So, our example

---

[2] Here is a short explanation of how the $\leftarrow$ key works. A computer's text editor starts with an empty screen and a cursor which we denote by "|". When you type a letter x, the cursor | is replaced by $x|$. So if the screen shows "*m|th*," and you press the "*o*" key, then the result is "*mo|th*."

The $\leftarrow$ key moves the cursor one space backward. That is, "*mo|th*" becomes "*m|oth*" and finally "*|moth*." If the cursor is at the beginning of the string, then the $\leftarrow$ key has no effect.

Finally, the cursor is not considered to be part of the final string. That is, if the string displays "*f |aecdb*," we just take the result to be *faecdb*.

shows that "*faecdb*" is reachable from "*abcdef.*" Prove that for any two strings A and B, A is reachable from B if and only if B is reachable from A.

1.3.3 Number Theory

In this section, by letting groups of transformations act on discrete geometric combinatorial structures, we will derive some classical results in number theory. We start with the famous proof of Wilson's Theorem discovered by Julius Petersen.

Theorem (Wilson) *If p be a prime number, then p divides $(p-1)! + 1$.*

Proof The result is obviously true for $p = 2$, so we only have to do the proof in the case where $p > 2$. Consider p equally spaced points on a circle, which form the vertices of a regular p-gon. Let us count the number of polygons (meaning closed polygonal lines) whose vertices are the p points. There are p ways to choose the first vertex, $p - 1$ ways to choose the second, $p - 2$ ways to choose the third, and so on, so the total number of polygons is $p!$. However, in a given polygon, the first vertex can be chosen in p ways, and the direction in which the polygon is traveled can be chosen in two ways. It follows that each polygon is counted $2p$ ways. Thus, the actual number of polygons is

$$\frac{p!}{2p} = \frac{(p-1)!}{2}.$$

If a polygon $A_1 A_2 \ldots A_p$ has the property that $A_1 A_2 = A_2 A_3 = \cdots = A_p A_1$, we call it a regular stellated polygon (it might be a true regular polygon, or it may have self-intersections). A regular stellated polygon is determined by the choice of two vertices, since all other vertices are determined by the condition that the sides are equal. Hence, there are $p(p - 1)$ regular stellated polygons. But we counted each regular stellated polygon $2p$ times, so the true number is

$$\frac{p(p-1)}{2p} = \frac{p-1}{2}.$$

The three regular stellated polygons for $p = 7$ are shown in Fig. 1.45.

We deduce that the number of polygons that are *not* regular stellated is

$$\frac{(p-1)!}{2} - \frac{p-1}{2} = \frac{[(p-1)! + 1] - p}{2}.$$

Now let us declare two polygons to be equivalent if they coincide under a rotation about the center of the circle. The regular stellated polygons are invariant under rotations by multiples of $2\pi/p$, so they are only equivalent to themselves. Conversely, assume that for some integer k, the rotation of angle $2k\pi/p$ keeps

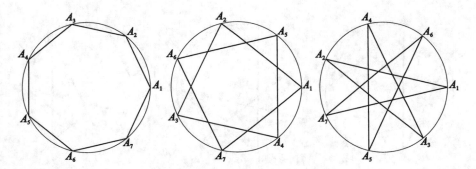

Fig. 1.45 Regular stellated polygons

the polygon $A_1 A_2 \ldots A_p$ invariant. Because p is prime, this rotation generates the group of all rotations, so $A_1 A_2 \ldots A_p$ is invariant under all rotations. The rotation that maps A_1 to A_2 should necessarily map A_2 to A_3 and so on, which implies that $A_1 A_2 \ldots A_p$ is regular stellated.

So if a polygon is not regular stellated, then it is not invariant under any nontrivial rotation. It follows that when rotating each such polygon by $2k\pi/p$, $k = 0, 1, 2, \ldots, p - 1$, p distinct polygons are obtained. An example is illustrated in Fig. 1.46. We deduce that the orbit of each polygon under the action of the group of rotations (namely, the polygon itself and its rotates) has p elements. It follows that the polygons that are not stellated can be divided into disjoint groups of p. So the number of equivalence classes of polygons that are not stellated is

$$\frac{1}{p} \left[\frac{(p-1)!}{2} - \frac{p-1}{2} \right] = \frac{1}{p} \cdot \frac{[(p-1)!+1] - p}{2},$$

and for this number to be an integer, the numerator $(p-1)!+1-p$ must be divisible by p. But then $(p - 1)! + 1$ itself is divisible by p and the theorem is proved. $\square$

Try similar ideas in order to solve the following problems.

73 *(Fermat's Little Theorem)* Let p be a prime number, and let n be a positive integer. Prove that

$$n^p - n \equiv 0 (\text{mod } p).$$

74 *(Lucas' Theorem)* Let p be prime, and let $a = a_0 + a_1 p + a_2 p^2 + \cdots + a_k p^k$, $b = b_0 + b_1 p + b_2 p^2 + \cdots + b_k p^k$ with a_j, b_j integers such that $0 \le a_j, b_j < p$, $j = 1, 2, \ldots, k$. Prove that

$$\binom{a}{b} \equiv \binom{a_0}{b_0} \binom{a_1}{b_1} \cdots \binom{a_k}{b_k} (\text{mod } p).$$

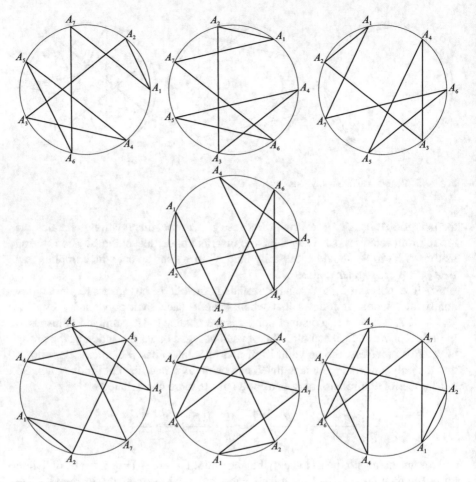

Fig. 1.46 The seven rotates of a 7-gon

75 *(Sylvester's Theorem)* Given two coprime positive integers a and b, show that $c = ab - a - b$ is the largest integer that cannot be represented as $ax + by$ with x, y non-negative integers. Also, show that for each positive integer $n < c$, exactly one of the numbers n or $c - n$ can be written as $ax + by$ with x, y non-negative integers.

76 *(The Cauchy-Davenport Theorem)* Let p be a prime number. Show that if A and B are two non-empty subsets of $\mathbb{Z}_p$ then

$$|A + B| \geq \min(|A| + |B| - 1, p).$$

1.3.4   Functions

We will conclude the first chapter with problems about functions. The example presented below is a problem from the 2013 Romanian Master of Mathematics, which was proposed by Alexander Betts from the United Kingdom. Its solution was found during the competition by Mark Sellke, member of the team that represented the United States.

Problem 1.8 Does there exist a pair of functions $g, h : \mathbb{R} \to \mathbb{R}$ such that the only function $f : \mathbb{R} \to \mathbb{R}$ satisfying $f(g(x)) = g(f(x))$ and $f(h(x)) = h(f(x))$ for all $x \in \mathbb{R}$ is the identity function $f(x) = x$?

Solution The answer is yes! This is the example

$$g(x) = \begin{cases} x + 1 & \text{if } \{x\} < \frac{1}{2} \\ x + \frac{1}{2} & \text{if } \{x\} \geq \frac{1}{2} \end{cases} \quad \text{and } h(x) = \sqrt{2}g\left(\frac{x}{\sqrt{2}}\right),$$

where $\{x\}$ denotes the *fractional part* of x. We have to prove that g and h have the required property. First, note that if x is in the range of g, then $g(x) = x + 1$, and if x is in the range of h, then $h(x) = x + \sqrt{2}$.

At the heart of the solution lies the study of the behavior of a function f satisfying the condition from the statement under *translations in the variable* by numbers of the form $m + n\sqrt{2}$, with m and n integers. We will make use of the following well-known result, which we now prove for sake of completeness.   □

Theorem (Kronecker) *A nontrivial subgroup of the additive group of real numbers is either cyclic (meaning that it has one element so that any other element is an integer multiple of it) or it is dense in the set of real numbers.*

Proof Let the group be G. It is either discrete, or it has an accumulation point on the real axis. If it is discrete, let a be its smallest positive element. Then any other element is of the form $b = ka + \alpha$ with $0 \leq \alpha < a$. But b and ka are both in G; hence α is in G is well. By the minimality of a, α can only be equal to 0, and hence, the group is cyclic.

If there is a sequence $(x_n)_n$ in G converging to some real number, then $\pm(x_n - x_m)$ approaches zero as $n, m \to \infty$. Choosing the indices m and n appropriately, we can find a sequence of positive numbers in G that converges to 0. Thus for any $\epsilon > 0$, there is an element $c \in G$ with $0 < c < \epsilon$. For some integer k, the distance between kc and $(k+1)c$ is less than ϵ; hence, any interval of length ϵ contains some multiple of c. Varying ϵ we conclude that G is dense in the real axis.   □

Applying this theorem we conclude that the set of numbers

$$\{m + n\sqrt{2} \mid m, n \in \mathbb{Z}\},$$

is dense, being a group that is not cyclic because 1 and $\sqrt{2}$ cannot both be integer multiples of the same number. Consequently, when n ranges among all positive integers, the numbers $\{n\sqrt{2}\}$ are dense in the unit interval.

Lemma 1 *The following properties hold:*

(g1) *If x is in the range of g then so is $f(x)$.*
(g2) *If x is in the range of g then for all positive integers m, $f(x+m) = f(x)+m$.*

Proof To prove (g1), let $x = g(x_0)$, then $f(x) = f(g(x_0)) = g(f(x_0))$, which is therefore in the range of g. For (g2) start with

$$f(x+1) = f(g(x)) = g(f(x)) = f(x) + 1,$$

then use induction to show that $f(x+m) = f(x)+m$.   □

A similar argument proves the next result.

Lemma 2 *The following properties hold:*

(h1) *If x is in the range of h then so is $f(x)$.*
(h2) *If x is in the range of h then for all positive integers n, $f(x+n\sqrt{2}) = f(x) + n\sqrt{2}$.*

We prove by contradiction that $f(x) = x$, by assuming that there are $a \neq b$ with $f(a) = b$.

Case 1: a is in the range of both g and h. Then so is b and $f(a+n) = b+n$ for all positive integers n. Since a and b are in the range of h, we have $\left\{\frac{a}{\sqrt{2}}\right\}, \left\{\frac{b}{\sqrt{2}}\right\} < \frac{1}{2}$. Then, unless the number $\frac{a-b}{\sqrt{2}}$ is an integer (and so $\left\{\frac{a}{\sqrt{2}}\right\} = \left\{\frac{b}{\sqrt{2}}\right\}$), by using Kronecker's Theorem, we can find a positive integer n such that $\left\{n \cdot \frac{1}{\sqrt{2}}\right\}$ is in one of the intervals

$$\left[\frac{1}{2} - \left\{\frac{b}{\sqrt{2}}\right\}, \frac{1}{2} - \left\{\frac{a}{\sqrt{2}}\right\}\right) \text{ or } \left[1 - \left\{\frac{a}{\sqrt{2}}\right\}, 1 - \left\{\frac{b}{\sqrt{2}}\right\}\right),$$

(depending on which of the fractional parts is larger, either the first or the second interval does not exist). If the first interval exists and $\left\{n \cdot \frac{1}{\sqrt{2}}\right\}$ belongs to it, then $\left\{\frac{a+n}{\sqrt{2}}\right\} < \frac{1}{2}$ and $\left\{\frac{b+n}{\sqrt{2}}\right\} \geq \frac{1}{2}$, so $a+n$ is in the range of h while $b+n$ is not. But $b+n = f(a+n)$, which contradicts (h1). The same reasoning rules out the other case, so $\frac{a-b}{\sqrt{2}}$ is an integer.
Similarly by Kronecker's Theorem, we obtain that there exists a positive integer m such that $\left\{a + m\sqrt{2}\right\} < \frac{1}{2}$ and $\left\{b + m\sqrt{2}\right\} \geq \frac{1}{2}$, which violates (g1) because $f(a + m\sqrt{2}) = b + m\sqrt{2}$ for all m, unless $a - b$ is an integer.
Thus, both $a - b$ and $\frac{a-b}{\sqrt{2}}$ should be integers, which is clearly impossible.

Case 2: a is in the range of *g* but not in the range of *h*. Then $f(a + n) = b + n$ for all positive integers n. There exists an n such that $\left\{ n \cdot \frac{1}{\sqrt{2}} \right\}$ is in the interval $\left(1 - \left\{ \frac{a}{\sqrt{2}} \right\}, \frac{3}{2} - \left\{ \frac{a}{\sqrt{2}} \right\} \right)$, so $\left\{ \frac{a+n}{\sqrt{2}} \right\} < \frac{1}{2}$ and $a + n$ is in the range of h. Then we can use $a + n$ as our a in Case 1 to rule out this case. The case where a is in the range of h but not g is similar.

Case 3: a is neither of the ranges of g and h. We have $f(g(a)) = g(a) = a + \frac{1}{2}$ and $f(g(a)) = g(f(a))$, which by the definition of g is equal to either $f(a) + 1$ or $f(a) + \frac{1}{2}$. But since $f(a) \neq a$, we must have $a + \frac{1}{2} = f(g(a)) = f(a) + 1$, so $f(a) = a - \frac{1}{2}$. And we also have $f(h(a)) = h(a) = a + \frac{\sqrt{2}}{2}$ and $f(h(a)) = h(f(a)) = f(a) + \sqrt{2}$ or $f(a) + \frac{\sqrt{2}}{2}$. By the same reasoning $f(a) = a - \frac{\sqrt{2}}{2}$, and this is a contradiction that rules out this case as well.

As we have ruled out all possible cases, we conclude that it is impossible for $f(a)$ not to be equal to a, and hence $f(a)$ is always equal to a. $\square$

Try your hand at the following problems.

77 Prove that any function $f : \mathbb{R} \to \mathbb{R}$ can be written as the sum of two functions whose graphs admit centers of symmetry.

78 Give an example of a function $f : \mathbb{R} \to \mathbb{R}$ whose graph is invariant under a $90°$ rotation about the origin. Show that for any such function $f(0) = 0$.

79 Let $n > 2$ be an integer, and let $f : \mathbb{C} \to \mathbb{C}$ be a function such that for every regular polygon with n vertices, $A_1 A_2 \ldots A_n$, in the plane

$$f(A_1) + f(A_2) + \cdots + f(A_n) = 0.$$

Prove that f is the zero function.

80 Let $f : \mathbb{R} \to \mathbb{R}$ be a continuous function whose graph has two axes of symmetry. Prove that

$$\lim_{x \to \infty} \frac{f(x)}{x}$$

exists and is finite.

81 Find all polynomial functions $P(x)$ with real coefficients that satisfy

$$P(x\sqrt{2}) = P(x + \sqrt{1 - x^2})$$

for all x with $|x| \leq 1$.

82 (a) Let $f : \mathbb{R} \to \mathbb{R}$ be a function such that $|f(x) - f(y)| = 1$ for every $x, y \in \mathbb{R}$ with $|x - y| = 1$. Is it necessarily true that f is an isometry?
(b) Let $f : \mathbb{C} \to \mathbb{C}$ be a function such that $|f(z) - f(w)| = 1$ for every $z, w \in \mathbb{C}$ with $|z - w| = 1$. Is it true that f is an isometry?

Chapter 2
Homotheties and Spiral Similarities

2.1 A Theoretical Introduction to Homotheties

This chapter is devoted to transformations that dilate or shrink geometric objects by a fixed factor. Such transformations occur naturally in real life, for example, when changing scale, with or without rotation. Several situations are depicted in Fig. 2.1.

It is customary, when presenting this subject, to split the narrative into two parts: first the story of the transformations that dilate without rotating, known as homotheties (from the Greek words *homo* meaning "similar" and *thesis* meaning "position") and then that of the spiral similarities, which dilate and rotate simultaneously.

2.1.1 Definition and Properties

We commence with the definition of homothety, which is illustrated in Fig. 2.2.

Definition Given a point O and a nonzero number k, the homothety of center O and ratio k is a transformation h of the plane that sends every point P to a point $P' = h(P)$ such that

$$\overrightarrow{OP'} = k\overrightarrow{OP}.$$

The homothety is called direct if $k > 0$, in which case P' is on the ray $|OP$, and inverse if $k < 0$, in which case O belongs to the segment PP'. Note that in both cases $OP'/OP = |k|$. The center O is a fixed point of the homothety, and it is the only fixed point (unless $k = 1$, when the homothety is the identity map).

Two geometric figures are called homothetic if there is a homothety that maps one into the other.

© The Author(s), under exclusive license to Springer Nature Switzerland AG 2022
R. Gelca et al., *Geometric Transformations*, Problem Books in Mathematics,
https://doi.org/10.1007/978-3-030-89117-6_2

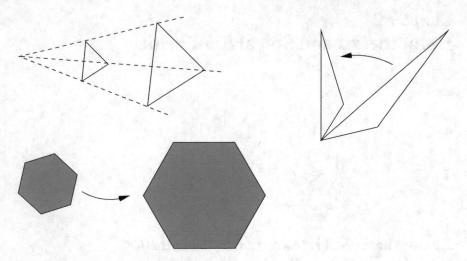

Fig. 2.1 Examples of polygons being shrunk or dilated

Fig. 2.2 The definition of homothety

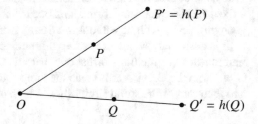

Being truthful to our plan to enlace synthetic and analytic reasoning, we rephrase homothety in the language of numbers. If the homothety h is centered at the origin and its ratio is $k \in \mathbb{R}$, then it is defined by the equation $h(z) = kz$. In general, if the complex coordinate of the center O is a, then the homothety is defined by

$$h(z) = a + k(z - a) = kz + (1 - k)a.$$

Indeed, this formula arises from the equation

$$\frac{h(z) - a}{z - a} = k,$$

and the latter expresses the fact that the segment $z - a$ is enlarged by a factor of k to the segment $h(z) - a$. Conversely, every map

$$h : \mathbb{C} \to \mathbb{C}, \quad h(z) = kz + b$$

for $k \in \mathbb{R} \setminus \{0, 1\}$ is a homothety of ratio k centered at the point $a = b/(1 - k)$.

Fig. 2.3 The image of a
segment through a homothety

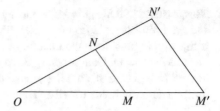

Most theoretical statements bellow are easier to attack analytically, since this avoids configuration-dependent proofs. But with the problems, it will be quite the opposite; geometric intuition yields sudden breakthroughs.

Theorem 2.1 *If M' and N' are the images of M and N through the homothety of center O and ratio k, then $M'N' = kMN$, and $M'N'$ is parallel to MN. Moreover, the image through the homothety of the segment MN is the segment $M'N'$, and the image of the line MN is the line $M'N'$.*

Proof The situation is exhibited in Fig. 2.3. Let a, m, n, m', n' be the complex coordinates of O, M, N, M', N', respectively. Then $m' = km + (1 - k)a$ and $n' = kn + (1 - k)a$. So $n' - m' = k(n - m)$, which expresses analytically the fact that $M'N'$ is parallel to MN and has length kMN.

If $z = tm + (1 - t)n$, then the image z' of z satisfies

$$z' = k[tm + (1 - t)n] + (1 - k)a = t[km + (1 - k)a] + (1 - t)[kn + (1 - k)a].$$

For $t \in [0, 1]$ we interpret this as saying that the segment MN is mapped to the segment $M'N'$, while for $t \in \mathbb{R}$, we interpret this as saying that the line MN is mapped to the line $M'N'$. And so homothety maps lines to lines.   □

An important corollary of this result is that homothety dilates (or shrinks) distances by a factor of $|k|$. The converse of Theorem 2.1 is also true, as the next result shows.

Proposition 2.2 *If a transformation h of the plane maps every segment to a segment parallel to it and of length r times the length of the original segment, with $r > 0$ and $r \neq 1$, then h is a homothety.*

Proof A solution in complex coordinates is again simpler, since it avoids the discussion of cases. You may want to try the synthetic solution and compare. Let $h(z) = z'$, and $h(w) = w'$. Then $h(z) - h(w) = \pm r(z - w)$ where the sign depends on how the two segments are oriented with respect to each other. Fix w (and then also w') and write

$$h(z) = \pm rz \mp rw + w'.$$

If the sign is always the same, then we are done, because we just obtained the formula for a homothety (direct if the sign is plus and inverse if the sign is minus).

Now, assume that $h(z_1) = rz_1 - rw + w'$ and $h(z_2) = -rz_2 + rw + w'$ for some z_1 and z_2. Then $h(z_2) - h(z_1) = -r(z_2 + z_1) + 2rw$. But $h(z_2) - h(z_1) = \pm r(z_2 - z_1)$. If the sign is plus, then we must have $z_2 = w$, while if the sign is minus, then we must have $z_1 = w$. Consequently there is a real number $k \neq 0$ such that $r = |k|$ and $h(z) = kz - kw + w'$ for all $z \neq w$. But then $h(w) = w' = kw - kw + w'$, so the formula actually holds for all z. Thus h is a homothety.   □

Theorem 2.3

(i) *Given a homothety, the image of a triangle is a triangle similar to it with the similarity ratio equal to the absolute value of the ratio of the homothety. The image of every polygon is a polygon similar to the original one with similarity ratio equal to the absolute value of the ratio of the homothety.*

(ii) *Homothety preserves angles.*

Proof Note that the sides of a triangle are mapped to segments parallel and in the ratio $|k|$ to them. Thus, a triangle is mapped to a triangle similar to it, with similarity ratio lkl. Since every polygon can be decomposed into triangles, the same is true for polygons. This proves (i).

For (ii), place the angle in a triangle. The property follows from the fact that similar triangles have equal angles.   □

As a corollary, if a homothety maps one polygon to another, then the ratio of the areas of the two polygons is equal to the square of the ratio of the homothety.

Theorem 2.4 *Homothety maps a circle to a circle of radius equal to the radius of the original circle multiplied by the absolute value of the ratio of the homothety.*

Proof Let the homothety be $h(z) = kz + (1 - k)a$. The equality

$$|z - \alpha| = R$$

that defines the circle is equivalent to

$$|(kz + (1 - k)a) - (k\alpha + (1 - k)a)| = kR.$$

This means that z satisfies $|z - \alpha| = R$ if and only if $z' = h(z)$ satisfies $|z' - h(\alpha)| = kR$. So z belongs to the circle of center α and radius R if and only if $h(z)$ belongs to the circle of center $h(\alpha)$ and radius kR.   □

The image ω' of a circle ω through a homothety of center O is illustrated in Fig. 2.4. Note that if the center of homothety is exterior to ω, it must also be exterior to ω', and in this case the common tangents of the circles pass through the center of homothety (this is a consequence of the fact that the line joining a point and its image passes through the center of homothety).

Fig. 2.4 The image of a
circle through a homothety

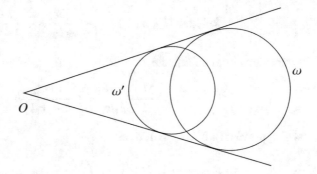

2.1.2   Groups Generated by Homotheties

Following the pattern from the previous chapter, we analyze the behavior of
homotheties under compositions and likewise interpret the results in the language
of group theory. First, a simple observation.

Theorem 2.5 *For a fixed point O, the homotheties with center O form a group,
the inverse of a homothety with ratio k is the homothety of ratio $1/k$, and the
composition of the homotheties of ratios k_1, k_2 is the homothety of ratio $k_1 k_2$.*

But if we look at all homotheties, then they do not form a group. In fact, we have
the following result.

Theorem 2.6 (Composition of Homotheties) *If h_1 and h_2 are homotheties of
centers A and B and ratios k_1 and k_2, respectively, then $h_2 \circ h_1$ is*

(i) *the homothety of ratio $k_1 k_2$ and center a point O on the line AB so that
$\overline{OA}/\overline{OB} = (1 - k_2)/[(1 - k_1)k_2]$, the bar denoting oriented segments*

(ii) *the translation of vector $(1 - k_2)\overrightarrow{AB}$ if $k_2 = 1/k_1$.*

*Consequently, if $h_2 \circ h_1$ is a homothety, then its center is collinear with the centers
of h_1 and h_2.*

Proof 1 Let a and b be the coordinates of A and B, respectively. The two
homotheties are

$$h_1(z) = k_1 z + (1 - k_1)a \text{ and } h_2(z) = k_2 z + (1 - k_2)b,$$

and their composition is

$$h_2 \circ h_1(z) = k_2[k_1 z + (1 - k_1)a] + (1 - k_2)b = k_1 k_2 z + (1 - k_1)k_2 a + (1 - k_2)b.$$

If $k_1 k_2 \neq 1$, this is further equal to

$$k_1 k_2 z + (1 - k_1 k_2) \left(\frac{(1 - k_1) k_2}{1 - k_1 k_2} a + \frac{1 - k_2}{1 - k_1 k_2} b \right),$$

whose center has coordinate

$$\frac{(1 - k_1) k_2}{1 - k_1 k_2} a + \frac{1 - k_2}{1 - k_1 k_2} b$$

which proves (i). If $k_1 k_2 = 1$, then

$$h_2 \circ h_1(z) = z + \frac{1 - k_1}{k_1} a + \left(1 - \frac{1}{k_1} \right) b = z + \frac{k_1 - 1}{k_1}(b - a)$$

$$= z + (1 - k_2)(b - a),$$

proving (ii).   □

Proof 2 Here is a synthetic proof, albeit without the precise location of the center of homothety. Let us first choose a point C that does not belong to the line AB. Let $B' = h_1(B)$ and $C' = h_1(C)$; then $B'C'$ is parallel to BC. Set $A'' = h_2(A)$, $B'' = h_2(B)$, and $C'' = h_2(C)$. Then the sides of $A''B''C''$ are parallel to the sides of $AB'C'$ and hence to the sides of ABC (Fig. 2.5).

Assume first that $k_1 k_2 \neq \pm 1$. Problem 84 below shows that there is a homothety h that maps ABC to $A''B''C''$. Its ratio is $k_1 k_2$. Note that A'' and B'' belong to the line AB, and so $h(AB) = AB$. This means that the center O of the homothety is on AB. The point O is a fixed point of $h = h_2 \circ h_1$. But h can only have one fixed point, since for any two points, the distance from $h(X)$ to $h(Y)$ is $|k_1 k_2| XY \neq XY$. Thus, the same homothety appears for all points C that are not on AB. If C is on AB, let D be a point that does not belong to AB. Set $C'' = h(C)$ and $D'' = h(D)$. Then the triangles $A'C'D'$ and ACD have parallel sides, so they are homothetic. And the homothety has ratio $k_1 k_2$ and center at the intersection of AA' and DD', so it must be h.

Fig. 2.5 The composition of two homotheties

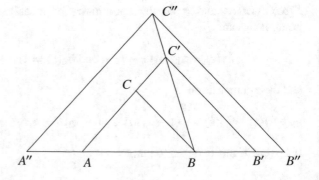

If $k_1 k_2 = \pm 1$, then the composition is an isometry, and we are in the environment of the previous chapter. The composition is therefore a translation (when $k_1 k_2 = 1$), or reflection over a point (when $k_1 k_2 = -1$), the latter case being an inverse homothety.   □

The collinearity proved in this theorem will play a major role in our story; we will highlight its significance later (stay alert!).

Theorem 2.7 *Let h be the homothety of center A and ratio $k \neq 1$, and let τ be the translation of vector $\vec{v}$. Then*

(i) $h \circ \tau$ is the homothety of ratio k and whose center is the translate of A by $\frac{k}{1-k}\vec{v}$,

(ii) $\tau \circ h$ is the homothety of ratio k and whose center is the translate of A by $\frac{1}{1-k}\vec{v}$.

Proof Let the complex coordinate of A be a, and let b be the complex number representing $\vec{v}$, so that $\tau(z) = z + b$. Also, let $h(z) = kz + (1 - k)a$. Then

$$h \circ \tau(z) = k(z + b) + (1 - k)a = kz + (1 - k)\left[a + \frac{k}{1 - k}b\right].$$

Also

$$\tau \circ h(z) = kz + (1 - k)a + b = kz + (1 - k)\left[a + \frac{1}{1 - k}b\right].$$

The proposition is proved.   □

You can also write a synthetic proof similar to that of Proposition 2.6. We are now able to state a fundamental result.

Theorem 2.8 *The group of transformations of the plane that take every line to a line parallel to it consists of all homotheties and all translations.*

Proof By Theorems 2.5–2.7, the set consisting of all homotheties and translations is a group. Since the image of a line through a translation or a homothety is a line parallel to it, every element of this group maps a line to a line parallel to it.

Conversely, let f be a transformation of the plane that maps every line to a line parallel to it, and let A and B be two distinct points. Compose f with a translation τ such that $\tau \circ f(A) = A$ (if $f(A) = A$ take τ to be the identity map). Then $\tau \circ f(B)$ belongs to the line AB and is a point different from A. Compose with the homothety h centered at A such that $h \circ \tau \circ f(B) = B$ (if $\tau \circ f(B) = B$ then h is just the identity map).

Now let $g = h \circ \tau \circ f$. Then $g(A) = A$, and $g(B) = B$, and g also maps every line to a line parallel to it. Let C be a point that does not belong to AB. Then each of the lines AC and BC is mapped into itself, and so the intersection of these lines must be mapped into itself. Thus $g(C) = C$ for all C that do not belong to AB.

Next, let $C \in AB$ and let D, E be points that are not on AB such that C does not belong to DE. Then, as seen above, $g(D) = D$, and $g(E)$, and applying the same reasoning for the triple D, E, C, we deduce that $g(C) = C$. Thus g is the identity map. This shows that $f = \tau^{-1} \circ h^{-1}$, so f is a composition of a translation and a homothety.   □

A consequence of this theorem is that if a transformation maps every line to a line parallel to it, and if one can find two points A and B with images A' and B', respectively, such that AA' and BB' intersect at some point O, then the transformation is a homothety of center O.

In view of Theorem 2.8 and the results preceding it, the smallest group that contains all homotheties necessarily contains all translations. Note also that direct homotheties and translations form a subgroup of this group.

2.1.3   Problems About Properties of Homotheties

We continue with a short list of theorems given as exercises, some of which will play an important role in what follows.

83 Given two distinct circles, how many homotheties map them one into the other?

84 (a) For two noncongruent triangles with parallel sides, show that there exists a homothety that maps one into the other.
(b) For two noncongruent polygons whose sides and diagonals are respectively parallel, show that there exists a homothety that maps one into the other.

85 What is the largest number of homotheties that can map two polygons into each other?

86 Let $\omega_1, \omega_2, \omega_3$ be three nonconcentric, noncongruent circles. Prove that:

(a) *(Monge's Theorem)* The centers of the direct homotheties of the pairs (ω_1, ω_2), (ω_1, ω_3), and (ω_2, ω_3) are collinear.
(b) *(d'Alembert's Theorem)* The six centers of the homotheties of the pairs (ω_1, ω_2), (ω_1, ω_3), and (ω_2, ω_3) lie on four lines.

2.2   Problems in Euclidean Geometry That Use Homothety

2.2.1   Theorems in Euclidean Geometry Proved Using Homothety

We now explain how to apply what we have learned about homotheties.

Fig. 2.6 Euler's line

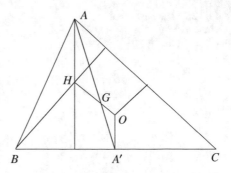

I. Euler's Line and the Nine-Point Circle
The following theorem was proved by Leonard Euler in 1765.

Theorem 2.9 (Euler's Line) *In a triangle the orthocenter, the centroid, and the circumcenter are collinear. Moreover, the centroid lies on the segment joining the circumcenter and the orthocenter and divides this segment in the ratio* 1 : 2.

Proof Figure 2.6 describes the situation. Let the triangle be ABC, and let G, H, O be its centroid, orthocenter, and circumcenter, respectively.

Consider the homothety h of center G and ratio $-\frac{1}{2}$. Then h maps the vertex A to the midpoint A' of BC, and it maps the altitude from A to the perpendicular bisector of BC. Reasoning similarly for the other altitudes, we deduce that h maps the intersection of the altitudes to the intersection of the perpendicular bisectors of the sides, thus $h(H) = O$. Since a point and its image are collinear with the center of homothety, it follows that H, O, and G are collinear. Moreover, since h is an inverse homothety, G is between H and O, and GO/GH is equal to the absolute value of the homothety ratio, which is $\frac{1}{2}$. The theorem is proved.   □

Remark If we place the origin of a complex coordinate system at the circumcenter O, with the coordinates of the vertices being a, b, c, then the coordinate of the orthocenter H is $a + b + c$. This is because H is the image of G under a homothety of center O and ratio 3, and the coordinate of G is $\frac{a+b+c}{3}$.

Theorem 2.10 (The Nine-Point Circle) *In a triangle, the midpoints of the sides, the feet of the altitudes, and the midpoints of the segments that join the orthocenter and the vertices are on a circle. Moreover, the center of this circle lies on Euler's line, and the orthocenter, the center of the nine-point circle, the centroid, and the circumcenter form a harmonic division.*

Four points X, Y, Z, W that appear in this order on a line form a harmonic division if

$$\frac{YX}{YZ} : \frac{WX}{WZ} = 1 \iff \frac{XY}{XW} : \frac{ZY}{ZW} = 1.$$

In complex coordinates, the four collinear points form a harmonic division if their coordinates satisfy

Fig. 2.7 The nine-point circle

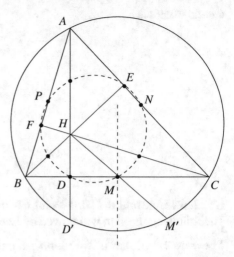

$$\frac{y-x}{y-z} : \frac{w-x}{w-z} = -1 \iff \frac{x-y}{x-w} : \frac{z-y}{z-w} = -1$$

namely, if their cross-ratio is -1.

Proof The proof of the theorem can be followed on Fig. 2.7. The midpoints of the segments that join the orthocenter with the vertices are on a circle that is mapped to the circumcircle by the homothety h centered at the orthocenter and of ratio 2. Thus it is natural to show that this homothety maps the other points in question to the circumcircle, too.

Let therefore the triangle be ABC, H its orthocenter, M, N, P the midpoints of BC, CA, AB, and D, E, F the feet of the altitudes from A, B, C, respectively. Consider a system of complex coordinates centered at the circumcenter O, and let a, b, c be the coordinates of A, B, C, respectively. In view of the above observation, the orthocenter has coordinate $a + b + c$. Then $M' = h(M)$ is the reflection of H over M, and since M has the coordinate equal to $\frac{b+c}{2}$, the complex coordinate of M' is $b + c - (a + b + c) = -a$. Thus, M' is the point diametrically opposite to A in the circumcircle.

The point $D' = h(D)$ is the reflection of H over BC, and this reflection is the composition of the reflection over the perpendicular bisector of BC with the reflection over M. Thus D' is the reflection of M' over the perpendicular bisector of the chord BC of the circumcircle, so it is itself on the circumcircle. We conclude that $h(M)$ and $h(D)$ are on the circumcircle, and the same is true for the other points. It follows that the nine points are on the image of the circumcircle through h^{-1}, so they are concyclic.

Moreover, the image of the circumcenter is the center of this nine-point circle, and it is the midpoint of the segment OH. It therefore has the complex coordinate $\frac{a+b+c}{2}$. Thus the orthocenter, center of the nine-point circle, the centroid, and the circumcenter have the coordinates

$$a+b+c, \quad \frac{a+b+c}{2}, \quad \frac{a+b+c}{3}, \quad 0.$$

We compute

$$\frac{(a+b+c)/2-(a+b+c)}{(a+b+c)/2-(a+b+c)/3} : \frac{0-a+b+c}{0-(a+b+c)/3} = \frac{-1/2}{1/6} : \frac{-1}{-1/3} = -1,$$

showing that the four points form a harmonic division.   □

And here is another proof of

Theorem 1.22 *Let ABC be a triangle. Then the symmedian from B passes through the point where the tangents to the circumcircle at A and C intersect.*

Proof Let W be the intersection point of the tangents. We step aside and explore the *nine-point circle* of some triangle ABC with orthocenter H. We have just learned that the nine-point circle passes through the midpoints M, N, and P of the sides BC, CA, and AB; the midpoints X, Y, and Z of AH, BH, and CH; and the feet D, E, and F of the altitudes through A, B, and C, respectively. Because $\angle YEN = 90°$, YN is a diameter. In particular, $\angle YDN = \angle YFN = 90°$ (Fig. 2.8).

Now consider the circle of diameter BH, whose center is Y. Because $\angle BDH = \angle BFH = 90°$, it is also the circumcircle of DBF. Since $YD \perp DN$ and $YF \perp FN$, ND and NF are tangents from N to this circle.

We are ready to finish the proof: since $\angle AFC = \angle ADC = 90°$, $AFDC$ is cyclic, and the power of the point B with respect to its circumcircle gives $BA \cdot BF = BC \cdot BD$. Hence $BD/BF = AB/BC$. This means that DBF is similar to ABC, and this similarity takes the median BN' to its reflection over the bisector of $\angle B$, and this reflection is BN. Consequently BN is a symmedian in the triangle DBF, and ND and NF are tangent to the circumcircle of BDF. Now just change the notation, trading D with A, F with C, and N with W.   □

Fig. 2.8 Symmedians and tangents

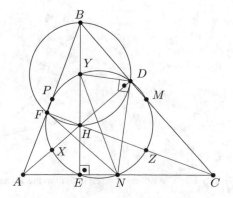

The nine-point circle will play an important role in many of the problems from this chapter. We will encounter it again briefly when we talk about inversion.

II. The Incircle and the Excircle

Homothety provides the natural setting for introducing the excircles of a triangle. Let ABC be the triangle, let I be its incenter, and let D, E, F be the points of tangency of the incircle with the sides BC, AC, AB, respectively.

Consider the point J on the incircle that lies diametrically opposite to D and, with it, the homothety h_a of center A that maps the tangent to the incircle at J to the line BC. Such a homothety exists because the two lines are parallel. Then the incircle is mapped to a circle that is exterior tangent to BC and is also tangent to AB and AC. This is one of the three *excircles* of the triangle, the other two are obtained by performing the same construction with respect to the sides AC and AB. The excircle corresponding to side BC is shown in Fig. 2.9. It is standard to denote its center by I_a, with the centers of the other two excircles being denoted by I_b and I_c.

The ratio of the homothety h_a is the result of a short computation. To do it, let $AB = c$, $AC = b$, and $BC = a$. Let also E' and F' be the images of E and F through the homothety, and let K be the point where the excircle is tangent to BC. Because the tangents from a point to a circle are equal, we can set $AE = AF = x$, $BD = BF = y$, $CD = CF = z$, $BK = BF' = u$, and $CK = CE' = v$. We have $x + y = c$, $x + z = b$, $y + z = a$, $u + v = a$, and $u + c = v + b$. Solving the system of equations, we obtain

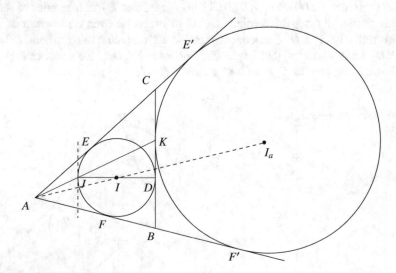

Fig. 2.9 The excircle corresponding to the vertex A

$$x = \frac{b+c-a}{2}, y = \frac{a+c-b}{2}, z = \frac{a+b-c}{2}, u = \frac{a+b-c}{2}, v = \frac{a+c-b}{2}.$$

The homothety ratio is therefore

$$\frac{AF}{AF'} = \frac{AF}{AB+BF'} = \frac{x}{c+u} = \frac{b+c-a}{b+c+a}.$$

This is the ratio of the inradius by the radius of the excircle. But from this computation, we can also infer that $BD = CK$, so D and K are symmetric with respect to midpoint of the side BC.

Proposition 2.11 *Let ABC be a triangle, let $D, K \in BC$ be the points where the incircle and the excircle are tangent to the side, let I and I_a be the centers of these circles, and let R be the midpoint of the altitude from A. Then R, I, K and R, D, I_a are collinear.*

Proof We add to the picture the foot N of the altitude from A, the point J that is diametrically opposite to D in the incircle, and the point L that is diametrically opposite to K in the excircle. Then the homothety h_a defined above maps DJ to LK, and so A, D, L and A, J, K are collinear. It follows that the triangles KDJ and KNA are homothetic under a homothety of center K, and so the midpoint R of AN, the midpoint I of JD, and K are collinear. Similarly, the triangles DAN and DKL are homothetic under a homothety of center K, so the midpoint R of AN, the midpoint I_a of LK, and D are collinear.   □

III. The Theorems of Pappus and Menelaus
We begin with the first version of Pappus' Theorem.

Theorem 2.12 (Pappus) *Let ℓ_1 and ℓ_2 be two parallel lines, let A_1, B_1, C_1 be distinct points on ℓ_1, and let A_2, B_2, C_2 be points on ℓ_2. If $P = B_1C_2 \cap C_1B_2$, $Q = C_1A_2 \cap A_1C_2$, and $R = A_1B_2 \cap A_2B_1$, then P, Q, R are collinear.*

Proof The theorem is illustrated in Fig. 2.10. We need the following lemma.

Fig. 2.10 Pappus' Theorem

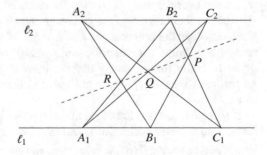

Lemma 2.13 *Let h_P, h_Q, h_R be three homotheties of centers P, Q, R and ratios p, q, r, respectively, such that*

$$h_P \circ h_Q \circ h_R = h_R \circ h_P \circ h_Q \ or \ h_P \circ h_Q \circ h_R = h_R \circ h_Q \circ h_P.$$

Assume additionally that $pq \neq 1$. Then the points P, Q, R are collinear.

Proof At the heart of the proof lies Theorem 2.6. Let us assume, for example, that

$$h_P \circ h_Q \circ h_R = h_R \circ h_Q \circ h_P.$$

Because $pq \neq 1$, by the abovementioned theorem, $h_P \circ h_Q$ and $h_Q \circ h_P$ are homotheties; denote their respective centers by S and T. By the same theorem, the points S, T are on the line PQ. We have two possibilities.

If $pqr = 1$, then the transformation $h_P \circ h_Q \circ h_R$ is the translation of vector

$$(1 - pq)\overrightarrow{RS} = (1 - r)\overrightarrow{TR},$$

and it follows that $R \in ST = PQ$.

If $pqr \neq 1$, let U be the center of the homothety $h_P \circ h_Q \circ h_R = h_R \circ h_Q \circ h_P$. Then U is on both lines SR and RT and if $R \notin ST = PQ$, then $R = U$. On the other hand, from Theorem 2.6, (i) we deduce that

$$pq(1 - r)\overrightarrow{UR} = -(1 - pq)\overrightarrow{US} \text{ and } r(1 - pq)\overrightarrow{UT} = -(1 - r)\overrightarrow{UR}.$$

But this can only happen if $U = S = T$, which is false. It follows that $R \in ST = PQ$, and the lemma is proved. $\qquad\square$

Returning to the proof of the theorem, let

$$p = \frac{\overline{PC_1}}{\overline{PB_2}} = \frac{\overline{PB_1}}{\overline{PC_2}}, \quad q = \frac{\overline{QA_2}}{\overline{QC_1}} = \frac{\overline{QC_2}}{\overline{QA_1}}, \quad r = \frac{\overline{RA_1}}{\overline{RB_2}} = \frac{\overline{RB_1}}{\overline{RA_2}},$$

where the overline notation stands for oriented segments (so that if P is between A_1 and B_2, then the ratio is negative; in the situation from Fig. 2.10 $p, q, r < 0$). Next, consider the homotheties h_P, h_Q, h_R of centers P, Q, R and ratios p, q, r, respectively. Note, for example, that $h_P(B_2) = A_1$ and $h_Q(A_1) = C_2$. If $pq = qr = 1$, then it follows easily that the lines PQ and QR are parallel to ℓ_1 and ℓ_2 respectively, thus P, Q, R are collinear. Otherwise, if say $pq \neq 1$, we have two cases.

If $pqr = 1$, then the transformations $h_P \circ h_Q \circ h_R$ and $h_R \circ h_Q \circ h_P$ have dilation factor 1 so they are translations, and they both take B_2 to B_1. So both must be the translation of vector $\overrightarrow{B_2B_1}$, and in particular they are equal.

$$h_P \circ h_Q \circ h_R = h_R \circ h_Q \circ h_P.$$

If $pqr \neq 1$, then $h_P \circ h_Q \circ h_R$ and $h_R \circ h_Q \circ h_P$ are homotheties of ratio pqr that again take B_2 to B_1. But there is only one point O on the line $B_1 B_2$ such that, with oriented segments, $\overline{OB_2}/\overline{OB_1} = pqr$. This point O must be the center of both homotheties, so again

$$h_P \circ h_Q \circ h_R = h_R \circ h_Q \circ h_P.$$

In both cases we can apply the lemma to conclude that P, Q, R are collinear.

If, instead, $qr = 1$, we are in the other hypothesis of the lemma, and we obtain the same conclusion. $\quad\square$

This proof requires that the lines are parallel, but the theorem is true even when the lines intersect.

Theorem 2.14 (Pappus) *Let ℓ_1 and ℓ_2 be two intersecting lines, let A_1, B_1, C_1 be distinct points on ℓ_1, and let A_2, B_2, C_2 be points on ℓ_2. If $P = B_1 C_2 \cap B_2 C_1$, $Q = A_1 C_2 \cap A_2 C_1$, and $R = A_1 B_2 \cap A_2 B_1$, then P, Q, R are collinear.*

The most natural approach is by two-dimensional real projective geometry. In that setting, the fact that the lines intersect is irrelevant, as we can move the intersection point at infinity. We land in the case we just proved, and we are done. But to be true to the scope of this book, we must write a proof that avoids transformations of the real projective plane.

The geometric configuration of the general Pappus' Theorem hides many homotheties in a subtle way. The geometry afficionado probably already thinks about Menelaus' Theorem, which is the most common tool for checking collinearity. And this theorem finds its most natural setting in the realm of homotheties.

Theorem 2.15 (Menelaus' Theorem) *Let ABC be a triangle, and let A_1, B_1, C_1 be points on BC, CA, AB, respectively. Then the points A_1, B_1, C_1 are collinear if and only if, with oriented segments, the following equality holds:*

$$\frac{\overline{A_1 B}}{\overline{A_1 C}} \cdot \frac{\overline{B_1 C}}{\overline{B_1 A}} \cdot \frac{\overline{C_1 A}}{\overline{C_1 B}} = 1,$$

We can rephrase this using complex numbers without mentioning oriented segments by saying that

$$\frac{b - a_1}{c - a_1} \cdot \frac{c - b_1}{a - b_1} \cdot \frac{a - c_1}{b - c_1} = 1,$$

with the lowercase letter being the coordinate of the uppercase point. The configuration of the Menelaus' Theorem is shown in Fig. 2.11.

Proof To avoid oriented segments, let us stay in the world of complex numbers. Consider the homothety h_a of center A_1 and ratio k_a that maps C to B, the homothety h_b of center B_1 and ratio k_b that maps A to C, and the homothety h_c of center C_1 and ratio k_c that maps B to A. Then

Fig. 2.11 Menelaus' Theorem

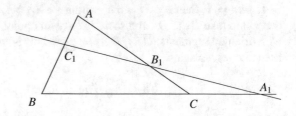

$$k_a = \frac{b - a_1}{c - a_1}, \quad k_b = \frac{c - b_1}{a - b_1}, \quad k_c = \frac{a - c_1}{b - c_1}.$$

By Theorem 2.6, the composition $h_a \circ h_b \circ h_c$ is either a homothety or a translation. It is not a translation, because it has a fixed point: $h_a \circ h_b \circ h_c(B) = B$. So it is a homothety. By the same theorem, the center of the homothety is on the line of the points A_1, B_1, C_1 and so is different from B and is another fixed point. A homothety with two fixed points is the identity map. Consequently $k_a k_b k_c = 1$, that is,

$$\frac{b - a_1}{c - a_1} \cdot \frac{c - b_1}{a - b_1} \cdot \frac{a - c_1}{b - c_1} = 1.$$

For the converse, consider the same homotheties, and notice that the given relation implies that $h_a \circ h_b \circ h_c$ has ratio 1. As this transformation has B as a fixed point, it is not a translation, so it is the identity map. This means that $h_c^{-1} = h_a \circ h_b$, and, by Theorem 2.6, h_a, h_b, h_c have collinear centers.   □

By applying repeatedly Menelaus' Theorem, we produce a proof of Pappus' Theorem with a large number of homotheties embedded in it.

Proof 1 Let X, Y, Z be the intersection points of the pairs of lines $(A_1 C_2, C_1 B_2)$, $(B_1 A_2, A_1 C_2)$, and $(C_1 B_2, B_1 A_2)$, respectively, as shown in Fig. 2.12. Using complex coordinates with the usual convention for upper-/lowercase letter, we write the relation from Menelaus' Theorem for the triangle XYZ and the transversals $A_1 B_2, B_1 C_2, C_1 A_2, \ell_1$, and ℓ_2

$$\frac{x - a_1}{y - a_1} \cdot \frac{y - r}{z - r} \cdot \frac{z - b_2}{x - b_2} = 1, \quad \frac{y - b_1}{z - b_1} \cdot \frac{z - p}{x - p} \cdot \frac{x - c_2}{y - c_2} = 1,$$

$$\frac{z - c_1}{x - c_1} \cdot \frac{x - q}{y - q} \cdot \frac{y - a_2}{z - a_2} = 1,$$

$$\frac{x - c_1}{z - c_1} \cdot \frac{z - b_1}{y - b_1} \cdot \frac{y - a_1}{x - a_1} = 1, \quad \frac{y - c_2}{x - c_2} \cdot \frac{x - b_2}{z - b_2} \cdot \frac{z - a_2}{y - a_2} = 1.$$

Multiply the five relations to obtain

$$\frac{y - r}{z - r} \cdot \frac{z - p}{x - p} \cdot \frac{x - q}{y - q} = 1,$$

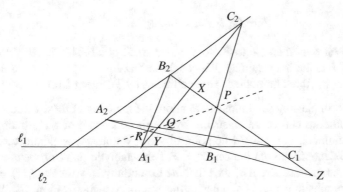

Fig. 2.12 The second version of Pappus' Theorem

which, by applying Menelaus' Theorem in reverse to the same triangle, implies that P, Q, R are collinear.   □

Proof 2 We can decrypt the meaning of these algebraic computations. Like in the proof of Menelaus' Theorem, behind each fraction lies a homothety. For example, behind the first fraction lies a homothety $h_{a_1}(w) = r_{a_1} w + s_{a_1}$ of center A_1 and that maps Y to X. Here r_{a_1} is a nonzero real number and s_{a_1} is a complex number. We use the same convention to associate to each fraction in the first three products a homothety indexed by the coordinate of the center written as a linear function with coefficients indexed by the same coordinate. The five equalities correspond, via the same argument as in the proof of the direct implication of Menelaus' Theorem, to

$$h_{a_1} \circ h_r \circ h_{b_2} = 1, \quad h_{b_1} \circ h_p \circ h_{c_2}, \quad h_{c_1} \circ h_q \circ h_{a_2} = 1,$$
$$h_{c_1}^{-1} \circ h_{b_1}^{-1} \circ h_{a_1}^{-1} = 1, \quad h_{c_2}^{-1} \circ h_{b_2}^{-1} \circ h_{a_2}^{-1} = 1,$$

where 1 stands for the identity transformation. By composing with the appropriate transformations on the left and right, we can change these equalities into

$$h_r = h_{a_1}^{-1} \circ h_{b_2}^{-1}, \quad h_p = h_{b_1}^{-1} \circ h_{c_2}^{-1}, \quad h_q = h_{c_1}^{-1} \circ h_{a_2}^{-1},$$
$$h_{c_1}^{-1} = h_{a_1} \circ h_{b_1}, \quad h_{c_2}^{-1} = h_{a_2} \circ h_{b_2}.$$

Substitute $h_{c_1}^{-1}$ and $h_{c_2}^{-1}$ in the first equations, and then compose to obtain

$$h_p \circ h_q \circ h_r = h_{b_1}^{-1} \circ h_{a_2} \circ h_{b_2} \circ h_{a_1} \circ h_{b_1} \circ h_{a_2}^{-1} \circ h_{a_1}^{-1} \circ h_{b_2}^{-1}.$$

For arbitrary functions, this is not the identity map. But here we work with affine transformations of the plane, and so the right-hand side is a affine transformation of the form $f(w) = rw + s$ with

$$r = r_{b_1}^{-1} r_{a_2} r_{b_2} r_{a_1} r_{b_1} r_{a_2}^{-1} r_{a_1}^{-1} r_{b_2}^{-1} = 1.$$

But $h_p \circ h_q \circ h_r$ maps Z to Z, so f has a fixed point. This implies $s = 0$, and therefore f is the identity map. We conclude that $h_p = h_r^{-1} \circ h_q^{-1}$, so the centers P, Q, R of the three homotheties are collinear, by Theorem 2.6.   □

It is worth pausing for a moment to allow the meaning of these examples to sink in. Homotheties can be parametrized by a pair consisting of a real number and a complex number. For a synthetic point of view, we choose as parameters the ratio k and the (coordinate of) the center a. For the analytic point of view, we choose different parameters, k and b, such that the homothety is written as $h(z) = kz + b$.

The composition rule is simpler in the second parametrization, for if $h_1(z) = k_1 z + b_1$, $h_2(z) = k_2 z + b_2$, and $h_2 \circ h_1(z) = kz + b$, then $k = k_1 k_2$, and $b = k_2 b_1 + b_2$. It is significantly more complicated to write the composition rule in the first parametrization, as Theorem 2.6 tells us that if the centers and ratios of h_1, h_2, and $h_2 \circ h_1$ are k_1, k_2, k and a_1, a_2, a, respectively, then $k = k_1 k_2$, while a is characterized by the geometric conditions of lying on the line passing through a_1 and a_2 and forming with a_1 and a_2 segments that are in a certain ratio. But it is *this* more geometrically intuitive parametrization, or rather the composition rule transcribed for the parameters, that empowers us with one of the best tools for proving collinearity, in the guise of Theorem 2.6. Both the Menelaus' Theorem and the Monge-d'Alembert Theorem (Problem 86) are immediate consequences of the composition rule for homotheties.

Groups that admit (local) parametrizations by real numbers in which the composition rule and the taking of the inverse are continuous maps are called Lie groups (after Marius Sophus Lie). The group of homotheties and translations is therefore a Lie group (parametrized by $\mathbb{R}^3$). We see how just by writing the composition rule in the appropriate parametrization we obtain powerful theorems in classical geometry.

2.2.2 Examples of Problems Solved Using Homothety

We start with an easy example.

Problem 2.1 The trapezoids $ABCD$ and $APQD$ share the base AD, while BC and PQ have different lengths. Show that the intersection points of the pairs of lines (AB, CD), (AP, DQ), and (BP, CQ) are collinear.

Solution Denote the intersections of the three pairs by K, L, M, respectively, as in Fig. 2.13. They are the centers of the direct homotheties h_1, h_2, and h_3 that transform the segments BC into AD, AD into PQ, and BC into PQ, respectively. Then $h_3 = h_2 \circ h_1$, so by Theorem 2.6, the centers K, L, M of these homotheties are collinear.   □

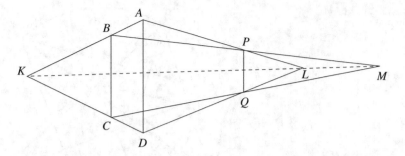

Fig. 2.13 Trapezoids sharing a side

Fig. 2.14 The orthic triangle of MNP

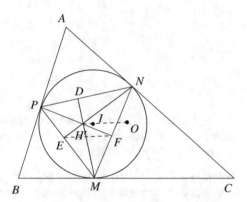

The next problem has appeared in the Iranian Mathematical Olympiad in 1995, the Kürschák Competition in 1997, and the Hungary-Israel Competition in 2000.

Problem 2.2 In a triangle ABC, let O be the circumcenter, let I be the incenter, and let M, N, P be the points where the incircle touches the sides BC, CA, and AB, respectively. Let also H' be the orthocenter of MNP. Prove that I, O, and H' are collinear.

Solution 1 The first solution that we present was found when this problem was discussed at the training program of the Bulgarian International Mathematical Olympiad Team in 2002. Let D, E, F be the feet of the altitudes from M, N, P, respectively, of the triangle MNP, as shown in Fig. 2.14. The points E and F are on the circle of diameter NP, so $\angle PNM = \angle MEF$. Using inscribed angles in the incircle, we have $\angle PNM = \angle BMP$. Hence $\angle BMP = \angle MEF$, showing that EF and BC are parallel. Similarly, $DF \| AC$ and $DE \| AB$. The triangles ABC and DEF have parallel sides, so by Problem 84, they are homothetic. Let S be the center of the homothety.

The incenter I of ABC is mapped by the homothety to the incenter of DEF, which is H', so S, I, H' are collinear. Since I is the circumcenter of MNP, the line IH' is the Euler line of this triangle, whose line contains also the circumcenter

Fig. 2.15 The triangle
$M'N'P'$

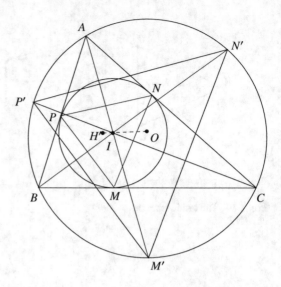

O' of this triangle. As the homothety maps O to O', S, O, O' are also collinear, showing that O belongs to the same Euler line. Thus I, O, H' are on the Euler line of MNP, so they are collinear.   □

Solution 2 Let M', N', P' be the points where the angle bisectors of $\angle A$, $\angle B$, $\angle C$ intersect the circumcircle, respectively (Fig. 2.15). Then

$$\angle(BN', M'P') = \frac{1}{2}\ \widehat{P'AN'} + \frac{1}{2}\ \widehat{B'M'} = \angle A/2 + \angle B/2 + \angle C/2 = 90°.$$

Hence $M'P'$ is perpendicular to the angle bisector of $\angle B$. But so is MP. So $M'P' \| MP$. Similarly, $M'N' \| MN$, $N'P' \| NP$. From here we also deduce that I is the orthocenter of $M'N'P'$.

The triangles MNP and $M'N'P'$ have parallel sides, so by Problem 84, they are homothetic. Let T be the center of the homothety. Then O (the circumcenter of $M'N'P'$), I (the circumcenter of MNP), and T are collinear. Also I (the orthocenter of $M'N'P'$), H' (the orthocenter of MNP), and T are collinear. Consequently, T, O, I, H' are collinear, and the problem is solved.   □

One should point out that in this configuration, the lines MM', NN', PP' intersect at the center S of the first homothety, while the lines AD, BE, CF intersect at the center T of the second homothety, and these intersection points are on the Euler line of MNP.

The method of these two solutions is to be remembered: If two triangles have parallel sides, then they are homothetic, and the lines connecting their corresponding vertices intersect at the center of homothety.

The next problem was given at the International Zhautykov Olympiad in 2011 in Kazakhstan.

Problem 2.3 Given a convex quadrilateral, draw two segments that join points on a pair of opposite sides and do not intersect. Draw two other segments that join points on the other pair of opposite sides and do not intersect. We know that the points of intersection of the first two segments with the last two segments lie on the diagonals of the quadrilateral, and that, of the 9 quadrilaterals they determine, those in three corners and the one in the middle admit inscribed circles. Show that the quadrilateral in the fourth corner also admits an inscribed circle.

Solution The solution needs two auxiliary results.

Lemma 2.16 *Let ABC be a triangle, and let ω and ω_1 be two circles that are tangent to the sides AB and AC and intersect the side BC. If the arcs on the circles ω and ω_1 that lie outside the triangle ABC have equal measures, then the two circles coincide.*

Proof Examining Fig. 2.16, we can see that there is a homothety of center A that maps ω and ω_1. Arguing by contradiction, let us assume that the circles do not coincide and that ω has smaller radius than ω_1. Then the arc of ω that lies outside of triangle ABC ($\overset{\frown}{DE}$ in the figure) is mapped by the homothety to an arc that lies strictly in the interior of the arc of ω_1 that lies outside of triangle ABC ($\overset{\frown}{D_1E_1}$ in the figure). Homothety preserves the measures of arcs (since it preserves the measure of angles), so the image has the same measure as the original arc, and hence the arc of ω that lies outside of the triangle ABC has measure strictly less than the arc of ω_1 that lies outside of the triangle. This is a contradiction, and the conclusion follows. □

Lemma 2.17 (Igor Voronovich) *Let $ABCD$ be a convex quadrilateral, and let ω_1 be a circle tangent to the sides AB and AD and ω_2 a circle tangent to the sides CB and BD. Then the quadrilateral $ABCD$ can be circumscribed to a circle if and only if the arcs of the circles ω_1 and ω_2 that lie on the same side of diagonal AC have equal measures.*

Fig. 2.16 Proof that ω and ω_1 coincide

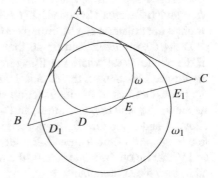

Fig. 2.17 Proof of the
Voronovich Lemma

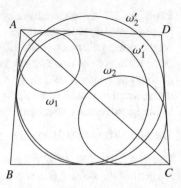

Proof Consider the homothety of center A that maps ω_1 to a circle ω_1' that is tangent
to BC and the homothety of center C that maps ω_2 to a circle ω_2' that is tangent to
AB. These new circles are shown in Fig. 2.17.

If $ABCD$ is a circumscribed quadrilateral, then $\omega_1' = \omega_2'$, so the arcs that lie
on the same side of AC on ω_1' and ω_2' are equal. Then, since homothety preserves
the measure of arcs, the arcs on the same side of AC of ω_1 and ω_2 have also equal
measures. This proves the direct implication.

For the converse implication, if the arcs of ω_1 and ω_2 that lie on the same side of
the diagonal AC as the vertex D have the same measure, then the same is true for
the circles ω_1' and ω_2', since again homothety preserves the measure of arcs. Using
Lemma 2.16 (applied to the triangle BAC and the circles ω_1' and ω_2'), we obtain
that $\omega_1' = \omega_2'$, and this is the circle inscribed in $ABCD$. The Voronovich Lemma is
proved.   □

Returning to the problem, we reason on Fig. 2.18 and notice that the circles 2 and
3 are inverse homothetic with respect to the point of intersection of two segments
that join opposite sides that lies on the diagonal between the two circles. The same
is true for the circles 3 and 4. So the circles 2 and 4 are directly homothetic, and
the center of the homothety lies on the line of support of the diagonal that joins
the vertices that correspond to these circles. This enforces the conditions of the
Voronovich Lemma (Lemma 2.17) for these two circles, and by the lemma, the
original quadrilateral can be circumscribed to a circle.

Draw the circles $5'$ and $5''$ so that each is tangent to a pair of adjacent sides
of the quadrilateral where the fifth circle is supposed to be inscribed, as shown in
the figure. Like before, the circles 1 and 3 are inverse homothetic with respect to
the point that lies at the intersection of the corresponding diagonal with the two
segments that join opposite sides, and the same is true for the circles 3 and $5'$, so the
circles 1 and $5'$ are directly homothetic, and the center of the homothety lies on the
line of support of the diagonal. It follows that the measures of the arcs of circles 1
and $5'$ that lie on the same side of the diagonal are equal, being the images of each
other by a direct homothety.

Fig. 2.18 Proof that the fifth
quadrilateral has an inscribed
circle

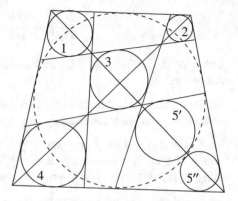

On the other hand, by applying the Voronovich Lemma to the big quadrilateral, we deduce that the measures of the arcs of circles 1 and $5''$ that lie on the same side of the diagonal are equal. Thus, the measures of the arcs of circles $5'$ and $5''$ that lie on the same side of the diagonal are equal. Now we can apply the Voronovich Lemma to the quadrilateral that arises in the lower right corner of the figure and conclude that this quadrilateral admits an inscribed circle. The problem is solved.

□

2.2.3   Problems in Euclidean Geometry to be Solved Using Homothety

This section lists problems that you should try to solve using homothety.

87 The circle ω_2 is interior tangent to the circle ω_1 at A. A line intersects ω_1 at P and Q and ω_2 at M and N. Show that $\angle PAM = \angle QAN$.

88 Show that the projections of the vertex A of the triangle ABC onto the interior and exterior bisectors of $\angle B$ and $\angle C$ are four collinear points.

89 Let I be the incenter of the triangle ABC, and let D be the point of tangency of the incircle with the side AC. Show that if B_1 is the midpoint of AC, then $B_1 I$ cuts the segment BD in half.

90 Let AB be a segment with K a fixed point on it. Let M be a variable point on the circle Ω of diameter AB, and let N be the point diametrically opposite to M. Find the locus of the intersection P of AN and KM when M varies on Ω.

91 *(Boutin's Theorem)* Let ABC be a triangle with circumcenter O. Let A', B', C' be the midpoints of BC, AC, and AB, respectively, and let A'', B'', C'' be points on the rays $|OA'$, $|OB'$, and $|OC'$ such that

$$\frac{OA''}{OA'} = \frac{OB''}{OB'} = \frac{OC''}{OC'}.$$

Show that AA'', BB'', and CC'' intersect at a point that lies on the Euler line of the triangle ABC.

92 Prove that if in a trapezoid the point of intersection of the diagonals is at equal distance from the two nonparallel sides, then the trapezoid is isosceles.

93 Let $A_1A_2 \ldots A_n$ be a polygon, $n \geq 3$, and let G_j be the centroid of the polygon with $n - 1$ sides obtained by removing the vertex A_j. Show that the polygons $A_1A_2 \ldots A_n$ and $G_1G_2 \ldots G_n$ are homothetic.

94 A polygon has the property that if its sides are pushed outward by one unit and then extended until they meet, the resulting polygon is similar to the original. Show that one can inscribe a circle in this polygon.

95 Let $ABCD$ be a convex quadrilateral, and let M, N, P, Q be the midpoints of the sides AB, BC, CD, DA, respectively. Construct the parallelogram $XYZW$ by taking the parallel lines through the vertices of the quadrilateral $ABCD$ to its diagonals. Let I be the intersection of the diagonals of $ABCD$, J the intersection of the diagonals of $MNPQ$, and K the intersection of the diagonals of $XYZW$. Prove that I, J, K are collinear.

96 Let ABC be a triangle, let $A_1B_1C_1$ be its orthic triangle, let $A'B'C'$ be the triangle determined by the midpoints of the sides, and let $A_2B_2C_2$ be the orthic triangle of $A'B'C'$. Assume that the nine-point circles of $A_1B_1C_1$ and $A_2B_2C_2$ are exterior to each other. Show that the interior tangents of these two nine-point circles intersect on the Euler line of the triangle ABC.

97 Let $ABCD$ be a quadrilateral. Consider the reflection of each of the lines AB, BC, CD, DA over the respective midpoints of the opposite sides. Prove that these four lines determine a quadrilateral that is homothetic to $ABCD$, and find the ratio and the center of the homothety.

98 The incircle of the triangle ABC touches the sides BC, CA, AB at the points A', B', C', respectively. Prove that the perpendiculars from the midpoints of $A'B'$, $B'C'$, and $C'A'$ to AB, BC, and CA, respectively, are concurrent.

99 Inside a triangle ABC, there are four circles $\alpha, \beta, \gamma, \delta$, of the same radius ρ, such that each of α, β, γ is tangent to two sides of the triangle and δ is tangent to α, β, and γ.

(a) Show that the center of δ lies on the line passing through the incenter and the circumcenter of ABC.
(b) Find ρ in terms of the circumradius and the inradius of ABC.

100 Prove that if the diagonals of a cyclic quadrilateral are perpendicular, then the midpoints of its sides and the feet of the perpendiculars drawn from the intersection point of the diagonals onto the sides lie on a circle.

101 Let ABC be a right triangle ($\angle A = 90°$) such that the vertex A is fixed and the vertices B and C vary on two fixed circles ω_b and ω_c that are exterior tangent at A. Find the locus of the foot of the altitude AD.

102 For an angle measure $\alpha > \pi/3$, consider all triangles ABC with $BC = 1$ and $\angle BAC = \alpha$. Among these triangles, find the one that minimizes the distance between the centroid and the incenter, and compute this distance in terms of α.

103 In the acute triangle ABC, let A_1, B_1, and C_1 be the feet of the altitudes from A, B, and C, respectively, and let H be the orthocenter. The perpendiculars from H onto A_1C_1 and A_1B_1 intersect the lines AB and AC at P and Q, respectively. Prove that the line that joins A with the midpoint of the segment PQ is orthogonal to B_1C_1.

104 Let ABC be a triangle, and let H, O, and R be its orthocenter, circumcenter, and circumradius, respectively. Let D be the reflection of A across BC, E that of B across CA, and F that of C across AB. Prove that D, E, and F are collinear if and only if $OH = 2R$.

105 Let ABC be a triangle, and let $A_2B_2C_2$ be the triangle formed by the tangents to the circumcircle at A, B, and C (with A_2 on the tangents at B and C, B_2 on the tangents at A and C, and C_2 on the tangents at A and B). Let also AA_1, BB_1, and CC_1 be the altitudes of the triangle ABC. Show that the lines A_1A_2, B_1B_2, and C_1C_2 intersect on the Euler line of the triangle ABC.

106 Let ABC be a scalene, acute triangle, and let N be the center of its nine-point circle. The lines r and s that are tangent to the circumcircle of the ABC at B and C, respectively, meet at the point D. Prove that A, D, and N are collinear if and only if $\angle BAC = 45°$.

107 A circle passing through the vertices A and C of the triangle ABC intersects the side AB at its midpoint D and the side BC at E. A circle that is tangent to AC at C and passes through E intersects the line DE again at F. Let K be the intersection of the lines AC and DE. Prove that AE, BK, and CF are concurrent.

108 Let ABC and DBC be two acute triangles, with $\angle A = \angle D$, such that the vertices A and D are separated by the line BC. Consider $E \in AC$ and $F \in BD$ such that $BE \perp AC$ and $CF \perp BD$, and let H_1 and H_2 be the orthocenters of the triangles ABC and DBC. Prove that the lines AD, EF, and H_1H_2 intersect at one point, and moreover, this point is on the nine-point circles of both triangles.

109 The trapezoid $ABCD$ has $AB \parallel CD$. A point P on the line BC, which does not coincide with B or with C, is joined with D and with the midpoint M of the segment AB. Let X be the intersection of PD and AB, Q the intersection of PM and AC, and Y the intersection of DQ and AB. Show that M is the midpoint of the segment XY.

110 Let ω be a circle, and let ℓ be a line tangent to it on which we fix a point M. Find the locus of the points P in the plane with the property that there exist points $Q, R \in \ell$ such that M is the midpoint of QR and ω is the incircle of PQR.

111 Let A, B, C, D be four collinear points (in this order). The circles of diameters AC and BD intersect at X and Y, and the line XY intersects the segment BC at Z. Let P be a point on the line XY different from Z. The line CP intersects the circle of diameter AC for the second time at M, and the line BP intersects the circle of diameter BD for the second time at N. Prove that the lines AM, DN, and XY intersect at one point.

112 In the triangle ABC, let A', B', and C' be the midpoints of the sides BC, CA, and AB, respectively, let H be the orthocenter, and let O_9 be the center of the nine-point circle. Let also A_1, B_1, C_1 be the intersections of the respective pairs of lines AO_9 and HA', BO_9 and HB', and CO_9 and HC'. Show that the triangles ABC and $A_1 B_1 C_1$ are homothetic and have the same Euler line.

113 Let ω be a circle, and let $ABCD$ be a square in its interior. Construct the circle ω_a that is tangent to the lines AB and AD, does not intersect the lines BC and CD, and is interior tangent to ω at A'. Construct analogously the points B', C', and D'. Prove that the lines AA', BB', CC' and DD' intersect at one point.

114 Let ABC be a triangle, and let AM and AN be the median and the angle bisector of $\angle A$ ($M, N \in BC$). Let P and Q be the points where the line perpendicular to AN at N intersects AB and AM, respectively, and let R be the point where the line perpendicular to AB at P intersects the line AN. Prove that RQ is perpendicular to BC.

115 *(Desargues' Theorem)* Let $A_1 B_1 C_1$ and $A_2 B_2 C_2$ be two triangles. Show that the lines $A_1 A_2$, $B_1 B_2$, and $C_1 C_2$ are concurrent or parallel if and only if the pairs of lines $(A_1 B_1, A_2 B_2)$, $(A_1 C_1, A_2 C_2)$, and $(B_1 C_1, B_2 C_2)$ intersect at three collinear points.

116 *(Ceva's Theorem)* Let ABC be a triangle, and let A_1, B_1, C_1 be points on the lines BC, CA, and AB, respectively. Give a proof based on homothety of the fact that AA_1, BB_1, and CC_1 are concurrent if and only if their complex coordinates satisfy

$$\frac{b - a_1}{c - a_1} \cdot \frac{c - b_1}{a - b_1} \cdot \frac{a - c_1}{b - c_1} = -1.$$

117 The circle ω is interior tangent to the circle Ω at K. The chords AB and BC of Ω are tangent to ω at M and N, respectively. Let P and Q be the midpoints of the arcs $\overset{\frown}{AB}$ and $\overset{\frown}{BC}$, respectively. The circumcircles of BPM and BQN intersect the second time at L. Prove that the quadrilateral $BPLQ$ is a parallelogram.

118 Let D be an arbitrary point on the side BC of the triangle ABC. Denote by I and J the incenters of the triangles ABD and ACD and by E and F the centers of the exscribed circles of these triangles tangent to BC. Prove that IJ and EF intersect on the line BC.

119 In the plane are given three circles ω_1, ω_2, and ω_3 that are exterior to each other, of radii r_1, r_2, r_3, such that r_1 is larger than both r_2 and r_3. From the point of intersection of the common exterior tangents of ω_1 and ω_2, one constructs the two tangents to ω_3, and from the point of intersection of the common exterior tangents of ω_1 and ω_3, one constructs the two tangents to ω_2. Prove that these two pairs of tangents determine a quadrilateral in which one can inscribe a circle, and then compute the radius of this circle in terms of r_1, r_2, and r_3.

120 (*Lucas' circles*) Let ABC be an acute triangle. On the side BC, construct in the exterior the square $BCDE$. Let M and N be the intersections of AE and AD with the side BC, respectively. Let Q be the intersection of the perpendicular at M to BC with AB, and let P be the intersection of the perpendicular at N to BC with AC.

(a) Prove that $MNPQ$ is a square.
(b) Let Ω_a be the circumcircle of APQ; construct similarly the circles Ω_b and Ω_c. Prove that the circles Ω_a, Ω_b, and Ω_c are pairwise tangent.

121 Let $A_1A_2A_3A_4$ be a quadrilateral with no parallel sides. For each $i = 1, 2, 3, 4$, define ω_i to be the circle tangent to the lines $A_{i-1}A_i$ and $A_{i+1}A_{i+2}$ and also tangent externally to the side A_iA_{i+1} of the quadrilateral, and let T_i be this point of tangency to the side (the indices are considered modulo 4, so $A_0 = A_4$, $A_5 = A_1$, and $A_6 = A_2$). Prove that the lines A_1A_2, A_3A_4, and T_2T_4 are concurrent if and only if the lines A_2A_3, A_4A_1, and T_1T_3 are concurrent.

122 Let $ABCD$ be a convex quadrilateral with $AB \neq BC$. Denote by ω_1 and ω_2 the incircles of ABC and ADC, respectively. Assume that there is a circle ω that is tangent to the ray $|BA$ at a point beyond A, to the ray $|BC$ at a point beyond C, as well as to the lines AD and CD. Prove that the common exterior tangents of ω_1 and ω_2 intersect at a point on ω.

123 Given is a convex quadrilateral $ABCD$ with an inscribed circle ω such that the rays $|AB$ and $|DC$ meet at the point P and the rays $|AD$ and $|BC$ meet at the point Q. The lines AC and PQ intersect at the point R. Let T be the point on the circle ω that is nearest to the line PQ. Prove that the line TR passes through the incenter of the triangle PQC.

2.3 Homothety in Combinatorial Geometry; Scaling

In this section, we demonstrate the presence of homothety in combinatorial geometry. The definition of homothety leads naturally to the idea of scaling, namely, of reducing or enlarging shapes or of changing the scale at which they are viewed. Scaling shows up in fractal geometry, where *fractals*, according to one of the

Fig. 2.19 The construction of the Sierpiński triangle

definitions given by Benoit Mandelbrot, are shapes made of parts similar to the whole in some way. An example is the *Sierpiński triangle*, named after Wacław Sierpiński. The Sierpiński triangle is some subset S of a closed equilateral triangular surface T, having the property that if h_1, h_2, h_3 are the homotheties of ratio $1/2$ centered at the vertices of T, then

$$S = h_1(S) \cup h_2(S) \cup h_3(S).$$

To construct S, you divide T into four equilateral triangles, remove the interior of the triangle in the middle, and call the resulting set S_1. Do the same with each of the four triangles in S_1, to obtain S_2. Repeat for S_2 to obtain S_3 and so on. The first three steps of the process are illustrated in Fig. 2.19. The Sierpiński triangle is the intersection of all the sets S_n, $n \geq 1$.

To illustrate how the idea of scaling can be used in solving mathematical Olympiad problems, we apply it to a question that was published by S.V. Konyagin in the journal *Kvant (Quantum)*.

Problem 2.4 In a kingdom, whose territory has the shape of a square with a side of 2 km, the king decides to invite all residents to his palace for a ball that starts at 7 pm. To this end he sends, at noon, a messenger who can give this information to any resident, who, in his turn, can take this information to any other resident, and so on. Every resident, before receiving the news, is located at home (in a known place) and can travel at 3 km/h in any direction. Show that the king can organize the transmission of messages so that all the residents can arrive at the court in time for the opening of the ball.

Solution The solution reminds of the construction of the Sierpiński triangle, and in general of the structure of fractals. To organize the notification, start by dividing the kingdom into four squares of side 1 km (the squares of first rank). Each of these squares is further divided into four squares of side 1/2 km (the squares of second rank). The process is continued until reaching the largest rank with the property that in each square there is no more than one residence of the kingdom (the ones on the boundary can be distributed to one of the neighboring squares).

The notification is organized in stages. At the first stage, one residence of each square of rank 1 is notified, after which all messengers return to the original position. This process, illustrated in Fig. 2.20, is done as follows: say the castle, which is

Fig. 2.20 The notification process

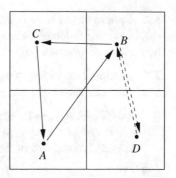

itself a residence, is at point A, and the other three residences are at B, C, D. The messenger from the castle travels to B then to C and then back to A; in the meanwhile, someone from B travels to D and back. In squares with no resident, no one is notified. The distances between any two of the points A, B, C, D are at most $2\sqrt{2} < 3$, so they can be traveled in less than 1 hour. Thus, the entire process took at most 3 hours.

At the next step, we repeat the same strategy in each of the four squares, at scale $1/2$. Then scale again, and repeat. After the nth stage, all residents will have been notified. Because of the scaling, the time is halved at each stage. Therefore, in at most

$$3\left(1 + \frac{1}{2} + \frac{1}{4} + \cdots + 1/2^n\right) < 3 \times 2 = 6$$

hours, everyone has been notified. In another hour every guest can reach the palace. So by $6 + 1 = 7$ o'clock, the ball can start. □

The following problems are based either on this idea of scaling, or just on plain homothety.

124 Let $P_1 = A_1 A_2 A_3 A_4 A_5$ be a convex pentagon, and let P_j be its translate by the vector $\overrightarrow{A_1 A_j}$, $j = 2, 3, 4, 5$. Show that the interiors of at least two of the pentagons P_1, P_2, P_3, P_4, P_5 overlap.

125 Let $\ell_1, \ell_2, \ldots, \ell_n$ be some lines in the plane, no two perpendicular and not all parallel. Show that for every k there is a unique point x_k on ℓ_k such that the perpendicular to ℓ_k at x_k passes through x_{k+1} for every $k = 1, 2, \ldots, n$ (where $x_{n+1} = x_1$).

126 It is given an equilateral triangle, with a flower and a grasshopper in its interior. Every second, the grasshopper picks a vertex and jumps in the direction of that vertex $1/10$ of the distance to it. Show that regardless of the initial position of the grasshopper, it can choose a sequence of jumps that will bring it arbitrarily close to the flower.

127 Several turtles walk in the plane with constant velocities. Their velocities are equal but have different directions. Show that after sometime, the turtles will be at the vertices of a convex polygon.

128 In an infinite lattice, N squares have been colored black. Prove that from the lattice one can cut a finite number of squares such that

(i) all the black squares lie in these squares
(ii) if K is one of the squares that was cut, the area of black squares inside K is between $1/5$ and $4/5$ of the area of K.

129 *(The Gohberg-Marcus Theorem)* Show that every convex polygon M that is not a parallelogram can be covered by three polygons homothetic to M and smaller than M. Show that for the parallelogram, the minimal number is four.

2.4   A Theoretical Study of Spiral Similarities

2.4.1   The Definition and Properties of Spiral Similarities

We now proceed to the more general situation that has been foretold at the beginning of the chapter.

Definition A *spiral similarity* about a point O by an angle α (counterclockwise) and scaling factor (i.e., ratio) $k > 0$ is the composition of a rotation about O by α and a homothety with ratio k and center O (Fig. 2.21).

We agree to include in the class of spiral similarities, as particular examples, rotations, when the homothety has scaling factor $k = 1$, and homotheties, when the rotation has angle $\alpha = 0$.

In complex numbers, a spiral similarity maps the point z to the point $f(z)$ defined by the equation

$$\frac{f(z) - z_0}{z - z_0} = k e^{i\alpha}.$$

Fig. 2.21 A spiral similarity with center O, angle $\alpha = 45°$, and scaling factor $k = 2\sqrt{2}$

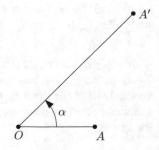

This gives the equation of the spiral similarity of center z_0, angle α, and ratio $k > 0$ as

$$f(z) = ke^{i\alpha}z + (1 - ke^{i\alpha})z_0.$$

Conversely, if $f(z) = rz + s$ with $r \neq 1$, then we can recover $z_0 = s/(1 - r)$, $\alpha = \arg(r)$, and $k = |r|$, so f is a spiral similarity.

Definition A transformation of the form $f : \mathbb{C} \to \mathbb{C}$, $f(z) = rz + s$ where $r, s \in \mathbb{C}$, $r \neq 0$ is called a *complex affine transformation*.

Adding the translations to the fold, we conclude that *the affine transformations of $\mathbb{C}$ represent all possible spiral similarities and translations*. This also proves that spiral similarities and translations together form a group. It is important to keep the distinction between the affine transformations of $\mathbb{C}$, the *complex line*, and the real affine transformations of the real two-dimensional plane, $\mathbb{R}^2$, which do not make their presence in this book.

We can derive immediately a number of useful properties of spiral similarities. If s is a spiral similarity of center O, (counterclockwise) angle α, and scaling factor k, and if we denote, as usual, $P' = s(P)$, then we have the following:

- (Automatic similarities) All triangles OPP' are similar, with the same orientation. In fact, they are all similar to a triangle with sides of lengths 1 and k that form an angle α; these similarities are "automatic" in the sense that you do not need to visualize them; the similar triangles are always there!
- (Conserved properties) A spiral similarity maps lines to lines, circles to circles, and segments to segments; it also preserves angles. If S is a polygon in the plane, S and $s(S)$ are similar, and corresponding sides in S and $s(S)$ form angles equal to α. These are consequences of the fact that a spiral similarity is the composition of a rotation and a homothety.

Endowed with the understanding of the analytic form of spiral similarities, we can state a principle that comes in handy when solving problems.

Theorem 2.18 (Averaging Principle) *Let F_1 and F_2 be two geometric figures with the property that there is a spiral similarity s such that $s(F_1) = F_2$. Let t be a fixed real number, and for each $P \in F_1$, consider a point Q in the plane such that $\overrightarrow{PQ} = t\overrightarrow{Ps(P)}$; let also F_3 be the geometric figure consisting of all such points Q when P varies in F_1. If F_3 does not consist of a single point, then there is a spiral similarity or translation that maps F_1 to F_3.*

Proof Let the spiral similarity be $s(z) = rz + s$. If p, q are the coordinates of P, Q, respectively, then the condition from the statement reads

$$q - p = t(s(p) - p) = trp + ts - tp,$$

that is,

$$q = (tr - t + 1)p + ts.$$

This is not a constant function of p when $tr - t + 1 \neq 1$, and in that case it is either a spiral similarity or a translation.   □

A simple observation, but a powerful problem-solving tool! The analytic proof can be adapted to several figures and to linear combinations of points, even with complex coefficients, giving rise to a more general statement.

Theorem 2.19 (Averaging Principle) *Let $F_1, F_2, \ldots, F_n$ be several figures that can be mapped into one another by spiral similarities, and let $t_1, t_2, \ldots, t_n$ be any complex numbers. Let also F be the figure consisting of the points of the form $t_1 z_1 + t_2 z_2 + \cdots + t_n z_n$, where $z_j \in F_j$ and z_j is mapped to z_k by the spiral similarity that maps F_j to F_k, $j, k = 1, 2, \ldots, n$, and assume that F is not a point. Then F can be mapped by spiral similarities or translations to each of the n figures $F_1, F_2, \ldots, F_n$.*

Proof Let $s_k(z) = r_k z + s_k$ be the spiral similarity that maps F_1 to F_k,, $k = 2, 3, \ldots, n$. Then

$$t_1 z_1 + t_2 z_2 + \cdots + t_n z_n = (t_1 + t_2 r_2 + \cdots + t_k r_k) z_1 + t_2 s_2 + t_3 s_3 + \cdots + t_k s_k,$$

showing that there is a spiral similarity mapping F_1 to F.   □

Remark The figure F defined by the numbers $t_1, t_2, \ldots, t_n$ depends on the coordinate system in use, except in the case where $t_1 + t_2 + \cdots + t_n = 1$.

The name of the principle comes from the case where $t_1, t_2, \ldots, t_n$ are real and add up to 1, for in that case F is the "average" (i.e., weighted mean) of the figures $F_1, F_2, \ldots, F_n$. This startling result is a straightforward consequence of the fact that spiral similarities are complex affine transformations.

2.4.2   The Center of a Spiral Similarity: The Generic Case

Now, suppose we run into a spiral similarity, but do not have its parameters (center O, angle α, and scale factor k) at hand. How do we recover them? The last two are easy to find: if we know the images A' and B' of two distinct points A and B, then $\alpha = \angle(AB, A'B')$, the angle between the lines AB and $A'B'$, and $k = A'B'/AB$. All that is left is to find the center O.

Theorem 2.20 *Given four points A, B, A', and B' in the plane such that $A \neq B$ and $A' \neq B'$, there is a unique spiral similarity s or translation that takes A to A' and B to B' (i.e., $A' = s(A)$ and $B' = s(B)$).*

Proof Complex numbers yield a short proof, as it usually happens with theoretical results. The center can be found using simple interpolation as follows. Let s be a spiral similarity such that $s(a) = a'$ and $s(b) = b'$. Then

$$s(z) - a' = \frac{b' - a'}{b - a}(z - a)$$

which gives

$$s(z) = \frac{a' - b'}{a - b}z + \frac{ab' - a'b}{a - b}.$$

Writing $s(z) = rz + (1 - r)z_0$, we have

$$1 - r = 1 - \frac{a' - b'}{a - b} = \frac{a - b - a' + b'}{a - b},$$

and hence

$$z_0 = \frac{ab' - a'b}{a - b - a' + b'},$$

For this to work, we should have $a - a' \neq b - b'$, but $a - a' = b - b'$ means that $\overrightarrow{AA'} = \overrightarrow{BB'}$, and then s is a translation. $\qquad \square$

There is an immediate corollary of this result. To formulate it, we need the concept of orientation of a polygon introduced in the first chapter. Recalling what was said in that chapter, a polygon is oriented counterclockwise if when traveling from one vertex to the next the polygon is on the left and clockwise if the polygon is on the right.

Definition Two polygons are called *directly similar* if they are similar and oriented the same way.

Proposition 2.21 *Given two directly similar polygons that are not translates of each other, there is a unique spiral similarity that maps one into the other.*

Proof Every polygon can be dissected into triangles whose vertices are also vertices of the polygon. If we perform the same dissection in each polygon (namely, we group corresponding vertices into triangles), then the polygons are similar if and only if the triangles from the dissection are similar. So it suffices to prove the result for triangles.

If ABC and $A'B'C'$ are two directly similar triangles, take the unique spiral similarity that maps the segment AB to the segment $A'B'$. Then the points C is mapped to some point C'' such that ABC and $A'B'C''$ are similar and oriented the same way. But there is only one such point C'' in the plane, namely, C'. Thus, the spiral similarity maps C to C', and we are done. $\qquad \square$

This proposition helps extend the concept of direct similarity to geometric figures other than polygons. Two geometric figures will be called directly similar if they are mapped into each other by a transformation of the form $f(z) = az + b, a, b \in \mathbb{C}$.

Simple as the complex number approach is, it gives no clue of how to locate geometrically the center of the spiral similarity. And we do want to know the answer to this geometric question. Two particular cases should be taken out of our way. The case where the line AB is parallel or coincides with $A'B'$ corresponds to AB being mapped to $A'B'$ by either a translation or a homothety (the latter is a spiral similarity with angle 0 or 180°). The translation has no center, and the center of the homothety is the intersection of AA' and BB'. The case of a true spiral similarity is clarified by the following result.

Theorem 2.22 (Center of Spiral Similarity) *Given four points A, B, A', and B' in the plane such that $A \neq B$ and $A' \neq B'$ and AB is not parallel to $A'B'$, the center of the unique spiral similarity that takes A to A' and B to B' is the second intersection point of the circumcircles of PAA' and PBB' where P is the intersection point of the lines AB and $A'B'$.*

Proof Let O be the center of the spiral similarity that maps AB to $A'B'$, and let us assume that O is different from P. Because the spiral similarity maps the line AB to the line $A'B'$ and the line OA to the line OA', and also because spiral similarities preserve angles

$$\angle(AB, OA) = \angle(A'B', OA').$$

It follows that P, A, A', O lie on a circle. For a similar reason, P, B, B', O lie on a circle, and so O is the second intersection point of the circumcircles of PAA' and PBB' (Fig. 2.22).

If O coincides with P, let us show that the two circles are tangent at P (in which case O is the "other" point of intersection because the tangency point is a double point). The segment AB is mapped to $A'B'$ by a rotation about P followed by a homothety about P. Consequently

$$\frac{PA}{PB} = \frac{PA'}{PB'}.$$

So the triangle PAA' is mapped to the triangle PBB' by a homothety of center P, and hence the circumcircle of PAA' is mapped to the circumcircle of PBB' by a

Fig. 2.22 Finding the center of a spiral similarity. Try to draw the points A, B, A', B' in other positions to see what happens!

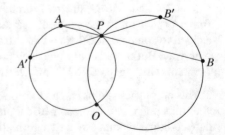

Fig. 2.23 Intuitive aid for
understanding how to find the
center of a spiral similarity

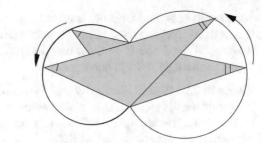

Fig. 2.24 Spiral similarity:
what if B' lies in AB?

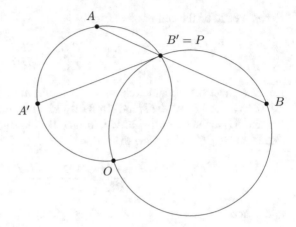

homothety of center P. Therefore, the two circles are tangent at P. The theorem is
proved.   □

This result can be easily remembered and understood by looking at Fig. 2.23.

There are some degenerate cases to consider. One such case is when three of the
four points are collinear. Suppose, without loss of generality, that A, B, and B' are
these points. Since AB and $A'B'$ are not parallel, A' does not lie on this line. In this
case, the intersection P of AB and $A'B'$ coincides with B', so the circumcircle of
PBB' degenerates to a circle that passes through B and B' and is tangent to the line
$A'B'$. You can interpret the equality $B' = P$ as a double root, so taking a tangent
circle makes sense. This situation is illustrated in Fig. 2.24

In fact, the angle equality $\angle(AB, OB) = \angle(A'B', OB')$ from Theorem 2.22 can
be interpreted exactly as saying that $A'B'$ is tangent to the circumcircle of OBB'.

2.4.3   The Center of a Spiral Similarity: The Case $A' = B$, Symmedians Revisited

Looking at Theorem 2.22, we realize that there is an ambiguity when $A' = B$, in
which case P also coincides with B and the construction degenerates. We could

address this case similarly with tangent lines, but instead, we study it separately, as it unveils a surprising relationship between spiral similarities and symmedians. Let therefore $A' = B$ and $B' = C$, so that the spiral similarity s takes A to B and B to C and the triangle ABC is nondegenerate.

Theorem 2.23 (Spiral Similarity and Symmedians) *Let ABC be a triangle. The center of the spiral similarity s that takes A to B and B to C lies on the B-symmedian of the triangle ABC, namely, on the symmetric of the median from B with respect to the bisector of $\angle ABC$.*

Proof Let K be the center of s. Then

$$\frac{KA}{KB} = \frac{KB}{KC} = \frac{AB}{BC}.$$

Let BK meet the circumcircle of ABC again at $L \neq B$ (Fig. 2.25). Notice that $\angle ALB = \angle ACB$, so the spiral similarity s' with center B that takes L to C takes the line AL to the line AC. Let $M = s'(A)$. By the automatic similarity, the triangles BAM and BLC are similar, so

$$\frac{AB}{BL} = \frac{AM}{LC} = \frac{BM}{BC},$$

and hence

$$AM = \frac{LC}{BC} \cdot BM.$$

By automatic similarity, the triangles ABL and MBC are also similar, so

$$\frac{AB}{BM} = \frac{BL}{BC} = \frac{AL}{MC},$$

and hence

$$MC = \frac{AL}{AB} \cdot BM.$$

Finally, using inscribed angles and s, $\angle LAC = \angle LBC = \angle KBC = \angle KAB$, and $\angle LCA = \angle LBA = \angle KBA$, so the triangles ALC and AKB are also similar, and

$$\frac{AL}{LC} = \frac{AK}{KB} = \frac{AB}{BC},$$

which implies

$$\frac{AL}{AB} = \frac{LC}{BC},$$

Fig. 2.25 Finding the center
of a spiral similarity, $A' = B$

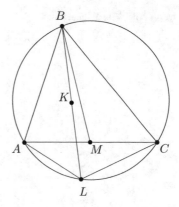

and using the relations derived above, we find that $AM = MC$. We have proved that M is the midpoint of AC and $\angle MBC = \angle ABK$. So the line BK is the symmetric of the median BM with respect to the internal bisector of $\angle B$, showing that it is the B-symmedian. □

Let us put symmedians more solidly in the context of spiral similarity. Because the spiral similarity s has ratio AB/BC, the ratio of the distances from K to AB and AC is $d_A/d_C = AB/BC$. This, of course, is true for every point lying on the symmedian, in particular for the point T where the symmedian BK meets the side AC. Then

$$\frac{AT}{TC} = \frac{\text{area } ATB}{\text{area } ATC} = \frac{AB \cdot d_A}{AC \cdot d_C} = \frac{AB^2}{BC^2}.$$

2.4.4 Spiral Similarities and Miquel's Theorem

We now interpret a classical result about complete quadrilaterals from the point of view of spiral similarities. A complete quadrilateral is the configuration formed by four lines, no two parallel. The name comes from the fact that it is the configuration obtained when extending the lines of support of the sides of a quadrilateral that is not a trapezoid.

Theorem 2.24 (Miquel's Theorem) *Consider four lines ℓ_1, ℓ_2, ℓ_3, and ℓ_4, no two of them parallel. For $i = 1, 2, 3, 4$, define C_i as the circumcircle of the triangle determined by the three lines different from ℓ_i. Then the four circles C_1, C_2, C_3, and C_4 have a common point, known as the Miquel point of ℓ_1, ℓ_2, ℓ_3, and ℓ_4.*

Proof There are several proofs using simple angle chasing (the reader is encouraged to try), but we present a very short approach based on spiral similarities, which has the advantage of avoiding configuration issues. One possible construction is shown in Fig. 2.26.

Fig. 2.26 The Miquel point

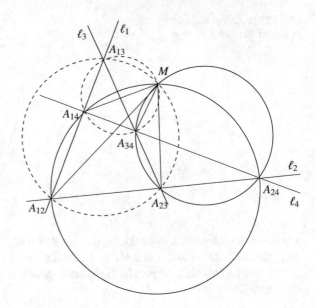

Let A_{ij} be the intersection of the lines ℓ_i and ℓ_j, for $1 \le i < j \le 4$. Consider the spiral similarity that takes A_{12} to A_{14} and A_{23} to A_{34}. By Theorem 2.22, its center is the second intersection point M of the circle that passes through A_{12}, A_{14}, and the intersection of $A_{12}A_{23} = \ell_2$ and $A_{14}A_{34} = \ell_4$, that is, the circumcircle C_3 of $A_{12}A_{14}A_{24}$, and the circle through A_{23}, A_{34}, and A_{24}, which is C_1.

Now the automatic similarity between the triangles $MA_{12}A_{14}$ and $MA_{23}A_{34}$ induces another spiral similarity with center M that takes A_{12} to A_{23} and A_{14} to A_{34}. Repeating the argument for these new pairs of points, we deduce that M is also the second intersection point of C_2 and C_4, and so M lies on all circles, as claimed.
$\square$

The trick of the proof can be phrased as follows.

Theorem 2.25 (Spiral Similarities Come in Pairs) *Let A, B, C, and D be points in the plane. Then the spiral similarity that takes A to C and B to D induces another spiral similarity with the same center that takes A to B and C to D.*

Proof The angle of the spiral similarity that maps A to B and C to D is the sum of $\angle COB$ and the angle of the initial spiral similarity. Its ratio is AO/BO, which is equal to CO/DO because $AO/CO = BO/DO$.
$\square$

We can rethink this result more intuitively and visualize it in Fig. 2.27: if the triangle OAB is mapped to the triangle OCD by a spiral similarity, then the triangle OAC is mapped to the triangle OBD by a spiral similarity.

We will revisit Miquel's Theorem at the end of Chap. 4, where we will employ a variety of geometric transformations to obtain detailed knowledge about complete quadrilaterals.

Fig. 2.27 Spiral similarities come in pairs

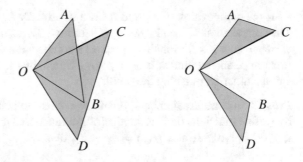

2.4.5   Compositions of Spiral Similarities

We want to understand compositions of spiral similarities. To this end, let us first look at the simpler case of the composition of a rotation and a homothety (both particular cases of spiral similarities), when the centers of rotation and homothety do not coincide.

Theorem 2.26 *The composition of a rotation and a homothety with different centers is a spiral similarity.*

Proof Let s be the resulting composition. It preserves the shape and orientation of geometric figures. We have learned that, given A and B, $A' = s(A)$, and $B' = s(B)$, there exists a unique spiral similarity or translation τ that takes A to A' and B to B'. Let P be any point in the plane, and let $P' = s(P)$. Then ABP and $A'B'P'$ are similar with the same orientation and thus $s(P) = \tau(P)$. We deduce that $s = \tau$.

We just have to rule out the possibility that s is a translation. If it were a translation, then $A'B' \parallel AB$, and $\overrightarrow{AB} = \overrightarrow{A'B'}$, so the homothety ratio is $k = 1$, and the rotation angle is 0; this means that s is the identity, which can be considered a spiral similarity with any center, angle 0, and ratio 1.   □

Theorem 2.27 (Composition of Spiral Similarities) *The composition of two spiral similarities is a spiral similarity or a translation.*

Proof 1 Surprisingly, the above particular case essentially solves the general case. The reason is that we now can break each spiral similarity into two transformations and handle them separately.

Specifically, let $s_1 = \rho_1 \circ h_1$ and $s_2 = h_2 \circ \rho_2$ be two spiral similarities, ρ_i being rotations and h_i being homotheties, $i = 1, 2$, whose centers coincide for each i (notice that we can commute ρ_i with h_i for each $i = 1, 2$ because of that). Then $s_1 \circ s_2 = \rho_1 \circ h_1 \circ h_2 \circ \rho_2$. The composition of two homotheties h_1 and h_2 is another homothety or a translation.

If the composition $h_1 \circ h_2$ were a translation, then $\rho_1 \circ h_1 \circ h_2 \circ \rho_2$ is an orientation preserving isometry, which is either a translation or a rotation (hence a spiral similarity).

If the composition of h_1 and h_2 is a homothety h, then $s_1 \circ s_2 = \rho_1 \circ h \circ \rho_2$. The composition of h and ρ_2 is, by Theorem 2.26, a spiral similarity $s' = \rho' \circ h'$, so $s_1 \circ s_2 = \rho_1 \circ \rho' \circ h'$. Finally, $\rho_1 \circ \rho'$ is a composition of rotations, which is another rotation ρ, and we find $s_1 \circ s_2 = \rho \circ h'$, which, by the previous theorem again, is a spiral similarity, finishing the proof.   □

Proof 2 In complex numbers, this is equivalent to stating that the composition of two affine functions on $\mathbb{C}$ is another affine function, which is immediate. Explicitly, if $f_1(z) = r_1 z + s_1$ and $f_2(z) = r_2 z + s_2$, then

$$(f_2 \circ f_1)(z) = f_2(r_1 z + s_1) = r_2(r_1 z + s_1) + s_2 = (r_1 r_2)z + (r_2 s_1 + s_2).$$

If $r_1 r_2 = 1$, then the composition is a translation; otherwise, it is a spiral similarity.   □

2.4.6   Groups Generated by Spiral Similarities

With the tools for studying spiral similarities that we have developed so far at hand, we are ready to prove a general result about geometric transformations of the plane.

Theorem 2.28 *Every transformation of the plane that maps lines to lines and circles to circles is either an isometry, a spiral similarity, or the composition of a spiral similarity and a reflection.*

Proof Let f be the transformation in question, and let A and B be two distinct points in the plane. Consider a translation τ that maps $f(A)$ to A and a spiral similarity s of center A that maps $\tau(f(B))$ to B. The transformation $g = s \circ \tau \circ f$ still maps lines to lines and circles to circles, and $g(A) = A$, $g(B) = B$, so g has two fixed points. We will show that g is either the identity map or the reflection over AB.

Note that because g is bijective, g maps the intersection of two lines to the intersection of their images. If the lines do not intersect, their images do not intersect either. So g maps parallel lines to parallel lines. For the same reason, g maps the intersection of two circles, to the intersection of their images. In particular, if the circles are tangent, their images must be tangent, too, and the tangency point of the original two circles is mapped to the tangency point of their images. The same is true if one of the circles is replaced by a line.

Take the lines ℓ_1 and ℓ_2 that are perpendicular to AB at A and B, respectively, and take also the circle ω of diameter AB (see Fig. 2.28). The images $g(\ell_1)$ and $g(\ell_2)$ are parallel lines through A and B as well, but they might be tilted, thus not forming an angle of $90°$ with AB. But $g(\omega)$ is a circle through A and B that is tangent to both $g(\ell_1)$ and $g(\ell_2)$, and this can only happen if these two lines are orthogonal to AB (as you can notice on Fig. 2.28). In particular $g(\ell_1) = \ell_1$, $g(\ell_2) = \ell_2$, and $g(\omega) = \omega$.

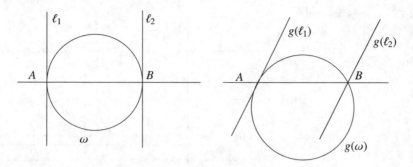

Fig. 2.28 The lines orthogonal to AB at A and B are mapped into themselves

Fig. 2.29 Construction of a
lattice whose nodes are
invariant under g

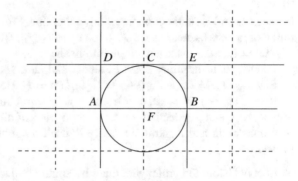

The line ℓ tangent to ω and parallel to AB is mapped to a line tangent to ω and parallel to AB, as well. There are two such lines: ℓ itself and its reflection over AB. By eventually replacing g by $\sigma \circ g$, where σ is the reflection over AB, we may assume that $g(\ell) = \ell$. Let C be the tangency point of ω and ℓ, D the intersection of ℓ and ℓ_1, and E the intersection of ℓ and ℓ_2. Let also F be the midpoint of AB (Fig. 2.29). Then $g(C) = C$, $g(D) = D$, and $g(E) = E$. But by replacing A and B by C and D in the above argument, and arguing that the circle of diameter CD is invariant under g, we conclude that $g(F) = F$ (as F plays the role of C in this new configuration).

A similar argument shows that the reflections of C and D over AB are invariant under g. Repeating the argument, we find that there exists a square lattice of size $AB/2$ whose nodes are invariant under g. Next, replace A and B by A and F, and repeat the argument. This proves that there is a square lattice of size $AB/4$ whose nodes are invariant under g. Inductively, we obtain that for every $n > 0$, there is a square lattice of size $AB/2^n$ whose sides are parallel to AB and whose nodes are invariant under g. The lines through these nodes that are parallel and perpendicular to AB are mapped into themselves by g.

Next, let Ω be an arbitrary circle. Let ℓ_u and ℓ_d be parallel to AB, ℓ_l and ℓ_r be perpendicular to AB, and all four tangent to Ω (Fig. 2.30). Then $g(\Omega)$ must intersect all lines that belong to each of the lattices and lie between ℓ_u and ℓ_d, respectively,

Fig. 2.30 g maps any circle into itself

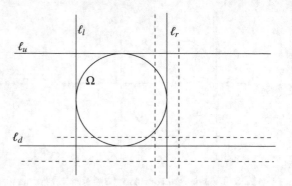

between ℓ_l and ℓ_r, and not intersect any other line. Since we can find lines from the union of lattices as close as we wish to any of ℓ_u, ℓ_d, ℓ_l, and ℓ_r, this is only possible if $g(\Omega) = \Omega$. So g maps any circle into itself.

Finally, for an arbitrary point P, let Ω_1 and Ω_2 be two circles that are tangent at P. Since $g(\Omega_1) = \Omega_1$ and $g(\Omega_2) = \Omega_2$, $g(P) = P$. Hence, g is the identity map.

Returning to f, it is either a translation, a spiral similarity, the composition of one of these and a reflection, or the composition of all three. But the composition of a translation and a spiral similarity is itself a spiral similarity. The theorem is proved.   □

In conclusion, the group generated by spiral similarities consists of spiral similarities (including rotations and homotheties as particular cases) and translations. In complex coordinates, it is

$$\{f : \mathbb{C} \to \mathbb{C} \mid f(z) = az + b, \quad a, b \in \mathbb{C}, a \neq 0\}.$$

The group of transformations that map lines to lines and circles to circles is generated by spiral similarities and reflections, and it is

$$\{f : \mathbb{C} \to \mathbb{C} \mid f(z) = az + b \text{ or } f(z) = a\bar{z} + b, \quad a, b \in \mathbb{C}, a \neq 0\}.$$

Remark It is important to compare Theorem 2.28 to Theorem 2.8. It is not true that every transformation that maps lines to lines is an isometry, a spiral similarity, or a spiral similarity composed with a reflection. In real coordinates, the map $(x, y) \mapsto (x + y, y)$ maps lines to lines, but it is unlike any transformation discussed in this book. It is what we call a real affine transformation; while such transformations are important in mathematics, we do not include them in our discussion.

2.4.7 Theoretical Questions About Spiral Similarities

Here are a few problems that will improve your understanding of these conceptual facts.

130 Let AB_1C_1, AB_2C_2, and AB_3C_3 be three directly similar triangles. Show that if B_1, B_2, B_3 are collinear, then C_1, C_2, C_3 are collinear.

131 Let ω be a circle, and let s be a spiral similarity whose center M is on ω. Show that all the lines XX' with $X \in \omega$ and $X' = s(X)$ pass through the second intersection point of the circles ω and $s(\omega)$.

132 Let ABC be a triangle, and let X be the intersection of the circle through B tangent to AC at A with the circle through C tangent to AB at A. Prove that AX is symmedian in the triangle ABC.

133 Let ℓ_1 and ℓ_2 be two lines that intersect at P, let A_1 be on ℓ_1, A_2 on ℓ_2, and let k be a positive number. Show that there are exactly two spiral similarities of ratio k that map ℓ_1 to ℓ_2 so that A_1 is mapped to A_2. Moreover, the centers of these spiral similarities are at the intersection of the circumcircle of the triangle PA_1A_2 and the Apollonian circle determined by the points A_1 and A_2 and the ratio k. (For the definition of the Apollonian circle, see Problem 165 in Sect. 3.1.10.)

134 Let ω_1 and ω_2 be two nonconcentric circles. How many spiral similarities that map one circle to the other and have the angle equal to $90°$ exist? (We allow both clockwise and counterclockwise $90°$ rotations.)

2.5 Spiral Similarity in Euclidean Geometry Problems

2.5.1 Similar Figures and the Circle of Similitude

We will demonstrate the power of spiral similarity by revisiting an exposition of the second author from *Transformari Geometrice. Omotetia si Inversiunea* (Matrixrom). While the theorems presented in this section are not necessarily useful in solving problems, the techniques employed in proving them are.

Proposition 2.21 has shown that two polygons that are directly similar and are not translates of each other can be mapped one into the other by a unique spiral similarity. And we have generalized the notion of direct similarity of figures by agreeing that it is decided by the existence of a spiral similarity or translation that maps one figure to the other. In this section we will assume that figures that are directly similar are *not* mapped into each other by translations, only by spiral similarities.

Definition Let F_1, F_2, F_3 be three directly similar figures; let O_1 be the center of the spiral similarity that maps F_2 to F_3, O_2 the center of the spiral similarity that maps F_3 to F_1, and O_3 the center of the spiral similarity that maps F_1 to F_2. If the points O_1, O_2, and O_3 are not collinear, the triangle they form is called the *triangle of similitude* of F_1, F_2, and F_3, and the circumcircle of $O_1 O_2 O_3$ is called the *circle of similitude* of these three figures.

If O_1, O_2, and O_3 coincide, we say that we have a center of similitude, and if they are collinear, then we have an axis of similitude. In what follows, we assume that neither of these two degenerate cases happen, so the three figures have a triangle of similitude.

Theorem 2.29 *Let $A_1 B_1$, $A_2 B_2$, and $A_3 B_3$ be three segments in the plane, and let P_1, P_2, P_3 be the intersection points of the pairs of lines $A_2 B_2$ and $A_3 B_3$; $A_1 B_1$ and $A_3 B_3$; and $A_1 B_1$ and $A_2 B_2$, respectively.*

(a) The circumcircles of $A_1 A_2 P_3$, $A_2 A_3 P_1$, and $A_3 A_1 P_2$ intersect at a point that lies on the circle of similitude of $A_1 B_1$, $A_2 B_2$, and $A_3 B_3$, and the same is true for the circumcircles of $B_1 B_2 P_3$, $B_2 B_3 P_1$, and $B_3 B_1 P_2$.

(b) Let O_i be the center of the spiral similarity that maps $A_j B_j$ to $A_k B_k$, $\{i, j, k\} = \{1, 2, 3\}$. Then $P_1 O_1$, $P_2 O_2$, $P_3 O_3$ intersect at a point that is on the circle of similitude of $A_1 B_1$, $A_2 B_2$, and $A_3 B_3$.

Part (b) is usually phrased as: In a system of three directly similar figures, the triangle formed by three homologous lines is in perspective with the triangle of similitude, with the center of perspective lying on the circle of similitude.

Proof We first prove an auxiliary result.

Lemma 2.30 *On the sides of the triangle ABC, we construct the triangles $A'BC$, $AB'C$, and ABC' such that the sum of the angles (or their supplements) at the vertices A', B', C' is a multiple of $180°$. Then the circumcircles of these triangles intersect at one point.*

Proof For simplicity, let us assume that the circumcircles of $A'BC$ and $AB'C$ are not tangent so that they intersect at a second point, P (Fig. 2.31).

Using directed angles modulo $180°$, we can write

$$\angle(PA, PB) = \angle(PA, PC) + \angle(PC, PB) = \angle(B'A, B'C) + \angle(A'C, A'B)$$

$$= \angle(C'A, C'B),$$

showing that P is also on the circumcircle of ABC'. The case where the circles are tangent at C is similar, with the line PC replaced by the tangent at C. The lemma is proved.   □

A particular case of the lemma is the configuration where $A \in B'C'$, $B \in A'C'$, and $C \in A'B'$, since in that case the three angles in question are $\angle BA'C = \angle B'A'C'$, $\angle AB'C = \angle A'B'C'$, and $\angle AC'B = \angle A'C'B'$, and, as angles of a triangle, they add up to $180°$. This particular case is used for proving the theorem. Such a

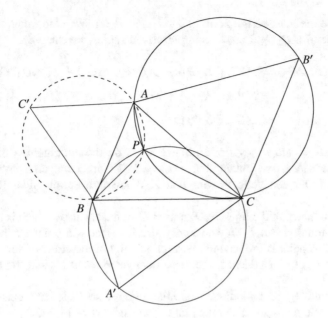

Fig. 2.31 The circumcircles of $A'BC$, $AB'C$, ABC' intersect

Fig. 2.32 Configuration with
the lines A_1B_1, A_2B_2, A_3B_3
intersecting at P_1, P_2, P_3

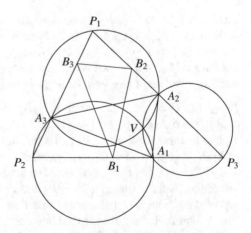

configuration can be discovered in Fig. 2.32, where the roles of A, B, C are played
by A_1, A_2, A_3 and the roles of A', B', C' are played by P_1, P_2, P_3.

Returning to the proof of the theorem, note first that by Theorem 2.22, the
points O_3, O_2, O_1 are on the circumcircles of the triangles $A_1A_2P_3$, $A_1A_3P_2$, and
$A_2A_3P_1$, respectively. Then, by applying Lemma 2.30 to the configuration where
$A_1 \in P_2P_3$, $A_2 \in P_1P_3$, and $A_3 \in P_1P_2$, we deduce that the circumcircles of
$A_1A_2P_3$, $A_1A_3P_2$, and $A_2A_3P_1$ intersect at some point V. For the same reason, the
circumcircles of $B_1B_2P_3$, $B_1B_3P_2$, and $B_2B_3P_1$ intersect at some point V'.

Let U be the intersection point of P_2O_2 and P_3O_3. We claim that V is on the circumcircle of UO_2O_3. Indeed, using inscribed angles, we can write

$$\angle(O_2V, VO_3) = \angle(VO_2, O_2P_2) + \angle(O_2P_2, P_3O_3) + \angle(P_3O_3, O_3V)$$
$$= \angle(VA_1, A_1P_2) + \angle(O_2U, UO_3) + \angle(P_3A_1, A_1V)$$
$$= \angle(O_2U, UO_3),$$

so O_2, O_3, U, V are concyclic. Similarly V' is on the circumcircle of UO_2O_3. As a consequence, we obtain that O_2, O_3, V, V' are concyclic, and similarly O_1, O_2, V, V' are concyclic, proving that V, V' are on the circle of similarity. This proves (a).

For (b), notice that the lines P_2O_2 and P_3O_3 intersect at U, which is also on the circle of similarity. But P_1O_1 and P_2O_2 also intersect at a point on the circle of similarity; this point is the second intersection point of the circle of similarity with P_2O_2, so it must be U. Hence, the three lines intersect at U, and the theorem is proved. □

Let F_1 and F_2 be two directly similar figures, so that there exists a spiral similarity s that maps F_1 to F_2. If S_1 and S_2 are subsets of F_1 and F_2, respectively, we say that S_1 and S_2 are *corresponding* subsets of F_1 and F_2 if $s(S_1) = s(S_2)$.

Theorem 2.31 *Let F_1, F_2, F_3 be three directly similar figures, and let $A_1B_1, A_2B_2, A_3B_3,$ and $A_1C_1, A_2C_2,$ and A_3C_3 be corresponding segments of F_1, F_2, F_3. Then the triangle formed by the lines $A_1B_1, A_2B_2,$ and A_3B_3 and the triangle formed by the lines $A_1C_1, A_2C_2,$ and A_3C_3 are directly similar, and the center of the spiral similarity that maps one into the other is on the circle of similitude of F_1, F_2, F_3.*

Proof Let P_i be the intersection of A_jB_j and A_kB_k, and let P'_j be the intersection of A_jC_j and A_kC_k, $\{i, j, k\} = \{1, 2, 3\}$ (Fig. 2.33). The spiral similarity that takes F_1 into F_2 takes the lines A_1B_1 and A_1C_1 into A_2B_2 and A_2C_2, respectively. Hence, $\angle(A_1B_1, A_2B_2) = \angle(A_1C_1, A_2C_2)$, as directed angles modulo π. This is the same as $\angle P_1P_3P_2 = \angle P'_1P'_3P'_2$, and we have similar equalities for the other two angles. We deduce that the triangles $P_1P_3P_2$ and $P'_1P'_3P'_2$ are similar and also (because we are working with directed angles) that they have the same orientation. Hence, by Proposition 2.21, they are mapped into each other by a spiral similarity s.

Applying Theorem 2.22 to P_2P_3 and $P'_2P'_3$, we deduce that the center of s is the second point of intersection of the circumcircles of $A_1P_3P'_3$ and $A_1P_2P'_2$. By the same argument applied to P_1P_3 and $P'_2P'_1$, we deduce that the center of s is the second intersection point of the circumcircles of $A_2P_3P'_3$ and $A_2P_1P'_1$. Thus, the circumcircles of $A_1P_3P'_3$ and $A_2P_3P'_3$ have three points in common: P_3, P'_3, and the center of s. But this can only happen if the circles coincide, and thus A_1, A_2, P_3, P'_3 are concyclic. It follows that the center of s is on the circumcircle of $A_1A_2P_3$. Similarly, the center of s is on the circumcircles of $A_2A_3P_1$ and $A_1A_3P_2$. But then by Theorem 2.29 (a) applied to the configuration consisting of A_1B_1, A_2B_2, A_3B_3, we deduce that the intersection point of these three circles, which is the center of s, is also on the circle of similitude of F_1, F_2, F_3. □

Fig. 2.33 Proof that $P_1 P_2 P_3$ and $P'_1 P'_2 P'_3$ are directly similar

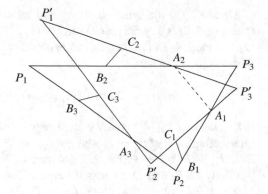

Fig. 2.34 Construction of the invariable points J_1, J_2, J_3

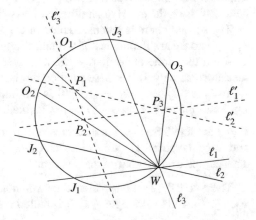

Theorem 2.32 *Let ℓ_1, ℓ_2, and ℓ_3 be three lines that intersect at a point W and are corresponding lines of the directly similar figures F_1, F_2, F_3. Then W is on the circle of similitude of F_1, F_2, F_3. If J_1, J_2, J_3 are the other points where ℓ_1, ℓ_2, and ℓ_3 intersect the circle of similitude, respectively, then J_1, J_2, J_3 depend only on F_1, F_2, F_3 and not on the choice of ℓ_1, ℓ_2, and ℓ_3.*

Before starting the proof, let us point out that the figures F_1, F_2, F_3 might not contain entire lines, such as in the case where they consist of polygons and circles. But we can choose any two points of F_1 and then add the line through them to F_1 and the corresponding lines to F_2 and F_3 to obtain larger figures that are still directly similar.

Proof The lines ℓ_1, ℓ_2, ℓ_3 are not very helpful, because they intersect at one point, so they do not allow us to use Theorem 2.29 or Theorem 2.31. To bring these theorems into the story, we add the lines ℓ'_1, ℓ'_2, ℓ'_3 to F_1, F_2, F_3, respectively, such that $\ell'_i \| \ell_i$, $i = 1, 2, 3$, and ℓ'_1, ℓ'_2, ℓ'_3 also correspond by the spiral similarities. We denote by P_i the intersection of ℓ'_j and ℓ'_k, $\{i, j, k\} = \{1, 2, 3\}$. The situation is shown in Fig. 2.34.

Denote by s_i the spiral similarity that takes F_j to F_k, and let O_i be its center, where (i, j, k) is a cyclic permutation of $(1, 2, 3)$. Let h_3 be the homothety of center O_3 that maps ℓ_1 to ℓ_1'. Then h_3 and s_3 commute (because they are spiral similarities of the same center), so

$$h_3(\ell_2) = (h_3 \circ s_3)(\ell_1) = (s_3 \circ h_3)(\ell_1) = s_3(\ell_1') = \ell_2'.$$

It follows that $h_3(W)$, which is the image of the intersection of ℓ_1 and ℓ_2, is the intersection of $h_3(\ell_1) = \ell_1'$ and $h_3(\ell_2) = \ell_2'$, and this intersection point is P_3. But $O_3, W, h(W) = P_3$ are then collinear, showing that $O_3 P_3$ passes through W. Similarly $O_2 P_2$ and $O_1 P_1$ pass through W, so the three lines intersect at W. But Theorem 2.31 shows that $O_1 P_1, O_2 P_2, O_3 P_3$ intersect on the circle of similitude; hence, W is on the circle of similitude of F_1, F_2, F_3.

The second claim is a consequence of the first. Note that the angles formed by ℓ_1, ℓ_2, and ℓ_3 are the angles of the spiral similarities. Because the lines intersect at W on the circle of similarity, the arcs determined by J_1, J_2, J_3 have the fixed measures, namely, twice these rotation angles. And the ratio of the distances from O_i to ℓ_j and ℓ_k, $\{i, j, k\} = \{1, 2, 3\}$ is the ratio of the spiral similarity with center O_i. Since the angle between ℓ_j and ℓ_k is fixed, this means that the angles $\angle O_j W J_k$, $k \neq j$ are fixed. So then the measures of the arcs $\overset{\frown}{O_j J_k}$ are also fixed for all $j \neq k$, and since the points O_j are fixed, the points J_k are fixed as well, for $j, k = 1, 2, 3$. This completes the proof of the theorem.   □

The points J_1, J_2, J_3 are called the *invariable points* of the directly similar figures F_1, F_2, F_3, and the triangle $J_1 J_2 J_3$ is called the invariable triangle.

Proposition 2.33 *Given three directly similar figures:*

(a) the triangle formed by three corresponding lines is similar to the invariable triangle but has opposite orientation;
(b) the three invariable points are corresponding points of the three figures.

Proof We reason once more on Fig. 2.34. We have

$$\angle(J_1 J_2, J_2 J_3) = \angle(J_1 W, W J_3) = \angle(P_3 P_2, P_2 P_1),$$

where for the first equality we used inscribed angles and for the second we used the fact that $\angle(J_1 W, W J_3)$ is the angle of the spiral similarity that maps F_1 to F_3. This proves (a). For (b), note that

$$\angle(J_2 O_1, O_1 J_3) = \angle(J_2 W, W J_3),$$

which is the angle of the spiral similarity of center O_1 that maps F_2 to F_3. By inscribed angles

$$\angle(O_1 J_2, J_2 W) = \angle(O_1 J_3, J_3 W),$$

which shows that $O_1 J_2 / O_1 J_3$ equals the ratio k of the distances from O_1 to ℓ_2 and ℓ_3. But k is the scaling factor of the spiral similarity of center O_1 that maps ℓ_2 to ℓ_3. Hence J_2 is mapped to J_3 by the spiral similarity that maps F_2 to F_3. The situation is the same for J_1, J_3 and J_1, J_2. The theorem is proved.   □

For a triangle ABC, the circle of similitude of the segments AB, BC, CA is called the circle of similitude of the triangle ABC.

Theorem 2.34 *The circle of similitude of a triangle is the circle of diameter KO where K is the Lemoine point (i.e., the intersection of the symmedians), and O is the circumcenter.*

Proof Let O_a, O_b, O_c be the centers of the spiral similarities that map AB to CA, BC to AB, and CA to BC, respectively. Then by Theorem 2.29, the lines AO_a, BO_b, and CO_c intersect at a point on the circle of similitude Σ.

Now we are in the situation described in Sect. 2.4.3, where the construction of the center of the spiral similarity degenerates, and O_a is the point where the circle that is tangent to AC at A and passes through B intersects for the second time the circle that is tangent to AB at A and passes through C. And we know that this is related to symmedians, more precisely; Theorem 2.23 shows that AO_a is a symmedian, and the same is true for BO_b and CO_c, so they intersect at the Lemoine point K (Fig. 2.35). Thus, we know that $K \in \Sigma$.

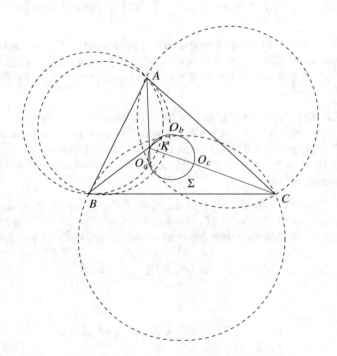

Fig. 2.35 Construction of the Lemoine point

The perpendicular bisectors of the sides are corresponding lines, and they intersect at one point, namely, O, and Theorem 2.32 implies that O is on the circle of similitude Σ, as well. The points A_1, B_1, C_1 where the perpendicular bisectors of BC, CA, AB intersect Σ are the invariable points of the three sides.

If we take the three lines ℓ_a, ℓ_b, ℓ_c through K parallel to AB, BC, CA, respectively, then we are in a converse situation of Theorem 2.32. We claim that these parallels correspond through the spiral similarities that map the sides of the triangle into each other. Indeed, consider the spiral similarities s_a, s_b, s_c of centers O_a, O_b, O_c that map ℓ_b to ℓ_c, ℓ_c to ℓ_a, and ℓ_a to ℓ_b, respectively, and let BC, $s_c(BC)$, and $s_a s_c(BC)$ be the corresponding lines that form the triangle A', B', C'. Then $B', C' \in BC$, and O_b, K, B' are collinear, and also O_c, K, C' are collinear. It follows that $B' = B, C' = C$. But because the triangles $A'B'C'$ and ABC have parallel sides and are oriented the same way, $A' = A$. Therefore, s_a, s_b, s_c are the spiral similarities that map the sides into each other, and so ℓ_a, ℓ_b, ℓ_c are corresponding lines. These lines must therefore pass through the invariable points A_1, B_1, C_1. The line OA_1, being the perpendicular bisector of BC, is perpendicular to ℓ_a. Thus, $\angle KA_1O = 90°$, showing that KO is a diameter in the circle of similitude, and we are done.   □

2.5.2 Examples of Problems Solved Using Spiral Similarities

Let us present some problems that use spiral similarities and have shown up in mathematics competitions. Our first example is a problem that was given at the Asian Pacific Mathematical Olympiad in 1998, being proposed by the first author of the book.

Problem 2.5 Let ABC be a triangle and let D be the foot of the altitude from A. Let E and F be two points on a line passing through D such that AE is perpendicular to BE, AF is perpendicular to CF, and E and F are different from D. Let M and N be the midpoints of the segments BC and EF, respectively. Prove that AN is perpendicular to NM.

Solution 1 The right angles imply that A, B, D, E lie on the circle of diameter AB and A, C, D, F lie on the circle of diameter AC (see Fig. 2.36). From here, using inscribed angles and working with directed angles modulo $180°$, we deduce that

$$\angle(AB, BC) = \angle(AB, BD) = \angle(AE, ED),$$

and

$$\angle(AC, CB) = \angle(AC, CD) = \angle(AF, FD).$$

It follows that the triangles ABC and AEF are directly similar, so there is a spiral similarity of center A taking B to E and C to F. This spiral similarity takes M,

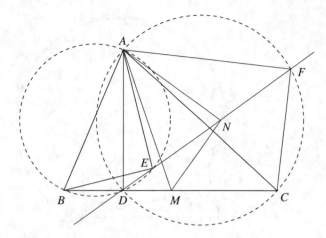

Fig. 2.36 Proof that $\angle ANM = 90°$

the midpoint of BC, to N, the midpoint of EF. So $B \mapsto E$, and $M \mapsto N$, but as spiral similarities come in pairs (Theorem 2.25), there is a spiral similarity of center A that maps the triangle ABE to the triangle AMN. We conclude that $\angle ANM = \angle AEB = 90°$, as required. □

Solution 2 We have seen in the first solution that ABC is mapped to AEF by a spiral similarity, and since spiral similarities come in pairs (Theorem 2.25), there is a spiral similarity of center A that takes ABE to ACF. Since M is the midpoint of BC and N the midpoint of EF, the Averaging Principle (Theorem 2.18) implies that the triangles ABE and ACF are also similar to AMN. Therefore, $\angle ANM = \angle AEB = 90°$. □

By comparing these two solutions, you should be able to derive a synthetic proof of the Averaging Principle (Theorem 2.18).

Our second example is a problem of Cosmin Pohoață that was given at the United States of America Mathematical Olympiad in 2014.

Problem 2.6 Let ABC be a triangle with orthocenter H, and let P be the second intersection point of the circumcircle of the triangle AHC with the internal bisector of the angle $\angle BAC$. Let X be the circumcenter of the triangle APB, and let Y be the orthocenter of the triangle APC. Prove that the length of the segment XY is equal to the circumradius of the triangle ABC.

Solution The configuration is shown in Fig. 2.37. Because Y is the orthocenter of APC, the circumcircles of APC and AYC are mapped into each other by the reflection over AC. But the circumcircle of APC is the same as the circumcircle of AHC, and it reflects to the circumcircle of ABC. Hence, the circumcircle of AYC coincides with the circumcircle of ABC, so Y is on the circumcircle of ABC.

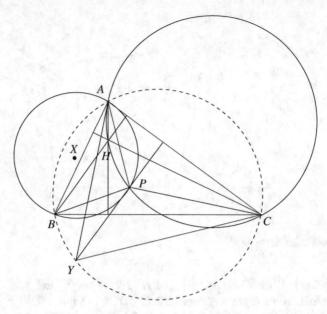

Fig. 2.37 Y is on the circumcircle of ABC

Next, let O be the circumcenter of ABC, and let O_1 be the circumcenter of APC, which is the reflection of O over AC. We will show that there is a spiral similarity that maps triangle O_1OX to triangle AOY. We continue our reasoning on Fig. 2.38.

Because OO_1 is orthogonal to AC and OX is orthogonal to AB, $\angle O_1OX$ is the supplement of $\angle BAC$. On the other hand, $\angle ACY = 90° - \angle CAP$, because AP is orthogonal to CY, so $\angle AOY = 2\angle CAP$ is the supplement of $\angle BAC$ as well. Therefore, $\angle O_1OX = \angle AOY$.

Also, XO perpendicular to AB and XO_1 perpendicular to AP implies $\angle XO_1O = \angle PAC = \angle A/2$. But then $\angle O_1XO = 180° - (180° - \angle A) - \angle A/2 = \angle A/2$, so the triangle OO_1X is isosceles. But $OA = OY$ because Y is on the circumcircle of triangle ABC, so the triangle AOY is also isosceles. We thus have two isosceles triangles OAY and OO_1X with equal angles at O, and they are mapped into each other by a spiral similarity.

But spiral similarities come in pairs (Theorem 2.25), so the triangle OO_1A is mapped into the triangle OXY by a spiral similarity. But $OO_1 = OX$, so the two triangles are congruent, and therefore $XY = O_1A$. Now the conclusion follows from the fact the circumcircles of AHC and ABC have the same radius, as they are reflections of each other over AC. Hence O_1A is equal to the circumradius of ABC, and so is XY.

It is worth observing that point Y can be located easily with complex numbers. Place the triangle in a coordinate system so that the origin is at the circumcenter O and A has coordinate 1. Let the coordinates of B and C be $e^{i\beta}$ and $e^{i\gamma}$, respectively. Then $\angle BAC = \angle BOC = (\gamma - \beta)/2$. By using inscribed angles in the circle of

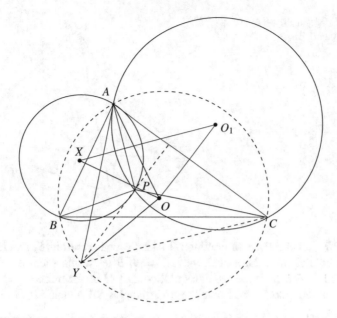

Fig. 2.38 The existence of the spiral similarity

center O_1, we obtain $\angle PO_1C = 2\angle PAC = \angle BAC$, so P is the rotate of C about O_1 by $-(\gamma-\beta)/2$. The reflection of P over AC, which we denote by P', is therefore obtained by rotating C about O by $(\gamma - \beta)/2$. Its coordinate is

$$p' = e^{i\frac{\gamma-\beta}{2}}e^{i\gamma} = e^{i\frac{3\gamma-\beta}{2}}.$$

We obtain P by reflecting P' over AC; by using Proposition 1.6 and performing the computations, we deduce that the coordinate of P is

$$p = -e^{i\frac{\beta-\gamma}{2}} - e^{i\gamma} - 1.$$

The triangle APC is inscribed in a circle centered at the origin, so the coordinate of the orthocenter is the sum of the coordinates of the vertices (see the discussion on the Euler line). Hence the coordinate of Y is

$$y = -e^{-i\frac{\gamma-\beta}{2}} = e^{i\left(\pi-\frac{\gamma-\beta}{2}\right)}.$$

From this formula we read that Y is on the circumcircle of ABC and that the angle $\angle AOY$ is the supplement of the angle $\angle BAC$.   □

The third example is from the short list of the 1978 International Mathematical Olympiad, proposed by Bulgaria.

Fig. 2.39 $SA'N$ and $SB'M$
are similar

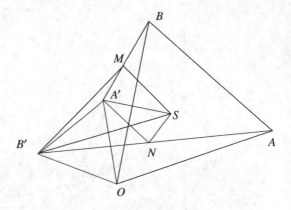

Problem 2.7 Let AOB be an equilateral triangle with centroid S, and let $A'OB'$ be another equilateral triangle such that $A' \neq S$, $B' \neq S$ and such that the angles $\angle A'OB'$ and $\angle AOB$ have the same orientation. Let M be the midpoint of $A'B$, and let N be the midpoint of AB'. Prove that the triangles $SA'N$ and $SB'M$ are similar.

Solution We consider the triangles placed as in Fig. 2.39. The composition of the homothety of center A and ratio 2 and the clockwise rotation about O is a spiral similarity that maps N to A' and keeps S fixed. The center of the spiral similarity is the fixed point S. Also, the homothety of center B and ratio 2 followed by the counterclockwise rotation about O maps M to B', and S is again its fixed point. Thus we have another spiral similarity, of center S, with the same ratio, but with opposite angle. We therefore have $SN/SA' = SM/SB' = 1/2$ and $\angle NSA' = \angle MSB' = 60°$. This proves that triangles $SA'N$ and $SB'M$ are similar, as desired.

□

We conclude the discussion with a simple example that illustrates how to use the general form of the Averaging Principle.

Problem 2.8 Two circles ω_1 and ω_2 intersect at M and N. Through M we take a line ℓ that intersects ω_1 and ω_2 again at A and B, respectively. In the half-plane determined by AB that does not contain N, construct a point C such that the triangle ABC is similar to a given triangle. What is the locus of C when the line ℓ varies?

Solution 1 The configuration is shown in Fig. 2.40. Assign to points complex coordinates using the convention that the lowercase complex number corresponds to the uppercase point. Because the triangle ABC is similar to some fixed triangle XYZ

$$\frac{c-a}{b-a} = w, \text{ where } w = \frac{z-x}{y-x}.$$

Solving for c we deduce that there exist complex numbers t_1 and t_2 that do not depend on ℓ such that $c = t_1a + t_2b$. By Problem 131, the spiral similarity that maps

Fig. 2.40 Finding the locus of C

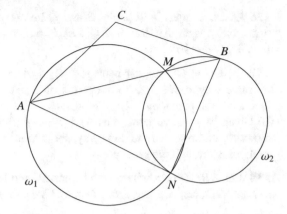

ω_1 to ω_2 maps A to B. Now we can apply the Averaging Principle in its general form, as stated in Theorem 2.19, to the figures ω_1 and ω_2 and the numbers t_1 and t_2 to deduce that as the point A traces the circle ω_1, the point C traces a figure similar to ω_1, hence a circle. This circle is the "weighted mean" of ω_1 and ω_2, with the complex weights t_1 and t_2. □

Solution 2 Of course, there is solution that does not use the Averaging Principle explicitly. Again by using the result proved in Problem 131, we deduce that the spiral similarity of center N that maps ω_1 to ω_2 maps A to B.

Since the ratios NA/AB and AC/AB do not depend on the position of ℓ, NA/AC does not depend on this position either. Combining this with the fact that $\angle NAC = \angle NAB + \angle BAC$ does not depend on the position of ℓ, we deduce that all triangles NAC are similar and oriented the same way. In particular, the angle $\angle CNA$ and the ratio CN/AN do not depend on the position of ℓ.

We deduce that the spiral similarity of center N, angle $\angle CNA$, and ratio CN/AN maps A to C. Hence the locus is the image of the circle ω_1 through this spiral similarity. Because the spiral similarity is a bijection, it establishes a one-to-one correspondence between the points of ω_1 and the locus, so the locus is actually the whole circle. □

2.5.3 Problems in Euclidean Geometry to be Solved Using Spiral Similarities

Endowed with the methods and ideas presented above, you are now invited to solve the following problems.

135 Given a triangle ABC and a polygon, show that there exist points M, N, P on the sides of the polygon such that the triangles MNP and ABC are similar.

136 Given a square $ABCD$, let P and Q be two points on the sides AB and BC, respectively, such that $BP = BQ$. Let H be the projection of B onto PC. Prove that $\angle DHQ = 90°$.

137 (a) Given two regular pentagons $A_1A_2A_3A_4A_5$ and $A_1'A_2'A_3'A_4'A_5'$ oriented the same way, prove that the midpoints of A_1A_1', A_2A_2', A_3A_3', A_4A_4', and A_5A_5' form a regular pentagon.
(b) Given two regular pentagons $A_1A_2A_3A_4A_5$ and $A_1A_2'A_3'A_4'A_5'$ with common vertex A_1 and oriented the same way, show that the lines A_2A_2', A_3A_3', A_4A_4', and A_5A_5' pass through the same point.

138 Let $ABCD$ be a square with center O, and let M and N be the midpoints of BO and CD, respectively. Prove that AMN is an isosceles right triangle.

139 Let ABC and $A'B'C'$ be two equilateral triangles such that the segments BC and $B'C'$ have the same midpoint. Find the ratio between the segments AA' and BB' and the angle that they form.

140 Given two directly similar triangles, ABC and $A_1B_1C_1$, and an arbitrary point O in the plane, consider the points A_2, B_2, and C_2 such that AA_1OA_2, BB_1OB_2, and CC_1OC_2 are parallelograms. Prove that the triangles ABC and $A_2B_2C_2$ are similar.

141 Given a convex polygon $A_1A_2 \ldots A_n$ of area S and a point M in the plane, find the area of the polygon $M_1M_2 \ldots M_n$, where M_j is the image of M under the rotation of center A_j and angle α, $j = 1, 2, \ldots, n$.

142 Two circles intersect at the points A and B, and the chords AM and AN are tangent to these circles. We consider the point C such that $AMCN$ is a parallelogram, and on the segments BN and CM, we consider the points P and Q, respectively, such that $BP/PN = MQ/QC$. Prove that $\angle APQ = \angle ANC$.

143 Let $\omega_1, \omega_2, \cdots, \omega_n$ be n circles passing through O. A grasshopper starts at a point $X_1 \in \omega_1$ and jumps to the points $X_2, X_3, X_4, \ldots$ such that $X_i \in \omega_i$ and X_iX_{i+1} passes through the other point of intersection of ω_i and ω_{i+1} for all i (where $\omega_{n+1} = \omega_1$). Show that the grasshopper returns to the initial position.

144 On the sides of the triangle ABC, construct the similar triangles BPA, AQC, and BRC such that the first two are in the exterior of the triangle ABC and the third is on the same side of BC as the point A. Prove that $APRQ$ is a parallelogram.

145 A circle passing through the vertex A and the circumcenter O of an acute triangle ABC intersects the sides AB and AC at the points P and Q, respectively. Prove that the orthocenter of the triangle POQ lies on BC.

146 Let $ABCD$ be a trapezoid with $AD\|BC$ and $\angle B > 90°$. Let M be a point on the side AB, let O_1 and O_2 be the circumcenters of the triangles MAD and MBC, and let N be the second point of intersection of the circumcircles of MO_1D and MO_2C. Prove that the line O_1O_2 passes through N.

147 Let ABC be a triangle, and let ℓ be a line that intersects the lines BC, CA, AB at D, E, and F, respectively. Construct the triangles A_1FE, FB_1D, and EDC_1 that are similar to the triangle ABC and oriented the same way as ABC. Show that the spiral similarities that map these triangles into one another have the same center.

148 The circles ω_1, ω_2 centered at O_1, O_2, respectively, intersect at P and S. The points A, B on ω_1 and C, D on ω_2 are chosen such that the segments AC and BD intersect at P. Denote the midpoints of the segments AC, BD, O_1O_2 by M, N, O, respectively. Prove that O is the circumcenter of the triangle MNP.

149 Let ABC be a triangle in the plane. In the exterior of this triangle, construct the triangles ABR, BCP, and CAQ such that

$$\angle PBC = \angle CAQ = 45°, \ \angle BCP = \angle QCA = 30°, \ \angle ABR = \angle RAB = 15°.$$

Prove that $\angle PRQ = 90°$ and that $PR = QR$.

150 Let ABC be a triangle, and let E and F be two arbitrary points on the sides AB and AC, respectively. The circumcircle of the triangle AEF meets the circumcircle of the triangle ABC again at the point M. Let D be the reflection of M over EF, and let O be the circumcenter of the triangle ABC. Prove that D is on BC if and only if O is on the circumcircle of the triangle AEF.

151 On the sides BC, CA, and AB of a triangle ABC, we consider the points A_1, B_1, C_1, respectively, such that the triangles ABC and $A_1B_1C_1$ are similar. Denote by A_2, B_2, C_2 the intersections of BB_1 and CC_1, CC_1 and AA_1, and AA_1 and BB_1, respectively, and assume that these three points are distinct. Prove that the circumcircles of the triangles ABC_2, BCA_2, CAB_2, $A_1B_1C_2$, $B_1C_1A_2$, and $C_1A_1B_2$ have a common point.

152 The points P and Q are chosen on the side BC of an acute-angled triangle ABC so that $\angle PAB = \angle ACB$ and $\angle QAC = \angle CBA$. The points M and N are taken on the rays AP and AQ, respectively, so that $AP = PM$ and $AQ = QN$. Prove that the lines BM and CN intersect on the circumcircle of the triangle ABC.

153 Let ABC be a triangle with $AB < AC$; let D be the point where the perpendicular bisector of the side BC intersects the side AC, and let E be the point where the angle bisector of $\angle ADB$ intersects the circumcircle of ABC. Prove that the angle bisector of $\angle AEB$ and the line that passes through the incenters of the triangles ADE and BDE are orthogonal.

154 Consider a convex pentagon $ABCDE$ such that

$$\angle BAC = \angle CAD = \angle DAE \text{ and } \angle CBA = \angle DCA = \angle EDA.$$

Let P be the point of intersection of the lines BD and CE. Prove that the line AP passes through the midpoint of the side CD.

155 Let ABC be a scalene triangle, and let X, Y, Z be points on the lines BC, CA, AB, respectively, such that $\angle AXB = \angle BYC = \angle CZA$. The circumcircles of the triangles BXZ and CXY meet again at $P \neq X$. Prove that P lies on the circle of diameter GH, where G and H are the centroid and the orthocenter of the triangle ABC, respectively.

156 Let ABC be an acute triangle with orthocenter H, and let W be a point on the side BC, lying strictly between B and C. The points M and N are the feet of the altitudes from B and C, respectively. Denote by ω_1 the circumcircle of BWN, and let X be the point on ω_1 such that WX is a diameter of ω_1. Analogously, denote by ω_2 the circumcircle of triangle CWM, and let Y be the point such that WY is a diameter of ω_2. Prove that $X, Y,$ and H are collinear.

Chapter 3
Inversions

3.1 Theoretical Results About Inversion

A great leap has to be taken from those geometric transformations of the first two chapters to inversion, and the intuition behind the latter is more difficult to explain. We will need a longer theoretical exposition to fully understand this transformation. Inversion should be thought of as reflection over a circle, with the points inside the circle being reflected outside and the points outside of the circle being reflected inside. With inversion, the point at infinity makes its appearance; it is the image of the center of the circle of inversion. There is just one point at infinity! The idea of inversion is suggested in Fig. 3.1.

3.1.1 The Definition of Inversion and Some of Its Properties

We state the following definition.

Definition Given a circle of center O and radius R, the *inverse* of a point $P \neq O$ with respect to the circle is a point P' such that P' lies on the ray from O through P and

$$OP \cdot OP' = R^2.$$

The transformation defined this way is called inversion with respect to the circle of center O and radius R, or inversion with center O and radius R. The quantity R^2 is called the ratio of the inversion.

Inversion is represented schematically in Fig. 3.2.

Inversion is not defined at its center O, which can be an issue when composing several inversions with different centers. The problem is solved by adding to the

© The Author(s), under exclusive license to Springer Nature Switzerland AG 2022   129
R. Gelca et al., *Geometric Transformations*, Problem Books in Mathematics,
https://doi.org/10.1007/978-3-030-89117-6_3

Fig. 3.1 The idea of
inversion

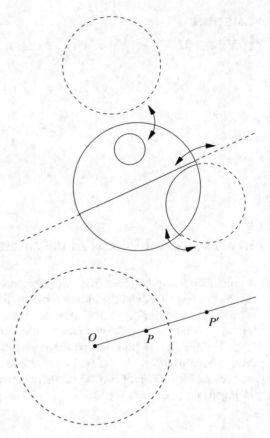

Fig. 3.2 The definition of
inversion with respect to a
circle

plane a single point at infinity, denoted by ∞. And this means that in whichever
direction of the plane you look, you see this point at infinity, all lines pass through
it. Then O is mapped to the point at infinity and vice versa. This renders inversion a
bijective transformation of the plane enlarged with the point at infinity.

Inversion shares several properties with reflection over a line:

- Composing an inversion with itself yields the identity map: if P is mapped to P',
  then P' must be mapped to P (because $OP \cdot OP' = OP' \cdot OP$). This means
  that inversion is an *involution*, exactly like reflection over a line is.
- The points on the circle of inversion are invariant under inversion, exactly how
  the points on the line of reflection are invariant under the reflection.
- Inversion maps points that are on one side of the circle of inversion to points
  on the other side (meaning that the interior is mapped to the exterior and the
  exterior is mapped to the interior), exactly in the same way as reflection maps
  one half-plane to the other.
- Inversion, like reflection, changes the orientation of figures.

We will see below that these similarities are not accidental and that in fact inversion should be thought of as *reflection over a circle*, as suggested in the preamble. We will arrive later there, and for that we will have to switch to coordinates.

The name "inversion" makes sense if you consider R as the unit, that is, $R = 1$: then the inverse P' of P satisfies $OP' = \frac{1}{OP}$. In this sense, inversion in a plane (or in space) generalizes the notion of the inverse (reciprocal) of a number on the real line.

It is quite easy to write in complex coordinates the formula for inversion over the unit circle centered at the origin

$$z \mapsto \frac{1}{\bar{z}}.$$

Indeed, if we write $z = re^{i\theta}$, then $1/\bar{z} = r^{-1}e^{i\theta}$, so on the one hand z and $1/\bar{z}$ are on the same ray from the origin (having the same argument) and on the other hand $|z| \cdot |1/\bar{z}| = r \cdot r^{-1} = 1$.

The inversion with respect to an arbitrary circle of center a and radius R maps a point $z = a + re^{i\theta}$ to the point $z^* = a + \frac{R^2}{r}e^{i\theta}$. This means that

$$\overline{(z - a)}(z^* - a) = R^2,$$

which gives

$$z^* = \frac{R^2}{\overline{z - a}} + a.$$

So in order to obtain the inverse of z with respect to the circle of center a and radius R^2, you should translate z by $-a$, take its inverse over the unit circle centered at the origin, map the new point by the homothety of center the origin and ratio R^2, and translate the result by a. Thus any inversion can be obtained by composing the inversion with respect to the unit circle with some translations and homotheties.

Proposition 3.1 *Let P be a point that does not lie on the circle of inversion, and let P' be its inverse. Let also M and N be the two points where PP' intersects the circle of inversion. Then P, M, P', N form a harmonic division.*

Proof We recall that the collinear points P, M, P', N form a harmonic division if exactly one of P and P' separates M and N and $PM/PN : P'M/P'N = 1$, which is equivalent, in complex coordinates, to the fact that their cross-ratio satisfies

$$\frac{p - m}{p - n} : \frac{p' - m}{p' - n} = -1,$$

where p, m, p', n are the complex coordinates of P, M, P', N, respectively.

For the proof, place the center of inversion at the origin of a system of coordinates for which PP' is the x-axis, the radius of inversion is 1, and M and N have

Fig. 3.3 Similarity of triangles obtained from inversion

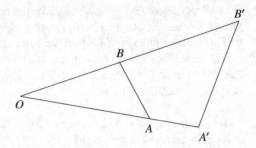

coordinates 1 and -1, respectively. If p is the real number which is the coordinate of P, then by the above result, P' has coordinate $1/\bar{p} = 1/p$. Then

$$
\frac{p-m}{p-n} : \frac{p'-m}{p'-n} = \frac{p-1}{p+1} : \frac{\frac{1}{p}-1}{\frac{1}{p}+1} = -1,
$$

and the proposition is proved. $\square$

Now that we understand what inversion does to a single point, we can study what it does to a pair of points.

Theorem 3.2 *Consider the inversion about the circle of center O and radius R. If A' is the inverse of A and B' is the inverse of B, then the triangles OAB and $OB'A'$ are (inverse) similar, and the similarity ratio is $\frac{R^2}{OA \cdot OB}$. In particular $\angle OAB = \angle A'B'O$.*

Proof Arguing on Fig. 3.3, we see that that $\angle AOB = \angle B'OA'$, and also the equality $OA \cdot OA' = R^2 = OB \cdot OB'$ gives

$$
\frac{OA'}{OB} = \frac{OB'}{OA} = \frac{R^2}{OA \cdot OB}.
$$

So the triangles are indeed similar, and the similarity ratio is as said. $\square$

We have written the equality of angles so that it holds for directed angles modulo π, too.

Remark As a corollary we obtain a formula that will be used extensively in what follows, a formula which describes how distances are distorted by inversion. If A' is the image of A and B' is the image of B through the inversion of center O and radius R, then

$$
A'B' = \frac{R^2}{OA \cdot OB} AB. \tag{3.1}
$$

Next, we learn how to construct the inverse of a point.

Fig. 3.4 Image of a point lying in the exterior of the circle of inversion

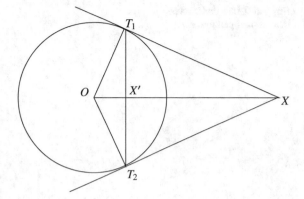

Proposition 3.3

(i) *If a point X lies outside of the circle of inversion, then its image X' is the midpoint of the segment formed by the tangency points of the two tangents from X to the circle of inversion.*

(ii) *If a point X lies inside the circle of inversion, then its image X' is the point obtained as the intersection of the tangents to the circle of inversion at the points where the perpendicular to OX intersects the circle of inversion.*

Proof We follow the proof of (i) on Fig. 3.4, where we have denoted the tangency points by T_1 and T_2. We notice that the midpoint X' of T_1T_2 satisfies $OX' \cdot OX = OT_1^2$, as given by the "Leg Theorem" in the right triangle T_1OX. So X' is the inverse of X. Part (ii) follows from (i) using the fact that inversion is an involution. □

3.1.2 Inverses of Lines and Circles

It is not hard to see that the image of a line that passes through the center of inversion is the same line. It is worth including the point at infinity, so that the image contains the center of inversion as well, a convention that we will tacitly follow. Throughout this section, the inversion has center O and radius R.

Theorem 3.4 (Inverse of a Line) *Let ℓ be a line that does not pass through the center O of inversion, A the projection of O onto ℓ, A' the image of A, and ω the circle of diameter OA'. Then the line ℓ and the circle ω are mapped into each other by the inversion.*

Proof We need not check the pairs of points (O, ∞) and (A, A'). Arguing on Fig. 3.5, let P and P' be two points that correspond through the inversion. Then $P \in \ell \backslash \{A\}$ is equivalent to $\angle OAP = 90°$. By Theorem 3.2, this is equivalent to

Fig. 3.5 Lines and circles
that are mapped into each
other

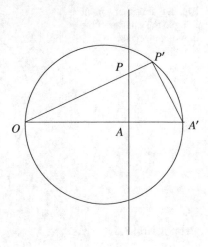

$\angle A'P'O = 90°$, and this is further equivalent to $P' \in \omega \backslash \{O, A'\}$. The theorem is proved.

□

Remark So a line that does not pass through the center of inversion is mapped to a circle that passes through the center of inversion and vice versa.

As a corollary, a line and a circle that are not tangent are mapped into each other by exactly two inversions, whose centers are the endpoints of the diameter that is perpendicular to the line.

Theorem 3.5 (Inverse of a Circle) *Let ω be a circle that does not pass through the center O of inversion, and let A and B be the endpoints of the diameter whose line of support passes through O. Let also A', B' be the images of A and B, respectively. Then the image of ω through the inversion is a circle ω' of diameter $A'B'$. In particular the center of inversion lies on the line connecting the center of the circle and the center of its image.*

Proof 1 This argument can be followed on Fig. 3.6. Certainly we do not have to discuss the pairs (A, A') and (B, B'). Assume that A is between O and B, and let P and P' be a point and its image through the inversion. Then

$$\angle APB = \angle OPB - \angle OPA = \angle P'B'O - \angle P'A'O = \angle B'P'A',$$

where for the last step we have used the Exterior Angle Theorem in the triangle $A'P'B'$. This shows that $\angle APB = 90°$ if and only if $\angle B'P'A' = 90°$. In other words, P is on ω if and only if P' is on ω'. The theorem is proved.   □

Proof 2 There exists also a short complex number argument. Choose a coordinate system with origin at the center of inversion and with the x-axis containing the

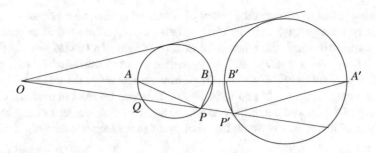

Fig. 3.6 Circles that are mapped into circles

center of the circle, and let $x \in \mathbb{R}$ be the coordinate of this center. The equation of the inversion is $z \mapsto R^2/\overline{z}$, where R is the radius of inversion. The equation of the circle is

$$|z - x| = r,$$

and if w is the image (and also the preimage) of z through the inversion, then

$$\left| \frac{R^2}{w} - x \right| = r.$$

So this is the equation of the image of the circle through the inversion. To check that this is the equation of a circle as well, we rewrite it as

$$R^4 - R^2 xw - R^2 x\overline{w} + (x^2 - r^2)w\overline{w} = 0,$$

and then as

$$w\overline{w} - \frac{R^2 x}{x^2 - r^2} w - \frac{R^2 x}{x^2 - r^2}\overline{w} + \frac{R^4 x^2}{(x^2 - r^2)^2} = \frac{R^4 x^2}{(x^2 - r^2)^2} - \frac{R^4}{x^2 - r^2}.$$

This is the equation of the circle

$$\left| w - \frac{R^2 x}{x^2 - r^2} \right| = \left(\frac{R^2 r}{x^2 - r^2} \right)^2,$$

with center $R^2 x/(x^2 - r^2)$ and radius $R^2 r/|x^2 - r^2|$.   □

Remark We have shown that a circle that does not pass through the center of inversion is mapped to a circle that does not pass through the center of inversion.

To summarize, while isometries and spiral similarities map lines to lines and circles to circles, inversion maps a line or a circle to a line or a circle.

It is important to note that the center of inversion is also the center of the unique homothety with positive ratio that maps the circles one into the other. Indeed, if in our figure we denote the other point of intersection of OP and ω with Q, then $OP \cdot OP' = R^2$ and $OP \cdot OQ = \rho$, where ρ is the power of O with respect to ω. Consequently $OP' = (R^2/\rho)OQ$, showing that P' is the image of Q under the homothety of center O and ratio R^2/ρ. This can be read immediately in the analytic proof. In particular, if the two circles admit common exterior tangents, then the center of inversion lies at the intersection of those tangents (see Fig. 3.6). And we have

Proposition 3.6 *If ω and ω' are circles of radii r and r', respectively, that are mapped into each other by an inversion of center O and radius R, and if the power of O with respect to ω is ρ, then*

$$r' = \frac{R^2}{\rho} r.$$

Unfortunately, the center of a circle is not mapped to the center of its image. This is to be expected because sometimes the image of the circle is a line, which has no center. Nevertheless we can still locate the image of the center of a circle through the inversion, as the following result shows.

Proposition 3.7 *The inverse of the center C of a circle ω through an inversion χ of center O is*

(i) *the reflection of O over $\chi(\omega)$ if $\chi(\omega)$ is a line;*
(ii) *the midpoint of the segment determined by the points of tangency to $\chi(\omega)$ of the tangents from O if $\chi(\omega)$ is a circle with O in its exterior;*
(iii) *the intersection of the tangents to $\chi(\omega)$ at the points where the perpendicular to OC at O intersects $\chi(\omega)$ if $\chi(\omega)$ is a circle and O is in its interior.*

Proof The three cases are shown in Fig. 3.7. For (i), denote by P the other end of the diameter of ω through the center O of the inversion, and let $C' = \chi(C)$ and $P' = \chi(P)$. Then $OP \cdot OP' = OC \cdot OC'$, and since $OP = 2OC$, it follows that $OC' = 2OP'$, as desired.

For (ii), with the same notation as above, if OTT' is a common tangent of ω and $\omega' = \chi(\omega)$ ($T \in \omega$, $T' \in \omega'$), then $\angle OTC = \angle OC'T'$, hence the conclusion.

Finally, for (iii), note that if T is at the intersection of ω with the perpendicular through O to OC, then the circumcircle of OCT has diameter OT, so it is tangent to ω. Its image through the inversion is a line that is tangent to ω', and the proposition is proved (please read in Sect. 3.1.6 the explanation of why inversion maps tangent curves to tangent curves). $\qquad\qquad\qquad\qquad\qquad\qquad\qquad\qquad\qquad\qquad\qquad\qquad\square$

On the other hand, given two circles ω_1 and ω_2, there is almost always a unique inversion χ that maps ω_1 to ω_2.

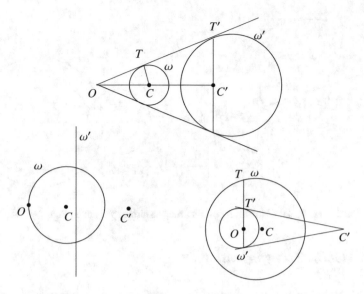

Fig. 3.7 The image through inversion of the center of a circle

Proposition 3.8 *Let ω_1 and ω_2 be two circles in the plane with distinct radii. There is a unique inversion χ that maps ω_1 to ω_2. The center of inversion coincides with the center of the unique homothety with positive ratio that maps ω_1 to ω_2, except when one of the circles is inside the other; in this case, the center of inversion coincides with the center of the unique homothety with negative ratio that maps ω_1 to ω_2.*

Proof We consider two cases, both shown in Fig. 3.8. If ω_1 is inside ω_2, let O and $-k$ be, respectively, the center and ratio of the inverse homothety h that maps ω_1 to ω_2. The point O must be inside both circles. Let ℓ be a line through O that meets ω_1 at A and B and ω_2 at $B' = h(A)$ and $A' = h(B)$ (yes, we reversed the points). Then, using oriented segments, $OA \cdot OA' = OA \cdot (-k \cdot OB) = k \cdot (-OA \cdot OB)$; this is k times the power of O with respect to ω_1, which is a positive constant. We can take χ with center O and radius $\sqrt{k|\rho|}$, and then ρ is the power of O with respect to ω_1.

The other case, where neither of the circles is inside the other, is handled analogously, with signs (doubly) inverted, as it follows from both the homothety and the fact that the center of the direct homothety that maps ω_1 to ω_2 is the intersection of the two common external tangent lines to both circles, which is outside both circles.

Theorem 3.2 shows that two points and their inverses determine uniquely the inversion, so the inversion χ is unique. $\square$

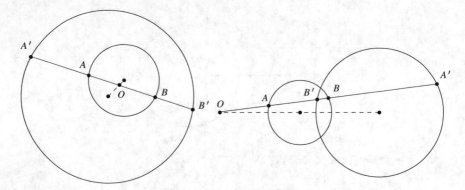

Fig. 3.8 Finding the center of the inversion that maps a circle to another circle

3.1.3   Möbius Transformations

As we have said in the first chapter, by introducing the concept of a group, we
introduced a new body of ideas, and one such idea is that we must understand the
compositions of geometric transformations. This was an easy task in the previous
chapters, but not so now, and to explain compositions of inversions, we must take a
step into a wider context. It will be the goal of the fourth chapter to truly take a global
look at all the transformations discussed in this book; here we intend to capture
only some essential facts about inversion that become clearer when discussed in
this general setting.

What we are doing now is to place ourselves in the framework of complex
projective geometry. To this end, as before, we add to the complex plane $\mathbb{C}$ one
point at infinity, denoted by ∞. The result, $\mathbb{C} \cup \{\infty\}$, is not the *real projective plane*
(in that case we have to add an entire line at infinity), but the *complex projective line*,
also called the *Riemann sphere* and denoted by $\mathbb{C}P^1$ by algebraic geometers. It is a
line in the sense that it has complex dimension equal to one, it is parametrized by
one complex number, and it is referred to as a sphere because to be able to add the
point at infinity you have to bend the plane into a sphere (we only bend it mentally;
the plane should stay straight so that we can talk about lines, segments, triangles,
etc.). The convention is that all real lines loop through the point at infinity.

Definition A *Möbius transformation* is a map $\phi : \mathbb{C} \cup \{\infty\} \to \mathbb{C} \cup \{\infty\}$ of the form

$$\phi(z) = \frac{az + b}{cz + d}, \quad \text{if } z \neq -d/c, \quad \phi(-d/c) = \infty, \text{ and } \phi(\infty) = a/c,$$

where a, b, c, d are complex numbers satisfying the condition $ad - bc \neq 0$.

Möbius transformations, named after August Ferdinand Möbius, are also known
as *linear fractional transformations* of $\mathbb{C}$, or as *homographic functions* (from the

French *fonctions homographiques*). Inversion is not here yet; it will show up in a few moments.

It is immediate to observe that a Möbius transformation is one-to-one and onto (from the Riemann sphere to itself) and that it does not change when multiplying a, b, c, d simultaneously by the same nonzero number.

Theorem 3.9 *The Möbius transformations form a group.*

Proof In other words, the composition of two Möbius transformations is a Möbius transformation, and the inverse of every Möbius transformation is a Möbius transformation. Indeed, if

$$\phi(z) = \frac{az+b}{cz+d} \text{ and } \phi'(z) = \frac{a'z+b'}{c'z+d'},$$

then

$$\phi' \circ \phi(z) = \frac{a'\dfrac{az+b}{cz+d} + b'}{c'\dfrac{az+b}{cz+d} + d'} = \frac{(aa'+cb')z + (a'b+db')}{(c'a+cd')z + (c'b+dd')},$$

and

$$\phi^{-1}(z) = \frac{1}{ad-bc} \cdot \frac{dz-b}{-cz+a}.$$

The reader familiar with the theory of matrices will immediately notice that if we arrange the four numbers defining the Möbius transformation in a 2×2 matrix, then composition corresponds to matrix multiplication, of course with a caveat that one Möbius transformation is represented by several matrices which can be transformed into one another by multiplication by nonzero scalars. □

The group of Möbius transformations plays a sufficiently important role to have a notation and a name attached to it; it is denoted by $PSL(2, \mathbb{C})$ and is called the complex projective special linear group. It is important to observe that this group is *not* commutative, namely, that in general $\phi \circ \psi \neq \psi \circ \phi$.

For a Möbius transformation that is not the identity map, the equation

$$\frac{az+b}{cz+d} = z$$

has at most two distinct solutions in $\mathbb{C} \cup \{\infty\}$. Indeed, ∞ is a fixed point if and only if $c = 0$, in which case any other fixed point is the solution to a linear equation, and if $c \neq 0$, then the above equation is equivalent to the quadratic equation $cz^2 + (d-a)z + b = 0$, which can have at most two distinct complex solutions. Hence, a Möbius transformation that is not the identity map has at most two fixed points. This means that such a transformation is determined by the images of three points.

Indeed, if ϕ and ψ are Möbius transformations that coincide at three points, then $\phi \circ \psi^{-1}$ has three fixed points, implying that it is the identity map, so $\phi = \psi$.

Another way to look at this fact is by noticing that a Möbius transformation is determined by four parameters, a, b, c, d, but one degree of freedom is lost since the multiplication by a nonzero constant does not change the transformation (this is one reason why the group has the word "projective" in its name.)

It is standard to look at the Möbius transformations that map three given points z_2, z_3, z_4 to $1, 0, \infty$, respectively. Such a transformation is defined by the formula

$$\phi_{z_2,z_3,z_4}(z) = \frac{z - z_3}{z - z_4} : \frac{z_2 - z_3}{z_2 - z_4}.$$

There are three degenerate versions of this formula, corresponding to the cases where $z_2, z_3,$ or z_4 is ∞, and they are, respectively,

$$\frac{z - z_3}{z - z_4} = \frac{z - z_3}{z - z_4} : \frac{\infty - z_3}{\infty - z_4}, \quad \frac{z_2 - z_4}{z - z_4} = \frac{z - \infty}{z - z_4} : \frac{z_2 - \infty}{z_2 - z_4}, \quad \frac{z - z_3}{z_2 - z_3} = \frac{z - z_3}{z - \infty} : \frac{z_2 - z_3}{z_2 - \infty}.$$

And we recognize the formula of $\phi_{z_2,z_3,z_4}(z)$ to be the *cross-ratio* (z, z_2, z_3, z_4) of z, z_2, z_3, z_4! As a rule, we equate all fractions of the form $(a - \infty)/(b - \infty)$ with 1.

Having the above standard transformations at hand, we can express every Möbius transformation as a composition of such standard transformations and their inverses. In fact, given three distinct points z_2, z_3, z_4 in $\mathbb{C} \cup \{\infty\}$ and three more distinct points w_2, w_3, w_4, there is a unique Möbius transformation ψ for which $z_2 \mapsto w_2$, $z_3 \mapsto w_3$, and $z_4 \mapsto w_4$ given by $\psi = \phi_{w_2,w_3,w_4}^{-1} \circ \phi_{z_2,z_3,z_4}$. In other words, we map $z_2, z_3,$ and z_4 to the points $1, 0,$ and ∞ and then map these to the desired images $w_2, w_3,$ and w_4.

Theorem 3.10 (Invariance of Cross-Ratio) *The cross-ratio is invariant under Möbius transformations, that is if ψ is a Möbius transformation, then*

$$(z_1, z_2, z_3, z_4) = (\psi(z_1), \psi(z_2), \psi(z_3), \psi(z_4)).$$

Proof It is worth changing back to the functional notation and writing this as

$$\phi_{z_2,z_3,z_4}(z_1) = \phi_{\psi(z_2),\psi(z_3),\psi(z_4)}(\psi(z_1)).$$

This equality states that z_1 is mapped by ϕ_{z_2,z_3,z_4} to the same point where $\psi(z_1)$ is mapped by $\phi_{\psi(z_2),\psi(z_3),\psi(z_4)}$. And this follows from

$$\phi_{\psi(z_2),\psi(z_3),\psi(z_4)} = \phi_{z_2,z_3,z_4} \circ \psi^{-1},$$

which is true because both transformations map $\psi(z_2), \psi(z_3), \psi(z_4)$ to $1, 0,$ and ∞, respectively. $\square$

Recall that the *cross-ratio of four points* in the plane is the cross-ratio of their complex coordinates, which is independent of the choice of coordinates. As coordinate changes are Möbius transformations, the cross-ratio does not depend on the coordinate system; it is an invariant of the configuration consisting of the four points. Let us place in our context Theorem 1.4 from Chap. 1.

Theorem 3.11 (Condition for Concyclicity) *Four points z_1, z_2, z_3, z_4 in $\mathbb{C} \cup \{\infty\}$ lie on a circle or on a line if and only if $(z_1, z_2, z_3, z_4) \in \mathbb{R}$.*

And we have the following result.

Theorem 3.12 *Möbius transformations map lines or circles into lines or circles.*

Proof Let ϕ be a Möbius transformation. By Theorem 3.10,

$$(z_1, z_2, z_3, z_4) = (\phi(z_1), \phi(z_2), \phi(z_3), \phi(z_4)).$$

The first of these is real if and only if z_1, z_2, z_3, z_4 are on a circle or line. The second is real if and only if $\phi(z_1), \phi(z_2), \phi(z_3), \phi(z_4)$ are on a circle or line. So z_1, z_2, z_3, z_4 are on a circle or line if and only if $\phi(z_1), \phi(z_2), \phi(z_3), \phi(z_4)$ are on a circle or line. Done. □

3.1.4 Möbius Transformations Versus Isometries, Spiral Similarities, and Inversions; Inversion and Circular Transformations

We now want to understand the so-called circular group, obtained by adding inversions to the group consisting of spiral similarities and translations. A good description of the elements of $PSL(2, \mathbb{C})$ is necessary. Some Möbius transformations we recognize immediately:

- $a = d = 1, c = 0, z \mapsto z + b$ is a translation;
- $b = c = 0, d = 1, z \mapsto az$ is a spiral similarity centered at the origin;
- $a = d = 0, b = c = 1, z \mapsto \frac{1}{z}$ is the inversion over the unit circle centered at the origin, composed with reflection over the x-axis.

We will call this last transformation *complex inversion*.

Theorem 3.13 *Every Möbius transformation is the composition of translations, spiral similarities, and the complex inversion.*

Proof If $c = 0$, the map is a spiral similarity, a translation, or just the identity map.

If $c \neq 0$, scale a, b, c, d so that $bc - ad = c$ (e.g., by multiplying each of them by $\frac{bc-ad}{c}$). Now take the composition

$$z \mapsto z + \alpha \mapsto \frac{1}{z + \alpha} \mapsto \frac{1}{\beta} \cdot \frac{1}{z + \alpha} \mapsto \frac{1}{\beta z + \alpha \beta} + \gamma.$$

Then $\beta = c$, $\alpha = \frac{d}{c}$, and $\gamma = \frac{a}{c}$, and we are done. □

Note that translations, spiral similarities, and the complex inversion preserve orientation, so Möbius transformations also preserve orientation.

We have seen in Sect. 3.1.1 that the inversion χ about the circle whose center has complex coordinate a and whose radius is R is given by the formula

$$\chi(z) = \frac{R^2}{\overline{z} - a} + a.$$

This formula shows that $\chi = \sigma_x \circ \phi$, where ϕ is the Möbius transformation

$$\phi(z) = \frac{\bar{a}z + (R^2 - a\bar{a})}{z - a},$$

and σ_x is the reflection over the x-axis. So you can obtain any inversion from a Möbius transformation and the reflection over the x-axis, $z \mapsto \bar{z}$. It is therefore natural to make the following definition.

Definition A *circular transformation* is a transformation of the Riemann sphere that is either a Möbius transformation or the composition of a Möbius transformation and the reflection over the x-axis.

We leave it to the reader to check that circular transformations form a group. This group contains, among many other transformations, the maps of the form $f(z) = az + b$ or $f(z) = a\bar{z} + b$ that were studied in the previous chapters.

By Theorem 3.12, Möbius transformations map circles or lines into circles or lines, and so does the reflection σ_x. We have thus produced the coordinate-based proof of the fact that circular transformations, and in particular inversions, map circles or lines to circles or lines.

We will now look at inversion from a modified point of view, which will then allow us to conclude that inversion plays indeed the role of reflection over a circle. We will conclude that the group of circular transformations is generated by inversions and reflections over lines, or, said differently, by reflections over circles and lines.

Let z_2, z_3, z_4 be three distinct points in the Riemann sphere (i.e., plane with the point at infinity). They can be either collinear, in which case they determine a line, or noncollinear, in which case they determine a circle. If one of the three points is at infinity, then they are collinear, since we have agreed that all lines pass through the point at infinity. In this sense we can think of lines as circles that have one point at infinity. With this convention, inversion always maps circles to circles.

Theorem 3.14 *Let z_2, z_3, z_4 be three distinct points in $\mathbb{C} \cup \{\infty\}$, and let χ be either the reflection over the line determined by z_2, z_3, z_4 if they are collinear or* *the inversion over the circle determined by them if they are not. Then*

$$(\chi(z), z_2, z_3, z_4) = \overline{(z, z_2, z_3, z_4)}.$$

Proof Let us verify first the case where z_2, z_3, z_4 are noncollinear. Because the cross-ratio is invariant under Möbius transformations (Theorem 3.10), it is invariant under isometries and spiral similarities. Inversion is well behaved under translations and spiral similarities (meaning that the images of a point and its inverse correspond through the inversion over the image of the circle). So we may assume that the circle of inversion is the unit circle centered at the origin, and so that $z_2\overline{z_2} = z_3\overline{z_3} = z_4\overline{z_4} = 1$. Applying Theorem 3.10 to the Möbius transformation $w \mapsto 1/w$, and using the fact that $\chi(z) = 1/\overline{z}$, we can write

$$(\chi(z), z_2, z_3, z_4) = \left(\overline{z}, \frac{1}{z_2}, \frac{1}{z_3}, \frac{1}{z_4}\right) = (\overline{z}, \overline{z_2}, \overline{z_3}, \overline{z_4}) = \overline{(z, z_2, z_3, z_4)}.$$

If z_2, z_3, z_4 are collinear, we may assume that they lie on the real axis, so $\overline{z_j} = z_j$, $j = 2, 3, 4$. In this case the equality is

$$(\chi(z), z_2, z_3, z_4) = (\overline{z}, z_2, z_3, z_4) = \overline{(z, z_2, z_3, z_4)}.$$

□

Remark In other words, the reflection over a line passing through z_2, z_3, z_4 is defined by the formula $(\chi(z), z_2, z_3, z_4) = \overline{(z, z_2, z_3, z_4)}$, and the same formula defines inversion over the circle passing through z_2, z_3, z_4. That is why we interpret inversion as *reflection over a circle*. And this result allows us to find the formula for the reflection $\chi(z)$ over a circle determined by three points z_2, z_3, z_4 by solving

$$\frac{\chi(z) - z_3}{\chi(z) - z_4} : \frac{z_2 - z_3}{z_2 - z_4} = \frac{\overline{z} - \overline{z_3}}{\overline{z} - \overline{z_4}} : \frac{\overline{z_2} - \overline{z_3}}{\overline{z_2} - \overline{z_4}}.$$

In light of this result, we say that $\chi(z)$ and z are symmetrical with respect to the line or circle determined by z_2, z_3, z_4.

Theorem 3.15 (The Symmetry Principle) *If f is a circular transformation, and if P and P' are points that are symmetrical with respect to a line or circle C, then $f(P)$ and $f(P')$ are symmetrical with respect to the line or circle $f(C)$.*

Proof This is easy to check if f is the reflection over the x-axis, so the problem reduces to checking this property for Möbius transformations. We do this with coordinates. Let f be a Möbius transformation, and let z and z' be symmetrical with respect to the circle or line passing through z_2, z_3, z_4. Then, by Theorem 3.14

$$(z', z_2, z_3, z_4) = \overline{(z, z_2, z_3, z_4)}.$$

This implies, by Theorem 3.10, that

$$(f(z'), f(z_2), f(z_3), f(z_4)) = \overline{(f(z), f(z_2), f(z_3), f(z_4))},$$

Hence, again, by Theorem 3.14, $f(z)$ and $f(z')$ are symmetrical with respect to the circle or line through $f(z_2)$, $f(z_3)$, $f(z_4)$, which is the image through f of the circle or line through z_2, z_3, z_4. The theorem is proved.   □

So, if two points are the image of each other through an inversion with respect to some circle, and you perform another inversion that maps this circle to some circle (or line), then the images of the points correspond under the inversion over this new circle (or under reflection over the line).

Remark The Symmetry Principle allows us to give another proof to Proposition 3.7. Assume that we are given an inversion χ of center O, and a circle ω of center C. Then C and ∞ are reflections of each other over ω. So $\chi(C)$ and $O = \chi(\infty)$ should be reflections of each other over $\chi(\omega)$. Using Proposition 3.3, and depending on whether $\chi(\omega)$ is a line or a circle, and in the latter case depending on the relative position of the circles, we obtain the three cases from Proposition 3.7.

Now we can describe the structure of the circular group.

Theorem 3.16 *Every circular transformation is the composition of reflections over lines and circles.*

Proof Every circular transformation is either a Möbius transformation or the composition of a Möbius transformation and the reflection over the x-axis. So it suffices to prove the property for Möbius transformations. By Theorem 3.13, every Möbius transformation is the composition of translations, spiral similarities, and the complex inversion, and spiral similarities are compositions of rotations and direct homotheties. Translations and rotations are compositions of reflections (Theorem 1.7), and so is the complex inversion. To complete the argument, we will prove synthetically that direct homotheties are compositions of inversions.

Let h be the homothety of center O and ratio $k > 0$. Consider the inversions χ_1 of center O and radius 1 and the inversion χ_2 of center O and radius $\sqrt{k}$. Let also $P \neq O$ be a point in the plane. Then $P' = \chi_1(P) \in |OP$ and $OP' = 1/OP$. If we let $P'' = \chi_2(P') = \chi_2(\chi_1(P))$, then $P'' \in |OP$ and $OP'' = k/OP' = kOP$. Consequently $\chi_2 \circ \chi_1 = h$, and the theorem is proved.   □

This result should be compared with Theorem 1.7.

While Möbius transformations preserve cross-ratios, the argument of the cross-ratio changes to its negative when reflecting over a line or circle. So although cross-ratio is not invariant under circular transformations, its absolute value

$$\frac{|z_1 - z_3|}{|z_1 - z_4|} : \frac{|z_2 - z_3|}{|z_2 - z_4|}$$

is. Real valued cross-ratios *are* invariant under circular transformations. So on the one hand, we have the following result:

Proposition 3.17 *If A, B, C, D are four points in the plane, and A', B', C', D' are their images through an inversion (or in general a circular transformation), then*

$$\frac{AC}{AD} : \frac{BC}{BD} = \frac{A'C'}{A'D'} : \frac{B'C'}{B'D'}.$$

On the other hand, if the four points lie on a circle or a line, then by combining Theorems 3.10, 3.11, and 3.14, we obtain the following result:

Theorem 3.18 (Invariance of Cross-Ratio) *If four points lie on a line or circle, then the cross-ratio of their complex coordinates is invariant under circular transformations.*

As a final observation, we recall that the condition for the points a, b, c, d to form a harmonic quadrilateral or a harmonic division on a line is $(a, c, b, d) = -1$. As a consequence of Theorem 3.18, circular transformations, and in particular inversion, map harmonic quadrilaterals to harmonic quadrilaterals (or to harmonic divisions when the image of the circle is a line). It is this invariance under all circular transformations of the property of a quadrilateral to be harmonic that makes harmonic quadrilaterals so organically related to the geometric transformations studied in this book and explains why they show up in so many problems.

3.1.5 Linear Fractional Transformations of the Real Line

Cross-ratios are probably a more familiar subject when talking about segments on a line, so let us open briefly a parenthesis, which will explain why we use the word "projective" when talking about Möbius and circular transformations. On a line we choose a coordinate system that identifies it with the x-axis and add to it the point at infinity to obtain what is called the *projective real line*. The linear fractional transformations[1] that map $\mathbb{R} \cup \{\infty\}$ into itself are of the form

$$\phi(x) = \frac{ax + b}{cx + d}, \quad a, b, c, d \in \mathbb{R}, ac - bd \neq 0.$$

These linear fractional transformations form what is called the real projective special linear group $PSL(2, \mathbb{R})$. As we have seen, the cross-ratio of four points on a line is invariant under linear fractional transformations of the line.

---

[1] The name Möbius transformation is restricted to the case of complex numbers.

Theorem 3.19 *Given two lines ℓ_1 and ℓ_2 in the plane and a point P on neither of them, define a map $f : \ell_1 \to \ell_2$ by associating to a point $X \in \ell_1$ the point $f(X)$ on ℓ_2 such that $P, X, f(X)$ are collinear; if no such point exists, associate the point at infinity of ℓ_2, and if for some $X' \in \ell_2$, PX' is parallel to ℓ_1, then X' is the image of the point at infinity of ℓ_1. If we fix an origin on each line, thus identifying each line with $\mathbb{R}$, then f is a linear fractional transformation of $\mathbb{R}$.*

Proof 1 If the lines do not intersect, then the map is just a homothety, so let us consider the case where the lines intersect (Fig. 3.9). As translations and dilations are linear fractional transformations, we may assume that on each line the origin is fixed at the intersection O of the two lines, some orientation is chosen, and the real coordinate of a point is specified by its (oriented) distance to the origin. Consider the points $X \in \ell_1$, $X' = f(X) \in \ell_2$ as given by the statement, and let $OP = a$, $OX = x$, $OX' = x'$, $\angle POX = \alpha$, $\angle POX' = \beta$, $\angle XPO = \theta$. Then the Law of Sines in the triangles POX and POX' gives

$$\frac{a}{\sin(\alpha + \theta)} = \frac{x}{\sin \theta} \text{ and } \frac{a}{\sin(\beta + \theta)} = \frac{x'}{\sin \theta}.$$

Using the addition formula the sine function we obtain

$$\frac{a}{x} = \frac{\sin \alpha \cos \theta + \cos \alpha \sin \theta}{\sin \theta} = \sin \alpha \cot \theta + \cos \alpha$$

and similarly

$$\frac{a}{x'} = \sin \beta \cot \theta + \cos \beta.$$

Substituting $\cot \theta$ from the first equation into the second, we obtain

Fig. 3.9 Definition and computation of the map f

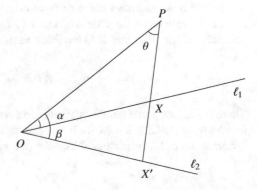

$$x' = f(x) = \frac{a}{\sin\beta \cot\theta + \cos\beta} = \frac{a}{\sin\beta \left(\dfrac{a}{x\sin\alpha} - \cot\alpha\right) + \cos\beta}.$$

Therefore

$$f(x) = \frac{(a\sin\alpha)x}{(\sin\alpha\cos\beta - \cos\alpha\sin\beta)x + a\sin\beta},$$

which is a linear fractional transformation of the real line.   □

The map $X \mapsto X'$ defined in this theorem is called a central projection, and the theorem can be phrased as follows: The central projection of one line to another defines a linear fractional transformation of $\mathbb{R} \cup \{\infty\}$. This is why the adjective "projective" is associated with linear fractional transformations.

Proof 2 We will now give an alternative proof borrowing some ideas from algebraic geometry. The real projective line $\mathbb{R} \cup \{\infty\}$ can be constructed as follows: place an observer at the origin O of the coordinate plane $\mathbb{R}^2$, whose points are parameterized by two real numbers. Each point $(x_1, x_2) \in \mathbb{R}^2$ different from the origin can be interpreted as a direction of view of the observer, and these directions of view allow the observer to perform a central projection from one line that lies in $\mathbb{R}^2$ and does not pass through the origin to another. Even the point at infinity of one line can be projected to the other. So the points of the real projective line are in one-to-one correspondence with the directions of view. To enhance your intuition, we have illustrated this in Fig. 3.10, with the directions of view represented as dotted lines (note the direction parallel to ℓ_1 that allows us to project $\infty \in \ell_1$ onto ℓ_2).

Each direction of view is a line passing through the origin, of the form $\{(tx_1, tx_2) \mid t \in \mathbb{R}\}$. Because of this, the real projective line can be identified with the set of equivalence classes of pairs of real numbers (x_1, x_2), with x_1 and x_2 not

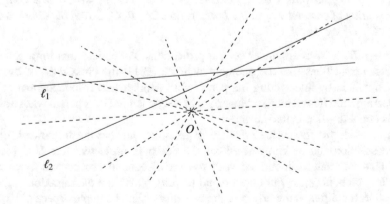

Fig. 3.10 A construction of the real projective line $\mathbb{R} \cup \{\infty\}$

simultaneously equal to zero, where (x_1, x_2) is equivalent to (x_1', x_2') if and only if there is a nonzero real number t such that $x_1' = tx_1$ and $x_2' = tx_2$.

The equivalence class of (x_1, x_2) is denoted by $[x_1 : x_2]$. With the reference line ℓ_1 of equation $x_2 = 1$, we notice that whenever $x_2 \neq 0$, the line $\{(tx_1, tx_2) \mid t \in \mathbb{R}\}$ intersects ℓ_1 at $(x_1/x_2, 1)$. So the map

$$[x_1 : x_2] \mapsto (x_1/x_2 : 1), \quad [1 : 0] \mapsto \infty$$

is a bijection from the set of directions of view to the real projective line, and it is realized geometrically by intersecting the directions of view with the line ℓ_1.

Now suppose that ℓ_2 is another line in $\mathbb{R}^2$ that does not pass through the origin, and let us perform the central projection from O of ℓ_1 onto ℓ_2. We can do this as follows: map ℓ_1 back to the set of directions of view, rotate the directions of view and dilate until ℓ_2 becomes the line $x_2 = 1$, and then project onto ℓ_2. The rotation+dilation is realized by a linear map

$$\mathbb{R}^2 \to \mathbb{R}^2, \quad (x_1, x_2) \mapsto (ax_1 + bx_2, cx_1 + dx_2), \text{ where } ad - bd \neq 1,$$

which realizes the central projection of ℓ_1 onto ℓ_2 as the map

$$\mathbb{R} \cup \{\infty\} \to \mathbb{R} \cup \{\infty\}, \quad [x_1 : x_2] \mapsto [ax_1 + bx_2 : cx_1 + dx_2].$$

For $x_2 \neq 0$ this map is $[x_1/x_2 : 1] \mapsto [(ax_1 + bx_2)/(cx_1 + dx_2) : 1]$, which after setting $x = x_1/x_2$ becomes $[x : 1] \mapsto [(ax + b)/(cx + d) : 1]$, a linear fractional transformation. $\qquad\qquad\square$

Here is an immediate corollary of the invariance of cross-ratios under linear fractional transformations.

Theorem 3.20 *Let ℓ_1 and ℓ_2 be two lines, and let P be on neither. Let also A, B, C, D be four points on ℓ_1, and let A', B', C', D' be their images through the projection of center P. Then the cross-ratio of A, B, C, D and the cross-ratio of A', B', C', D' are equal.*

Cross-ratio is invariant under central projections. We deduce that for a pencil of four concurrent lines like the one shown in Fig. 3.11, the cross-ratio of the four points obtained by intersecting the pencil with another line is independent of this intersecting line. It is an intrinsic invariant of the pencil itself. A pencil whose cross-ratio is equal to -1 is called *harmonic*.

As an aside for the curious reader, let us point out that the framework of the second solution can be constructed for the complex projective line $\mathbb{C} \cup \{\infty\}$ as well. This line can be identified with the set of equivalence classes of pairs of complex numbers (z_1, z_2) not both equal to zero, which are the directions of view of an observer sitting at the origin of the two-dimensional complex space $\mathbb{C}^2$, where (z_1, z_2) is equivalent to (z_1', z_2') if and only if there is a nonzero complex number λ such that $z_1' = \lambda z_1$ and $z_2' = \lambda z_2$. The equivalence class of (z_1, z_2) is denoted by

Fig. 3.11 A pencil of four
lines crossed by two lines

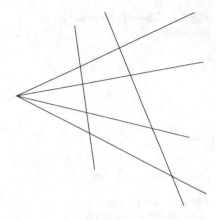

$[z_1 : z_2]$, and it is important to notice that whenever $z_2 \neq 0$, $[z_1 : z_2] = [z_1/z_2 : 1]$. Hence, $\mathbb{C}$ is identified with the set of points of the form $[z : 1]$ and ∞ is the point $[1 : 0]$. In this framework, the Möbius transformation $z \mapsto \frac{az+b}{cz+d}$ assumes the form $[z_1 : z_2] \mapsto [az_1 + bz_2 : cz_1 + dz_2]$. It arises from "projecting" a linear transformation of the two-dimensional complex space onto the complex projective line!

3.1.6 The Invariance of Angles

Like isometries and spiral similarities, inversions preserve angles. But because lines are usually transformed into circles, we have to look at angles in a more general sense. Given two curves intersecting at some point, if the tangent lines to the curves at that point exist, then the angle between the curves is, by definition, the angle formed by their tangents. For two intersecting lines, this recovers the classical definition of angles. And the angle between two intersecting circles is the angle between the tangents at any of the two intersection points (the two angles are equal by symmetry). This is illustrated in Fig. 3.12.

In general, one defines the angle between two intersecting curves using coordinates as follows. Parametrize the curves as $\gamma_1, \gamma_2 : [0, 1] \to \mathbb{C}$, so they intersect at some $t_0 \in (0, 1)$, meaning that $\gamma_1(t_0) = \gamma_2(t_0)$. The curves have tangents at t_0 if they are differentiable with respect to t and have nonzero derivative at t_0, meaning that

$$\frac{d\gamma_j}{dt}(t_0) = \lim_{h \to 0} \frac{\gamma_j(t_0 + h) - \gamma(t_0)}{h}, \quad j = 1, 2,$$

exist and are nonzero numbers. The angle between the curves is the angle between the tangents and is

Fig. 3.12 The angle between
two circles

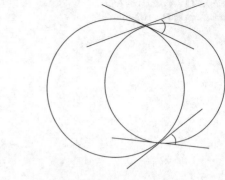

Fig. 3.13 Inversion
preserves the angle between a
curve and a ray through O

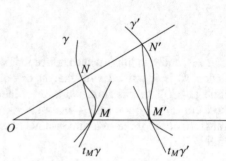

$$\arg \frac{d\gamma_2}{dt}(t_0) - \arg \frac{d\gamma_1}{dt}(t_0).$$

In this book we are only concerned with the angles formed by circles and lines, in
which case we can rely on synthetic geometry; we use this general definition only
for the proof of the results in this section. We denote the tangent line to the curve γ
at the point X by $t_X\gamma$.

Theorem 3.21 (Invariance of Angles) *The angle formed by two curves is invariant
under inversion, but the orientation of angles is reversed.*

Proof Let the center of inversion be O. Let also $\gamma : [0, 1] \to \mathbb{C}$ be a curve that has
a tangent at $M = \gamma(t_0)$ for some $t_0 \in (0, 1)$. We will show that the angle between γ
and OM is invariant under inversion. The argument can be followed on Fig. 3.13.

Using dashes for inverted points and curves, we have

$$\angle(OM, t_M\gamma) = \lim_{N \to M} \angle(OM, MN) = \lim_{N \to M} \angle OMN = \lim_{N' \to M'} \angle ON'M'$$

$$= \lim_{N' \to M'} (ON', N'M') = \angle(OM', t_{M'}\gamma').$$

Here we have used tacitly the property that if γ is differentiable, then its image γ'
is differentiable as well; in other words, if γ admits a tangent line at a point, the so

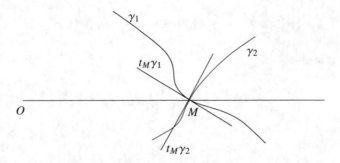

Fig. 3.14 Proof that inversion preserves angles

does its image. And we used the fact that if $N \in \gamma$ and N approach M, then $N' \in \gamma'$ and N' approach M'.

Next, consider the line OM connecting the center of inversion with the intersection point of the two curves, as shown in Fig. 3.14. Using what we just proved, we obtain

$$\angle(t_M \gamma_2, t_M \gamma_1) = \angle(t_M \gamma_2, OM) + \angle(OM, t_M \gamma_1)$$
$$= \angle(t_{M'} \gamma_2', OM') + \angle(OM', t_{M'} \gamma_1') = \angle(t_{M'} \gamma_2', t_{M'} \gamma_1').$$

$\square$

If an inversion maps M, N, P to M', N', P', is it true that $\angle MNP = \angle M'N'P'$? Absolutely not. The segments MN and NP are mapped into arcs of circles. What we have proved is that the angle between these arcs is equal to the angle $\angle MNP$.

As a corollary of Theorems 3.21 and 3.16, we obtain the following.

Theorem 3.22 (Invariance of Angles) *Circular transformations preserve angles (if the orientation of angles is ignored).*

Here is a short proof of this result using complex analysis; the reader unfamiliar with this technique can skip it.

Proof Since every circular transformation is either a Möbius transformation, or the composition of a Möbius transformation and the reflection over the x-axis, it suffices to check the result for Möbius transformations.

Let $\gamma_1, \gamma_2 : [0, 1] \to \mathbb{C}$ be the parametrization of two curves that intersect at $z_0 = \gamma_1(t_0) = \gamma_2(t_0)$, $t_0 \in (0, 1)$. Let also ϕ be a Möbius transformation. The angle between γ_1 and γ_2 at t_0 is

$$\arg \frac{d\gamma_2}{dt}(t_0) - \arg \frac{d\gamma_1}{dt}(t_0),$$

while the angle between their images is

$$\arg \frac{d\phi \circ \gamma_2}{dt}(t_0) - \arg \frac{d\phi \circ \gamma_1}{dt}(t_0).$$

The chain rule gives

$$\frac{d\phi \circ \gamma_j}{dt}(t_0) = \frac{d\phi}{dz}(z_0)\frac{d\gamma_j}{dt}(t_0), \quad j = 1, 2,$$

so

$$\arg \frac{d\phi \circ \gamma_j}{dt}(t_0) = \arg \frac{d\phi}{dz}(z_0) + \arg \frac{d\gamma_j}{dt}(t_0), \quad j = 1, 2.$$

Thus ϕ rotates both the tangent to γ_1 and the tangent to γ_2 by the same angle $\arg \frac{d\phi}{dz}(z_0)$. This means that the angle between the tangents is preserved under ϕ.

$\square$

We point out that if

$$\phi(z) = \frac{az + b}{cz + d},$$

then

$$\frac{d\phi}{dz} = \frac{ad - bc}{(cz + d)^2}.$$

For Möbius transformations, we have the stronger result.

Theorem 3.23 *Möbius transformations preserve angles, including their orientation.*

Circular transformations, and in particular inversion, map tangent curves to tangent curves and orthogonal curves to orthogonal curves. Two curves are called orthogonal if their tangents at the intersection point form a right angle.

3.1.7 Inversion with Negative Ratio

Let us point to a particular type of circular transformation, known as *inversion with negative ratio*. For $R > 0$, the inversion of ratio $-R^2$ and center O is the transformation that maps every point P in the plane to a point P' such that O belongs to the segment PP' and $OP \cdot OP' = R^2$. This circular transformation is the composition of

- the inversion of center O and radius R, which maps P to $P'' \in |OP$ such that $OP \cdot OP'' = R^2$

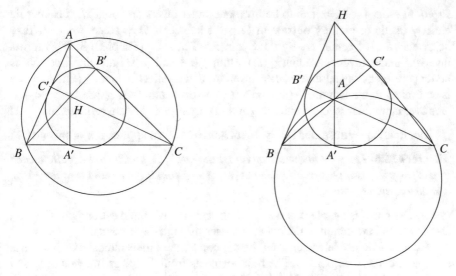

Fig. 3.15 The nine-point circle is the image of the circumcircle through an inversion

- the reflection over O.

This transformation is of interest because it appears in the geometry of the triangle. Let ABC be a triangle that is not right, and let H be its orthocenter. Then

$$AH \cdot HA' = HB \cdot HB' = HC \cdot HC' = 4R^2 \cos A \cos B \cos C,$$

where R is the circumradius. Thus A', B', C' are the images of A, B, C, respectively, through an inversion of center H and ratio equal to $-4R^2 \cos A \cos B \cos C$. This ratio is negative if the triangle is acute and positive if the triangle is obtuse. We thus have the following result (see also Fig. 3.15).

Proposition 3.24 *The nine-point circle of a triangle that is not right is the image of the circumcircle through an inversion whose center is the orthocenter and whose ratio is negative if the triangle is acute and positive if the triangle is obtuse.*

3.1.8   Circles Orthogonal to the Circle of Inversion

As said above, two circles are orthogonal if their tangent lines at either of the two intersection points form a right angle. Two circles are orthogonal if the radii of one circle at the points of intersection are tangent to the other circle.

Proposition 3.25 *A circle that is not the circle of inversion is mapped into itself by the inversion if and only if it is orthogonal to the circle of inversion.*

Proof In order for a circle ω to be invariant under inversion about Ω, it cannot lie entirely inside or entirely outside of Ω, so it intersects Ω at some point O. Take an inversion χ of center O. By the Symmetry Principle (Theorem 3.15), $\chi(\omega)$ is invariant under inversion about $\chi(\Omega)$. Both $\chi(\omega)$ and $\chi(\Omega)$ are lines, so this is better phrased as saying that $\chi(\omega)$ is invariant under reflection over $\chi(\Omega)$. And this is equivalent to the fact that $\chi(\omega)$ and $\chi(\Omega)$ are orthogonal. As χ preserves angles, ω is invariant under inversion about Ω if and only if ω and Ω are orthogonal. $\square$

The situation is illustrated in Fig. 3.16. And here is an application of this fact.

Theorem 3.26 *Let χ be an inversion about the circle Ω, and let ω be a circle such that $\omega' = \chi(\omega)$ is also a circle. Then Ω, ω, ω' are coaxial, meaning that they have a common radical axis.*

Proof The proof is simple if ω and Ω intersect. The two intersection points belong also to ω', and the common radical axis passes through these points.

If Ω and ω do not intersect, let T be the point where their radical axis intersects the line of centers (Fig. 3.17). The four segments from T that are tangent to Ω and ω have equal lengths $\sqrt{p}$, where p is the power T with respect to the two circles. So the four tangency points lie on a circle γ that is centered at T and orthogonal to Ω and ω. This circle is invariant under inversion and remains orthogonal to ω' because inversion preserves angles. Hence, the segment from T that is tangent to ω'

Fig. 3.16 Circle orthogonal to the circle of inversion

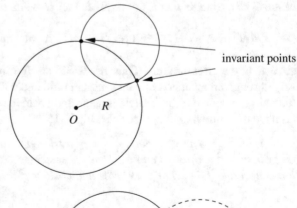

invariant points

Fig. 3.17 The circle of inversion, a circle, and its inverse are coaxial

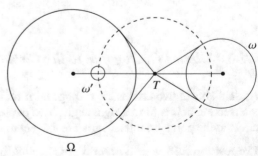

is a radius in γ, so it has length $\sqrt{p}$. This implies that the power of T with respect to ω' is also p, and so T lies on the common radical axis of the three circles. □

3.1.9 The Limiting Points of Two Circles

A surprising result:

Theorem 3.27 *Given two nonintersecting circles, there is an inversion that maps them into two concentric circles.*

Proof 1 Two concentric circles are characterized by the fact that there are infinitely many lines but no circles that are orthogonal to both. The lines orthogonal to the two circles pass through the common center, and they actually have two common points: the center of the circles and the point at infinity. From this observation, we deduce that it is possible to map a pair of nonintersecting circles ω_1 and ω_2 by an inversion into a pair of concentric circles if and only if *all* circles orthogonal to both ω_1 and ω_2 have two common points (in which case one of the points is the center of inversion). A configuration is shown in Fig. 3.18.

We will thus prove that for every two nonintersecting circles, ω_1 of center O_1 and ω_2 of center O_2, the circles that are orthogonal to both have two common points. Any circle that is orthogonal to both ω_1 and ω_2 must intersect the line $O_1 O_2$ at two points, and for such a circle ω, whose center we denote by O, let P and Q be its intersections with $O_1 O_2$ (Fig. 3.19). Then OP^2 is the power of O with respect to both circles. This is because the radius of ω at the point where it intersects ω_1 is tangent to ω_1 (by the orthogonality condition), and the same is true for ω_2. Hence, O is on the radical axis of ω_1 and ω_2. This radical axis is therefore OM, where M is the projection of O onto $O_1 O_2$. So the projection M of O on $O_1 O_2$ does not

Fig. 3.18 The circles that are orthogonal to two nonintersecting circles have two common points

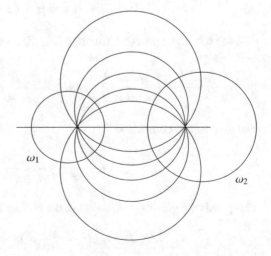

Fig. 3.19 Finding the center
of inversion that makes two
circle concentric

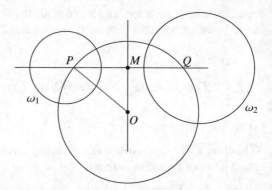

depend on ω, only on the original circles, because it is where the radical axis of the two circles intersects the line of centers.

The power of O with respect to ω_1 and ω_2, being equal to OP^2, is also the power of O with respect to the degenerated circle of center P and radius 0. So OM is in fact the radical axis of the triple of circles ω_1, ω_2, and P. This means that MP^2 is the power of M with respect to both ω_1 and ω_2, and the same is true if we replace P by Q. Hence, the pair $\{P, Q\}$ is uniquely determined by ω_1 and ω_2, and so all orthogonal circles pass through both these points, and the theorem is proved.   □

Proof 2 Let us see the proof by complex numbers. Instead of varying the center of inversion, we keep it fixed at the origin 0 of the coordinate system and vary the centers of the circles on the x-axis while keeping the distance between their centers fixed. Let the distance between the centers be d, and let the first circle have center x and radius r_1 and the second circle have center $d + x$ and radius r_2. Let the radius of inversion be R. Then the inversion is the map $z \mapsto R^2/\bar{z}$. The two circles have the equations

$$|z - x| = r_1 \text{ and } |z - (x + d)| = r_2$$

and, as seen in the proof of Theorem 3.5, their inverses have the equations

$$\left| w - \frac{R^2 x}{x^2 - r_1^2} \right| = \left(\frac{R^2 r_1}{x^2 - r_1^2} \right)^2 \text{ and } \left| w - \frac{R^2(x + d)}{(x + d)^2 - r_2^2} \right| = \left(\frac{R^2 r_2}{(x + d)^2 - r_2^2} \right)^2.$$

The two centers coincide if

$$\frac{R^2 x}{x^2 - r_1^2} = \frac{R^2(x + d)}{(d + x)^2 - r_2^2}.$$

This yields the quadratic equation in the real variable x

$$dx^2 - (d^2 + r_1^2 - r_2^2)x + dr_1^2 = 0.$$

The discriminant of this equation is

$$(d^2 + r_1^2 - r_2^2) - 4d^2 r_1^2 = (d^2 + r_1^2 - r_2^2 + 2dr_1)(d^2 + r_1^2 - r_2^2 - 2dr_1)$$
$$= [(d + r_1)^2 - r_2^2][(d - r_1)^2 - r_2^2]$$
$$= (d + r_1 + r_2)(d + r_1 - r_2)(d - r_1 + r_2)(d - r_1 - r_2)$$
$$= (d + r_1 + r_2)[d^2 - |r_1 - r_2|^2](d - r_1 - r_2).$$

The circles do not intersect if either $d > r_1 + r_2$ or $d < |r_1 - r_2|$. In the first case, all factors are positive; in the second case, the last two factors are negative. So if the circles do not intersect, the equation has exactly two solutions. As the center of inversion must be on the line of circles, this algebraic computation shows that there are exactly two centers of inversion that map the circles into concentric circles. Any radius of inversion would work. This concludes the analytic proof.   □

Definition The two points that are the centers of the inversions that map the two nonintersecting circles to concentric circles are called the *limiting points* of the two circles.

This theorem has an important corollary. Given two nonintersecting circles ω_1 and ω_2, a Steiner chain is a sequence of circles tangent to both ω_1 and ω_2, and such that each circle is also tangent to the previous one. If the sequence is finite, and the first and the last circles are also tangent, the Steiner chain is called closed. An example is shown in Fig. 3.20.

Theorem 3.28 (Steiner's Porism) *If for two nonintersecting circles ω_1 and ω_2 a closed Steiner chain of n circles exists, then any circle that is tangent to both ω_1 and ω_2 is part of such a closed Steiner chain of n circles.*

Proof We can map ω_1 and ω_2 by an inversion to concentric circles, without changing the conclusion. But now the statement is trivial, since the closed Steiner chains are mapped into one another by rotations about the common center of ω_1 and ω_2.   □

Fig. 3.20 Example of a closed Steiner chain

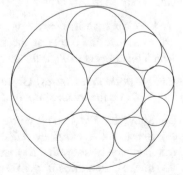

3.1.10 Problems with Theoretical Flavor About Properties of Inversion and Möbius Transformations

The understanding of inversion, and more generally of Möbius and circular transformations, can be enhanced by solving the following problems.

157 (a) What is the composition of two inversions with the same center?
(b) What is the composition of an inversion and a homothety with the same center?
(c) What is the composition of a homothety and an inversion with the same center?

158 Show that two points are reflections of each other over a circle if and only if the circle is orthogonal to every circle passing through the two points.

159 Let A, B, P be distinct points in the plane such that P is not the midpoint of the segment AB. Show that there is an inversion that maps A, B, P to A', B', P', respectively, such that P' is the midpoint of $A'B'$.

160 Let AB be a chord of a circle ω, and let χ be an inversion. Set $\omega' = \chi(\omega)$, $A' = \chi(A)$, and $B' = \chi(B)$. Show that AB and $A'B'$ intersect on the radical axis of ω and ω'.

161 Show that if two circles are tangent to each other and orthogonal to a third circle, ω, then their tangency point lies on ω.

162 Given the pairwise exterior tangent circles ω_1, ω_2, and ω_3, show that one and only one of the following situations occurs:

 (i) there is a unique circle which is exterior tangent to all three circles and a unique circle to which the three circles are interior tangent;
 (ii) ω_1, ω_2, and ω_3 have a common tangent and there is a unique circle that is tangent to all three circles, this circle being exterior tangent to them;
(iii) there are exactly two circles that are exterior tangent to the three circles and no circle is interior tangent.

163 Let A, B, C, D be four points, no three of them collinear. Prove that the angle between the circumcircles of ABC and ABD is equal to the angle between the circumcircles of ACD and BCD.

164 Prove that if a circle is orthogonal to two circles ω_1 and ω_2, then it is orthogonal to the circle of the inversion that maps ω_1 to ω_2 (or to the line over which ω_1 reflects to ω_2 if the circles have equal radii).

165 *(Apollonian Circles)* Given two points A and B and a positive number k, let $\Gamma_k(A, B)$ be the locus of points P in the plane such that $PA/PB = k$.

(a) Show that $\Gamma_k(A, B)$ is a line for $k = 1$ and a circle for $k \neq 1$.
(b) Show that A and B are reflections of each other over $\Gamma_k(A, B)$.
(c) Conversely, show that if A and B are reflections of each other over the circle (or line) Γ, then there is $k > 0$ such that $\Gamma = \Gamma_k(A, B)$.

166 Find all Möbius transformations that map the upper half-plane $\{z \mid \text{Im } z > 0\}$ to itself.

167 Prove that every transformation of $\mathbb{C} \cup \{\infty\}$ that maps every line and circle to a line or a circle is a circular transformation. (It is understood that every line contains $\{\infty\}$).

168 To a polynomial $P(x) = ax^3 + bx^2 + cx + d$ with real coefficients, of degree at most 3, one can apply the following two operations: (i) swap simultaneously a and d, respectively, b and c, and (ii) translate the variable x to $x + t$, where t is a real number.

(a) Can one transform, by a successive application of these operations, the polynomial $P_1(x) = x^3 + x^2 - 2x$ into the polynomial $P_2(x) = x^3 - 3x - 2$?
(b) What if we replace $P_2(x)$ by $x^3 - 3x - 3$?
(c) Allow a third operation in which all coefficients of a polynomial can be multiplied simultaneously by the same nonzero real number, and replace $P_2(x)$ by $x^3 - 3x - 1$. Can we now transform $P_1(x)$ into this polynomial?

3.2 Inversion in Euclidean Geometry Problems

3.2.1 Applications of Inversion to Proving Classical Results

We present one important construction and two groupings of theorems, as excursions through old-fashioned geometry, guided by inversion. While the results do have well-known proofs by other means, we think that the inversive proofs shown here can deepen their understanding.

I. Pole and Polar with Respect to a Circle
In this section we need a result which, by itself, does not involve inversion.

Lemma 3.29 (Poncelet) *Let Ω be a circle of center O, and let M be a point that does not belong to the circle. Through M, we take a variable line that intersects the circle at M_1 and M_2, and let P be the harmonic conjugate of M with respect to M_1 and M_2 (meaning that the points M, P, M_1, M_2 form a harmonic division). Then the locus of P is*

(i) a line p_M perpendicular to OM if M lies inside Ω;
(ii) the segment that lies inside Ω of a line p_M that is perpendicular to OM.

Proof The two situations are shown in Fig. 3.21. Let N be the midpoint of the segment MP. Then the relation that expresses the fact that M, P, M_1, M_2 form a harmonic division

$$\frac{MM_1}{MM_2} = \frac{PM_1}{PM_2}$$

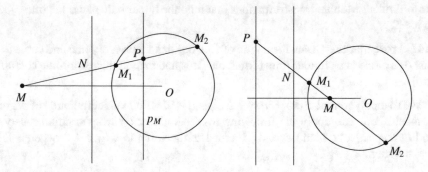

Fig. 3.21 Proof of Poncelet's lemma

can be rewritten as

$$\frac{MN \pm NM_1}{NM_2 \pm MN} = \frac{PN \mp NM_1}{NM_2 \mp PN},$$

with the signs depending on whether M is outside or inside the circle Ω.

Using the fact that $MN = PN$, we obtain $NM^2 = NM_1 \cdot NM_2$, showing that N is on the radical axis of the circle Ω and the point M (the point is a degenerate circle). But P is the image of N under a homothety of center M and ratio 2, so the locus is the image p_M of this radical axis, or more precisely the intersection of p_M with the interior of the circle if M is exterior to Ω, and the entire line if M is exterior. And p_M is orthogonal to OM, as the radical axis is. $\qquad\square$

Definition The line p_M is called the polar of M with respect to Ω, and M is called the pole of p_M with respect to Ω. If M is on the circle, its polar with respect to the circle is the tangent at M.

It is not hard to see that given a circle every line has a pole and every point has a polar with respect to the circle. It is at this moment that we exhibit the connection with inversion, with the following straightforward corollary to Proposition 3.1.

Proposition 3.30 *The polar of a point M with respect to a circle Ω of center O is the line perpendicular to OM passing through the inverse of M with respect to Ω.*

From here we deduce immediately that the polar of a point that lies exterior to a circle passes through the tangency points of the tangents from the point to the circle. Here is another consequence.

Theorem 3.31 (La Hire) *Considering polars with respect to a circle, $A \in p_B$ if and only if $B \in p_A$.*

Proof The proof can be followed on Fig. 3.22. Let A', B' be the inverses of A and B with respect to Ω, respectively. Then the similarity of the triangles OAB and $OB'A'$ implies that the quadrilateral $AA'B'B$ is cyclic. Then $\angle AB'B = \angle BA'A$, so

Fig. 3.22 Proof of La Hire's theorem

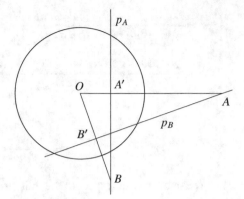

the fact that one of these angles is right is equivalent to the fact that the other is right. But $\angle AA'B = 90°$ means that $B \in p_A$ and $\angle BB'A = 90°$ means that $A \in p_B$. The theorem is proved. □

Proposition 3.32 *Three points are collinear if and only if their polars with respect to a circle are concurrent.*

Proof Using La Hire's Theorem, we deduce that the polars of three collinear points pass through the pole of the line determined by the points. Conversely, if three lines are concurrent, the polar of their intersection passes through their poles. □

Theorem 3.33 *Let A, B, C, D be four concyclic points, and let $M = AC \cap BD$, $N = AB \cap CD$, and $P = AD \cap BC$. Then $p_M = NP$, $p_N = MP$, and $p_P = MN$, the polars being considered with respect to the circle passing through the four points.*

Proof 1 To prove that $p_M = NP$ amounts to show that $N \in p_M$ and $P \in p_M$. Moreover, showing that $N \in p_M$ is the same as showing that NM' is perpendicular to OM, where M' is the inverse of M with respect to the circle, and O is the center of the circle.

Consider a system of coordinates with O at the origin so that the circle passing through A, B, C, D is the unit circle. The coordinates of these four points satisfy

$$a\bar{a} = b\bar{b} = c\bar{c} = d\bar{d} = 1.$$

In complex coordinates, the point of intersection of the lines passing through the pairs of points (z_1, z_2) and (z_3, z_4) has the coordinate

$$\frac{((\bar{z_2} - \bar{z_1})z_1 - (z_2 - z_1)\bar{z_1})(z_4 - z_3) - ((\bar{z_4} - \bar{z_3})z_3 - (z_4 - z_3)\bar{z_3})(z_2 - z_1)}{(z_4 - z_3)(\bar{z_2} - \bar{z_1}) - (z_2 - z_1)(\bar{z_4} - \bar{z_3})}.$$

Using the fact that a, b, c, d have absolute value 1, we can simplify the formulas for m and n as

$$m = \frac{(\bar{b}a - b\bar{a})(d-c) - (\bar{d}c - d\bar{c})(b-a)}{(d-c)(\bar{b}-\bar{a}) - (b-a)(\bar{d}-\bar{c})} = \frac{\left(\frac{a}{b} - \frac{b}{a}\right)(d-c) - \left(\frac{c}{d} - \frac{d}{c}\right)(b-a)}{(d-c)\left(\frac{1}{b} - \frac{1}{a}\right) - \left(\frac{1}{d} - \frac{1}{c}\right)(b-a)}$$

$$= \frac{(a-b)(c-d)(c+d-a-b)\frac{1}{abcd}}{(a-b)(c-d)(ab-cd)\frac{1}{abcd}} = \frac{c+d-a-b}{ab-cd},$$

and

$$n = \frac{b+d-a-c}{ac-bd}.$$

The inverse of m with respect to the unit circle is

$$m' = \frac{1}{\bar{m}} = \frac{\bar{a}\bar{b} - \bar{c}\bar{d}}{\bar{c}+\bar{d} - \bar{a} - \bar{b}}.$$

To show that NM' is perpendicular to OM is equivalent to showing that the ratio of $n - m'$ and m is imaginary, which is the same as showing that the product of $n - m'$ and $\bar{m}$ is imaginary (because $(n-m')/m = (n-m')\bar{m}/(m\bar{m})$, whose denominator is real). We compute

$$(n-m')\bar{m} = n\bar{m} - \frac{1}{m} \cdot \bar{m} = \frac{b+d-a-c}{ac-bd} \cdot \frac{\bar{c}+\bar{d}-\bar{a}-\bar{b}}{\bar{a}\bar{b}-\bar{c}\bar{d}} - 1$$

$$= \frac{b\bar{c} + b\bar{d} - b\bar{a} + d\bar{c} - d\bar{a} - d\bar{b} - a\bar{c} - a\bar{d} + a\bar{b} - c\bar{d} + a\bar{c} + c\bar{b}}{c\bar{b} - a\bar{d} - d\bar{a} + b\bar{c}} - 1,$$

where we have used the fact that a, b, c, d have absolute value 1. Now bring to the common denominator, reduce terms, and rearrange to obtain

$$\frac{b\bar{d} - \bar{b}d + a\bar{b} - \bar{a}b + d\bar{c} - \bar{d}c + c\bar{a} - \bar{c}a}{c\bar{b} + \bar{c}b - a\bar{d} - \bar{d}a}.$$

The numerator is imaginary because it is a sum of numbers of the form $z - \bar{z}$, while the denominator is real being the sum of numbers of the form $z + \bar{z}$. Hence, the quotient is imaginary, and we are done.   □

Proof 2 There is a quick proof that has homothety at its heart. Let us assume that we are in the configuration from Fig. 3.23, and let us show that MP is the polar of N (the other two cases are similar). Denote by R and Q the intersections of PM with AB and CD, respectively. Using again complex coordinates, the theorems of Menelaus (Theorem 2.15) and Ceva (Problem 116) imply that

$$\frac{n-b}{n-a} = -\frac{q-b}{q-a},$$

Fig. 3.23 MP is the polar of
N

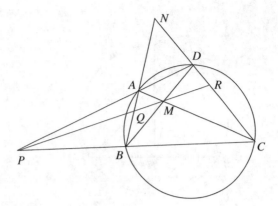

which in the context of those theorems means that the direct homothety of center N
that maps A to B and the inverse homothety of center Q that maps A to B have the
same ratio in absolute value. This means that N, A, Q, B form a harmonic division,
and the same is true for N, D, R, C. By definition, QR is the polar of N with respect
to the circle, and we are done.   □

The second proof shows that N, A, Q, B form a harmonic division on AB
regardless of whether the quadrilateral is cyclic or not. It is worth mentioning that a
more appropriate placement of the notion of pole and polar, as well as of the above
theorem, is within the field of real two-dimensional projective geometry. But that
lies outside of our discussion.

II. The Theorems of Simson-Wallace and Salmon
Let us turn our attention to some classical theorems, which we place in the context
of inversion.

Theorem 3.34 (Simson-Wallace) *Let ABC be a triangle and let M be an arbitrary point in the plane. Let also N, P, Q be the projections of M on the lines BC,
AC, and AB, respectively. Then the points N, P, Q are collinear if and only if M
is on the circumcircle of ABC.*

Proof In the proof, which can be followed on Fig. 3.24, we use the fact that
inversion preserves the angles but changes their orientation. In this figure, on the
left is the original configuration, and on the right is the inverted one. To avoid
case dependency, everywhere in the proof we will be working with directed angles
modulo 180°.

Consider an inversion of center M and arbitrary radius, and use the standard
notation for the inverse of a point. We have

$$\angle N'C'P' = \angle N'C'M + \angle MC'P' = -\angle CNM - \angle MPC = 0,$$

which shows that the points C', N', P' are collinear. Similarly the points A', P', Q'
are collinear. The fact that N, P, Q is collinear is equivalent to $\angle NPQ = 0$. We

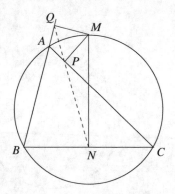 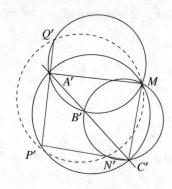

Fig. 3.24 Proof of the Simson-Wallace theorem

have the following sequence of equivalences:

$$\angle NPQ = 0 \iff \angle MPN = \angle MPQ \iff \angle MN'P' = \angle MQ'P'$$
$$\iff \angle(MN', MC') + \angle(MC', C'P') = \angle(MQ', MA') + \angle(MA', A'P')$$
$$\iff \angle N'MC' = \angle Q'MA' \iff \angle NMC = \angle QMA$$
$$\iff \angle(MN, BC) + \angle(BC, MC) = \angle(MQ, AB) + \angle(AB, MA)$$
$$\iff \angle BCM = \angle BAM.$$

The last condition is equivalent to the fact that A, B, C, M are concyclic, as these points are not collinear.   □

If M is on the circumcircle of ABC, so that the points N, P, Q are collinear, then the line they determine is called the *Simson line* corresponding to M with respect to the triangle ABC.

This proof is more complicated than the standard proof based on cyclic quadrilaterals and angle chasing. But, by examining the inverted figure, we are led naturally to another theorem. In the inverted figure, N', P', and Q' lie diametrically opposite to M in the circles that are the images of BC, AC, and AB, respectively. This is because MN, MP, and MQ are orthogonal to the respective sides, and so MN', MP', and MQ' are orthogonal to the respective circles, and hence are diameters in these circles.

Examining thus the figure, we observe the four endpoints of the diameters MN', MP', and MQ' of some three circles that lie on a circle themselves, and also we notice that the pairwise intersection points of these three circles are collinear. This brings us to the next theorem.

Theorem 3.35 (Salmon) *Let $ABCD$ be a cyclic quadrilateral, and let M, N, P be the points where the circles of diameters AB and AC, AB and AD, and AC and AD intersect the second time, respectively. Then M, N, P are collinear.*

Fig. 3.25 Proof of Salmon's
theorem

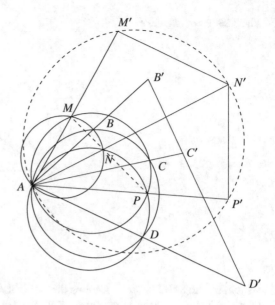

Proof Of course this is the inverted figure of the Simson-Wallace Theorem, and so
the result should follow immediately by inversion from that theorem. Arguing on
Fig. 3.25, take an inversion of center A and arbitrary radius. Then the circumcircle
of $ABCD$ transforms into the line $B'C'D'$, and the circles of diameters AB, AC,
and AD transform into the lines $M'N'$, $M'P'$, and $N'P'$, respectively, and these
lines are orthogonal to the corresponding diameters. The perpendiculars from A to
the sides of the triangles $M'N'P'$, which are the points B', C', D', are collinear. By
the Simson-Wallace Theorem, A is on the circumcircle of triangle $M'N'P'$, which
implies that M, N, P are collinear, and we are done.   □

III. The Theorems of Euler, Tzitzeica, and Poncelet
We commence our next essay with Euler's relation.

Theorem 3.36 (Euler's Relation) *Let* R, r, O, *and* I *be the circumradius, the*
inradius, the circumcenter, and the incenter of a triangle. Then

$$OI^2 = R^2 - 2rR.$$

Proof In the triangle ABC, let D, E, F be the tangency points of the incircle and
the sides BC, CA, AB, respectively (see Fig. 3.26). Consider the inversion of center
I and radius r. We use the convention that the inverse of a point X is denoted by X'.
   In the right triangle EAI, we have

$$IA \cdot IA' = r^2 = IE^2,$$

Fig. 3.26 Proof of Euler's relation

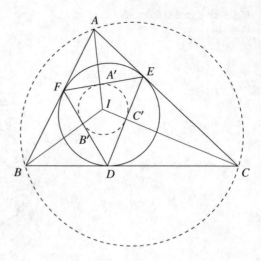

which, by the Leg Theorem, implies that A' is the foot of the altitude from E in this triangle. Thus A' is the midpoint of EF. Similarly B' is the midpoint of FD, and C' is the midpoint of ED. From here we deduce that the inverse of the circumcircle of ABC is the nine-point circle of DEF. The radius of the nine-point circle is $r/2$, namely, half of the circumradius. Using Proposition 3.6, we obtain

$$\frac{r}{2} = \frac{r^2}{\rho} R,$$

where ρ is the power of I with respect to the circumcircle of ABC. By computing the power of I using the diameter of the circumcircle that passes through it, we obtain $\rho = (R + OI)(R - OI) = R^2 - OI^2$. Therefore

$$\frac{r}{2} = \frac{r^2}{R^2 - OI^2} R,$$

from which we derive $R^2 - OI^2 = 2rR$, and finally Euler's relation: $OI^2 = R^2 - 2rR$.   □

Here is one of the many corollaries of this result.

Proposition 3.37 *Let ABC be a triangle, let I be its incenter, and let A', B', C' be the points where the angle bisectors of the triangle intersect again the circumcircle. Then the circumcircles of the triangles $IA'B'$, $IB'C'$, and $IA'C'$ have the same radius as the circumcircle of the triangle ABC.*

Proof We consider the circular transformation that is the composition of the reflection over I and the inversion with center I and radius $\rho = \sqrt{R^2 - OI^2}$, where O is the circumcenter of ABC. (Sometimes this transformation is referred to as an inversion of center I and negative ratio $OI^2 - R^2$.) A point P on the circumcircle of ABC is mapped to the point P' in such a way that I is between P and P' and

$$IP \cdot IP' = R^2 - OI^2.$$

And this quantity is the power of I with respect of the circumcircle (see the proof of the previous result). Consequently P' is on the circumcircle, and so the circular transformation maps the circumcircle of ABC into itself. The points A', B', C' are therefore the images of A, B, C, respectively, so the notation is appropriate!

We now follow the argument on Fig. 3.27. As a corollary of what we have just shown is the fact that the circumcircles of $IA'B'$, $IB'C'$, $IC'A'$ are the images of the lines AB, BC, AC, respectively. The lines AB, BC, AC are at distance r from the center of inversion, so the diameters of the circles that are their images are given by the formula

$$d = \frac{\rho^2}{r} = \frac{R^2 - OI^2}{r} = \frac{R^2 - R^2 + 2Rr}{r} = 2R.$$

The last expression is the diameter of the circumcircle of triangle ABC, and the proposition is proved.   □

Remark We have solved this problem by means of a special circular transformation. This transformation is of interest in itself, so let us examine it in more detail. Let ω be a circle and let O be a point that does not belong to ω. Using O we can define a circular transformation that maps ω to itself:

• If O is outside the circle, then the transformation is the inversion with center O and radius equal to the square root of the power of O with respect to ω.
• If O is inside the circle, then the transformation is the composition of the reflection over O with the inversion of center O whose radius is the square root

Fig. 3.27 Proof of the equality of the three circles

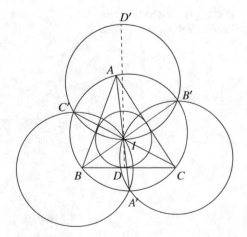

of the power of O with respect to ω (which therefore is an inversion with negative ratio).

We will use this circular transformation to prove the two results below.

Theorem 3.38 (Tzitzeica's Five-Lei Coin Problem) *Three circles of equal radii have a common point and intersect pairwise at three distinct points. Then the circle passing through these three points is equal to any of the three circles.*

Proof 1 Let I be the intersection of the three circles, and let A', B', C' be the pairwise intersection points. The reason for the notation is that if we consider the circular transformation ϕ that is obtained by composing the reflection over I with the inversion of center I and radius the square root of the power of I with respect to the circumcircle of $A'B'C'$, we arrive at the configuration of Proposition 3.37, which can be seen by examining Fig. 3.28.

Indeed, the respective images A, B, C of A', B', C' are on the circumcircle of $A'B'C'$. Also, the images of the three equal circles are the lines AB, BC, and AC, and because the circles are equal, these lines are at equal distance from I. Hence I is the incenter of ABC, and by Proposition 3.37, the circumcircle of ABC (which is the same as the circumcircle of $A'B'C'$) is equal to any of the three circles. $\square$

Proof 2 With the notation from the first proof, consider an inversion of center I, and let A_1, B_1, and C_1 be the images of A', B', C', respectively. Because the circles have equal radii and pass through the center of inversion, their images, which are the lines A_1B_1, B_1C_1, C_1A_1, are at equal distance from I, so I is the incenter of the triangle $A_1B_1C_1$. Let us focus on this triangle and study Fig. 3.29. An easy computation yields

$$\angle B_1 I C_1 = 180° - \angle A_1 B_1 C_1/2 - \angle A_1 C_1 B_1/2 = 90° + \angle B_1 A_1 C_1,$$

and for a similar reason

Fig. 3.28 First proof of Tzitzeica's theorem

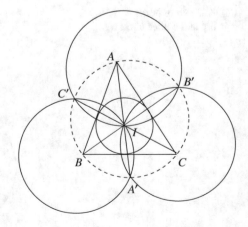

Fig. 3.29 Second proof of Tzitzeica's theorem

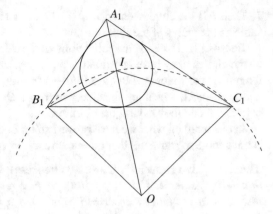

$$\angle A_1 I C_1 = 180° - \angle B_1 A_1 C_1/2 - \angle A_1 C_1 B_1/2 = 90° + \angle A_1 B_1 C_1/2.$$

Let O be the circumcenter of $B_1 I C_1$. Then, in the circumcircle of this triangle,

$$\angle B_1 O C_1 = 360° - 2\angle B_1 I C_1 = 180° - \angle B_1 A_1 C_1,$$

and hence in the isosceles triangle $B_1 O C_1$, $\angle O C_1 B_1 = \angle B_1 A_1 C_1/2$. Then in the isosceles triangle $O I C_1$

$$\angle O I C_1 = \angle O C_1 I = \angle O C_1 B_1 + \angle B_1 C_1 I = \angle B_1 A_1 C_1/2 + \angle A_1 C_1 B_1/2$$
$$= 180° - \angle A_1 I C_1.$$

It follows that A_1, I, O are collinear, so $A_1 I$ passes through the circumcenter of $B_1 I C_1$. From here we deduce that $A_1 I$ and the circumcircle of $B_1 I C_1$ are orthogonal. But then, so are their preimages, which are the lines $A' I$ and $B' C'$. Thus, $A' I$ is an altitude in the triangle $A' B' C'$, and for a similar reason, so are $B' I$ and $C' I$. Then I is the orthocenter of $A' B' C'$, and in this situation, the circumcircles of $A' I B'$, $B' I C'$, and $C' I A'$ are equal to the circumcircle of the triangle $A' B' C'$. The theorem is proved. □

The name of this result comes from the fact that the Romanian mathematician George Tzitzeica (also spelled Gheorghe Țițeica) discovered it by playing with a 5-lei coin (*leu*, plural *lei*, being the currency of Romania). The first proof demonstrates that this theorem is a direct consequence of Euler's relation. We suggest that you try to work backward from Tzitzeica's Theorem via Proposition 3.37 to prove Euler's relation.

Tzitzeica's Theorem can be proved easily using complex numbers, albeit without inversion. Here is how. Place the origin of the coordinate system at I, and let w_1, w_2, w_3 be the coordinates of the centers O_1, O_2, O_3, of the three circles, respectively; $|w_1| = |w_2| = |w_3|$ because O_1, O_2, O_3 are at equal distance from

1. Then I is the circumcenter of $O_1O_2O_3$, and the circumradius of $O_1O_2O_3$ is the same as the radius of the three circles.

Because A, B, C are the reflections of I over the lines that connect the centers of the circles, they have coordinates $w_1 + w_2$, $w_1 + w_3$, and $w_2 + w_3$, so the side lengths of the triangle ABC are $|w_1 + w_2 - w_1 - w_3| = |w_2 - w_3|$, $|w_1 - w_3|$, and $|w_2 - w_3|$, which are the side lengths of $O_1O_2O_2$. Thus the two triangles are congruent, and they have equal circumradii.

Let us build on what we have proved so far and apply the circular transformation ϕ once more to prove another classical result in elementary geometry.

Theorem 3.39 (Poncelet's Closure Theorem) *Let Ω and ω be two circles such that ω is inside Ω. If it is possible to find one triangle that is simultaneously inscribed in Ω and circumscribed to ω, then it is possible to find infinitely many such triangles, and each point of Ω or ω is a vertex or tangency point, respectively, of one such triangle.*

Proof Let ABC be the triangle inscribed in Ω and circumscribed to ω, and let I be the center of ω. We will show that by starting with any point A_1 on Ω, we can construct a triangle $A_1B_1C_1$ inscribed in Ω and circumscribed to ω. The proof can be followed on Fig. 3.30.

We start by choosing A_1 arbitrarily on Ω and constructing the tangents A_1B_1 and A_1C_1 to ω, with $B_1, C_1 \in \Omega$. We then consider the circular transformation ϕ obtained by composing the reflection over I with an inversion of center I that keeps Ω invariant, and let $A_1' = \phi(A_1)$, $B_1' = \phi(B_1)$, and $C_1' = \phi(C_1)$.

By applying Proposition 3.37 to the triangle ABC, we deduce that the images of the sides are three equal circles through I that are tangent to the circle of center I and have radii equal to the diameter of Ω. Now take another line ℓ tangent to ω. Because ℓ is the rotate about I of one of the sides of ABC, and because ϕ has circular symmetry, being the composition of a reflection over I with an inversion centered at I, it follows that $\phi(\ell)$ is a circle through I of diameter equal to the

Fig. 3.30 Proof of Poncelet's closure theorem

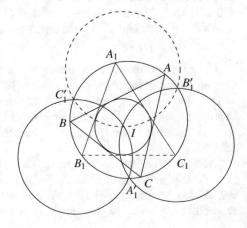

diameter of Ω. We have therefore shown that ϕ maps lines tangent to ω to circles through I that are equal to Ω, and vice versa.

Then the circumcircles Ω_1 and Ω_2 of $IA'_1B'_1$ and $IA'_1C'_1$ are the images of A_1B_1 and A_1C_1, respectively, and hence are equal to Ω. We can apply Tzitzeica's Theorem to the circles Ω, Ω_1, and Ω_2, which pass through A'_1 and intersect pairwise at I, B'_1, and C'_1, to deduce that the circumcircle Ω_3 of $IB'_1C'_1$ is equal to the three circles Ω, Ω_1 and Ω_2. Hence, the line $B_1C_1 = \phi(\Omega_3)$ is tangent to ω. We conclude that the triangle $A_1B_1C_1$ is also inscribed in Ω and circumscribed to ω, and the theorem is proved. □

This result is also known as Poncelet's porism. It has the following general form.

Let Γ_1 and Γ_2 be two plane conics. If it is possible to find, for a given $n > 2$, an n-sided polygon that is simultaneously inscribed in Γ_1 (meaning that all of its vertices lie on Γ_1) and circumscribed to Γ_2 (meaning that all of its edges are tangent to Γ_2), then it is possible to find infinitely many such n-sided polygons, and each point of Γ_1 or Γ_2 is a vertex or tangency point, respectively, of such a polygon.

We will not prove this general result in this book, as its proof uses advanced tools of algebraic geometry. One should note the similarity between the Poncelet and Steiner porisms.

IV. The "$\sqrt{bc}$ Inversion" and the Mixtilinear Incircles

In this section we introduce a special Möbius transformation associated to a triangle and a vertex and present a construction motivated by this transformation. To define the transformation, which we denote by ϕ_A, we let the triangle be ABC, with $BC = a$, $AC = b$, and $AB = c$, and let the chosen vertex be A. Then ϕ_A is the composition of the inversion of center A and radius $\sqrt{bc}$ with the reflection over the internal bisector of $\angle BAC$. This is indeed a Möbius transformation because it is the composition of two reflections, one over a circle the other over a line. By abuse of language, it is called the "$\sqrt{bc}$ inversion." As we will see in what follows, this Möbius transformation combining the reflections over a circle and a line passing through its center is a mighty tool for solving problems.

Considering a system of coordinates with A at the origin and such that $B = ce^{i\theta}$ and $C = be^{-i\theta}$ (so that $AB = c$ and $AC = b$), the $\sqrt{bc}$ inversion is given by

$$\phi_A : \mathbb{C} \cup \{\infty\} \to \mathbb{C} \cup \{\infty\}, \quad \phi_A(z) = \frac{bc}{z}.$$

Proposition 3.40 *The transformation ϕ_A has the following properties:*

(i) *$\phi_A(A) = \infty$, $\phi_A(B) = C$, and $\phi_A(C) = B$;*

(ii) *the circumcircle of ABC and the line BC are mapped into one another, and consequently the arc $\overset{\frown}{BC}$ and the segment BC are mapped into each other;*

(iii) *if P is a point that is not on AB or AC, then the triangle ABP is similar to the triangle $AP'C$, and consequently $P' = \phi_A(P)$ is the image of B under the spiral similarity of center A that maps P to C;*

(iv) *if P and Q are two points such that A, P, Q are not collinear, then the triangles APQ and AQ'P' are similar, where $P' = \phi_A(P)$ and $Q' = \phi_A(Q)$;*

(v) *the image of the circumcenter O of ABC through ϕ_A is the reflection of A over BC, and the image of the orthocenter H of ABC is the second intersection point of AO with the circumcircle of BOC.*

Proof Property (i) follows from the fact that $AB \cdot AC$ equals the square of the radius of inversion. For (ii) note that $\phi_A(BC)$ is a circle that passes through A, and by (i) it also passes through B and C. Parts (iii) and (iv) are direct consequences of Theorem 3.2 (see Fig. 3.31). In fact (iii) can be checked quickly with complex coordinates, if we place A at the origin, with $B = ce^{i\theta}$, $C = be^{-i\theta}$, and $b, c > 0$. If P has coordinate z, then $\phi_A(z) = \frac{bc}{z}$, and

$$\frac{c}{\phi_A(z)} = \frac{c}{\dfrac{bc}{z}} = \frac{z}{b},$$

showing that triangles ABP' and APC are directly similar, and hence are mapped into each other by a spiral similarity.

For (v), note that O is the reflection of ∞ over the circumcircle, so, by the Symmetry Principle (Theorem 3.15), $O' = \phi_A(O)$ is the reflection of $A = \phi_A(\infty)$ over the line BC which is the image of the circumcircle. H is at the intersection of AO' and the circumcircle of $BO'C$ (because this circle is the reflection of the circumcircle of ABC over BC), so $H' = \phi_A(H)$ is at the intersection of the image of AO', which is AO, and the image of the circumcircle of $BO'C$, which is the circumcircle of BOC.   □

Fig. 3.31 The image of P through the $\sqrt{bc}$ inversion

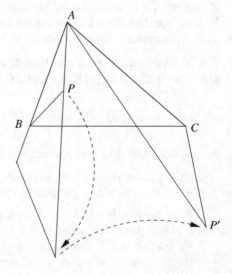

What happens if we compose two $\sqrt{bc}$ inversions of the same triangle? The result is surprising.

Proposition 3.41 *Let ϕ_B and ϕ_C be the $\sqrt{bc}$ inversions determined by triangle ABC in vertices B and C, respectively. Then $\phi_B \circ \phi_C = \phi_A$.*

Proof Let $\phi = \phi_B \circ \phi_C$. This is a Möbius transformation, so we have to find the images of three points. The natural choices are A, B, and C. Indeed

$$\phi(A) = \phi_B(\phi_C(A)) = \phi_B(B) = \infty, \quad \phi(B) = \phi_B(\phi_C(B)) = \phi_B(A) = C,$$

$$\phi(C) = \phi_B(\phi_C(C)) = \phi_B(\infty) = B.$$

These coincide with $\phi_A(A)$, $\phi_A(B)$, and $\phi_A(C)$, respectively, hence the conclusion.
□

The $\sqrt{bc}$ inversion allows for the following slick proof of

Proposition 1.23 *The point where the symmedian of a triangle intersects the circumcircle forms with the vertices a harmonic quadrilateral.*

Proof Let the triangle be ABC, let P be the midpoint of BC, and let P' be its image through the $\sqrt{bc}$ inversion corresponding to the vertex A in the triangle ABC. Denote the complex coordinates of points by the same letter, but lowercase. Because the $\sqrt{bc}$ inversion is a Möbius transformation, it preserves the cross-ratio. As A is mapped to ∞ and B is mapped to C, we have

$$(a, p', b, c) = (\infty, p, c, b).$$

The fact that $ABP'C$ is harmonic is equivalent to $(a, p', b, c) = -1$, and the fact that $(\infty, p, c, b) = -1$ is equivalent to the fact that P is the midpoint of BC. The proposition is proved.
□

Inversion and Möbius transformations become more interesting when they involve circles and tangencies, so let us look at incircles and excircles. To recapitulate: if ϕ_A is the $\sqrt{bc}$ inversion determined by ABC at the vertex A, the side AB is mapped to the ray r of the line AC that starts at C and does not contain A, the side AC is mapped to a similar ray s of the line AB, and the side BC is mapped to the arc $\overset{\frown}{BC}$ of the circumcircle of ABC that does not contain A. Then the incircle ω is mapped to a circle that is tangent to another circle that is tangent to the rays r and s and the circumcircle of ABC. And this is the *A-mixtilinear excircle*, illustrated in Fig. 3.32.

If we switch to the A-excircle ω_A, then $\phi_A(\omega_A)$ is a circle that is tangent to sides AB and AC and to the circumcircle internally. This is the *A-mixtilinear incircle* (see Fig. 3.33).

Let us first find a suitable way to construct the A-mixtilinear incircle Ω_A with ruler and compass. One possible way to do that is to locate the tangency points of Ω_A to AB, AC, and the circumcircle Ω of ABC.

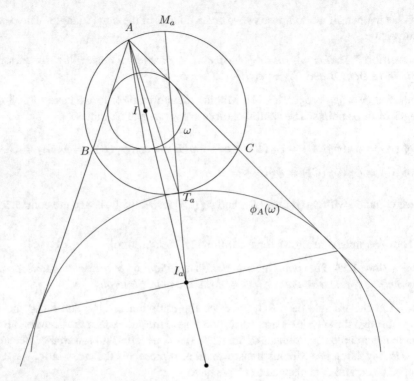

Fig. 3.32 The A-mixtilinear excircle

Proposition 3.42 (Finding Tangency Points, Part I) *Let Ω_A be the A-mixtilinear incircle of triangle ABC. Let Ω be the circumcircle of ABC, and let I be the incenter of this triangle. Let Ω_A touch AB at D, AC at E, and Ω at U_a. Let also Y_a be the tangency point of the A-excircle to BC. Then the lines AY_a and AU_a are isogonal and I is the midpoint of DE.*

Proof You can follow our reasoning on Fig. 3.33. Let ϕ_A be the $\sqrt{bc}$ inversion determined by the triangle ABC and the vertex A. The first part is immediate from the fact that $\phi_A(Y_a) = U_a$ and the fact that ϕ_A inverts about a circle centered at A and then reflects across the bisector of $\angle BAC$.

The second part is more interesting: Proposition 3.7 implies that the A-excenter I_a of ω_A is mapped to the midpoint of the segment determined by the points of tangency of the tangents to $\phi_A(\omega_A) = \Omega_A$ through A, that is, I_a is mapped to the midpoint of DE. It remains to show that $\phi_A(I_a) = I$, which is true because

$$\angle ACI_a = \angle ACB + \frac{1}{2}(180° - \angle ACB) = 90° + \frac{1}{2}\angle ACB = \angle AIB,$$

which implies that the triangles AIB and ACI_a are similar. And then we apply Proposition 3.40, part (iii).   □

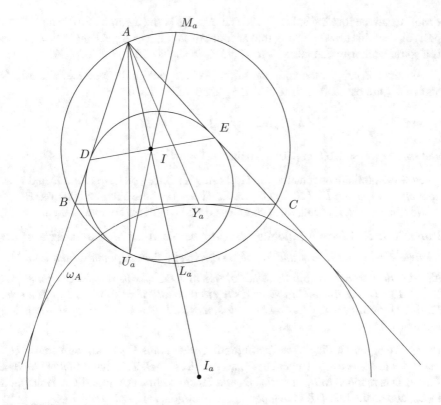

Fig. 3.33 The A-mixtilinear incircle

It should be mentioned that proving that I is the midpoint of the segment DE has made the object of a problem given at the International Mathematical Olympiad in 1978. It is also true that each excenter is the midpoint of the segment determined by the tangency points of the corresponding mixtilinear excircle.

We can now see how to construct Ω_A: find the incenter I (which can be done by drawing the bisector of $\angle BAC$ and then drawing the circle with center at the midpoint L_a of the arc $\overset{\frown}{BC}$ of Ω that does not contain A) and then draw the perpendicular to AI through I to obtain the points D and E. Unfortunately, locating U_a requires too much work, because we need to find the A-excircle, or, at least, Y_a. When we are using pencil and paper, this leads to loss of precision in the construction. But the following result makes the construction of U_a easy and precise.

Proposition 3.43 (Finding Tangency Points, Part 2) *Let Ω_A be the A-mixtilinear incircle of triangle ABC. Let Ω be the circumcircle of ABC, and let I be the incenter. Let Ω_A touch Ω at U_a, and let M_a be the midpoint of $\overset{\frown}{BC}$ that contains A. Then U_a lies on the line IM_a.*

Proof Continue looking at Fig. 3.33. The proof is based again on Proposition 3.40, part (iii), but this time by taking into account that $\phi(Y_a) = U_a$ and $\phi(I) = I_a$, there is a spiral similarity that takes Y_a to I and I_a to U_a, so $\angle AU_aI = \angle AI_aY_a$.

As above, let L_a be the midpoint of the $\overset{\frown}{BC}$ that does not contain A. Since I_aY_a and L_aM_a are perpendicular to BC, they are parallel. Then

$$\angle AI_aY_a = \angle AL_aM_a = \angle AU_aM_a,$$

and so $\angle AU_aI = \angle AU_aM_a$. The conclusion follows.   □

Now the problem of constructing Ω_A is simple: Once you have found D and E as above, draw the line M_aI to find U_a on Ω. Then Ω_A is the circumcircle of DEU_a.

Mixtilinear circles bring forth countless little facts. Here are two of them.

Proposition 3.44 (Arc Midpoints) *Let Ω_A be the A-mixtilinear incircle of the triangle ABC, and let L_b and L_c be, respectively, the midpoints of the arcs $\overset{\frown}{CA}$ and $\overset{\frown}{AB}$ that do not contain the other vertex of ABC. As before, let Ω_A touch the line AB at D, the line AC at E, and the circumcircle Ω of ABC at U_a. Then the line U_aD passes through L_c and (analogously) the line U_aE passes through L_b. Moreover, DE is parallel to L_bL_c.*

Proof We argue on Fig. 3.34. The homothety that maps Ω_A to Ω, with center U_a, maps AB to a line that is tangent to Ω and parallel to AB. This line is tangent to Ω at L_c. So D is mapped to L_c, and U_a, D, and L_c are collinear. Similarly, E is mapped to L_b, and so the line DE is mapped to the line L_cL_b. Hence, $DE \parallel L_bL_c$.   □

Proposition 3.44 can also be used to obtain another proof of the fact that I is the midpoint of DE: by Pascal's Theorem applied to the hexagon $U_aL_cCABL_b$, the

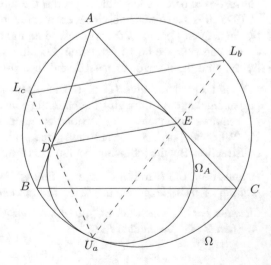

Fig. 3.34 Midpoints of arcs and a mixtilinear circle

points $U_a L_c \cap AB = \{D\}$, $L_c C \cap B L_b = \{I\}$, and $L_c C \cap L_b U_a = \{E\}$ are collinear, and since AI is the angle bisector of $\angle DAE$ in the isosceles triangle ADE, I is the midpoint of DE.

It is natural to interpret the lines AU_a, BU_b, and CU_c as cevians. The symmetry of the configuration makes us expect that these lines are concurrent. This is the case, and the common point is actually special!

Proposition 3.45 (Concurrent Cevians) *Let Ω_A be the A-mixtilinear incircle of the triangle ABC. Let Ω_A touch the circumcircle Ω of ABC at U_a, and define U_b and U_c similarly. Then AU_a passes through the center of the direct homothety that maps the incircle of ABC to the circumcircle. Consequently, the lines AU_a, BU_b, CU_c are concurrent.*

Proof Arguing on Fig. 3.35, let h be the (direct) homothety that takes Ω_A to Ω, and let h' be the (direct) homothety that takes the incircle ω to Ω_A. Then h has center X_a, h' has center A, and the composition $h' \circ h$ takes ω to Ω, because $h'(h(\omega)) = h'(\Omega_A) = \Omega$. This composition is the direct homothety between ω and Ω, so we are done by Proposition 2.6. You can also use Monge's Theorem (see Problem 86 (a)) to prove this result. □

Analogous results are true for the mixtilinear excircles. The reader should explore, discover, and prove these results.

Fig. 3.35 Lines with the same opinion; they concur!

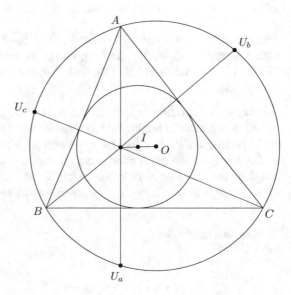

3.2.2   Examples of Problems Solved Using Inversion

The first example of our discussion is a problem given at the Israel Mathematical Olympiad in the year 1995, whose first two solutions that are presented below were suggested to us by Leandro Maia.

Problem 3.1 Let PQ be the diameter of a semicircle Γ. The circle ω is tangent internally to Γ and is also tangent to the segment PQ at C. Let A be a point on Γ, and let B be a point on PQ such that AB is tangent to ω and perpendicular to PQ. Prove that AC is the angle bisector of $\angle PAB$.

Solution 1 There is an inversion present in the picture, which can be revealed by examining closely Fig. 3.36. The Leg Theorem in the right triangle APQ gives $QB \cdot QP = QA^2$. So the inversion χ_1 of center Q and radius QA maps P and B into each other. But there is more to it. To discover this hidden fact, note first that $\chi_1(\Gamma)$ is a ray that is perpendicular to PQ, and this ray is $|BA$. It follows that $\chi_1(\omega)$ is a circle that is tangent to $\Gamma = \chi_1(|AB)$, $|AB = \chi_1(\Gamma)$, and $PQ = \chi_1(PQ)$. There are four such circles, but because the image of χ lies entirely in the union of rays $|QX$, with $X \in \omega$, it follows that $\chi_1(\omega)$ must be ω itself. But then $\chi_1(C) = C$; this is the hidden fact that solves the problem. From here we obtain $QA = QC$, so the triangle QAC is isosceles. We thus have

$$\angle CAB = \angle CAQ - \angle BAQ = \angle CAQ - \angle APB = \angle ACQ - \angle APB = \angle PAC,$$

where for the second equality we have used the fact that AB is altitude in the right triangle APQ and for the fourth equality we have used the Exterior Angle Theorem in the triangle ABC. This proves that AC is the angle bisector of $\angle PAB$, as required.   □

Solution 2 An inversion χ_2 of center A (Fig. 3.37) solves the problem with a "position argument," meaning that the conclusion is derived from the understanding of the positions of the images of various objects with respect to one another.

   The image of the circle of which Γ is a semicircle is the line $P'Q'$, while the image of the line PQ is the circumcircle Ω of $P'AQ'$. And because $\angle P'AQ' = \angle PAQ = 90°$, Ω has diameter $P'Q'$, so its center O is on $P'Q'$.

Fig. 3.36 ω is tangent to AB, PQ, and Γ

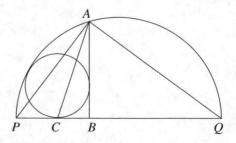

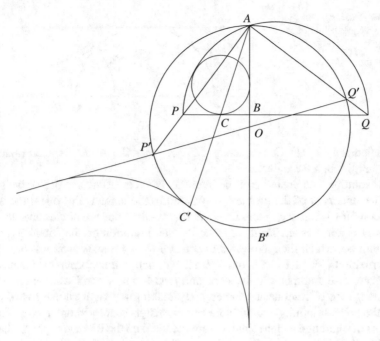

Fig. 3.37 The inversion χ_2

Additionally, $\chi_2(\omega)$ is a circle that is (exterior) tangent to $\Omega = \chi_2(PQ)$ and is also tangent to $AB = \chi(AB)$ and $P'Q'$. Note that because AB is perpendicular to PQ, the center O of $\chi_2(PQ) = \Omega$ is on AB as well (Theorem 3.4). So the lines $P'Q'$ and AB pass through the center O of Ω, and $\chi_2(\omega)$ is tangent to Ω at $C' = \chi_2(C)$ and is also tangent to AB and $P'Q'$. By the symmetry of the figure formed by the two lines and the two circles, C' must be the midpoint of the arc $P'B'$ ($B' = \chi_2(B)$), and consequently $\angle P'AC' = \angle C'AB'$, that is, $\angle PAC = \angle CAB$, as desired.   □

Solution 3 Let χ_3 be an inversion of center C. As $\angle CAP = \angle CP'A'$ and $\angle CAB = \angle CB'A'$ (Theorem 3.2), the line AC bisects $\angle PAB$ if and only if $\angle CP'A' = \angle CB'A'$.

The inversion maps the circle ω to a line ω' that is parallel to the line PQ, while the semicircle Γ is mapped to a semicircle Γ' that is tangent to ω' and whose endpoints are $P' = \chi_3(P)$ and $Q' = \chi_3(Q)$. We want to determine the image of the segment AB. For that, let $A' = \chi_3(A)$ and $B' = \chi_3(B)$. Because AB is perpendicular to PQ, by Theorem 3.2, $\angle CA'B' = \angle CBA = 90°$, so the segment AB is mapped into an arc on the semicircle with diameter $B'C$. Also, as inversion preserves tangencies, this semicircle is tangent to the line ω'. So we are in the situation from Fig. 3.38. The two semicircles in this figure are equal, and they are mapped into each other by the reflection over the line through A' that is

Fig. 3.38 The inversion χ_3
yields two equal semicircles

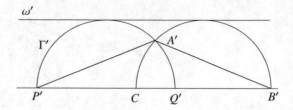

perpendicular to $P'Q'$. Consequently $\angle CP'A' = \angle CB'A'$, as they are images of
each other through the reflection.

This solution is an instance of the interplay between affine and projective geometry. The statement of the problem is phrased in the language of one-dimensional
affine complex geometry, where lines are lines and circles are circles and the point
at infinity is not present. Then we place the configuration on the Riemann sphere,
by adding ∞, which then allows us to act by Möbius transformations, or circular
transformations, or just inversion. We act by such a transformation to move ∞
somewhere else and then eliminate ∞ and read the new configuration once more
in the language of Euclidean geometry. By making the right choice for where to
move the point at infinity, the problem becomes significantly simpler. Note that the
original configuration and the final one are identical on the Riemann sphere, the only
difference is that we postulate a different point at infinity for each of them. When
we remove these different points at infinity and look at the problem in the Euclidean
setting, we see different things.   □

Solution 4 This last solution emphasizes once more this method of "moving the
point at infinity," but we go further, by stating the conclusion of the problem so
that it is invariant under Möbius or circular transformations, and then transform the
configuration to one where the conclusion is easy to verify. And indeed, using the
Bisector Theorem, we write the conclusion of the problem as saying that

$$\frac{AB}{AP} = \frac{CB}{CP}.$$

In complex coordinates this states that the absolute value of the cross-ratio of
A, C, B, P is equal to 1, that is, $|(a, c, b, p)| = 1$.

The cross-ratio is invariant under Möbius transformations, so we consider the
Möbius transformation $\phi = \chi_2 \circ \chi_1$, where χ_1 is the inversion of center P and
radius PA and χ_2 is the inversion of center Q and radius QA. Then

$$\phi(P) = \chi_2(\chi_1(P)) = \chi_2(\infty) = Q \text{ and } \phi(A) = \chi_2(\chi_1(A)) = \chi_2(A) = A.$$

The Leg Theorem applied twice in the right triangle APQ yields $PA^2 = PB \cdot PQ$
and $QA^2 = QB \cdot QP$, so

$$\phi(Q) = \chi_2(\chi_1(Q)) = \chi_2(B) = P, \text{ and } \phi(B) = \chi_2(\chi_1(B)) = \chi_2(Q) = \infty.$$

Let $C' = \phi(C)$. Then, in coordinates

$$|(a, c, b, p)| = 1 \iff |(a, c', \infty, q)| = 1 \iff |(\infty, q, a, c')| = 1.$$

We are left with showing that

$$\left| \frac{\infty - a}{\infty - c'} : \frac{q - a}{q - c'} \right| = 1.$$

The first fraction is 1, so we have to check that the second fraction has absolute value equal to 1, and geometrically this means that $QA = QC'$.

Since $\phi(P) = Q, \phi(Q) = P, \phi(A) = A$, we have $\phi(\Gamma) = \Gamma$. Note also that A and the reflection of A over PQ are invariant under ϕ and also $\phi(B) = \infty$ which means that $\phi(AB)$ is a line and this line is AB. Thus, $\phi(AB) = AB$ and $\phi(|AB) = |AB$. And because ϕ maps the line PQ into itself, $\phi(\omega)$ is tangent to Γ, the ray $|BA$, and the line PQ. But $\chi_1(\omega)$ satisfies the same tangency conditions.

Note that χ_1 maps ω to the other side of Γ, and there is only one circle on that side of Γ tangent to Γ, the ray $|BA$, and the line PQ simultaneously, namely, the circle denoted by ω' in Fig. 3.39. And χ_2 maps $\omega' = \chi_1(\omega)$ to the same side of Γ. But then $\chi_2(\omega') = \omega'$. In particular, $\chi_2(\chi_1(C)) = \chi_1(C)$, which means that C' is invariant under χ_2. This can only happen if C' is on the circle of inversion, that is, if $AQ = QC'$. The problem is solved. □

Here is a short-listed problem of the 1995 International Mathematical Olympiad, proposed by Turkey.

Problem 3.2 The incircle of the triangle ABC touches the sides BC, CA, and AB at D, E, and F, respectively. A point X is chosen inside the triangle ABC such that the incircle of XBC touches BC at D as well, and let Y and Z be the points where this incircle touches CX and BX, respectively. Prove that $EFZY$ is cyclic.

Solution The solution presented below was published by Dan Brânzei in *Gazeta Matematică (Mathematics Gazette, Bucharest)* in 1996. It can be followed on Fig. 3.40.

Fig. 3.39 The solution based on an invariant

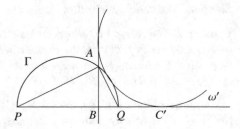

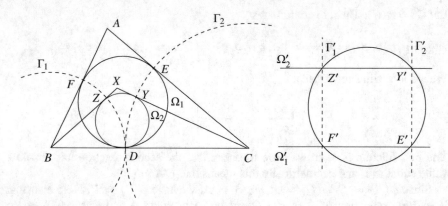

Fig. 3.40 Proof that $EFZY$ is cyclic

Denote the incircles of ABC and XBC by Ω_1 and Ω_2, respectively. Because the tangents from a point to a circle are equal, $BZ = BD = BF$ and $CY = CD = CE$, so there are two more circles in the picture, namely, the circle Γ_1 of center B and radius BD and the circle Γ_2 of center C and radius CD. Because the radii of Γ_1 and Γ_2 are tangent to Ω_1 and Ω_2 at the points of contact, the circles Γ_1 and Γ_2 are orthogonal to both Ω_1 and Ω_2.

Consider an inversion of center D. The circles $\Omega_1, \Omega_2, \Gamma_1, \Gamma_2$ are mapped into four lines $\Omega_1', \Omega_2', \Gamma_1', \Gamma_2'$ such that Ω_1' is parallel to Ω_2' and Γ_1' is parallel to Γ_2' and Ω_j' makes an angle of $90°$ with Γ_k', $j, k = 1, 2$. Consequently, the images E', F', Z', Y' of E, F, Z, Y form a rectangle, which is cyclic. As inversion maps the circumcircle of $E'F'Z'Y'$ to a line or a circle, and as the points E, F, Z, Y are not collinear, these points must therefore be concyclic.   □

The third example is a problem that was given at the Romanian Mathematical Olympiad in 1997.

Problem 3.3 Given n circles ($n \geq 3$) that have a common point O, if the circles cut on two rays starting at O equal segments, then they cut on any ray starting at O equal segments.

Solution The conclusion follows from the following.

Lemma *Let $\omega_1, \omega_2, \omega_3$ be three circles that pass through a point O, and let $|OX$ and $|OY$ be two rays that cut the circles a second time at A_1, A_2, A_3 and B_1, B_2, B_3, respectively (with A_2 between A_1 and A_3 and B_2 between B_1 and B_3), such that $A_1A_2/A_2A_3 = B_1B_2/B_2B_3$. Then*

(a) $\omega_1, \omega_2, \omega_3$ have a second common point or are tangent at O;
(b) if $|OZ$ is a ray that cuts $\omega_1, \omega_2, \omega_3$ at C_1, C_2, C_3, respectively, then $C_1C_2/C_2C_3 = A_1A_2/A_2A_3$.

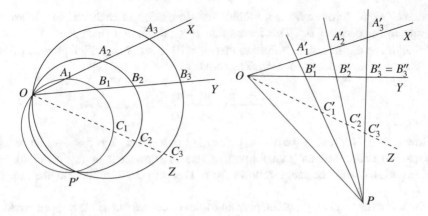

Fig. 3.41 Möbius transformation acting on the configuration of rays and circles

Proof The proof of the lemma combines linear fractional transformations in both one-dimensional real and one-dimensional complex projective geometry.

First, we use a Möbius transformation ϕ to map O (where we locate the origin of the coordinate system) to the point at infinity and the point at infinity to O, and then $\omega'_1 = \phi(\omega_1)$, $\omega'_2 = \phi(\omega_2)$ and $\omega'_3 = \phi(\omega_3)$ are lines, while the rays $|OX$, $|OY$ and $|OZ$ are still mapped into rays that start at $\phi(\infty) = O$. The configurations before and after the Möbius transformation are shown on the left and on the right of Fig. 3.41.

Denote the complex coordinate of a point by the corresponding lowercase letter, and use a dash to denote the image of a point through ϕ. The collinearity and order of the points allow us to write the metric relation from the statement as

$$\frac{a_1 - a_2}{a_3 - a_2} = \frac{b_1 - b_2}{b_3 - b_2},$$

both quotients being negative real numbers. Because the cross-ratio is invariant under Möbius transformations (Theorem 3.10), we have

$$(a'_1, a'_3, a'_2, 0) = (a_1, a_3, a_2, \infty) = \frac{a_1 - a_2}{a_3 - a_2} : \frac{a_1 - \infty}{a_3 - \infty} = \frac{a_1 - a_2}{a_3 - a_2}$$

$$= \frac{b_1 - b_2}{b_3 - b_2} = \frac{b_1 - b_2}{b_3 - b_2} : \frac{b_1 - \infty}{b_3 - \infty} = (b_1, b_3, b_2, \infty) = (b'_1, b'_3, b'_2, 0).$$

Let P be the intersection point of $A'_1 B'_1$ and $A'_2 B'_2$. Let also B''_3 be the intersection of PA'_3 with OY. Then by Theorem 3.20 applied to the pencil of four lines $PO, \omega'_1, \omega'_2, PA'_3$

$$(b'_1, b'_3, b'_2, 0) = (a'_1, a'_3, a'_2, 0) = (b'_1, b''_3, b'_2, 0),$$

hence $b_3'' = b_3'$ (cross-ratio is a Möbius transformation in each variable; hence it is injective), so $B_3' = B_3''$. This means that $P \in A_3' B_3'$, and since $P \in A_1' B_1'$ and $P \in A_2' B_2'$ by construction, it follows that $\phi^{-1}(P) \in \omega_j$, $j = 1, 2, 3$, proving (a).

Part (b) is now straightforward. The equality

$$\frac{c_1 - c_2}{c_3 - c_2} = \frac{a_1 - a_2}{a_3 - a_2}$$

follows from $(c_1, c_3, c_2, \infty) = (a_1, a_3, a_2, \infty)$, since by the invariance of cross-ratio under Möbius transformations this is equivalent to $(c_1', c_3', c_2', 0) = (a_1', a_3', a_2', 0)$, and the latter follows from Theorem 3.20 applied to the pencil $PO, \omega_1', \omega_2', \omega_3'$. $\qquad\square$

To solve the original problem, pick all triples of circles out of the n given circles, and apply the lemma. $\qquad\square$

This is a common technique for proving the existence of common intersection points of curves: map them by a Möbius (or circular) transformation, and check that their images have a common point. This is a consequence of the fact that Möbius (and circular) transformations are injective. In this problem we had to show that the circles have two common points, and this is equivalent to the fact that their images have two common points on the Riemann sphere, which are P and the point at infinity.

The next example is a problem proposed by Poland which was given at the 59th International Mathematical Olympiad in 2018.

Problem 3.4 A convex quadrilateral $ABCD$ satisfies $AB \cdot CD = BC \cdot DA$. A point X is chosen inside the quadrilateral so that $\angle XAB = \angle XCD$ and $\angle XBC = \angle XDA$. Prove that $\angle AXB + \angle CXD = 180°$.

Solution 1 The metric relation from the statement can be rewritten as

$$\frac{AB}{AD} : \frac{CB}{CD} = 1,$$

which is a condition about the cross-ratio of segments, and in view of the discussion from Sect. 3.1.4, we should think about inversion and circular transformations. A circular transformation will preserve this cross-ratio, and it might yield a nicer quadrilateral, such as one with equal angles placed in better locations. Inversion centered at X does the job, and the second official solution starts with an inversion of center X and radius 1.

With the usual convention that $A \mapsto A'$, $B \mapsto B'$, $C \mapsto C'$, and $D \mapsto D'$, we immediately have that

$$\frac{A'B'}{A'D'} : \frac{C'B'}{C'D'} = 1,$$

and this yields $A'B' \cdot C'D' = B'C' \cdot D'A'$.

Fig. 3.42 The image of
$ABCD$ through an inversion
of center X

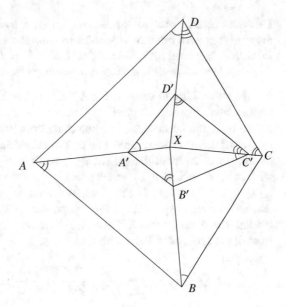

As for the angles, Theorem 3.2 gives

$$\angle XB'A' = \angle XAB = \angle XCD = \angle XD'C'$$

and

$$\angle XC'B' = \angle XBC = \angle XDA = \angle XA'D'.$$

Nothing seems to have changed! But by examining Fig. 3.42 carefully, we notice
also that $\angle XB'C' = \angle XCB$, and hence

$$\angle A'B'C' = \angle A'B'X + \angle XB'C' = \angle BAX + \angle BCX = \angle BCD.$$

Similarly $\angle B'C'D' = \angle CDA$, $\angle C'D'A' = \angle DAB$, and $\angle D'A'B' = \angle ABC$. So not
only that two quadrilaterals satisfy the same properties specified in the statement of
the problem, but they have equal angles, too.
   The metric relations give

$$\frac{AB}{AD} : \frac{CB}{CD} = 1 = \frac{D'C'}{D'A'} : \frac{B'C'}{B'A'},$$

which combined with the equalities of angles $\angle DAB = \angle C'D'A'$ and $\angle BCD = \angle A'B'C$ implies that the cross-ratio of the points A, B, C, D is equal to the cross-
ratio of the points D', A', B', C'. But then the Möbius transformation that maps
A, B, C to D', A', B' also maps D to C'. And we have the following result.

Lemma *Assume that in the convex quadrilaterals $XYZT$ and $X'Y'Z'T'$ are mapped into each other by a Möbius transformation, and assume that we have the following equalities of angles: $\angle X = \angle X'$, $\angle Y = \angle Y'$, $\angle Z = \angle Z'$, and $\angle T = \angle T'$. Then the Möbius transformation is a spiral similarity or a translation.*

Proof All we have to show is that $XYZT$ and $X'Y'Z'T'$ are directly similar, since in that case, there is a spiral similarity or a translation that maps the first into the second, and because the Möbius transformation is uniquely determined by the images of three vertices, it must be this spiral similarity or translation.

Construct the quadrilateral XYZ_1T_1 similar to $X'Y'Z'T'$ and sharing side XY with $XYZT$ such that Z_1 and T_1 lie on the rays YZ and XT, respectively, and $Z_1T_1 \parallel ZT$, as shown in Fig. 3.43. We need to prove that $Z_1 = Z$ and $T_1 = T$. Assume the contrary. Without loss of generality $TX > T_1X$ (the way Fig. 3.43 is drawn). Let the segments XZ and Z_1T_1 intersect at U. Using the similarities of the triangles XT_1U and XTZ we obtain

$$\frac{T'X'}{T'Z'} : \frac{Y'X'}{Y'Z'} = \frac{T_1X}{T_1Z} : \frac{YX}{YZ_1} < \frac{T_1X}{T_1U} : \frac{YX}{YZ} = \frac{TX}{TZ} : \frac{YX}{YZ},$$

which contradicts the fact that the cross-ratio is invariant under Möbius transformations. The lemma is proved. □

It follows from the lemma that the quadrilaterals $ABCD$ and $D'A'B'C'$ are similar. Using this similarity and the fact that, by formula (3.1)

$$A'B' = \frac{AB}{XA \cdot XB} \text{ and } D'A' = \frac{DA}{XD \cdot XA},$$

we have

$$\frac{AB}{BC} = \frac{D'A'}{A'B'} = \frac{DA}{XD \cdot XA} \cdot \frac{XA \cdot XB}{AB} = \frac{DA}{AB} \cdot \frac{XB}{XD}.$$

Therefore

Fig. 3.43 The construction of the quadrilateral XYZ_1T_1

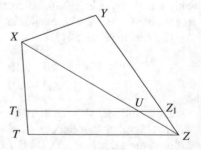

$$\frac{XB}{XD} = \frac{AB^2}{BC \cdot DA} = \frac{AB^2}{AB \cdot CD} = \frac{AB}{CD}.$$

A similar computation shows that

$$\frac{XA}{XC} = \frac{DA}{BC}.$$

Now using the Law of Sines in the triangles AXB and CXD and the first of these two relations, we can write

$$\frac{\sin \angle AXB}{\sin \angle XAB} = \frac{AB}{XB} = \frac{CD}{XD} = \frac{\sin \angle CXD}{\sin \angle XCD}.$$

The denominators of the first and last fraction are equal by hypothesis, hence so are the numerators: $\sin \angle AXB = \sin \angle CXD$. The other relation yields $\sin \angle DXA = \sin \angle BXC$. If at least one of the pairs $(\angle AXB, \angle CXD)$ and $(\angle BXC, \angle DXA)$ consists of supplementary angles, we are done. Otherwise, $\angle AXB = \angle CXD$ and $\angle DXA = \angle BXC$. In this case $X = AC \cap BD$, and the conditions about angles from the statement imply that $ABCD$ is a parallelogram. The metric relation shows that it is in fact a rhombus, so the diagonals make a right angle, and the sum of the two angles is again $180°$.   □

Solution 2 Inversion about X is not the only possibility. The contestant Jonas Walter of the German team used the inversion whose center is the point K where the perpendicular bisector of the segment BD intersects the line AC and whose radius is KB. This is motivated by the properties of the Apollonian circles (see Problem 165). Let us explain this approach, with the aid of Fig. 3.44.

The metric relation from the statement can be rewritten as

$$\frac{AB}{AD} = \frac{CB}{CD},$$

and from it we read that A and C lie on the same Apollonian circle defined by B and D and this ratio. Let us assume that we are not in the degenerate case where AC, and thus the Apollonian circle, is the perpendicular bisector of BD, and let us also assume that B is inside the Apollonian circle (otherwise swap B and D). Call this Apollonian circle ω, and let its intersections with the line BD be U and V, with the points U, B, V, D appearing in this order on the line BD. By Proposition 3.1 and by Problem 165, the points U, B, V, D form a harmonic division on the line BD.

Consider the circle γ of diameter BD, and take an inversion χ of center D. Then $U^* = \chi(U)$, $B^* = \chi(B)$, $V^* = \chi(V)$, and ∞ still form a harmonic division (by the invariance of real valued cross-ratio), so B^* is the midpoint of the segment U^*V^*. Note that $\chi(\gamma)$ is the perpendicular bisector of U^*V^*, and U^* and V^* are reflections of each other over $\chi(\gamma)$. By the Symmetry Principle (Theorem 3.15), U and V are

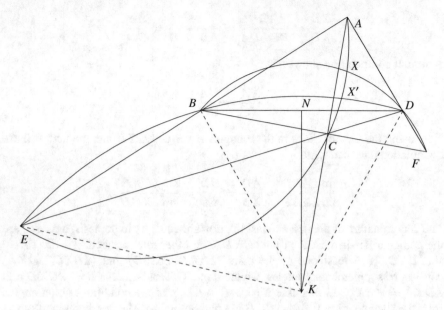

Fig. 3.44 The inversion of $ABCD$ with center at K

symmetric with respect to γ. So if the center of γ is N, then $NU \cdot NV = NB^2$, the later being the ratio of the inversion χ.

From this metric relation we deduce that the perpendicular bisector of BD is the radical axis of ω and the degenerate circle of center B and radius 0. It follows that the point K defined above satisfies

$$KB^2 = KA \cdot KC.$$

This equality implies that the inversion of center K and radius $KB = KD$ maps any circle through A and C into itself. We now choose a particular such circle. For that we make a second assumption that $ABCD$ is not a trapezoid. Let the intersection of AB and CD be E and the intersection of AD and BC be F. The circumcircle of ACE is the circle we have in mind. Working with directed angles modulo 180°, so that the argument does not depend on the configuration, we translate the condition $\angle XAB = \angle XCD$ from the statement into $\angle XAE = \angle XCE$, which then implies that X belongs to the circumcircle of ACE.

So we have determined that the circumcircle of AEC contains X and is invariant under the inversion. The other angle equality from the statement implies that X also lies on the circumcircle of BDF. An important observation is that the circumcircle of BDF is the image through the inversion of the circumcircle of BDE. Indeed, if E' is the image of E, then by working once more with directed angles modulo 180° and using Theorem 3.2, we have

$$\angle BE'D = \angle BE'K + \angle KE'D = \angle KBE + \angle EDK = \angle KBA + \angle CDK$$

By the same proposition the triangles KCD and KDA are similar, and KBC and KAB are also similar, and so this is further equal to

$$\angle BCK + \angle KAD = \angle FCA + \angle CAD = \angle FCA + \angle CAF = \angle CFA = \angle BFD.$$

So, as directed angles modulo $180°$, $\angle BE'D = \angle BFD$ showing that B, F, E', D are concyclic, which proves our claim.

Because X lies at the intersection of the circumcircles of ACE and BDF, its image X' must lie at the intersection of the circumcircles of ACE and BDE. This point cannot be E, for X and E lie on different sides of line AC which passes through the center of inversion. Hence X' is the point of intersection of the circumcircles of ACE and BDE different from E. Using the formula (3.1) we compute

$$X'B = \frac{KB^2}{KX \cdot KB} XB, \quad X'D = \frac{KB^2}{KX \cdot KD} XD = \frac{KB^2}{KX \cdot KB} XD,$$

and so

$$\frac{XB}{XD} = \frac{X'B}{X'D}.$$

Examining the configuration consisting of the circumcircles of ACE and BDE, we see that these circles intersect at E and X, and we know that the points E, D, C are collinear and the points E, B, A are collinear. Using the result proved in Problem 131 we deduce that there is a spiral similarity of center X' that maps the circles into each other such that A is mapped to B and C is mapped to D. Then the triangles $X'AB$ and $X'CD$ are similar, so

$$\frac{X'B}{X'D} = \frac{AB}{CD}.$$

We have obtained that

$$\frac{XB}{XD} = \frac{AB}{CD}.$$

Switching A, C with B, D we deduce that

$$\frac{XA}{XC} = \frac{DA}{DC}.$$

And we have seen in the first solution that this yields the conclusion.

Of course, we have left out a few cases. If BC and AD are parallel, then F lies at infinity, so the circumcircle of BDF becomes the line BF. The argument adapts *mutatis mutandis*. It might happen that AB is parallel to CD. Again, circumcircles become lines, and the argument works. Finally, it might happen that AC is the perpendicular bisector of BD, so that $ABCD$ is a kite. We can switch A, C with B, D if BD is not the perpendicular bisector of AC. If it is, then we have a rhombus, and X is necessarily its center, and then the conclusion is obvious.   □

Solution 3 During the discussions of the jury, a solution arose that uses inversion with respect to the Apollonian circle ω defined by the points B and D and the ratio $AB/AD = CB/CD$. Let O be the center of ω. As the circle of diameter BD is orthogonal to the Apollonian circle, by Proposition 3.25, it is invariant under the inversion, so the inversion maps

$$A \mapsto A, \quad C \mapsto C, \quad B \mapsto D.$$

From here, the argument can be followed on Fig. 3.45. Again, let $E = AB \cap CD$, $F = AD \cap BC$, and we have seen above that X is on the circumcircles of both ACE and BDF. The other intersection point of this two circles is F', the inverse of F. Indeed, by $OF' \cdot OF = OB \cdot OD$, we have that F' is on the circumcircle of BDF (by power-of-a-point). To show that it is on the circumcircle of ACE, we notice first that Theorem 3.2 gives $\angle BCO = \angle ODC$, $\angle OAD = \angle ABO$, $\angle AF'O = \angle OAF$, and $\angle OF'C = \angle CFO$. In the following computation, and for the rest of the solution, we work with directed angles modulo $180°$

$$\angle AF'C = \angle AF'O + \angle OF'C = \angle OAF + \angle FCO = \angle OAD + \angle BCO$$

$$= \angle ABO + \angle ODC = \angle ABD + \angle BDC = \angle AEC,$$

where for the last step we have used the Exterior Angle Theorem in the triangle EBD. So F' is indeed on the circumcircle of AEC.

Let T be the intersection of AC and BD, and let T' be its image through the inversion. Define X_1 as the intersection of the circumcircles of BTC and ATD. We will show that the image X_1' of X_1 through the inversion is X.

Indeed, using inscribed angles and the Exterior Angle Theorem in ACF

$$\angle BX_1D = \angle BX_1T + \angle TX_1D = \angle BCT + \angle TAF = \angle AFC = \angle DFB.$$

Therefore X_1 is on the circumcircle of BDF, and so is X_1', its image through the inversion, because inversion preserves this circle (as $OB \cdot OD = OA^2$). A similar angle chasing argument gives

$$\angle CX_1A = \angle CX_1T + \angle TX_1A = \angle CBT + \angle TDA = \angle CBD + \angle BDA = \angle BFA$$

$$= \angle CFA.$$

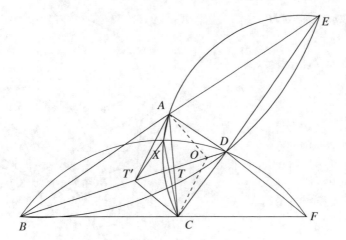

Fig. 3.45 The inversion about the circle of Apollonius

From here we deduce that A, F, C, X_1 lie on a circle, and using the inversion, we obtain that A, F', C, X_1' lie on a circle as well. So X_1' lies on the circle through A, C, F', which we have seen is the circumcircle of ACE. We conclude that X_1' is the intersection point of ACE and BDF that is distinct from F', and thus it is X.

As the quadrilateral $DX_1T'A$ is cyclic, so is the quadrilateral formed by the images of its vertices through the inversion, and the latter is $ABT'X$. Exchanging the roles of A and C, as well as B and D, we deduce that $DCT'X$ is also cyclic. Using the fact that OAC is isosceles and that $\angle OT'A = \angle TAO$ by the inversion, together with the two quadrilaterals that were proved to be cyclic, we can write

$$\angle BXA = \angle BT'A = \angle OT'A = \angle TAO = \angle OCT = \angle CT'O = \angle CT'D = \angle CXD,$$

We are working with directed angles modulo $180°$, so

$$\angle AXB = -\angle BXA = -\angle CXD = 180° - \angle CXD,$$

and the problem is solved. □

The lesson learned from this problem is that whenever you are in the presence a configuration of four points such that the cross-ratio of their coordinates has absolute value 1 (which is the case with the vertices of the quadrilateral), it is worth thinking about the circles of Apollonius.

Next, we present an example from a 2003 Team Selection Test for the Moldovan International Mathematical Olympiad Team because it makes use of the polar of a point with respect to a circle.

Problem 3.5 Let $ABCD$ be a quadrilateral inscribed in a circle of center O, and let M and N be the midpoints of the diagonals AC and BD, respectively. Prove

Fig. 3.46 The images of M'
and N'

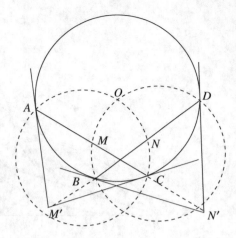

that the points O, M, B, D are concyclic if and only if the points O, N, A, C are
concyclic.

Solution 1 Consider the inversion with respect to the circumcircle of $ABCD$. Then
the circumcircle of OAC becomes the line AC, and the circumcircle of OBD
becomes the line BD. Let M' and N' be the images of M and N, respectively
(Fig. 3.46), which, as we know from Proposition 3.3, lie the first at the intersections
of the tangents to the circle of inversion at A and C and the second at the intersection
of the tangents at B and D. Then M is on the circumcircle of OBD if and only if
M' is on the line BD and N is on the circumcircle of OAC if and only if N' is on
the line AC.

But the line BD is the polar $p_{N'}$ of N' with respect to the circle of inversion, and
the line AC is the polar $p_{M'}$ of M' with respect to the same circle. By La Hire's
Theorem (Theorem 3.31), $M' \in p_{N'}$ if and only if $N' \in p_{M'}$, and the problem is
solved.   □

Solution 2 It is worth showing the complex number solution just because it is
thoroughly elementary. Four points lie on a line or circle if and only if their cross-
ratio is real. Let A, B, C, D have coordinates a, b, c, d of absolute value 1, so that
O is at the origin. The complex number translation of the equivalence from the
statement is

$$\frac{\frac{a+c}{2} - b}{\frac{a+c}{2} - d} : \frac{b}{d} \in \mathbb{R} \iff \frac{\frac{b+d}{2} - a}{\frac{b+d}{2} - c} : \frac{a}{c} \in \mathbb{R}.$$

Setting $x = a/b$, $y = c/b$, $z = a/d$, and $w = c/d$, we translate this equivalence
into

$$\frac{x + y - 2}{z + w - 2} \in \mathbb{R} \iff \frac{\dfrac{1}{x} + \dfrac{1}{z} - 2}{\dfrac{1}{y} + \dfrac{1}{w} - 2} \in \mathbb{R}.$$

Using the fact that x, y, z, w have absolute value one, by taking the complex conjugate of the right-hand side, we can transform this equivalence into

$$\frac{\dfrac{x + y}{2} - 1}{\dfrac{z + w}{2} - 1} \in \mathbb{R} \iff \frac{\dfrac{x + z}{2} - 1}{\dfrac{y + w}{2} - 1} \in \mathbb{R}.$$

Note that $xw = yz$, so the points X, Y, Z, W with coordinates x, y, z, w form an isosceles trapezoid inscribed in the unit circle with XZ parallel to YW. The first condition expresses the fact that the line joining the midpoints of XY and ZW passes through 1, while the second condition expresses the fact that the line joining the midpoints of XZ and YW passes through 1. But the two lines coincide with the midline of the trapezoid, and the problem is solved. $\square$

The next example uses two inversions simultaneously. It is a short-listed problem which was proposed by Poland for the 1998 International Mathematical Olympiad, and the first solution that we present belongs to Octav Drăgoi.

Problem 3.6 Let $ABCDEF$ be a convex hexagon such that $\angle B + \angle D + \angle F = 360°$ and

$$\frac{AB}{BC} \cdot \frac{CD}{DE} \cdot \frac{EF}{FA} = 1.$$

Prove that

$$\frac{BC}{CA} \cdot \frac{AE}{EF} \cdot \frac{FD}{DB} = 1.$$

Solution 1 Using the hypothesis of the problem, we can rewrite the conclusion as the more complicated equality

$$\frac{AB}{CA} \cdot \frac{CD}{DB} = \frac{DE}{FD} \cdot \frac{FA}{AE}.$$

We now consider two inversions, the first of center A and the second of center D, and both of arbitrary radii. With the convention that we denote, the image of a point P under the first inversion by P' and under the second inversion by P'' (Fig. 3.47), by applying the formula (3.1), we obtain

$$\frac{AB}{CA} \cdot \frac{CD}{DB} = \frac{C'D'}{D'B'} \quad \text{and} \quad \frac{DE}{FD} \cdot \frac{FA}{AE} = \frac{F''A''}{A''E''}.$$

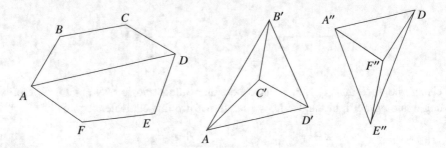

Fig. 3.47 The hexagon $ABCDEF$ and the two inversions

The problem reduces to showing that

$$\frac{C'D'}{D'B'} = \frac{F''A''}{A''E''},$$

and this is somewhat simpler. We prove this equality by placing the segments in the two similar triangles $B'C'D'$ and $E''F''A''$.

So let us show that the two triangles are similar. Rewrite the hypothesis as

$$\frac{AB \cdot CD}{BC} = \frac{DE \cdot FA}{EF},$$

and apply again formula (3.1) to obtain

$$\frac{AB \cdot CD}{BC \cdot AD} = \frac{C'D'}{B'C'} \quad \text{and} \quad \frac{DE \cdot FA}{EF \cdot AD} = \frac{F''A''}{E''F''}.$$

We obtain the desired equality of the ratios of the sides.

$$\frac{C'D'}{B'C'} = \frac{F''A''}{E''F''}.$$

As for the angles formed by those sides, we first use Theorem 3.2 to obtain

$$\angle B'C'D' = \angle B'C'A + AC'D' = \angle B + \angle ADC$$

and

$$\angle E''F''A'' = \angle E''F''D + \angle DF''A'' = \angle E + \angle DAF = 2\pi - \angle F - \angle ADE.$$

And then we use the angle equality from the hypothesis of the problem to conclude that $\angle B'C'D' = \angle E''F''A''$. This combined with $C'D'/B'C' = F''A''/E''F''$ proved above implies that the triangles $B'C'D'$ and $E''F''A''$ are similar, as claimed, which then implies

$$\frac{C'D'}{D'B'} = \frac{F''A''}{A''E''},$$

completing the proof.   □

Solution 2 The above solution can be encoded in complex numbers, where it becomes, surprisingly, more natural. Using lowercase letters for the coordinates of the uppercase points, we translate the hypothesis of the problem into the fact that the number

$$\frac{a-b}{c-b} \cdot \frac{c-d}{e-d} \cdot \frac{e-f}{a-f}$$

is both real positive (because the arguments of the three fractions add up to 2π) and has absolute value equal to 1 (the metric relation from the statement). Therefore this number is equal to 1, and we can rewrite this fact in terms of cross-ratios as

$$\frac{b-a}{b-c} : \frac{d-a}{d-c} = \frac{e-d}{e-f} : \frac{a-d}{a-f}.$$

The conclusion of the problem, as rewritten in the first solution, states that the cross-ratios

$$\frac{a-b}{a-c} : \frac{d-b}{d-c} \quad \text{and} \quad \frac{d-e}{d-f} : \frac{a-e}{a-f}$$

have equal absolute values. In fact

$$\frac{a-b}{a-c} : \frac{d-b}{d-c} = \frac{d-e}{d-f} : \frac{a-e}{a-f};$$

these cross-ratios are equal! To be able to see this, we examine separately the left-hand sides and the right-hand sides of the two equalities of cross-ratios that comprise the hypothesis and the conclusion of the problem. To this end we choose a Möbius transformation that maps a to ∞, and by using the invariance of the cross-ratios under Möbius transformations, we infer that the two left-hand sides transform into

$$\frac{d'-c'}{b'-c'} \quad \text{and} \quad \frac{d'-c'}{d'-b'}.$$

If we call the first fraction p, then the second is $\frac{p}{p-1}$. Similarly, if we choose a Möbius transformation that maps d to ∞, the cross-ratios on the right sides become

$$\frac{a'-f'}{e'-f'} \quad \text{and} \quad \frac{a'-f'}{a'-e'},$$

and if we denote the first fraction by q, the second is $\frac{q}{q-1}$. Of course $p = q$ implies $\frac{p}{p-1} = \frac{q}{q-1}$, so by using the invariance of cross-ratios under Möbius transformations, we deduce the desired equality of cross-ratios. Certainly, we could have used the more general circular transformations (and in particular inversions), as we did in the first solution, and work with complex conjugates.   □

Let us illustrate the use of the $\sqrt{bc}$ inversion with two Olympiad problems. The first has appeared in the 2015 United States of America Junior Mathematical Olympiad, being proposed by Kim Sungyoon.

Problem 3.7 Let $ABCD$ be a cyclic quadrilateral. Prove that there exists a point X on the segment BD such that $\angle BAC = \angle XAD$ and $\angle BCA = \angle XCD$ if and only if there exists a point Y on segment AC such that $\angle CBD = \angle YBA$ and $\angle CDB = \angle YDA$.

Solution We argue on Fig. 3.49. Because AC and AX are isogonals in the triangle ABD, the $\sqrt{bc}$ inversion corresponding to the vertex A in this triangle swaps B and D as well as C and X. Hence the triangles ABC and AXD are similar, which gives

$$\frac{AD}{XD} = \frac{AC}{BC}.$$

Analogously, the $\sqrt{bc}$ inversion corresponding to vertex C in the triangle CBD swaps B and D as well as A and X. Hence the triangles ABC and DXC are similar and

$$\frac{CD}{XD} = \frac{CA}{BA}.$$

Substituting XD from one equation into the other, we obtain

$$\frac{AD}{CD} = \frac{AB}{CB}.$$

Thus, the quadrilateral $ABCD$ is harmonic. Moreover, in a harmonic quadrilateral, the midpoint of BD is swapped with point C by both $\sqrt{bc}$ inversions, so it plays the role of the point X. Thus, the existence of X is equivalent to $ABCD$ being harmonic. But so is the existence of Y, so the two conditions from the statement are equivalent.   □

We have seen here yet another characterization of harmonic quadrilaterals.

The second example that illustrates the use of $\sqrt{bc}$ inversion is a problem that was given at the Romanian Master of Mathematics in 2011, being proposed by Vasily Mokin and Feodor Ivlev from Russia.

Problem 3.8 The triangle ABC is inscribed in the circle ω. A variable line ℓ parallel to BC intersects the segments AB and AC at D and E, respectively, and intersects ω at K and L such that D is between K and E. The circle γ_1 is tangent to

Fig. 3.48 Finding the locus
of the intersection point of the
interior tangents

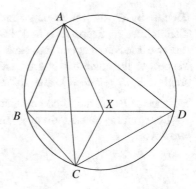

Fig. 3.49 The quadrilateral
$ABCD$ is harmonic

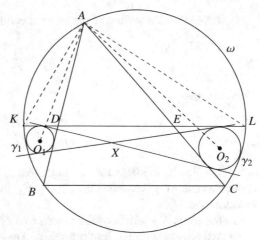

the segments KD and BD and to ω, and the circle γ_2 is tangent to the segments LE
and CE and to ω. Find the locus of the point of intersection of the common interior
tangents of γ_1 and γ_2 when ℓ varies.

Solution Denote the intersection of the interior tangents of γ_1 and γ_2 by X. The
situation is illustrated in Fig. 3.48.

Let ϕ_A be the $\sqrt{bc}$ inversion determined by the triangle AKL and the vertex A.
Then

$$\phi_A(\omega) = KL, \quad \phi_A(KL) = \omega, \quad \phi_A(AB) = AB, \quad \phi_A(AC) = AC.$$

Also $\phi_A(C) = D$ (the intersection of the image of ω with AB), and $\phi(B) = E$.
Because tangencies are preserved

$$\phi_A(\gamma_1) = \gamma_2 \text{ and } \phi_A(\gamma_2) = \gamma_1.$$

Denote by O_1 the center of γ_1 and by O_2 the center of γ_2. Because the inversion that defines ϕ_A keeps AO_1 and AO_2 invariant, after composing with the reflection, the two lines are mapped into each other. Hence, AO_1 and AO_2 are symmetric with respect to the angle bisector of $\angle KAL$, which is the same as the angle bisector of angle $\angle BAC$ (since KL and BC are parallel, the bisectors of $\angle KAL$ and $\angle BAC$ cut the arc $\overset{\frown}{BC}$ at the same point, so they coincide).

Let r_1 and r_2 be the radii of γ_1 and γ_2, respectively. Because $\angle BAO_1 = \angle CAO_2$, we have

$$\frac{AO_1}{AO_2} = \frac{r_1}{r_2}.$$

Also, if X is the intersection of the interior common tangents of γ_1 and γ_2, then

$$\frac{XO_1}{XO_2} = \frac{r_1}{r_2}.$$

Consequently

$$\frac{AO_1}{AO_2} = \frac{XO_1}{XO_2},$$

which implies, by the Bisector Theorem, that X is on the bisector of $\angle O_1AO_2$. But, again because AO_1 is mapped to AO_2 by ϕ_A, this angle bisector coincides with the bisector of $\angle BAC$.

So the locus is a subset of the bisector of $\angle ABC$. To understand which part of the locus, we examine the limiting cases. These limiting cases are when ℓ passes through A, when $X = A$, and when $\ell = BC$, when X is on BC. And, by continuity, any point on the angle bisector of $\angle BAC$ that lies on the open segment of this bisector with one endpoint A and the other endpoint on BC is part of the locus. So the locus is this open segment.   □

3.2.3   Problems in Euclidean Geometry to be Solved with Inversion (or with Möbius Transformations)

And now, a rich list of practice problems.

169 Let ω_1 and ω_2 be two circles that intersect at A and B. The tangents at A to the two circles intersect these circles a second time at M and N. Let C be the reflection of A over B. Show that the quadrilateral $AMCN$ is cyclic.

170 (*Steiner's Theorem*) Let ABC be a triangle, and let D and E be two points on the side BC such that $\angle DAB = \angle EAC$. Prove that

$$\frac{DB}{DC} \cdot \frac{EB}{EC} = \frac{AB^2}{AC^2}.$$

171 Let $ABCD$ be a quadrilateral that admits an inscribed and a circumscribed circle. Let M, N, P, Q be the tangency points of the inscribed circle with the sides AB, BC, CD, DA, respectively. Prove that MP is perpendicular to NQ.

172 Consider two lines ℓ_1 and ℓ_2 that intersect at P and the circles $\omega_1, \omega_2, \omega_3, \omega_4$ through P such that ω_1 and ω_3 are tangent to ℓ_1 and ω_2 and ω_4 are tangent to ℓ_2. Let A_j be the intersection of ω_j and ω_{j+1}, $j = 1, 2, 3, 4$ (with $\omega_5 = \omega_1$).
(a) Prove that the following relation holds:

$$\frac{A_1 A_2}{A_1 A_4} \cdot \frac{A_3 A_2}{A_3 A_4} = \frac{PA_2^2}{PA_4^2}.$$

(b) Prove that A_1, A_2, A_3, A_4 are concyclic if and only if ℓ_1 and ℓ_2 are orthogonal.

173 Let ω be a circle and let ST be a chord. The circle ω_1 is orthogonal to ω and has the center on ST. Let ω_1 intersect the line ST at A_1 and A_2 and the circle ω at B_1 and B_2, such that A_1 lies on the chord ST. Let also M be the midpoint of the arc $\overarc{ST}$ of ω that does not contain the point B_1. Prove that the lines $A_1 B_1$ and $A_2 B_2$ intersect at M.

174 *(Shoemaker's Knife Problem)* Let Ω_1 and Ω_2 be two circles that are interior tangent, and let ω_n, $n \geq 0$, be circles such that the center of ω_0 is collinear with the centers of Ω_1 and Ω_2, and for each $n \geq 1$, the circle ω_n is tangent to Ω_1, Ω_2, and ω_{n-1}. Denote by h_n the distance from the center of ω_n to the line of centers of Ω_1 and Ω_2 and by d_n the diameter of ω_n. Prove that

$$\frac{h_n}{d_n} = n.$$

175 Given a non-isosceles triangle ABC, consider the medians AA_1, BB_1, CC_1 and the circumcenter O. Show that the circumcircles of $AA_1 O$, $BB_1 O$, and $CC_1 O$ have a second intersection point besides O.

176 Let Ω and ω be two concentric circles with common center O, with ω inside Ω. Consider a point A on Ω, and let B and C be the points of intersection of ω with the circle of diameter OA. Let also D be the point where the ray $|OA$ intersects ω, and let E and F be the points where the rays $|OB$ and $|OC$ intersect Ω, respectively. Show that D, E, F are collinear.

177 Show that the three perpendiculars taken from the vertices of a triangle onto the corresponding sides of its orthic triangle intersect at the circumcenter of the triangle.

178 Let Ω be a circle of diameter AB, let P be a point outside of Ω that does not lie on the tangent at A, and let t be the tangent at B. Consider the points C and D on the circle Ω such that PC and PD are tangent to the circle, and let C', D', P' be the intersections of AC, AD, and AP with the line t, respectively. Prove that P' is the midpoint of the segment $C'D'$.

179 Two points, A and B, are chosen on a circle k of center S such that $\angle ASB = 90°$. The circles k_1 and k_2 are internally tangent to k at A and B, respectively, and are also externally tangent to each other. The circle k_3 lies in the interior of the angle $\angle ASB$ and is internally tangent to k at C and externally tangent to k_1 and k_2 at X and Y, respectively. Prove that $\angle XCY = 45°$.

180 (a) Let ω and ω_1 be two circles such that ω_1 is interior tangent to ω at A. A chord ST of ω is tangent to ω_1 at B. Let M be the midpoint of the arc $\overset{\frown}{ST}$ of ω that does not contain A. Prove that A, B, M are collinear.
(b) Let ω be a circle, ST a chord of ω, and ω_1 and ω_2 two circles that are tangent to both ω and ST and intersect at P and Q. Prove that the line PQ passes through the midpoint of the arc $\overset{\frown}{ST}$ of ω that lies on the other side of ST from ω_1 and ω_2.
(c) Let ABC be a triangle, D a point on the side BC, ω_1 a circle tangent to the circumcircle and to the segments AD and BD, and ω_2 a circle tangent to the circumcircle and to the segments AD and CD. Prove that ω_1 and ω_2 are tangent if and only if $\angle BAD = \angle DAC$.

181 A circle ω is tangent to the parallel lines ℓ_1 and ℓ_2, and the circles ω_1 and ω_2 are exterior tangent to each other and to ω, also ω_1 is tangent to ℓ_1, while ω_2 is tangent to ℓ_2. Let A be the point where ω_1 and ω are tangent, and let B be the point where ω_2 and ℓ_2 are tangent. Prove that the line AB is the common interior tangent of ω and ω_1.

182 Let ABC be a triangle with circumcenter O. On the sides, AC and BC are constructed; in the exterior, the rectangles $ACDE$ and $BCFG$ of equal areas. Let M be the midpoint of the segment DF. Prove that the points O, C, M are collinear.

183 A right triangle ABC ($\angle B = 90°$) is inscribed in a circle. We let K and N be the midpoints of the arc $\overset{\frown}{BC}$ and the segment AC, respectively, and let M be the second intersection point of KN with the circle. The tangents to the circle at B and C intersect at E. Prove that $\angle EMK = 90°$.

184 Let ABC be a triangle that is not isosceles and let H be its orthocenter. Denote by A_1, B_1, C_1 the feet of the altitudes AH, BH, CH, respectively. Let A_2, B_2, C_2 be the projections of H onto B_1C_1, C_1A_1, and A_1B_1, respectively. Prove that:

(a) the circumcircles of the triangles HA_1A_2, HB_1B_2, HC_1C_2 have a second common point besides the orthocenter H.
(b) the circumcircles of the triangles HAA_2, HBB_2, HCC_2 have a second common point besides the orthocenter H.

185 Let C, C_1, C_2, C_3 be four circles in the plane, and let ℓ be a line that does not intersect C, such that each of the circles C_1, C_2, C_3 is tangent to the other two, as well as to the circle C and to the line ℓ. Knowing that the radius of C is 1, find the distance from its center to ℓ.

186 Let ABC be a triangle for which $\angle A < \angle C$, with two points, D and E, chosen on the sides AC and AB, respectively, so that $\angle BED = \angle C$. Let also F be a point inside the quadrilateral $BCDE$ such that the circumcircles of BCF and DEF are tangent and the circumcircles of BEF and CDF are tangent. Prove that the quadrilateral $ACFE$ is cyclic.

187 In the triangle ABC, consider the excircle tangent to side BC at A_1 and to the lines AC and AB at B_1 and C_1, respectively, and let I_a be the center of this circle. Denote by A', B', C' the midpoints of the segments B_1C_1, A_1C_1, and A_1B_1, respectively. Let ℓ_a be the line that passes through I_a and the circumcenter of $A'B'C'$. Using the excircles corresponding to the other two sides, define similarly the lines ℓ_b and ℓ_c. Prove that the lines ℓ_a, ℓ_b, and ℓ_c are concurrent.

188 In a triangle ABC, D is the foot of the altitude from A, E is the point where the symmedian from A intersects the circumcircle, and F is the point where the diameter from A of the circumcircle intersects the side BC. Prove that the circumcircles of the triangles ABC and DEF are tangent.

189 In the triangle ABC, consider a point X on the altitude from A. The circles C_1 and C_2 have centers on the line BC, and C_1 passes through B and X, while C_2 passes through C and X. Let M and N be the intersections of C_1 with the side AB and with the altitude from B, and let P and Q be the intersections of C_2 with the altitude from C and with the side AC. Prove that the points M, N, P, Q are collinear.

190 Let ABC be a triangle, and let K and L be the midpoints of the sides AB and AC, respectively. Let P be the second intersection point of the circumcircles of the triangles ABL and ACK, and let Q be the second intersection point of the line AP with the circumcircle of the triangle AKL. Prove that $2AP = 3AQ$.

191 Let ω be the circumcircle of the triangle ABC, and let ℓ be the line through A that is tangent to ω. The circles ω_1 and ω_2 are tangent to ℓ and to the line BC and are exterior tangent to ω. Let D and E be the points where ω_1 and ω_2 are tangent to BC, respectively. Prove that the circumcircles of ABC and ADE are tangent.

192 Let $ABCD$ be a quadrilateral satisfying:

(i) $\angle ABC = \angle ADC = 135°$
(ii) $AC^2 \cdot BD^2 = 2AB \cdot BC \cdot CD \cdot DA$.

Prove that the diagonals of the quadrilateral are orthogonal.

193 Connect a point M with the vertices of the triangle ABC. Prove that the perpendiculars from the orthocenter H onto MA, MB, and MC meet BC, CA, and AB, respectively, at three collinear points. Show, additionally, that the line determined by these three points is orthogonal to MH.

194 Let ABC be a triangle, and let D be the second intersection point of the circle that passes through B and is tangent to the line AC at A with the circle that passes through C and is tangent to the line AB at A. Let E be a point on the line AB such that B is the midpoint of AE, and let F be the second point of intersection of the ray $|CA$ with the circumcircle of ADE. Prove that $AF = AC$.

195 Let ABC be a triangle and let O be its circumcenter. The lines AB and AC intersect the circumcircle of the triangle BOC again at B_1 and C_1, respectively. Let D be the intersection of BC and B_1C_1. Prove that the circle tangent to AD at A and with center on B_1C_1 is orthogonal to the circle of diameter OD.

196 In the acute triangle ABC, let M be the midpoint of the angle bisector AD ($D \in BC$). Consider a point X on the segment BM such that $\angle MXA = \angle DAC$. Show that $\angle AXC = 90°$.

197 Can one find 1975 points on the unit circle such that the distance between any two is rational?

198 Let P be a point inside the triangle ABC such that

$$\angle APB - \angle C = \angle APC - \angle B.$$

Let D and E be the incenters of the triangles APB and APC, respectively. Prove that AP, BD, CE meet at one point.

199 Let BD be the internal bisector of $\angle B$ in the triangle ABC ($D \in AC$). The line BD intersects the circumcircle Ω of the triangle ABC at $E \neq B$. The circle ω of diameter DE intersects Ω again at F. Prove that BF is a symmedian in the triangle ABC.

200 Let Ω be the circumcircle of the triangle ABC, and let ω be a circle that passes through B and C, whose center we denote by O_1. We also denote by E and F the second intersection points of ω with the lines AB and AC, respectively, by Q the intersection of the lines EF and BC, and by P the second intersection point of AQ with Ω. Prove that $\angle APO_1 = 90°$.

201 Let ABC be a triangle, let Ω be its circumcircle, let ω be the C-mixtilinear incircle, and let ℓ be the line that is parallel to the side AB, is tangent to the circle ω, and crosses the sides AC and BC. Denote by P and Q the tangency points of the circle ω with the circle Ω and the line ℓ, respectively. Prove that $\angle ACP = \angle BCQ$.

202 Prove that the perpendiculars from the vertices of a triangle onto the lines determined by the orthocenter and the midpoints of the opposite sides meet the corresponding sides along a line orthogonal to Euler's line.

203 Let $ABCD$ be a cyclic quadrilateral, and let M and N be the midpoints of the diagonals AC and BD. Prove that AC is the angle bisector of $\angle BMD$ if and only if BD is the angle bisector of $\angle ANC$.

204 Let Ω_A be the A-mixtilinear incircle of the triangle ABC, and let Ω_A touch the circumcircle Ω of ABC at U_a. Invert the circumcircle of ABC about its incircle, obtaining Ω'. Prove that A and U_a are mapped to diametrically opposite points in Ω'.

205 Let Ω and ω be two circles interior tangent at P (with ω inside Ω), and let AB be a chord of Ω that is tangent at some point C to ω. The line PC intersects a second time the circle Ω at Q, and the tangents from Q to ω intersect again Ω at R and S. Let I, X, Y be the incenters of triangles ABP, ABR, and ABS, respectively. Show that $\angle PXI + \angle PYI = 90°$.

206 A circle with center O passes through the points A and C and intersects the sides AB and BC of the triangle ABC at the points K and N, respectively. The circumcircles of the triangles ABC and KBN intersect at two distinct points B and M. Prove that $\angle OMB = 90°$.

207 Let ABC be an acute triangle with circumcenter O, and let BE be the altitude from B ($E \in AC$). The circumcircle of AOC intersects AB at $P \neq A$ and BC at $Q \neq C$. Let ω be the circle of center E and radius EB. Prove that P, Q, E are collinear if and only if the circumcircle of AOC is tangent to ω.

208 Two circles Γ_1 and Γ_2 intersect at the points X and Y. The line through Y that is parallel to the closer common tangent to Γ_1 and Γ_2 intersects Γ_1 and Γ_2 a second time at A and B, respectively. Let O be the center of the circle that is exterior tangent to Γ_1 and Γ_2 and interior tangent to the circumcircle of XAB. Prove that the line XO is the angle bisector of $\angle AXB$.

209 Let ω be a circle in the plane, and let A and B be two points on it. Let M be the midpoint of the chord AB, and let P be another point on this chord. Construct the circles γ and δ tangent to AB at P and to ω at C and D, respectively. Let E be the point that is diametrically opposite to D in ω. Prove that the circumcenter of the triangle BMC lies on the line BE.

210 Let ABC be an acute triangle, with $AB > AC$, let Γ be its circumcircle, let H be the orthocenter, and let F be the foot of the altitude from the vertex A. Let also M be the midpoint of the side BC, and let Q, K be the points on the circle Γ for which $\angle HQA = \angle HKQ = 90°$. We assume that the points A, B, C, K, Q are distinct and show up on the circle Γ in this order. Prove that the circumcircles of triangles KHQ and FKM are tangent to each other.

211 Let $ABCD$ be a convex quadrilateral with $\angle B = \angle D = 90°$. The point H is the foot of the perpendicular from A to BD. The points S and T are chosen on the sides AB and AD, respectively, in such a way that H lies inside triangle SCT and

$$\angle SHC - \angle BSC = 90°, \qquad \angle THC - \angle DTC = 90°.$$

Prove that the circumcircle of triangle SHT is tangent to the line BD.

212 Let ABC be a triangle with incenter I and circumcenter O. A circle is tangent to the side AB at point K, to the side BC at point L, and is externally tangent to the circumcircle of AOC. Prove that the midpoint of BI lies on the line KL.

213 Let ABC be a triangle with circumcenter O. The circle tangent to the side AB that passes through the points A and C meets the circumcircle of BOC at $T \neq C$. The lines TO and BC meet at the point K. Show that the line AK is tangent to the circumcircle of the triangle ABC.

214 Let P be a point on the circumcircle ω of the triangle ABC. The lines PA and BC meet at A_1, the lines PB and CA meet at B_1, and the lines PC and AB meet at C_1. The inversion about ω maps A_1 to A_2, B_1 to B_2, and C_1 to C_2. Prove that the lines AA_2, BB_2, and CC_2 have a common point whose isogonal conjugate is on the nine-point circle of ABC.

215 Let ABC be a triangle with AD as altitude ($D \in BC$). Let M be the midpoint of AD. The points X and Y are the orthogonal projections of D onto CM and BM, respectively. The lines BX and CY meet at Z. Show that the circumcircle of the triangle XYZ and the circle with diameter AD are tangent.

216 Let ω be the circumcircle of the triangle ABC. The mixtilinear incircle γ is tangent to the sides AB and AC at the points P and Q, respectively, and to the circumcircle ω at the point S. The lines AS and PQ meet at T. Prove that $\angle BTP = \angle CTQ$.

217 Let O_b and O_c be the centers of the B- and C-mixtilinear excircles of the triangle ABC, respectively. These mixtilinear circles touch the circumcircle of ABC at R and S, respectively, and the line BC at D and E, respectively. The lines O_bS and O_cR meet at X, and the lines O_bE and O_cD meet at Y. Prove that $\angle BAX = \angle CAY$.

218 The incircle of a scalene triangle ABC has center I and touches the side BC at the point D. The point X lies on the arc $\overset{\frown}{BC}$ of the circumcircle of ABC that does not contain A, and is such that if E and F are the orthogonal projections of X onto BI and CI, and if M is the midpoint of EF, then $MB = MC$. Prove that $\angle BAD = \angle CAX$.

219 Let ABC be a scalene triangle with circumcircle Ω, and suppose the incircle of ABC touches BC at D. The angle bisector of $\angle A$ meets BC and Ω at E and F, respectively. The circumcircle of the triangle DEF intersects the A-excircle at the points S_1 and S_2 and the circumcircle Ω the second time at T. Prove that the line AT passes through either S_1 or S_2.

220 *(Descartes' Theorem)* Let $\omega_1, \omega_2, \omega_3, \omega_4$ be four circles that are pairwise exterior tangent, and let r_1, r_2, r_3, r_4 be their radii, respectively. Show that

$$2\left(\frac{1}{r_1^2} + \frac{1}{r_2^2} + \frac{1}{r_3^2} + \frac{1}{r_4^2}\right) = \left(\frac{1}{r_1} + \frac{1}{r_2} + \frac{1}{r_3} + \frac{1}{r_4}\right)^2.$$

221 Let ABC be a triangle, and let D, E, and F be the tangency points of the incircle with BC, CA, and AB, respectively. Let EF meet the circumcircle Γ of ABC at X and Y. Furthermore, let T be the second intersection point of the circumcircle of DXY with the incircle. Prove that AT passes through the tangency point of the A-mixtilinear incircle with Γ.

Chapter 4
A Synthesis

4.1 Bringing Together All Transformations

In the first three chapters, we have put each geometric transformation in its own spotlight. And we have observed that each has its own character, leading to a specific mindset. Now that we understand the transformations well enough, it is the time to allow them to work together. As different as they look synthetically, in the analytic perspective, the main actors of this book are very much alike. Just take a look at the list of orientation-preserving transformations:

Transformation	General equation in $\mathbb{C}$
Translation	$f(z) = z + b$
Rotation	$f(z) = az + b, |a| = 1, a \neq 1$
Homothety	$f(z) = az + b, a \in \mathbb{R} \setminus \{0, 1\}$
Spiral similarity	$f(z) = az + b, a \in \mathbb{C} \setminus \mathbb{R}$
Möbius transformation	$f(z) = \dfrac{az + b}{cz + d}, ad - bc \neq 0$

The first four are particular cases of the affine transformation $f(z) = az + b$, and *all* transformations are, in fact, Möbius transformations! And we have seen in Sect. 3.1.5 that Möbius transformations themselves are closely related to linear transformations. If you throw in reflections over lines and circles, you have the complete picture of all the transformations discussed in this book. The point is that all these transformations are naturally intertwined; this is why so many problems admit solutions based on different transformations. We should not think of a problem as a "translation problem" or an "inversion problem"; but as a "transformation problem"!

Then the reader might ask, "why think of synthetic geometry at all and not focus on the analytic?" The answer is that, while complex geometry is potentially effective, it masks the geometric insight under lengthy, and sometimes fruitless,

© The Author(s), under exclusive license to Springer Nature Switzerland AG 2022
R. Gelca et al., *Geometric Transformations*, Problem Books in Mathematics,
https://doi.org/10.1007/978-3-030-89117-6_4

computations. Moreover, a good synthetic observation leads organically to the right transformation, and points to a more concise and elegant solution.

Finally, there is nothing wrong with integrating synthetic observations with complex computations. Some of these computations have synthetic meanings that are harder to convey through synthetic means. Better be pragmatic than purist.

4.1.1   Some Examples

Let us visit mathematics' most ubiquitous theorem.

Theorem 4.1 (Pythagorean Theorem) *In a right triangle, the square of the length of the hypotenuse is equal to the sum of the squares of the lengths of the other two sides.*

Proof 1 We present first the proof by *rotation* published by Euclid in his *Elements*. Let the right triangle be ABC with $\angle A = 90°$. Construct in the exterior of the triangle the squares $ABMN$, $BCPQ$, and $CARS$ as shown in Fig. 4.1. We are supposed to prove that the areas of $ABMN$ and $CARS$ add up to the area of $BCPQ$.

Construct the altitude AD and let E be the intersection of the lines AD and PQ.

The triangles MBC and ABQ are mapped into each other by a 90° rotation about B, so they have the same area. Similarly, the triangles SCB and ACP are mapped into each other by a 90° rotation about C, so they have the same area. The area of the triangle MBC is half the area of the square $ABMN$, because they have the same base MB and the same height MN. For a similar reason, the area of the triangle ABQ is half the area of the rectangle $BDEQ$. Thus, the area of the square $ABMN$ is equal to the area of the rectangle $BDQE$. The same argument shows that the area of the square $CARS$ is equal to the area of the rectangle $CDEP$. But the rectangles

Fig. 4.1 Proof of the Pythagorean Theorem using rotations

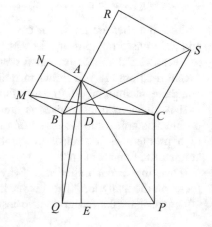

$BDQE$ and $CDEP$ make up the rectangle $BCPQ$. Hence, the area of the latter is equal to the sum of the areas of $ABMN$ and $CARS$, and the Pythagorean Theorem is proved. ☐

Proof 2 The second proof, by *translation*, has appeared in Roger B. Nelsen's book *Proofs without Words*. It can be followed on Fig. 4.2. On the sides BC, AC, and AB of the right triangle ABC ($\angle A = 90°$), construct in the exterior the squares S_1, S_2, S_3, respectively. We want to show that S_1 can be dissected into pieces that can be translated so as to produce the squares S_2 and S_3. That would imply that the area of S_1 is the sum of the areas of S_2 and S_3, which proves the Pythagorean Theorem.

To this end, translate the sides of S_1 to segments whose midpoints coincide with the center of S_3. They divide S_3 into four quadrilaterals, α, β, γ, and δ, which are equal because they map into one another by rotations about the center of S_3. Let x and y be the lengths of the two segments that the side of S_3 is divided into (see Fig. 4.2). Translate α, β, γ, and δ inside S_1 as shown in the figure. A square S is left uncovered at the center of S_1. We are to show that S is equal to S_2. The side-length of S is $y - x$. By examining the parallelogram having BC as one side that appears in the figure (this parallelogram has a side equal to y and the opposite side equal to $x + AC$), we deduce that the side-length of S_2 is also equal to $y - x$. Thus, S and S_2 are equal, and the theorem is proved. ☐

Fig. 4.2 Proof of the Pythagorean Theorem using translations

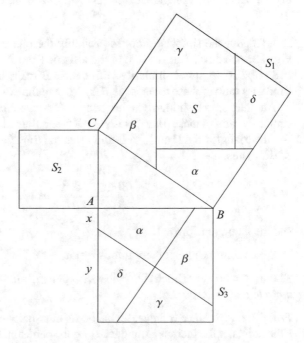

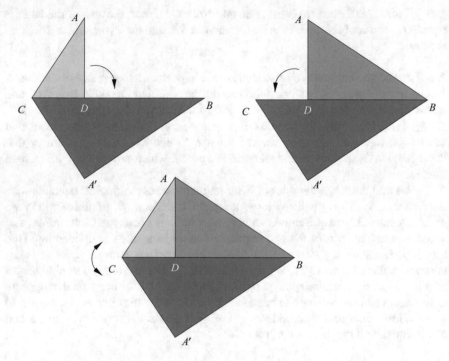

Fig. 4.3 Proof of the Pythagorean Theorem using spiral similarity

Proof 3 But the simplest proof is probably the one by *spiral similarity* shown in Fig. 4.3. As above, D is the foot of the altitude from A, and let us reflect A over BC to a point A'. A spiral similarity with center B maps triangle BAD to BCA'. The similarity ratio of these triangles is BA/BC and the ratio of their areas is BA^2/BC^2.

Another spiral similarity with center C maps triangle CAD to CBA'. The similarity ratio of these triangles is BA/BC and the ratio of their areas is CA^2/BC^2. But the triangles BAD and CAD add up to the triangle ABC which is congruent to $A'BC$. Thus,

$$\frac{BA^2}{BC^2} + \frac{CA^2}{BC^2} = 1,$$

and the theorem is proved.   □

We continue with another famous result.

Theorem 4.2 (Feuerbach) *In every triangle, the nine-point circle is tangent to the incircle and the excircles.*

Proof Let ABC be a triangle, let ω_9 be its nine-point circle, let I be the center and r the radius of the incircle ω, and let I_a be the center and r_a the radius of the excircle ω_a corresponding to the vertex A. We denote by D and D_a the tangency points of

Fig. 4.4 Proof of
Feuerbach's Theorem

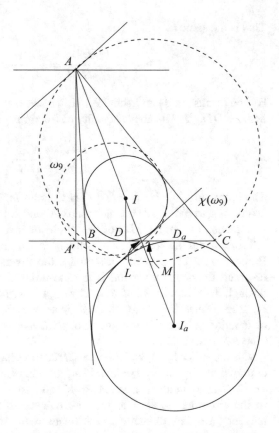

ω and ω_a with the side BC, respectively. Finally, let $A', M, L \in BC$ be such that AA', AM, AL are the altitude and median from A and the angle bisector of $\angle BAC$, respectively. The proof can now be followed on Fig. 4.4.

There is a *direct homothety* of center A and an *inverse homothety* of center L that map ω to ω_a, and consequently

$$\frac{LD}{LD_a} = \frac{r}{r_a} = \frac{AI}{AI_a}.$$

The latter is further equal to the ratio of the distances from A' to ID and $I_a D_a$, hence to $A'D/A'D_a$. From the equality,

$$\frac{LD}{LD_a} = \frac{A'D}{A'D_a}$$

we can derive

$$\frac{LD_a + LD}{LD_a - LD} = \frac{A'D_a - A'D}{A'D_a + A'D}.$$

This is the same as

$$\frac{DD_a}{2LM} = \frac{2A'M}{DD_a}.$$

By the discussion from Sect. 2.2.1 II, we know that M is the midpoint of DD_a, so $MD = DD_a/2$. The above equality of fractions yields

$$\frac{DM}{LM} = \frac{A'M}{DM}.$$

Therefore, $MA' \cdot ML = MD^2$. There is an *inversion* χ hidden in the figure, of center M and radius MD, which maps A' and L into each other.

Because the radius MD of the circle of inversion is orthogonal to the radius ID of ω and the circle ω is orthogonal to the circle of inversion, hence by Proposition 3.25, ω is invariant under the inversion. For the same reason, ω_a is invariant under the inversion. The nine-point circle ω_9 on the other hand passes through the midpoint M of BC, so $\chi(\omega_9)$ is a line that is parallel to the tangent to ω_9 at M. Since $A' \in \omega_9$, $L = \chi(A')$ is on $\chi(\omega_9)$. We deduce that the image $\chi(\omega_9)$ of the nine-point circle is a line through L that is parallel to the line tangent to the nine-point circle at M.

On the other hand, by Theorem 2.10, the nine-point circle is the image of the circumcircle by an *inverse homothety* of center the centroid of ABC and of ratio $-1/2$. This homothety maps A to M, and consequently it maps the line tangent to the circumcircle at A to the line tangent to the nine-point circle at M. The tangent ℓ to the circumcircle at A forms with AC an angle equal to $\angle ABC$, and consequently its *reflection* ℓ' over the angle bisector AI of $\angle BAC$ is parallel to BC (the antiparallel is mapped to the parallel). The line through L that is parallel to ℓ' is therefore just BC, and this is the common tangent of ω and ω_a. The line through L that is the reflection of BC with respect to the angle bisector AI (which is also the line of centers of ω and ω_a) is therefore the other common tangent of the two circles.

So $\chi(\omega_9)$ is a line tangent to $\omega = \chi(\omega)$ and $\omega_a = \chi(\omega_a)$. Because inversion preserves tangencies, ω_9 is tangent to both ω and ω_a. By symmetry, the nine-point circle is tangent to the other two excircles, and the theorem is proved.   □

The geometric transformations mindset brings on a system of coordinates associated with a (convex) cyclic quadrilateral that can be quite handy. If $ABCD$ is the cyclic quadrilateral and P is the intersection of the diagonals AC and BD, then the triangles APB and DPC are similar. The triangle DPC can be obtained from APB by an *inverse homothety* of center P (and ratio $-k$, for some $k > 0$) followed by a *reflection* over the angle bisector of $\angle APB$ (Fig. 4.5). We can therefore choose a system of coordinates with P at the origin, such that the coordinate of A is 1, while the coordinate of B is $a\omega$, for some $a > 0$ and $|\omega| = 1$. Then the coordinates of C and D are $-ka$ and $-k\omega$, respectively. This is illustrated in Fig. 4.6.

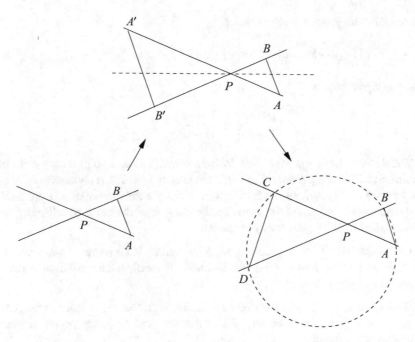

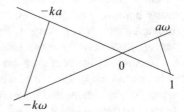

Fig. 4.5 Constructing a cyclic quadrilateral from a triangle using homothety and reflection

Fig. 4.6 Coordinates for a
cyclic quadrilateral

Let us show how this coordinate system simplifies the solution to a problem of
Titu Andreescu that was given at the USA Mathematical Olympiad in 1999.

Problem 4.1 Let $ABCD$ be a cyclic quadrilateral. Prove that

$$|AB - CD| + |AD - BC| \geq 2|AC - BD|.$$

Solution In this system of coordinates, we compute

$$AB = |1 - \omega a| = |\omega| \cdot |\overline{\omega} - a| = |\omega - a|, \quad CD = k|\omega - a|,$$
$$AD = |1 + k\omega| = |\omega| \cdot |\overline{\omega} + k| = |\omega + k|, \quad BC = a|\omega + k|,$$
$$AC = 1 + ka, \quad BD = a + k.$$

We are left with proving the inequality

$$|1 - k| \cdot |\omega - a| + |1 - a| \cdot |\omega + k| \geq 2|1 - a| \cdot |1 - k|,$$

which is equivalent to

$$\frac{|\omega - a|}{|1 - a|} + \frac{|\omega + k|}{|1 - k|} \geq 2.$$

The distance from a point on the positive x-semiaxis to a point on the unit circle is minimized when the point on the unit circle is at 1, which is a consequence, as it can be seen on Fig. 4.7, of the fact that in a triangle the longer side is opposed to the larger angle. So each of the terms on the left side of the last inequality is greater than or equal to 1. The inequality is proved.   □

Let us present now a problem of Vyacheslav Viktorovich Proizvolov that has appeared in the Russian journal *Kvant (Quantum)*, whose solution combines *inversion* and *homothety*.

Problem 4.2 Let $\omega_1, \omega_2, \omega_3, \omega_4$ be four circles such that ω_j is exterior tangent to ω_{j+1}, $j = 1, 2, 3, 4$, ($\omega_5 = \omega_1$). Show that the four tangency points are either concyclic or collinear.

Solution Let A_j be the tangency point of ω_j and ω_{j+1}, $j = 1, 2, 3, 4$. There are many possible configurations, some of which are illustrated in Fig. 4.8. What is important is that for every j, the circles ω_{j-1} and ω_{j+1} are outside ω_j, so the tangency points of these two latter circles with ω_{j+2} are outside ω_j.

We simplify the configuration by an *inversion* of center A_1, and use the standard convention that the inverse of a geometric object is denoted by adding a dash to its name. The circles ω_1 and ω_2 are mapped to two parallel lines ω_1' and ω_2', which are tangent to the circles ω_3' and ω_4'. Because the circles are exterior tangent, the point A_3' lies between the two lines. Two possible configurations are shown in Fig. 4.9.

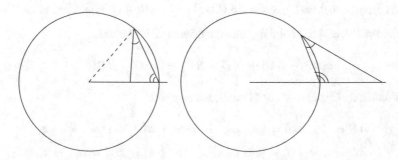

Fig. 4.7 Minimizing the distance from a point on the positive x-semiaxis to a point on the unit circle

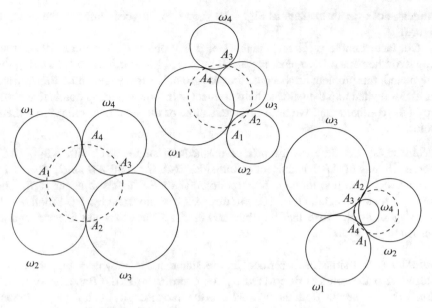

Fig. 4.8 Concyclic tangency points

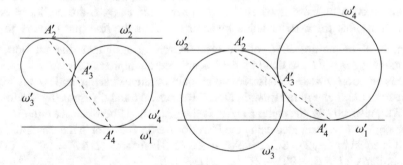

Fig. 4.9 Collinear tangency points

The *inverse homothety* of center A'_3 that maps ω'_3 to ω'_4 must map A'_2 to A'_4 because it maps tangents to tangents and parallel lines to parallel lines. Thus, A'_2 and A'_4 are collinear with the center of homothety. In other words, A'_2, A'_3, A'_4 are collinear, and as inversion maps, a line to a line through the center of inversion or a circle through the center of inversion, A_1, A_2, A_3, A_4, is either collinear or concyclic.   □

Certainly, this problem can also be solved by an angle chasing argument. But there are many possible configurations and the solution is case-dependent. The combined use of inversion and homothety addresses all cases at once. The property does not necessarily remain true if we relax the condition for the circles to only be tangent, not exterior tangent, as Fig. 4.10 shows. What goes wrong in the inverted figure?

Our last example is the most elaborate; it is Problem 6 from the 2019 International Mathematical Olympiad, proposed by Anant Mudgal from India. During the Olympiad, this problem has become the playground of many geometric transformations, all invited into the game to help connect somehow the many points that show up in the configuration. We have collected some of the solutions, and present them below.

Problem 4.3 Let I be the incenter of the acute triangle ABC with $AB \neq AC$. The incircle ω of ABC is tangent to the sides BC, CA, and AB at D, E, and F, respectively. The line through D perpendicular to EF meets ω again at R. The line AR meets ω again at P. The circumcircles of the triangles PCE and PBF meet again at Q. Prove that the lines DI and PQ meet on the line through A perpendicular to AI.

Solution 1 The situation described in the statement is shown in Fig. 4.11. We present first the solution discovered by the coordinator Žarko Ranđelović, which uses *spiral similarity*. Both the line DI and the line through A that is perpendicular to AI are easy to understand; the first is perpendicular to BC at D, and the second is the exterior angle bisector of $\angle A$. It is therefore natural to start by letting L be the intersection of these lines and try to prove that P, Q, L are collinear. Note additionally that the exterior angle bisector of $\angle A$ is the line that joins A to the midpoint Y of the arc $\overset{\frown}{BAC}$ of the circumcircle. Throughout this solution, we assume that $\angle B > \angle C$, so that Fig. 4.12 is an accurate representation.

This solution will use a considerable amount of angle chasing, and so it is worth computing the angles of the triangle DEF. Because AI and CI *rotate* by $90°$ to EF and ED, respectively, the angle $\angle DEF$ is the acute angle between AI and CI. The latter angle can be computed using the Exterior Angle Theorem in the triangle AIC, and it is equal to $\angle A/2 + \angle C/2 = 90° - \angle B/2$. Therefore, $\angle DEF = 90° - \angle B/2$, and similarly $\angle DFE = 90° - \angle C/2$, $\angle EDF = 90° - \angle A/2$.

If we add to the figure the circumcircle of BIC, then it appears that Q lies on it (see Fig. 4.12). This is indeed true, so let us prove it with the help of angle chasing. Because B, F, Q, P are on a circle, $\angle BQP = \angle BFP$, and because C, E, Q, P are on a circle, $\angle PQC = \angle PEC$. Combining these facts and using inscribed angles in the circumcircles of BPF, CPE, and DEF, we obtain

Fig. 4.10 A counterexample

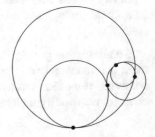

Fig. 4.11 Statement of
Problem 4.3

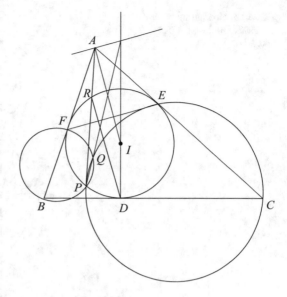

$$\angle BQC = \angle BQP + \angle PQC = \angle BFP + \angle PEC = \frac{1}{2}\,\overset{\frown}{BP} + \frac{1}{2}\,\overset{\frown}{PE}$$

$$= \angle FEP + \angle PFE = 180° - \angle EPF = 180° - \angle EDF = 90° + A/2 = \angle BIC.$$

This shows that $BQIC$ is a cyclic quadrilateral, as claimed.

Next, note that $\angle ALD$ is the angle between AL and LD, which is the same as
the angle between AI and BC because AL and LD rotate by $90°$ to AI and BC.
And the angle bisector AI forms with the side BC an angle equal to $\angle A/2 + \angle C$.
Therefore,

$$\angle ALD = \angle A/2 + \angle C.$$

Also, using inscribed angles in the incircle and the above equalities, we can compute

$$\angle APD = \angle RPD = \angle RFD = \angle RFE + \angle EFD = \angle RDE + \angle EFD$$

$$= 90° - \angle DEF + \angle EFD = 90° - (90° - \angle B/2) + 90° - \angle C/2$$

$$= 90° + \angle B/2 - \angle C/2 = \angle A/2 + \angle B.$$

Hence, $\angle ALD + \angle APD = \angle A + \angle B + \angle C$, so $APDL$ is a cyclic quadrilateral.

Proving that $BQIC$ and $APLD$ are cyclic is one of two major steps toward our
goal of connecting P and Q to the rest of the configuration. The other step involves
a geometric transformation, and for that we let T and X be the second intersection
points of PQ and ID with the circumcircle of BIC, respectively.

Fig. 4.12 Configuration with
the circumcircles of ABC
and BIC

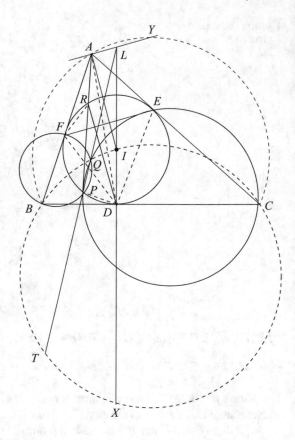

We will show that there is a *spiral similarity* that takes

$$X \mapsto D, \ T \mapsto P, \ B \mapsto F, \ I \mapsto R, \ C \mapsto E, \ Y \mapsto A.$$

We proceed to Fig. 4.13, from which we read

$$\angle FER = \angle FDR = 90° - \angle DFE = \angle C/2 = \angle BCI,$$

and similarly $\angle EFR = \angle CBI$, so that the triangles BIC and FRE are directly
similar. By Proposition 2.21, there is a spiral similarity s that maps the triangle
BIC to the triangle FRE, that is, $s(B) = F$, $s(I) = R$, $s(C) = E$. The spiral
similarity s is the one we have predicted.

Because IX is perpendicular to BC, and RD is perpendicular to EF, the point
X plays in the configuration consisting of the triangle BIC and its circumcircle the
same role that D plays in the configuration consisting of the triangle FRE and its
circumcircle. From this we deduce that $s(X) = D$.

Also, both triangles BYC and EAF are isosceles, and $\angle BYC = \angle A = \angle EAF$,
so they are directly similar. Hence, they are mapped one into the other by a spiral

Fig. 4.13 Proof that
X, T, B, I, C, Y are mapped
to D, P, F, R, E, A

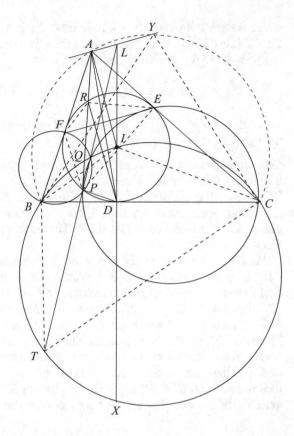

similarity, and this spiral similarity maps BC to EF, so it coincides with s. Thus, $s(Y) = A$.

Finally, using inscribed angles in the circumcircles of PFB and BIC, we have

$$\angle PEF = \angle PFB = \angle PQB = \angle TQB = \angle TCB.$$

Similarly, using the circumcircles of PCE and BIC, we obtain

$$\angle PFE = \angle TBC.$$

It follows that the triangles TBC and PFE are similar, the spiral similarity that maps the first into the second is s, and hence $s(T) = P$. Therefore, s maps the points as specified.

Let us see how we can use this spiral similarity to complete the solution. Spiral similarities preserve angles, so

$$\angle YTX = \angle s(Y)s(T)s(X) = \angle APD = 180° - \angle ALD = \angle YLX,$$

where for the third equality, we have used the fact that $APDL$ is cyclic. Hence, $YLTX$ is cyclic too. Using the fact that these two quadrilaterals are cyclic, and that spiral similarities preserve angles, we can write

$$\angle TLX = \angle TYX = \angle s(T)s(Y)s(X) = \angle PAD = \angle PLD = \angle PLX.$$

We deduce that T, P, L are collinear. But Q belongs to the line PT, so PQ passes through L. The problem is solved.   □

Solution 2 Here is a solution that combines *reflections* over lines and circles, which was found by Ilya Igorevich Bogdanov, member of the Problem Selection Committee. The midpoint A' where AI intersects EF and the point K where DI intersects ω appear naturally in the figure. Drawing several situations, we may guess that K, A', P are collinear (Fig. 4.14). The solution begins with the proof of this fact.

First, notice that K is the image of R under the reflection over AI. Indeed $\angle DRK = 90°$, because KD is a diameter in ω, and so RK is parallel to EF. And since AI is perpendicular to the chord EF, it is also perpendicular to the chord RK; thus, it is the perpendicular bisector of RK. From here we deduce that PR and PK are isogonal in the triangle PRK. But PR is symmedian in this triangle, by Theorem 1.22, so PK is median, and therefore it passes through A'.

We *invert* with respect to the incircle ω, and use the convention that the images of the points bear dashes (Fig. 4.15). There are some points that are fixed by the inversion: D, E, F, P, R, K. By Proposition 3.3, A' is the image of A, which explains the notation. For the same reason, B' is the midpoint of DF and C' is the

Fig. 4.14 K, A', P are collinear

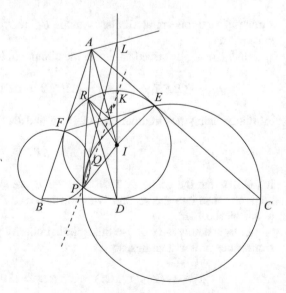

Fig. 4.15 Proof that
$\angle PQ'I = \angle PL'I$

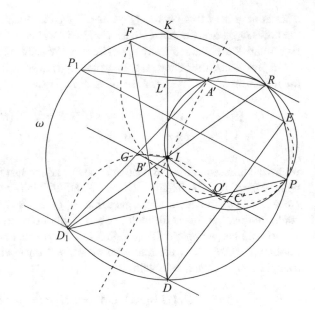

midpoint of DE. Some cyclicities are automatic: (E, P, C', Q'), (F, P, B', Q'), (I, R, P, A'). Additionally, $B'C'$ is parallel to EF, being midline in DEF.

The collinearity of P, Q, L is equivalent to the cyclicity of P, Q', L', I. To avoid configuration issues, we will work with directed angles modulo $180°$. We are left with showing that

$$\angle PQ'I = \angle PL'I.$$

The latter angle is the same as $\angle PL'D$.

The point L' is easy to locate; it is the projection of A' onto ID, because $\angle A'L'I = \angle LAI = 90°$ (Theorem 3.2). Notice that this fact combined with $\angle DPA' = \angle DPK = 90°$ implies that $A'L'DP$ is cyclic. So we obtain $\angle PA'D = \angle PL'D$. This reduces the problem to showing that

$$\angle PQ'I = \angle PA'D.$$

Locating the point Q', which lies at the intersection of the circumcircles of FPB' and EPC', is slightly more difficult. We use Fig. 4.15, from where we guess that $Q' \in B'C'$. This is equivalent to the fact that B, Q, I, P are concyclic, which was proved in the first solution. Or we can compute directly with inscribed angles

$$\angle B'Q'P = \angle B'FP = \angle DFP = \angle DEP = \angle C'EP = \angle C'Q'P,$$

a faster argument that proves $Q' \in B'C'$. We can locate Q' even better. Introduce the point $D_1 \in \omega$ such that $DD_1 \| EF$. Then $D_1 R$ is the reflection of DK over $A'I$, so D_1 and R are diametrically opposite in ω. We claim that $Q' \in PD_1$.

To prove this, introduce also the point $P_1 \in \omega$ such that $PP_1 \| EF$. Then $P_1 R$ is the reflection of PK over $A'I$, so $A' \in P_1 R$. Then

$$\angle(PP_1, Q'P) = \angle(B'Q', Q'P) = \angle B'Q'P = \angle DEP = \angle DP_1 P = \angle P_1 P D_1$$
$$= \angle(PP_1, D_1 P),$$

where the third equality was checked above, and the last follows from the symmetry with respect to $A'I$. Hence, $Q' \in PD_1$, as claimed. It should be noted that $Q' \in PD_1$ is equivalent to the fact that P, Q, I, D_1 are concyclic in the original configuration, but this is difficult to check without inversion.

Now we can complete the solution. Let $A'D_1$ meet $B'C'$ at G. Because $B'C'$ is midline in DEF, G is the midpoint of $A'D_1$. Consequently, GI is midline in the triangle $D_1 RA'$. We thus obtain

$$\angle D_1 IG = \angle D_1 RA' = \angle D_1 RP_1 = \angle PKD,$$

where for the last equality, we use reflection over $A'I$. Now using the midline GI in the triangle $D_1 RA'$ as well as inscribed angles in the circle ω and the circumcircle of $PB'F$, we have

$$\angle D_1 IG = \angle PKD = \angle PFD = \angle PFB' = \angle PQ'B' = \angle PQ'G$$

showing that D_1, G, I, Q' are concyclic. This yields

$$\angle PQ'I = \angle D_1 Q'I = \angle D_1 GI = \angle D_1 A'R = \angle D_1 A'P_1.$$

Using the reflection over $A'I$, we deduce that $\angle D_1 A'P_1 = \angle PA'D$. Combining, we obtain that $\angle PQ'I = \angle PA'D$, which is the equality that we had to prove.   □

Solution 3 And here is a solution that combines *inversion* and *homothety*, also due to Ilya Igorevich Bogdanov. Like in the previous solution, we introduce the midpoint A' of EF, which is the image of A through the inversion over ω, K which is the second intersection point of DI with ω, and L which is the intersection point of the external bisector of $\angle A$ with DI, and with this auxiliaries aim to prove that P, Q, L are collinear.

Let DA' intersect ω for the second time at S, as shown in Fig. 4.16. Let us take a look at the cyclic quadrilateral $SKDP$ with Theorem 3.33 in mind. First, note that EF is the polar of A with respect to ω, and since the polar of A' with respect to ω is perpendicular to AA', La Hire's Theorem (Theorem 3.31) implies that the polar of A' must be AL. Note that A' is the intersection of PK and DS (see the second solution), so by Theorem 3.33 the lines PS and DK intersect on the polar of the point A', which is AL. But DK intersects AL at L, so L is the intersection

Fig. 4.16 The construction
of S

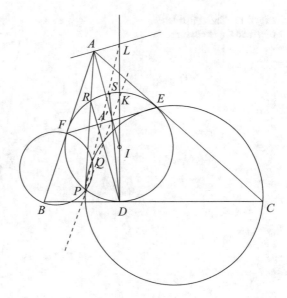

of PS and DK. We deduce that P, L, S are collinear, and the problem is reduced to
showing that P, Q, S are collinear, an easier task since S lies on ω.

Now we perform the inversion with respect to ω. As explained in the second
solution, the inverses A', B', C' of A, B, C are the midpoints of EF, DF, and DE,
respectively, and Q' is at the intersection of the circumcircles of $FB'P$ and $EC'P$
and lies on $B'C'$. We maintain the convention about denoting inverses by adding
dashes.

Let T be the second intersection point of $B'C'$ with the circumcircle of $Q'PI$
(Fig. 4.17). When working with directed angles modulo π, we can phrase the angle
chasing in terms of rotating lines (as explained in Chap. 1); let us adopt this style
here. On the one hand I, T, Q', P are concyclic, and on the other hand, E, P, C', Q'
are concyclic, so we can write

$$\angle(TI, IP) = \angle(TQ', Q'P) = \angle(Q'C', Q'P) = \angle(EC', EP) = \angle(ED, EP).$$

Furthermore, because K, E, P, D are concyclic,

$$\angle(ED, EP) = \angle(KD, KP) = \angle(IP, KP),$$

the latest because IKP is isosceles. Combining these equalities, we obtain
$\angle(TI, IP) = \angle(IP, KP)$; therefore, TI and KP are parallel.

Consider the homothety of center D and ratio 2. It maps I to K, and hence it
maps the line IT to KP. Also $B'C'$ is mapped to EF. Consequently, T is mapped
to the intersection of EF and KP, which is A'. So $T \in DS$. Now using the fact that
K, S, P, D lie on a circle, we deduce that $\angle(SD, SP) = \angle(KD, KP)$, and we can
read above that the latter is equal to $\angle(TQ', Q'P)$. We conclude that I, T, Q', P, S

Fig. 4.17 The construction of T and angle chasing

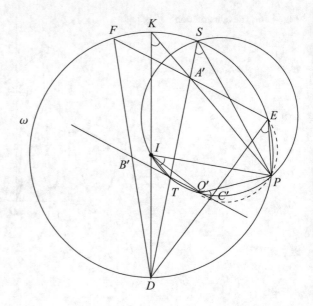

lie on a circle, in particular S is on the circumcircle of $IQ'P$. And this is equivalent to the fact that P, Q, S are collinear, as desired.   □

Solution 4 The fourth solution that we present was discovered by Régis Prado Barbosa, member of the Brazilian delegation, and is based on *inversion* and *spiral similarity*. This solution can be followed on Fig. 4.18. It starts with the construction of the points L, A', R, and S as before, and, like in the third solution, proves that B, Q, I, C are concyclic, and by proving that P, S, L are collinear, reduces the problem of showing that P, Q, L are collinear to that of showing that P, Q, S are collinear. Again we can assume without loss of generality that $\angle B > \angle C$, so that the figure is as drawn.

Next, we introduce X as the second intersection point of AD and ω, Y as the intersection of DR and EF, and Z as the second intersection point of the circle passing through B, Q, I, C with QP (point T in the first solution). For simplicity, let $\alpha = \angle PFE, \beta = \angle PEF, \gamma = \angle SDE, A = \angle BAC$. There are some automatic angle equalities obtained using inscribed angles:

$$\angle CBZ = \angle CQZ = \angle CQP = \angle CEP = \angle EFP = \alpha,$$
$$\angle BCZ = \angle BQZ = \angle BQP = \angle BFP = \angle FEP = \beta,$$

and we have seen in the first solution that

$$\alpha + \beta = 180° - \angle EDF = 180° - (90° - A/2) = 90° + A/2.$$

Fig. 4.18 Proof that P, Q, S are collinear

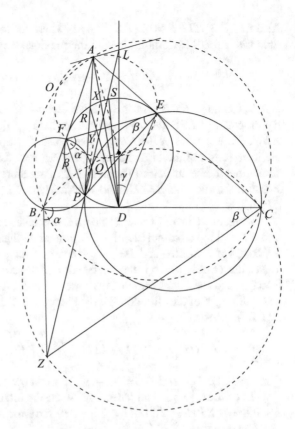

One more angle can be computed easily. By the symmedian construction (Theorem 1.22), DA is a symmedian in the triangle DEF. Also the line DS is the same as DA', hence is a median in the same triangle. Thus, $\gamma = \angle SDE = \angle XDF$.

Let us focus for a moment on the quadrilateral $PRXD$. By Theorem 3.33, the polar of the point that is the intersection of the diagonals DR and PX passes through A, thus by La Hire's Theorem (Theorem 3.31), the polar of A passes through this point. But the polar of A is EF, and DR intersects EF at Y, so Y is the intersection of DR and PX. Therefore, P, Y, X are collinear.

A few more computations of angles before we switch to geometric transformations:

$$\angle PIX = \angle PIF + \angle FIX = \overset{\frown}{PF} + \overset{\frown}{FX} = 2\angle PEF + 2\angle XDF = 2\beta + 2\gamma,$$

and since the triangle IPX is isosceles, we have

$$\angle IPY = \angle IPX = \frac{180° - 2\beta - 2\gamma}{2} = 90° - \beta - \gamma = \alpha - A/2 - \gamma.$$

Also $\angle IEY = \angle IEF = \angle IAE = A/2$, where for the second equality we used the right triangle EAI. Combining with the previous computation, we obtain

$$\angle IPY + \angle IEY = \alpha - \gamma.$$

The triangles PFE and ZBC have equal angles, so they are directly similar. By Proposition 2.21, there is a spiral similarity s that takes the triangle PFE to ZBC. Since s takes EF to BC, Theorem 2.22 implies that the center O of s is the intersection of the circumcircles of AEF and ABC (see Fig. 4.18). But spiral similarities come in pairs (Theorem 2.25), so there is a second spiral similarity s' that maps the triangle OPZ to OEC. Hence, $\angle(PZ, EC) = \angle(OP, OE) = \angle POE$.

And again we invert over ω. This inversion takes the circumcircle of AEF to the line EF and the circumcircle of ABC to a circle that passes through the midpoints A', B', C' of EF, DF, and DE, respectively. This is the nine-point circle of DEF. The center O of s, which is the intersection of the circumcircles of AEF and ABC, is mapped to the intersection of the nine-point circle with EF other than A'. And this is the leg Y of the altitude DY. By Theorem 3.2, $\angle POI = \angle YPI$ and $\angle IOE = \angle IEY$. Therefore,

$$\angle POE = \angle POI + \angle IOE = \angle YPI + \angle IEY = \alpha - \gamma.$$

Consequently, $\angle(PZ, EC) = \alpha - \gamma$. If we show that PS forms the same angle with EC, we are done. And indeed, let W be the intersection of PS with AC. Then $\angle WPE = \angle SPE = \angle SDE = \gamma$ and the Exterior Angle Theorem in the triangle WPE shows that

$$\angle PWC = \angle CEP - \angle WPE = \alpha - \gamma.$$

The problem is solved.   □

Solution 5 The last solution that we present has been put together by the US contestants Luke Robitaille and Brandon Wang, by the member of the US delegation Evan Chen, and by Michael Ren. It uses a *Möbius transformation* $\phi = \chi_2 \circ \chi_1$, where χ_1 is the inversion about ω and χ_2 is an inversion of center P and arbitrary radius. We mark by a dash the images of points under χ_1 and by a star the images under ϕ.

As in the second solution, we let K be the intersection of ID with ω and let A' be the midpoint of EF; we know that $A' \in PK$; hence, $A^* \in PK^*$. Note that $R^* = \chi_2(R)$ and $K^* = \chi_2(K)$ are both on $\phi(\omega) = \chi_2(\omega) = E^*F^*$. We claim that $PK^* = PA^*$ is a symmedian in the triangle PE^*F^*. Indeed, by Proposition 1.23, the cyclic quadrilateral $PERF$ is harmonic, so, in coordinates,

$$\frac{p - e}{p - f} : \frac{r - e}{r - f} = -1.$$

As Möbius transformations preserve cross-ratios, and as P is mapped to ∞, we have

$$\frac{r^* - e^*}{r^* - f^*} = \frac{\infty - e^*}{\infty - f^*} : \frac{r^* - e^*}{r^* - f^*} = -1.$$

So R^* is the midpoint of $E^* F^*$, that is, PR^* is a median in $PE^* F^*$. But the lines PR^* and PK^* (which are the same as PR and PK) form equal angles with PF^* and PE^*, respectively, showing that PK^* is a symmedian in the triangle $PE^* F^*$.

Also, the first inversion maps A to A', and the second inversion maps A' to A^* which is on the image of EF through this second inversion. We deduce that A^* is at the intersection of the line PK^* with the circumcircle of $PE^* F^*$.

Because D and K are diametrically opposite in ω and they are invariant under χ_1, and because the lines PD and PK are invariant under χ_2, $\angle D^* P K^* = 90°$.

As above, PB^* and PC^* are symmedians in the triangles $PD^* F^*$ and $PD^* E^*$, respectively, and the quadrilaterals $PD^* B^* F^*$ and $PD^* C^* E^*$ are harmonic. Additionally, as Q is the intersection of the circumcircles of BPF and CPE, Q^* is the intersection of the lines $C^* E^*$ and $B^* F^*$, which are the images of these two circles.

Let us add to the picture the point L where $DI = DK$ intersects the exterior angle bisector of $\angle A$; if $L' = \chi_1(L)$, then $\angle A' L' D = \angle LAD = 90°$; thus, L' is the projection of A' on DK. Applying Theorem 3.2 to χ_2 and working with directed angles modulo $180°$, we obtain

$$\angle K^* L^* D^* = \angle K^* L^* P + \angle P L^* D^* = \angle P K L' + \angle L' D P = \angle P K D + \angle K D P$$
$$= \angle K P D = 90°,$$

where the fourth equality uses the sum of (directed angles) in the triangle KDP. Also, $PA'L'D$ is cyclic ($\angle DPA' = \angle DL'A' = 90°$), so $L^* \in A^* D^*$. We deduce that L^* is the projection of K^* onto $D^* A^*$.

In the perspective of the inversion χ_1, showing that that P, Q, L are collinear is equivalent to showing that P, Q', I, L' are concyclic. In the perspective of the second inversion, showing that P, Q', I, L' are concyclic is equivalent to showing that $I^* = \chi_2(I), Q^*, L^*$ are collinear. Note that I^* is the reflection of P over EF (Proposition 3.7). Dropping the stars, we have reduced the problem to the following:

Problem In the triangle PEF, the P-symmedian meets the side EF at K and the circumcircle at A. Let D be a point on the line EF such that $\angle DPK = 90°$ and let L be the foot of the perpendicular from K to AD. Denote by I the reflection of P over EF. Construct also the cyclic harmonic quadrilaterals $PDCE$ and $PDBF$. Prove that the lines EC, FB, and LI are concurrent.

The solution to this new problem can be followed on Fig. 4.19. Suppose that the line through A perpendicular to EF meets the line EF at W and the circumcircle of PEF again at Z.

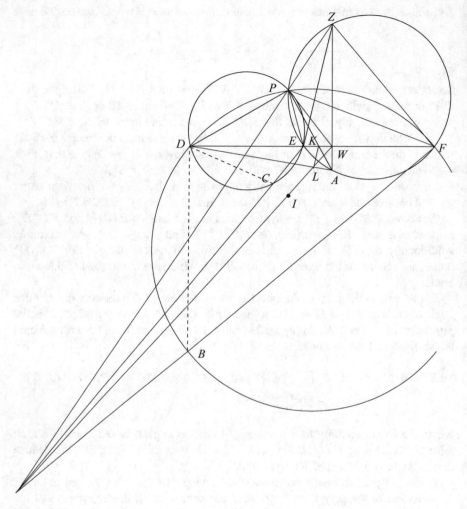

Fig. 4.19 Proof that CE, BF, and IL are concurrent

Because $\angle DPA = \angle DWA = 90°$, the points D, P, W, A are concyclic, so $\angle EDP = \angle WDP = \angle WAP = \angle ZAP$. But because Z, P, E, A are concyclic, the latter angle is equal to $\angle ZEP$. Hence, $\angle EDP = \angle ZEP$, showing that ZE is tangent to the circumcircle of $PDCE$. Similarly, ZF is tangent to the circumcircle of $PDBF$.

The orthogonalities $DP\perp KA$, $KL\perp DA$, $AW\perp DK$ show that PLW is the orthic triangle of DKA. So WD is the angle bisector of $\angle PWL$. Consequently, the line WL is the reflection of the line WP over EF, and so $I \in WL$.

The lines WZ, WP, WD, WI form a harmonic pencil, meaning that they determine on any line that intersects them four points whose cross ratio is -1. Indeed, in view of what was discussed in Sect. 3.1.5, we have to check that one particular line intersects this pencil at four points that form a harmonic division, and this is the line PI, which intersects WD at the midpoint of DI and WZ at the point at infinity.

We will discover two more harmonic pencils in the figure with the aid of a lemma.

Lemma 4.3 *Let $ABCD$ be a cyclic quadrilateral and let M be a point on the circumcircle. Then the quadrilateral $ABCD$ is harmonic if and only if the lines MA, MB, MC, MD form a harmonic pencil.*

Proof Consider an inversion of center M. Then the four lines in question are mapped into themselves, while the circumcircle of the quadrilateral becomes a line parallel to the tangent at M (Fig. 4.20). As Theorem 3.18 shows, the cross-ratio of the points A, B, C, D is equal to that of their images A', B', C', D'. So A, B, C, D form a harmonic quadrilateral if and only if A', B', C', D' form a harmonic division on the line that is the image of the circumcircle. But that is what characterizes the pencil as being harmonic. It should be noted that when $A = M$, then the line AM becomes the tangent to the circle at A. □

As a consequence of Lemma 4.3, EZ, EP, ED, EC form a harmonic pencil ($PDCE$ is harmonic and EZ is tangent to its circumcircle) and so do FZ, FP, FD, FB ($PDBF$ is harmonic and FZ is tangent to its circumcircle). Examining how these pencils intersect the line PZ, we notice that three of the intersection points are the same, namely, P, Z, and the point where PZ intersects EF. Because all three pencils determine harmonic divisions on PZ, it follows that the fourth point is the same. This means that PZ, WI, EC, FB meet at one point. But since $L \in WI, WI = IL$, so EC, FB, LI meet at one point, and the problem is solved. □

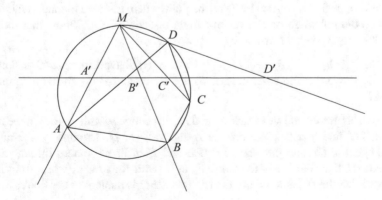

Fig. 4.20 Harmonic quadrilaterals yield harmonic pencils

4.1.2 Some Problems

Discover which geometric transformations solve the following problems. Let us point out that for each of these problems, we were able to write either several solutions that employ different geometric transformations or one solution that combines geometric transformations of different types. These types of transformations are highlighted in italics in the solutions printed at the back of the book, in order to emphasize their diversity.

Can you find more than one solution, and can you use more than one technique? Which is the most elegant solution?

222 On the sides of a triangle ABC construct in the exterior the equilateral triangles ABC_1, BCA_1, and CAB_1.

(a) *(Napoleon's Theorem)* Prove that the centers of the triangles ABC_1, BCA_1, and CAB_1 form an equilateral triangle.
(b) Prove that the segments AA_1, BB_1, and CC_1 are equal and their lines of support are concurrent.

223 *(Ptolemy's Theorem)* Show that in a cyclic quadrilateral $ABCD$

$$AB \cdot CD + AD \cdot BC = AC \cdot BD.$$

224 Let A, B, C, D be four points on the circle ω, in this order, and let S be a point inside ω such that $\angle SAD = \angle SCB$ and $\angle SDA = \angle SBC$. The angle bisector of the angle $\angle ASB$ intersects the circle ω at P and Q. Prove that $SP = SQ$.

225 Let $A_1A_2A_3$ be a scalene triangle, and, for $1 \le i \le 3$, let M_i be the midpoint of the side opposite to A_i, and let T_i be the point where this side touches the incircle. Let S_i be the reflection of T_i over the angle bisector of $\angle A_i$. Show that the lines M_iS_i, $1 \le i \le 3$, are concurrent.

226 Let OA_1B_1, OA_2B_2, and OA_3B_3 be three equilateral triangles that share the vertex O. Prove that the midpoints of B_1A_2, B_2A_3, and B_3A_1 form an equilateral triangle.

227 Two circles Γ_1 and Γ_2 of radii r_1 and r_2, respectively, with $r_1 < r_2$, are exterior tangent. The line t_1 is tangent to Γ_1 at A and to Γ_2 at D. The line t_2 is parallel to t_1, is tangent to Γ_1, and intersects Γ_2 at E and F. A line t_3 passes through D and intersects the line t_2 at B and the circle Γ_2 at C (with B and C different from E and F). Prove that the circumcircle of the triangle ABC is tangent to the line t_1.

228 Let $ABCD$ be a quadrilateral inscribed in the circle ω, with $AC \perp BD$. Let E and F be the reflections of D over the lines BA and BC, respectively, and let P be the intersection point of the lines BD and EF. Suppose that the circumcircle of the triangle EPD meets ω at D and Q, and the circumcircle of the triangle FPD meets ω at D and R. Show that $EQ = FR$.

229 On the sides AB and AC of the triangle ABC, construct in the exterior the equilateral triangles ABN and ACM. Let P, Q, R be the midpoints of BC, AM, and AN, respectively. Prove that the triangle PQR is equilateral.

230 The points A, B, C, D are chosen such that C, D are on the same side of the line AB, $AC \cdot BD = AD \cdot BC$, and $\angle ADB = 90° + \angle ACB$. Find the ratio

$$\frac{AB \cdot CD}{AC \cdot BD}$$

and prove that the circumcircles of ACD and BCD are orthogonal.

231 Let A be one of the two intersection points of the circles ω and ω', whose centers are O and O', respectively. The common tangents touch ω at P and Q and ω' at P' and Q', with P and P' on the same side of OO'. Let M and M' be the midpoints of the segments PQ and $P'Q'$. Prove that $\angle OAO' = \angle MAM'$.

232 Let ABC be an acute scalene triangle, and let H and O be its orthocenter and circumcenter, respectively. The line OA crosses the altitudes from B and C of the triangle ABC at P and Q, respectively. Show that the circumcenter of the triangle HPQ lies on one of the medians of the triangle ABC.

233 (a) Consider five points A, B, C, D, E such that $ABCD$ is a parallelogram and $BCED$ is a cyclic quadrilateral. A line ℓ through A intersects the line BC at K and the side DC at L so that $EL = EK = EC$. Prove that ℓ is the angle bisector of $\angle BAD$.
(b) Conversely, let the bisector of the angle $\angle BAD$ of the parallelogram $ABCD$ intersect the lines BC and CD at K and L, respectively. Prove that the circumcenter of the triangle CKL lies on the circumcircle of the triangle BCD.

234 Let ω_1 and ω_2 be two circles that intersect at A and B. A line passing through A intersects ω_1 and ω_2 at C and D, respectively. Let M be the midpoint of the segment CD, P the midpoint of the arc $\overarc{BC}$ that does not contain A, and Q the midpoint of the arc $\overarc{BD}$ that does not contain A. Prove that $\angle PMQ = 90°$.

235 (a) *(Steiner's Theorem)* Prove that the Simson line of a point on the circumcircle of a triangle passes through the midpoint of the segment that joins the point with the orthocenter of the triangle.[1]
(b) *(Lemoine's Theorem)* Consider for each vertex of a cyclic quadrilateral its Simson line with respect to the triangle formed by the other three vertices. Show that the four Simson lines obtained this way intersect at one point.

---

[1] The Simson line of a point on the circumcircle of a triangle is the line determined by the projections of the point onto the sides.

236 Let $ABCD$ be a cyclic quadrilateral, and let M and N be the midpoints of the sides AB and CD, respectively. Let E be the point of intersection of the lines AD and BC and let F be the point of intersection of the diagonals AC and BD. Let also P and Q be the projections of F onto BC and AD, respectively.

(a) Prove that

$$\frac{MN}{EF} = \frac{1}{2}\left|\frac{AB}{CD} - \frac{CD}{AB}\right|.$$

(b) Prove that PQ is perpendicular to MN.

237 The diagonals AC and BD of a quadrilateral $ABCD$ are perpendicular. Four squares, $ABEF$, $BCGH$, $CDIJ$, and $DAKL$, are erected externally on its sides. The intersection points of the pairs of straight lines CL, DF; DF, AH; AH, BJ; BJ, CL are denoted by P_1, Q_1, R_1, S_1, respectively, and the intersection points of the pairs of straight lines AI, BK; BK, CE; CE, DG; DG, AI are denoted by P_2, Q_2, R_2, S_2, respectively. Prove that the quadrilaterals $P_1Q_1R_1S_1$ and $P_2Q_2R_2S_2$ are congruent.

238 Let $ABCD$ be a parallelogram that is not a rhombus. The point Q is the intersection of the reflections of the lines AB and CD across the diagonals AC and BD, respectively. Prove that Q is the center of the spiral similarity that maps the segment AO to OD, where O is the center of the parallelogram.

239 Let ABC be a triangle and let M be the midpoint of the side BC. The circle centered at M and of radius MA intersects a second time the lines AB and AC at B' and C', respectively. Let D be the point where the tangents to this circle at B' and C' intersect. Prove that the perpendicular bisector of the side BC passes through the midpoint of the segment AD.

240 Let $A_1A_2A_3$ be a scalene triangle, and let I be its incenter. Let C_i be the smaller circle through I that is tangent to the sides A_iA_{i+1} and A_iA_{i+2}, $i = 1, 2, 3$ (where indices are taken modulo 3). Let B_i be the second point of intersection of C_{i+1} and C_{i+2}, $i = 1, 2, 3$. Prove that the circumcenters of A_1B_1I, A_2B_2I, and A_3B_3I are collinear.

241 Let Γ be a circle, let AB be a diameter, and let ℓ be a line perpendicular to AB and exterior to the circle so that B is between A and ℓ. Let C be a point of Γ, different from A and B, and let D be the intersection point of the lines AC and ℓ. The line DE is tangent to the circle Γ at E so that E and B are on the same side of the line AC. The line BE intersects the line ℓ at F, and the line AF intersects the circle Γ at G. Prove that the reflection of G over AB is on the line CF.

242 Let ABC be a triangle, let P, Q, and R be the tangency points of the incircle with the sides BC, AC, and AB, respectively, and let I be the incenter. The lines PQ and AB intersect at D and the lines QR and BC intersect at E. Prove that the circumcircles of QEP, QDR, and QBI have a second common point.

243 Let R and S be distinct points on the circle Ω, and let t denote the tangent line to Ω at R. The point T is the reflection of R over S. A point J is chosen on the smaller arc $\overset{\frown}{RS}$ of Ω so that the circumcircle Γ of the triangle JST intersects t at two different points. Denote by A the common point of Γ and t that is closest to R. The line AJ meets Ω again at K. Show that the line KT is tangent to the circle Γ.

244 Let ABC be a triangle and let I and O denote its incenter and circumcenter, respectively. Let ω_A be the circle through B and C that is tangent to the incircle of the triangle ABC; the circles ω_B and ω_C are defined similarly. The circles ω_B and ω_C meet at a point A' different from A; the points B' and C' are defined similarly. Prove that the lines AA', BB', and CC' are concurrent at a point on the line IO.

245 Two circles Γ_1 and Γ_2 touch internally the circle Γ at M and N, respectively, and the center of Γ_2 is on Γ_1. The common chord of the circles Γ_1 and Γ_2 intersects Ω at A and B. The lines MA and MB intersect Γ_1 a second time at C and D, respectively. Prove that Γ_2 is tangent to CD.

246 Let ABC be a given triangle. Let X be a variable point on the arc $\overset{\frown}{AB}$ of the circumcircle Ω of ABC that does not contain C, and let O_1 and O_2 the incenters of the triangles CAX and CBX. Prove that the circumcircle of XO_1O_2 meets the circle Ω at a fixed point.

247 Let M be an arbitrary point in the circumcircle of ABC. Suppose that the lines through M that are tangent to the incircle of ABC meet BC at K_1 and K_2. Prove that the circumcircle of MK_1K_2 meets the circumcircle of ABC again at the tangency point with the A-mixtilinear incircle.

248 A circle Γ centered at the vertex A of the triangle ABC intersects the side BC at D and E so that B, D, E, C are distinct and lie on the line BC in this order. Let F and G be the intersection points of Γ with the circumcircle Ω, so that A, F, B, C, G are on Ω in this order. Let K and L be the points at which the circumcircles of the triangles BDF and CEG intersect the second time the sides AB and AC. Suppose that the lines FK and GL are distinct and intersect at the point X. Prove that X lies on the line AO, where O is the circumcenter of ABC.

249 Fix a circle Γ, a line ℓ tangent to Γ, and another circle Ω disjoint from ℓ such that Γ and Ω lie on opposite sides of ℓ. The tangents to Γ from a variable point X on Ω cross ℓ at Y and Z. Prove that, as X traces Ω, the circumcircle of the triangle XYZ is tangent to two fixed circles.

250 Let $ABCD$ be a cyclic quadrilateral whose diagonals AC and BD intersect at E. Show that the Euler lines of the triangles AEB, BEC, CED, and DEA are concurrent.

251 For every triangle T that is not right, we will denote by $H(T)$ its orthic triangle (whose vertices are the feet of the altitudes of T). Starting with a triangle T_0, we define recursively $T_{n+1} = H(T_n)$, $n \geq 0$. For a given $n > 1$, find how many triangles T_0, not similar to each other, exist such that T_n is similar to T_0.

4.2 A Story of Complete Quadrilaterals

Complete quadrilaterals have their place in two-dimensional real projective geometry, but we can already establish a number of beautiful facts about them with our own tools. And we will see all the stars of this book performing on the stage of this rich configuration.

We recall that a *complete quadrilateral* is any set of four lines, no three passing through the same point, and no two parallel.[2] An example of a complete quadrilateral, determined by the lines $\ell_1, \ell_2, \ell_3, \ell_4$, is shown in Fig. 4.21. We agree to define the points A_{ij} as the intersections on the lines ℓ_i and ℓ_j, $1 \le i < j \le 4$, and call them the vertices of the complete quadrilateral.

4.2.1 Miquel's Theorem and $\sqrt{bc}$ Inversion

We apply a $\sqrt{bc}$ inversion to... a quadrilateral! We have seen in Chap. 2 that complete quadrilaterals have a lot to do with spiral similarities, Miquel's Theorem (Theorem 2.24) states that all four circumcircles of the triangles determined by three of four lines $\ell_1, \ell_2, \ell_3, \ell_4$, no two of them parallel, pass through a common point: the *Miquel point M*. In that context, we have uncovered the following transformation-focused version of this result:

Theorem 4.4 (Miquel's Theorem, Rephrased) *Let $\ell_1, \ell_2, \ell_3, \ell_4$ be the lines that define a complete quadrilateral, and let A_{ij} be its vertices. Then there is a spiral similarity that maps A_{ij} to A_{ik} and A_{mj} to A_{mk}, for $\{i, j, k, m\} = \{1, 2, 3, 4\}$,*

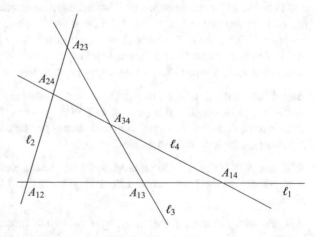

Fig. 4.21 A complete quadrilateral

---

[2] The latter condition is dropped in real projective geometry, since parallelism does not exist in real projective geometry.

whose center is the intersection of the four circumcircles of the triangles determined by three of the four lines.

On the other hand, Proposition 3.40 establishes a connection between $\sqrt{bc}$ inversions and spiral similarities. In fact, putting together the aforementioned proposition and Miquel's Theorem gives rise to $\sqrt{bc}$ inversion on complete quadrilaterals.

Proposition 4.5 (Inverting Complete Quadrilaterals) *Let ℓ_1, ℓ_2, ℓ_3, and ℓ_4 be the four lines of a complete quadrilateral, whose vertices are A_{ij}, and let M be its Miquel point. Then the $\sqrt{bc}$-inversion determined by M in the triangle $MA_{12}A_{34}$ maps A_{13} to A_{24} and A_{14} to A_{23}.*

Proof Let ϕ be the aforementioned $\sqrt{bc}$ inversion. Combining the proof of Miquel's Theorem given in Chap. 2 with Proposition 3.40, (iii), we find that there is a spiral similarity centered at M that maps A_{12} to A_{14} and A_{23} to A_{34}, which implies that $\phi(A_{14}) = A_{23}$ and $\phi(A_{23}) = A_{14}$. Analogously, there is a spiral similarity, also centered at M, that maps A_{12} to A_{13} and A_{24} to A_{34}, which implies that $\phi(A_{13}) = A_{24}$ and $\phi(A_{24}) = A_{13}$. This proof can be followed on Fig. 4.22. □

A quick corollary of this proposition is that the diagram formed by the lines ℓ_1, ℓ_2, ℓ_3, and ℓ_4, and the four circles determined by three of these four lines, is fixed by the Möbius transformation ϕ. Lines and circles are swapped; more specifically, each line is mapped to the circle corresponding to the other three lines.

A consequence of this result is that one can assume that the complex coordinates of four vertices of the complete quadrilateral are $a, b, 1/a$, and $1/b$, with the Miquel point being at the origin of the coordinate system. The coordinates of the other two vertices, with less elegant formulas, are

$$x = \frac{(ab - \overline{ab})(\overline{a} - \overline{b})}{(\overline{ab} - 1)(a\overline{b} - \overline{a}b)} \text{ and } 1/x.$$

4.2.2   Some Classical Results

We will now prove some well-known results about complete quadrilaterals.

Theorem 4.6 *If the lines ℓ_1, ℓ_2, ℓ_3, and ℓ_4 determine a complete quadrilateral with Miquel point M, then*

(i)   *the four orthogonal projections of M onto all four lines ℓ_i are collinear;*
(ii) **(The Orthocentric Line)** *The orthocenters of all four triangles determined by the four lines are collinear;*
(iii) *the circumcenters of all four triangles determined by the four lines are concyclic, and M lies on this circle.*

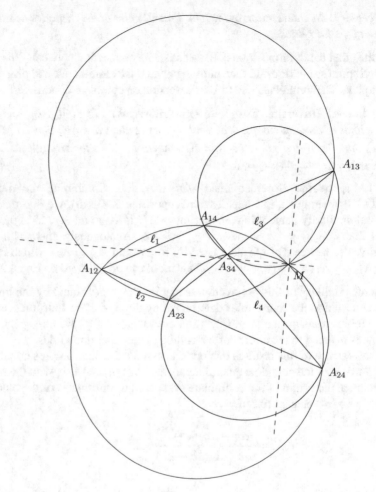

Fig. 4.22 Complete quadrilaterals and $\sqrt{bc}$ inversion. The common angle bisectors are represented by dashed lines

Proof (i) Any two triangles formed in the complete quadrilateral share two sides, so they share the Simson line of the Miquel point. So the Simson line of the Miquel point is the same for all four triangles, and this is the line of the four orthogonal projections.

(ii) By Steiner's Theorem (Problem 235), the orthocenters are on the line that is the image of the above Simson line by a homothety with center M and ratio 2. So the orthocenters are collinear.

(iii) Consider the same homothety centered at M and with ratio 2, and let P_i be the image of the projection of M onto ℓ_i under this homothety, $i = 1, 2, 3, 4$; in other words, P_i is the reflection of M across ℓ_i. By Proposition 3.7, the $\sqrt{bc}$ inversion ϕ

that fixes the configuration maps the points P_i to the circumcenters O_i of T_i; since the points P_i are collinear, the points O_i are concyclic; moreover, the center M of ϕ lies on this circle. $\qquad\square$

Here is another result focused on transformations.

Theorem 4.7 *Let ℓ_1, ℓ_2, ℓ_3, and ℓ_4 be four lines that form a complete quadrilateral, and let M be the Miquel point. Denote by T_i the triangle formed by all lines except ℓ_i, and by O_i the circumcenter of T_i. Then there exists a spiral similarity, centered at M, that takes the triangle $O_j O_k O_m$ to the triangle T_i, where $\{i, j, k, m\} = \{1, 2, 3, 4\}$.*

Proof Let O_i be the circumcenter of T_i. We prove that there is a spiral similarity with center M that takes $O_1 O_2 O_3$ to T_4. Specifically, we show that $O_1 O_2 O_3$ and $A_{23} A_{13} A_{12}$ are similar.

Consider the spiral similarity with center M that takes $A_{13} A_{14}$ to $A_{23} A_{24}$. It also maps the circumcircle of $M A_{13} A_{14}$, to the circumcircle of $M A_{23} A_{24}$; in particular, O_2 is mapped to O_1. So we have the automatic (spiral) similarity s between the triangles $M O_1 O_2$ and $M A_{23} A_{13}$. This transformation has center M, and is such that $s(O_1) = A_{23}$ and $s(O_2) = A_{13}$. We can prove in the same way that triangles $M O_1 O_3$ and $M A_{23} A_{12}$ are similar; furthermore, it follows that the spiral similarity s' with center M that takes O_1 to A_{23} also takes O_3 to A_{12}. Since a spiral similarity is determined by its center and the image of some other point, $s' = s$. So $s(O_3) = A_{12}$ and we are done. $\qquad\square$

For the next result, we define the *diagonals* of a complete quadrilateral to be the line segments that connect two of the vertices that are not on the same line ℓ_i.

Theorem 4.8 (The Newton-Gauss Line) *The midpoints of the three diagonals of a complete quadrilateral are collinear.*

Proof We will use a fact that is proved easily:

Lemma 4.9 *In a parallelogram $XYZW$, a point M is on the diagonal XZ if and only if the two parallelograms that have Y and W as vertices and are determined by the lines through M parallel to the sides have equal areas.*

With this result at hand, let ℓ_1, ℓ_2, ℓ_3, and ℓ_4 be the four lines, and, with the usual conventions about vertices, let N_1, N_2, and N_3 be the midpoints of $A_{12} A_{34}$, $A_{13} A_{24}$, and $A_{14} A_{23}$, respectively. Without loss of generality, we may assume that the four lines form the configuration from Fig. 4.23. Through the points A_{13}, A_{34}, A_{24}, A_{23}, and A_{14}, take parallels to the sides $A_{12} A_{13}$ and $A_{12} A_{24}$; let Q, R, S, T, U, V be the points shown in the figure.

The homothety of center A_{12} and ratio 2 maps:

- N_1 to A_{34},
- the midpoint N_2 of the diagonal $A_{13} A_{24}$ of the parallelogram $A_{12} A_{13} Q A_{24}$ to the opposite vertex Q, and

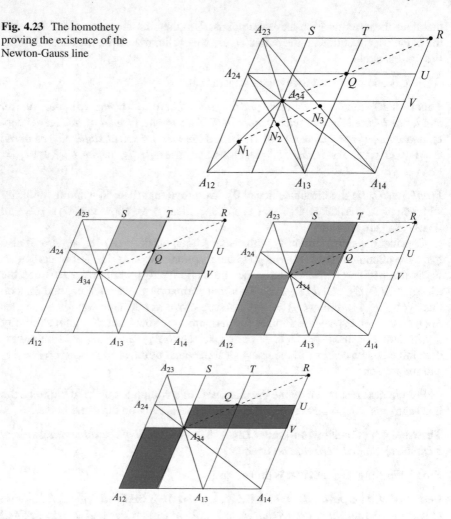

Fig. 4.23 The homothety proving the existence of the Newton-Gauss line

Fig. 4.24 The area argument proving the existence of the Newton-Gauss line

- the midpoint N_3 of the diagonal $A_{14}A_{23}$ of the parallelogram $A_{12}A_{14}RA_{23}$ to the opposite vertex R.

As homothety maps lines to lines, the collinearity of the midpoints N_1, N_2, N_3 is equivalent to the collinearity of their images A_{34}, Q, and R.

Applying Lemma 4.9 to the parallelogram $A_{34}VRS$, we deduce that the fact that Q is on the diagonal $A_{34}R$ is equivalent to the equality of the areas of the shaded parallelograms from the top left configuration in Fig. 4.24.

We can apply the direct implication of Lemma 4.9 to the parallelogram $A_{12}A_{13}TA_{23}$ because A_{34} is on the diagonal $A_{13}A_{23}$, so the shaded parallelograms from the top right configuration in Fig. 4.24 have equal areas. Also, in the parallelogram $A_{12}A_{14}UA_{24}$, A_{34} is on the diagonal $A_{14}A_{24}$, and so, by the same lemma, the shaded parallelograms from the configuration at the bottom of Fig. 4.24 have equal areas. The two equalities of areas that we have proved imply, by subtraction, the equality of areas for the shaded parallelograms on the top left, and this implies that the points A_{34}, Q, R are collinear. The theorem is proved. □

We have seen in Theorem 4.6 that the orthocenters of the four triangles formed in a complete quadrilateral are collinear; they determine what is called the orthocentric line. A proof of this collinearity based on power-of-a-point is also possible, and it proceeds as follows. Let H_i be the orthocenter of T_i, and let $A_{pq}B_r$ be the altitudes. We know that $A_{pq}H_i \cdot H_iB_r$ is the same for all three choices of indices. But these products are the powers of H_i with respect to the three circles that have the diagonals as diameters. Consequently, the orthocenters have the same power with respect with the three lines, so on the one hand, the circles with the diagonals as diameters are coaxial (the Gauss-Bodenmiller Theorem), and on the other hand, the orthocentric line is the common radical axis, which is therefore orthogonal to the Newton-Gauss line.

After this discussion, let us solve an enticing problem from the 2016 Brazilian Mathematical Olympiad. This problem was proposed by the third author of the book and, as we will discover, is about complete quadrilaterals.

Problem 4.4 Let $ABCD$ be a noncyclic, convex quadrilateral with no parallel sides. The lines AB and CD meet at the point E. The circumcircles of the triangles ADE and BCE meet again at a point $M \neq E$. The internal bisectors of $ABCD$ determine a cyclic convex quadrilateral with circumcenter I, and the external bisectors of $ABCD$ determine a cyclic convex quadrilateral with circumcenter J. Prove that I, J, and M are collinear.

Solution 1 We argue on Fig. 4.25, in which we define E and F to be the intersection points of the pairs of lines (AB, CD) and (AD, BC), respectively. We have constructed a complete quadrilateral! Our first solution uses the aforementioned $\sqrt{bc}$ *inversion* and some facts about *Möbius transformations*.

Denote by I_{XY} the intersection of the internal bisectors of the angles $\angle X$ and $\angle Y$, and by J_{XY} the intersection of the external bisectors of the angles $\angle X$ and $\angle Y$, where $\angle X$ and $\angle Y$ range among the consecutive angles of the original quadrilateral. Notice that I_{AB}, I_{CD}, J_{AB}, and J_{CD} are incenters and excenters of the triangles ABF or CDF, so they lie on the internal bisector of $\angle F$; similarly, I_{BC}, I_{DA}, J_{BC}, and J_{DA} are the incenters and excenters of the triangles ADE and BCE, so they lie on the internal bisector of $\angle E$.

Consider the $\sqrt{bc}$ inversion ϕ that fixes the complete quadrilateral $ABCDEF$. We will prove a stronger statement:

The circumcircles ω of $I_{AB}I_{BC}I_{CD}I_{DA}$ and Ω of $J_{AB}J_{BC}J_{CD}J_{DA}$ are fixed by ϕ, so the centers of these circles lie on the common bisector of $\angle AMC$, $\angle BMD$, and $\angle EMF$.

As a consequence of this fact, since ω and Ω are fixed by ϕ, both are orthogonal to the circle of inversion, and M lies on the radical axis of ω and Ω. This radical axis is therefore the common external bisector of $\angle AMC$, $\angle BMD$, and $\angle EMF$.

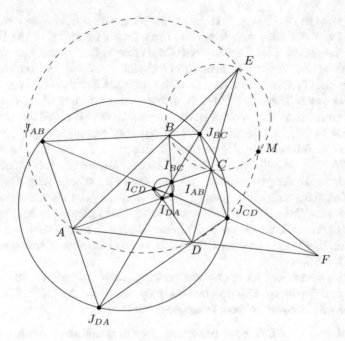

Fig. 4.25 Incenters and excenters

We now proceed with proving this stronger statement. To this end, let us find $J'_{AB} = \phi(J_{AB})$. Observe that J_{AB} is the excenter of triangle ABF, and focus on this triangle. At this moment, we apply the fact that the cross-ratio is invariant under Möbius transformations, and we use complex coordinates. Recall that

$$\angle XYZ = \arg \frac{x - y}{z - y} = \arg \frac{y - x}{y - z}.$$

From $\phi(A) = C, \phi(B) = D, \phi(C) = A, \phi(D) = B, \phi(E) = F$, and $\phi(F) = E$, it follows that $(j_{AB}, a, f, b) = (j'_{AB}, c, e, d)$, which is the same as

$$\frac{j_{AB} - f}{j_{AB} - b} : \frac{a - f}{a - b} = \frac{j'_{AB} - e}{j'_{AB} - d} : \frac{c - e}{c - d}.$$

Taking arguments, this yields

$$\angle EJ'_{AB}D + \angle ECD = \angle FJ_{AB}B - \angle FAB,$$

where the angles are directed and taken modulo π. Because C, D, E are collinear, $\angle ECD \in \{0, \pi\}$. Working with directed angles modulo π and using the cyclic quadrilaterals $I_{AB}AJ_{AB}B$ and $I_{DA}AJ_{DA}D$ (the incenter, excenter, and adjacent vertices form a cyclic quadrilateral), we obtain

$$\angle E J'_{AB} D = \angle F J_{AB} B - \angle FAB = \angle I_{AB} AB - \angle FAB$$

$$= -\angle FAI_{AB} = -\angle D J_{DA} E = \angle E J_{DA} D.$$

This means that J'_{AB} lies on the circumcircle of DEJ_{DA}. Analogously, by inter-changing A and B, and C and D, we find that J'_{AB} lies on the circumcircle of CEJ_{BC}. This implies that J'_{AB} is the Miquel point of the complete quadrilateral obtained by completing $CD J_{DA} J_{BC}$, and thus it also lies on the circumcircle of the triangle determined by the lines $J_{BC} J_{DA}$, $C J_{BC}$, and $D J_{DA}$, that is, on the circumcircle of the triangle $J_{BC} J_{DA} J_{CD}$. This circumcircle is... Ω. So J'_{AB} lies on Ω, and similarly J'_{BC}, J'_{CD}, and J'_{DA} lie on Ω too. This means that $\phi(\Omega) = \Omega$.

For the proof that $\phi(\omega) = \omega$, we could invoke the same argument, but instead let us show how to perform the angle chase with complex numbers instead. Again we have an equality of cross-ratios: $(i'_{AB}, c, e, d) = (i_{AB}, a, f, b)$. In all that follows, we work with angles modulo π. The points C, D, E are collinear, so the equality of cross-ratios implies

$$\arg \frac{e - i'_{AB}}{d - i'_{AB}} = \arg \left(\frac{e - i'_{AB}}{d - i'_{AB}} : \frac{e - c}{d - c} \right) = \arg \left(\frac{f - i_{AB}}{b - i_{AB}} : \frac{f - a}{b - a} \right).$$

From the fact that F, I_{AB}, J_{AB} are collinear and A, I_{AB}, B, J_{AB} are concyclic, it follows that

$$\arg \frac{f - i_{AB}}{b - i_{AB}} = \arg \frac{j_{AB} - i_{AB}}{b - i_{AB}} = \arg \frac{j_{AB} - a}{b - a}.$$

So

$$\arg \left(\frac{f - i_{AB}}{b - i_{AB}} : \frac{f - a}{b - a} \right) = \arg \left(\frac{j_{AB} - a}{b - a} : \frac{f - a}{b - a} \right) = \arg \frac{j_{AB} - a}{f - a}.$$

Because A, D, F are collinear, and J_{AB}, A, J_{DA} are collinear,

$$\arg \frac{j_{AB} - a}{f - a} = \arg \frac{j_{AB} - a}{d - a} = \arg \frac{j_{DA} - a}{d - a}.$$

Because A, I_{AD}, D, J_{AD} are concyclic, I_{AD}, J_{AD}, E are collinear,

$$\arg \frac{j_{DA} - a}{d - a} = \arg \frac{j_{DA} - i_{DA}}{d - i_{DA}} = \arg \frac{e - i_{DA}}{d - i_{DA}}.$$

Consequently,

$$\arg \frac{e - i'_{DA}}{d - i'_{DA}} = \arg \frac{e - i_{DA}}{d - i_{DA}} \iff \frac{e - i'_{AB}}{d - i'_{AB}} : \frac{e - i_{DA}}{d - i_{DA}} \in \mathbb{R}.$$

So (i'_{AB}, i_{DA}, e, d) is a real number, proving that i'_{AB} lies on ω. From here the solution continues as above to conclude that ω, like Ω, is fixed by ϕ, and the conclusion follows.   □

Solution 2 The second solution, discovered by Régis Prado Barbosa, is perhaps a bit more elementary, and is based on *homotheties*, namely, on the two homotheties between the circumcircles ω of $I_{AB}I_{BC}I_{CD}I_{DA}$ and Ω of $J_{AB}J_{BC}J_{CD}J_{DA}$. We argue on Fig. 4.26.

We use the same notation as for the previous solution and start by proving some basic facts. Because J_{BC} and I_{BC} are the incenter and the E-excenter of the triangle BCE, J_{BC} and I_{BC} lie on the internal bisector of $\angle BEC$, and the quadrilateral $I_{BC}BJ_{BC}C$ is cyclic, having as circumcenter the midpoint K of $I_{BC}J_{BC}$. Reasoning in a similar way for the triangle ADE, we deduce that the points I_{DA} and J_{DA} lie on the internal bisector of $\angle AED$ (which angle coincides with $\angle BEC$), and the quadrilateral $I_{DA}AJ_{DA}D$ is cyclic as well, having as circumcenter the midpoint L of $I_{DA}J_{DA}$. The common bisector of $\angle AED$ and $\angle BEC$ seems important for the solution, so let us denote it by ℓ_E. We define ℓ_F analogously—notice that the points I_{AB}, J_{AB}, I_{CD}, and J_{CD} all lie on ℓ_F.

Recall from the statement of the problem that I is the center of ω and J is the center of Ω. Our strategy is to use the centers O_+ and O_- of the two homotheties

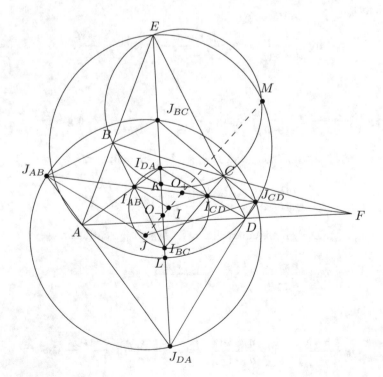

Fig. 4.26 Homotheties

that map these circles into each other. First, note that if we are in the degenerate situation with $O_- = O_+$, then $I = J$ and there is nothing to prove. Else, since both O_+ and O_- lie on the line IJ, the lines O_+O_- and IJ coincide, and we are left with proving that O_+, O_- and M are collinear.

It seems unnatural to replace I and J with O_- and O_+, but the latter two points are easier to locate! To see why this is so, note that since $\angle J_{AB}JJ_{CD}$ is an angle at the center of the circle and $I_{BC}BJ_{BC}C$ is a cyclic quadrilateral, we have

$$\angle J_{AB}JJ_{CD} = 2\angle J_{AB}J_{DA}J_{CD} = 2(180° - \angle J_{CD}J_{BC}J_{AB}) = 2(180° - \angle BJ_{BC}C)$$

$$= 2\angle CI_{BC}B = 2\angle I_{CD}I_{BC}I_{AB} = \angle I_{CD}II_{AB}.$$

Moreover, because I_{AB}, J_{AB}, I_{CD}, J_{CD} lie in ℓ_F, $II_{CD} \parallel JJ_{CD}$ and $II_{AB} \parallel JJ_{AB}$. We conclude that the direct homothety that takes ω to Ω takes I_{AB} to J_{AB} and I_{CD} to J_{CD}. From here we infer that the center O_+ of this homothety lies on the line $I_{AB}J_{AB}$, that is, on ℓ_F. Analogously, O_- lies on ℓ_E and the inverse homothety centered at O_- that maps the circles into each other takes I_{BC} to J_{BC} and I_{DA} to J_{DA}.

Now that we can work with the lines ℓ_E and ℓ_F, how do we bring the Miquel point M into the picture? Let us first connect the inverse homothety with M. The key elements are the midpoints K and L of $I_{BC}J_{BC}$ and $I_{DA}J_{DA}$, which are on the circumcircles of BCE and ADE, and these circumcircles intersect at M. We will show that O_-M bisects the angle $\angle BMD$.

Notice first that, in the circumcircles of $BEMCK$ and $AEMDL$,

$$\angle BMK = \angle BEK = \frac{1}{2}\angle BEC = \frac{1}{2}\angle AED = \angle LED = \angle LMD.$$

We now show that MO_- bisects $\angle KML$, because then $\angle BMO_- = \angle BMK + \angle KMO_- = \angle LMD + \angle O_-ML = \angle O_-MC$, and so MO_- bisects $\angle BMD$.

To prove that MO_- bisects $\angle KML$, we use ratios, which are intrinsic to homotheties, and the Bisector Theorem. From the inverse homothety, we obtain

$$\frac{O_-J_{BC}}{O_-J_{DA}} = \frac{O_-I_{BC}}{O_-I_{DA}} = \frac{O_-J_{BC} + O_-I_{BC}}{O_-J_{DA} + O_-I_{DA}} = \frac{I_{BC}J_{BC}}{I_{DA}J_{DA}} = \frac{KJ_{BC}}{LJ_{DA}},$$

so

$$\frac{I_{BC}J_{BC}}{I_{DA}J_{DA}} = \frac{O_-J_{BC}}{O_-J_{DA}} = \frac{KJ_{BC}}{LJ_{DA}} = \frac{O_-J_{BC} - KJ_{BC}}{O_-J_{DA} - LJ_{DA}} = \frac{O_-K}{O_-L}.$$

We are almost done: the *spiral similarity* centered at M that takes BC to AD also takes the circumcircle of BCE to the circumcircle of ADE. Now, $I_{BC}J_{BC}$ is the diameter of the circumcircle of $BJ_{BC}CI_{BC}$, since $\angle J_{BC}BI_{BC} = 90°$ (angle between internal and external bisectors). Also, the midpoint K of arc $\overarc{BC}$ is taken to the midpoint L of arc $\overarc{AD}$. Thus,

$$\frac{O_-K}{O_-L} = \frac{I_{BC}J_{BC}}{I_{DA}J_{DA}} = \frac{MK}{ML},$$

which is the relation from the Bisector Theorem, and so MO_- bisects $\angle KML$. A similar argument shows that MO_+ bisects the angle $\angle BMD$, and so the lines MO_+ and MO_- coincide. The conclusion follows.   □

4.2.3   Problems About Complete Quadrilaterals

We end our book with some questions involving complete quadrilaterals that can be solved with the geometric transformations techniques that we have developed and with the aid some of the theorems proved above.

252 Let $ABCD$ be a cyclic quadrilateral with no parallel sides. Let E be the intersection point of the lines AB and CD, and let F be the intersection point of the lines AD and BC (i.e., we complete the quadrilateral.) Let M be the Miquel point of this complete quadrilateral. Prove that:

(a) the point M lies on the line EF;
(b) the circumcircle Ω of $ABCD$ is fixed by the $\sqrt{bc}$ inversion that fixes the complete quadrilateral;
(c) if P is the intersection of diagonals AC and BD and O is the circumcenter of $ABCD$, then O and P lie on the common internal bisector of $\angle AMC$, $\angle BMD$, and $\angle EMF$;
(d) the line OP is perpendicular to the line EF;
(e) the circle ω_{AC} through M, A, and C passes through the circumcenter O of $ABCD$; moreover, ω_{AC} meets again each of the perpendicular bisectors of the sides of $ABCD$ on points belonging to the sides of the complete quadrilateral.

253 Let $ABCD$ be a quadrilateral with no parallel sides. Let E be the intersection point of the lines AB and CD, and let F be the intersection point of the lines AD and BC (so we complete the quadrilateral). Let M be the Miquel point of the resulting complete quadrilateral and let P be the intersection point of the diagonals AC and BD. Prove that if P lies on the common internal bisector of $\angle AMC$, $\angle BMD$, and $\angle EMF$, then $ABCD$ is cyclic.

254 Let $ABCD$ be a quadrilateral inscribed in the circle ω, and let K and N be the intersections of the pairs of lines (AB, CD) and (AD, BC), so that $ABCDKN$ forms a complete quadrilateral. Prove that the circumcircle of AKN is tangent to ω if and only if the circumcircle of CKN is tangent to ω.

255 Let $ABCD$ be a quadrilateral, and let E and F be points on the sides AD and BC, respectively, such that

$$\frac{AE}{ED} = \frac{BF}{FC}.$$

The ray $|FE$ meets the rays $|BA$ and $|CD$ at S and T, respectively. Prove that the circumcircles of the triangles SAE, SBF, TCF, and TDE pass through a common point.

256 Let $ABCD$ be a fixed convex quadrilateral with $BC = AD$ and BC not parallel to AD. Two variable points E and F lie on the sides BC and AD, respectively, such that $BE = DF$. The lines AC and BD meet at P, the lines AC and EF meet at Q, and the lines BD and EF meet at R. Prove that the circumcircles of the triangles PQR, as E and F vary, have a common point other than P.

257 Let MN be a line parallel to the side BC of the triangle ABC, with M on the side AB and N on the side AC. The lines BN and CM meet at the point P. The circumcircles of the triangles BMP and CNP meet at two distinct points P and Q. Prove that $\angle BAQ = \angle CAP$.

258 The points P and Q lie on the diagonals AC and BD, respectively, of a quadrilateral $ABCD$ such that

$$\frac{AP}{AC} + \frac{BQ}{BD} = 1.$$

The line PQ meets the sides AD and BC at the points M and N. Prove that the circumcircles of the triangles AMP, BNQ, DMQ, and CNP intersect at one point.

259 The circles ω and Ω meet at the points A and B. Let M be the midpoint of the arc $\overset{\frown}{AB}$ of the circle ω (M lies inside Ω). A chord MP of the circle ω intersects Ω at Q (Q lies inside ω). Let ℓ_P be the tangent line to ω at P, and let ℓ_Q be the tangent line to Ω at Q. Prove that the circumcircle of the triangle formed by the lines ℓ_P, ℓ_Q, and AB is tangent to Ω.

260 Show that in the four triangles formed by a complete quadrilateral, the perpendiculars to the Euler lines at the centers of the nine-point circles intersect at one point.

261 Let D be an interior point of the acute triangle ABC with $AB > AC$ so that $\angle DAB = \angle CAD$. The point E on the segment AC satisfies $\angle ADE = \angle BCD$, the point F on the segment AB satisfies $\angle FDA = \angle DBC$, and the point X on the line AC satisfies $CX = BX$. Let O_1 and O_2 be the circumcenters of the triangles ADC and EXD, respectively. Prove that the lines BC, EF, and O_1O_2 are concurrent.

Part II
Hints

Chapter 5
Isometries

1 What is the composition of two reflections? Where does it map the first triangle?

2 The set is invariant under the composition of the reflections over the centers of symmetry.

3 The isometry has fixed points. To prove this, examine the behavior under the isometry of a segment determined by a point and its image.

4 The composition of two reflections is a translation.

5 Use (and prove) the fact that if $\sigma_1, \sigma_2, \sigma_3, \sigma_4$ are the reflections over four points in the plane, then

$$\sigma_1 \circ \sigma_2 \circ \sigma_3 \circ \sigma_4 = \sigma_3 \circ \sigma_4 \circ \sigma_1 \circ \sigma_2.$$

6 Look at the images of the points of pairwise intersections of a, b, and c under the reflections σ_a, σ_b, and σ_c.

7 What kind of transformation is $(\sigma_a \circ \sigma_b \circ \sigma_c)^2$, where σ_a, σ_b, σ_c are the reflections over the three lines?

8 Multiply the relation from the statement on the left by $\sigma_a \circ \sigma_b$ and on the right by $\sigma_d \circ \sigma_c$. Examine the relative position of the lines a and b.

9 By examining what types of isometries the compositions should be, deduce that $\sigma_a \circ \sigma_{\ell_1} \circ \sigma_b = \sigma_A \circ \sigma_{\ell_1} \circ \sigma_B$. Then check that $\sigma_{\ell_2} = \sigma_a \circ \sigma_{\ell_1} \circ \sigma_b$.

10 Look at a pair of axes of symmetry that minimize the angle between them. Create new axes of symmetry by reflecting one over the other.

11 What geometric transformation is the composition of the five reflections over the midpoints (taken in order) of the pentagon?

© The Author(s), under exclusive license to Springer Nature Switzerland AG 2022
R. Gelca et al., *Geometric Transformations*, Problem Books in Mathematics,
https://doi.org/10.1007/978-3-030-89117-6_5

12 Consider rotations that map one trajectory into the other. Can you find such a rotation that maps the starting points into each other?

13 Look at the 90° rotation about the center of the square. Where is P mapped?

14 Translate the segment CD into a segment AE. Look at the triangle EAB.

15 Construct the parallelogram $ABJC$. Do you notice that it rotates by 90° into $EAGI$? You can also try to use complex numbers, the vertices of the squares being 90° rotates of the vertices of the triangle. For (d), find a metric relation that characterizes the symmedian.

16 Consider the 90° rotation about the center of the square.

17 The answer is yes, and the easiest construction uses a translation.

18 Try shearing, that is, try to slide the first two parallelograms, without changing their areas, into two parallelograms that partition the third parallelogram.

19 Turn the relation into the Pythagorean Theorem in some right triangle.

20 Examine the proof of Theorem 1.24.

21 Use Theorem 1.26.

22 For (a), use complex coordinates with the origin at the circumcenter, which yield a nice expression for the coordinate of the orthocenter. For (b), reflect over the circumcenter.

23 Reflect D over AB.

24 What happens when you reflect the squares over the angle bisectors of the triangle?

25 What happens when you translate ℓ? Translate ℓ so that A_0 lies on the circumcircle of ABC.

26 Look at compositions of the 90° rotations about the centers of the squares.

27 You can use either a 60° rotation or a 120° rotation.

28 Look at the 60° rotation about the center of the hexagon. What happens to the quadrilateral $ABCM$?

29 Parallelograms go hand in hand with translations!

30 Reflect A over the side BC and over the midpoint of this side. The images are on the circumcircle of HBC, where H is the orthocenter of ABC. What else is on this circle?

31 Consider either the reflection over the perpendicular bisector of the chord c, or the reflection over A.

32 The angle is 60°, and a 60° rotation about A might help you prove it.

33 Look at $60°$ rotations about A and B.

34 Examine the figure formed by $\mathcal{P}$ and its reflection over the point O.

35 Use the fact that AN is mapped to CM by a $120°$ rotation.

36 Reflect over the midpoint M of CE. If A maps to A' and B to B', what kind of quadrilateral is $ABA'B'$?

37 Use Problem 6. Or, for a different solution, use the observation that if two lines ℓ and ℓ' are the reflections of each other over the point X, then they are reflections over any point that lies on the parallel through X to ℓ and ℓ'.

38 Can you dissect the triangle ABC into some triangles congruent to ABC_1, AB_1C, and A_1BC?

39 Think "cut and paste"! You can translate AB_2C_1, A_1BC_2, A_2B_1C so that A_1 coincides with A_2, B_1 coincides with B_2, and C_1 coincides with C_2. What figure do you get? Can you use this configuration to deduce that $A_1B_1C_1$ is equilateral?

40 There is a rotation about I that takes the triangle DBF to the triangle GCE.

41 Study the compositions of the clockwise $120°$ rotations about A, B, C. Or if you use complex numbers, you will see that the coordinates of the successive locations of the man after $3n$ steps have a nice formula.

42 Look at the translation that maps one circle into the other. This translation will map some points that are important in the solution into one other. Also, $30°$ angles can be obtained from equilateral triangles!

43 The minimum should be a billiard path. An elegant approach uses five reflections of the triangle over its sides that turn the perimeter of the orthic triangle into a straight line.

44 The main step of the solution does not explicitly use isometries; it uses properties of the isogonal conjugate of a point. Let Q be the isogonal conjugate of P with respect to triangle AED. Show that QE is parallel to CD.

45 Use the same idea as for the proof of the Pappus' Area Theorem.

46 Show that if ABC is a triangle, then you can construct a point D such that $ABCD$ is a parallelogram (it is not hard to do this if ABC is equilateral with side-length equal to the opening of the compass, but by using compositions of translations, you can do it in general). Show also that if A, B are two points, then you can construct a point C such that ABC is equilateral (again, this is not hard if AB has length equal to the opening of the compass, but by using compositions of isometries, you can do it in general as well). Try to combine these two constructions.

47 Use "inscribed circles" and put the nails at the tangency points.

48 Use three cuts that meet at a (special) interior point.

49 One of the squares can be obtained from the other by a rotation about the center followed by a translation. Show that the property holds true when the squares have the same center. Check that the total length of the red (or blue) segments does not change under translations.

50 Consider the rotations of the polygon by the angles $2k\pi/n$, $k = 1, 2, \ldots, n-1$, and examine the overlaps of red vertices with black vertices. What is the average number of red-black overlaps?

51 For part (a), use two translations of the set of points so that the resulting sets and the original set do not overlap, and so that the three sets lie in a region whose area is as small as possible. For (b) you can allow overlaps.

52 If such a set did not exist, then every reflection over a point would change the color of all but finitely many points of the plane. Consider the group generated by two reflections.

53 Tessellate the plane by regular hexagons with the distance between the centers of two neighboring hexagons slightly greater than 4. Let H be one of the hexagons. Look at the problem "modulo H."

54 Add to the figure the reflections of the chords over the center of the circle. Then use the Pigeonhole Principle.

55 Reflect "arcs" of people in order to reduce the number of neighboring enemies.

56 Imagine instead that the figure is fixed and the points move on the cylinder, all rigidly linked to each other. Now rotate and translate the cylinder so that the points avoid the surface. What region does one of the points trace?

57 The "reversed" configuration is the reflection over a line of the original configuration.

58 Each of the equilateral triangles is partitioned into four equal triangles: medial, top, bottom left, and bottom right. Focus just on the union of the top triangles and show that the part of it, A, that is in the complement of T, has area smaller than T. Note that a top triangle turns into the medial triangle by a reflection over the common side. Slice A by the lines of support of these common sides, and then reflect the slices.

59 Use the following Lemma: Let Γ be a polygonal line in the plane with endpoints A and B, and consider a number $\alpha \in (0, 1)$. Then among all segments with endpoints in Γ that are parallel to AB, there is either one of length αAB or one of length $(1 - \alpha)AB$.

60 Call the angle between the lines α. The trajectory of the flea is determined by the vectors that describe the jumps. Can you relate the angle between the first and the third vector to α? What about the angle between the second and the fourth vector?

61 Play symmetrically.

62 Think about invariants! Color the chessboard by elements of the group of symmetries of a nonsquare rectangle in such a way that at every jump the color of the final square is obtained by multiplying the color of the initial square by the same element.

63 Can you reduce the problem to symmetric play?

64 There are two key observations: (1) if two 1×4 rectangles coincide by a horizontal or vertical translation by 3 units in the direction of their shorter side, then they must contain an equal number of a's, b's, and c's, and (2) if two 1×3 rectangles coincide by a horizontal or vertical translation by 4 units in the direction of their shorter side, then they must contain an equal number of a's, b's, and c's.

65 For (a) consider the two orthogonal vectors that define the lattice of one given color. Factor out the plane by identifying two points if and only if one is mapped into the other by a linear combination with integer coefficients of the two vectors. You obtain a torus. What is the surface area of the torus? What region of that torus is colored by each color?

66 Arrange the trees at the nth roots of unity. When a bird flies from the kth to mth tree, its position rotates by $\frac{2\pi i(m-k)}{n}$. Define an invariant associated with each configuration of birds using powers of the rotation about the center of the circle by $\frac{2\pi i}{n}$.

67 Translate A and B such that $\max A = \min B = 0$.

68 Translate $A \mapsto A' = A + \{-a\}$ and $B \mapsto B' = B + \{b\}$, and then compare $A \cup B$ to $A' \cup B'$.

69 In other words, we are supposed to show that the union of finitely many nondegenerate intervals with total length less than 1 may be translated so as not to intersect $\mathbb{Z}$. Can the translates of this union by integers cover the entire interval $[0, 1]$?

70 The number is 2.

71 Use binary expansions to construct an example.

72 Encode the strings by geometric objects, and then use symmetry.

73 Count the number of non-monochromatic colorings of the vertices of a regular p-gon by n colors.

74 Do the case $k = 1$ first. Follow the idea in Problem 73 where you consider a_1 regular p-gons that rotate about their centers, together with a_0 fixed points that do not belong to these polygons. Pick $b_0 + b_1 p$ points among the a_1 vertices and the a_0 fixed points. Count configurations that are not rotation invariant.

75 Represent the family of equations $ax + by = n$ as lines in the plane.

76 Think about Problem 67, but this time with rotations about a point since you work modulo p.

77 Write $f = g + h$ such that the center of symmetry of the graph of g is $(0, 0)$ and the center of symmetry of the graph of h is $(1, 0)$. You can start by constructing f and g on one small interval and then extend this construction inductively.

78 Make $f(x) = x + 1$ on $(0, 1]$.

79 To prove $f(A_0) = 0$, choose a regular n-gon, $A_0 A_1 A_2 \cdots A_{n-1}$, and consider the rotations ρ_k of center A_0 and angles $\frac{2k\pi}{n}$, $k = 0, 1, \ldots, n - 1$. Define $A_{kj} = \rho_k(A_j)$, $k, j = 0, 1, \ldots, n - 1$. Examine all regular polygons that arise in the figure after the rotations. There are more regular polygons than just the rotations of the original polygon.

80 What is the composition of two reflections?

81 By a change of variable, you can turn P into a polynomial in z and z^{-1} that is invariant under rotation of z by $\frac{\pi}{4}$ about the origin $z \mapsto e^{\frac{\pi i}{4}} z$.

82 The answers to (a) and (b) are different. For (b), show that if $|z - w| = \sqrt{3}$, then $|f(z) - f(w)| = \sqrt{3}$. From this and the hypothesis, deduce that if z and w are such that $|z - w| = |m + n\sqrt{3}|$ with $m, n \in \mathbb{Z}$, then $|f(z) - f(w)| = |z - w|$.

Chapter 6
Homotheties and Spiral Similarities

83 The answer is 1 or 2. There is a simple solution using coordinates. For a synthetic solution, you can use the group of homotheties and translations.

84 Beware of the many possible configurations, and of the fact that the homothety can be direct or inverse! Think about the group of homotheties and translations.

85 There can be one or two such homotheties. Use composition to rule out the existence of a third homothety.

86 Let h_{12}, h_{13}, h_{23} be the direct homotheties of the pairs (ω_1, ω_2), (ω_1, ω_3), (ω_2, ω_3), respectively. Then $h_{13} = h_{23} \circ h_{12}$.

87 What is the image of the line through the homothety that maps ω_2 to ω_1?

88 There is a special line on which these projections lie. Use a homothety of center A and ratio $1/2$ to find it.

89 Use the homothety that maps the incircle to the excircle.

90 Can you spot an inverse homothety of center K?

91 If h_1, h_2, h_3 are homotheties such that $h_3 = h_2 \circ h_1$, then their centers are collinear.

92 Use homothety to show that, in a trapezoid, the line determined by the midpoints of the parallel sides passes through the intersection of the diagonals and also through the intersection of the nonparallel sides.

93 The center of homothety is the centroid of the polygon.

94 The two polygons are homothetic.

95 A homothety of center I maps $MNPQ$ to $XYZW$.

96 The two orthic triangles are homothetic. What is the center of homothety?

© The Author(s), under exclusive license to Springer Nature Switzerland AG 2022   255
R. Gelca et al., *Geometric Transformations*, Problem Books in Mathematics,
https://doi.org/10.1007/978-3-030-89117-6_6

97 In complex coordinates, the line through a and b reflects to the line through $c + d - a$ and $c + d - b$, whose parametric equation is $t \mapsto c + d - a + t(a - b)$, $t \in \mathbb{R}$. Similarly, the line through a and d reflects to the line $s \mapsto b + c - d + s(a - d)$, $s \in \mathbb{R}$. Can you guess the intersection point?

98 The perpendiculars through A', B', C' to the sides BC, AC, AB, respectively, are concurrent.

99 The triangle formed by the centers of three of the circles is homothetic to ABC. The two triangles have the same incenter.

100 A homothety whose center is the intersection of the diagonals maps the circle through the eight points to a circle centered at the circumcenter of the original quadrilateral.

101 How are B and C connected by the direct homothety that maps the circles into each other?

102 Fix B and C in the plane, and let A vary so that ABC has the property from the statement. What are the loci of the centroid and the incenter?

103 A homothety of center H and ratio 2 can help.

104 Use a homothety centered at the centroid and of ratio $-1/2$. The Simson-Wallace Theorem (Theorem 3.34) might be helpful.

105 Given a homothety, the line joining a point with its image passes through the center of homothety.

106 Consider a homothety of center A and ratio $1/2$.

107 There are two parallel lines in the figure! Consider homotheties of centers K and E.

108 Use the homotheties centered at the orthocenters that map the circumcircles to the nine-point circles.

109 Consider the homotheties h_Q of center Q and which maps A to C, and h_P of center P and which maps C to B. What kind of map is the composition $h_Q \circ h_P$?

110 Think about the relationship between the incircle and the excircle.

111 Use a homothety of center Z and ratio ZB/ZA.

112 Find a better description of the points A_1, B_1, C_1 using the triangles HBC, HAC, HAB.

113 Construct the incircle of $ABCD$ and consider homotheties that map it to ω and ω_a.

114 Let D be the point where AN intersects the circumcircle. Consider a homothety of center A that maps D to R.

115 Use the idea from the proofs of Menelaus' and Pappus' Theorems.

116 To prove that if the lines intersect then the identity holds, use homotheties that map B_1 to C, C to B, B to A, and A to B_1.

117 Prove that each of the triples (K, M, P), (K, N, Q), and (M, L, N) consists of collinear points.

118 Use the Monge-d'Alembert Theorem (Problem 86).

119 Let A be the intersection of the common exterior tangents of ω_1 and ω_2, and let B be the intersection of the common exterior tangents of ω_1 and ω_3. You should show that there is a circle ω that is mapped to ω_2 by a homothety of center B and to ω_3 by a homothety of center A.

120 The two squares are homothetic. Use the homothety to compute the radius of Ω_a. Show that the radii of Ω_a and Ω_b add up to the distance between their centers.

121 Let P be the intersection of A_1A_4 and A_2A_3 and assume that P, T_1, T_3 are collinear. Use the Monge-d'Alembert Theorem (Problem 86) to conclude that there is a circle ω that is tangent to A_1A_2 at T_1 and to A_3A_4 at T_3. Use the same theorem in reverse to prove the other collinearity.

122 Add to the picture the B-excircle of the triangle ABC and the D-excircle of the triangle ACD. Use homotheties that map the circles of this enriched configuration into one another.

123 The previous problem is actually a good hint. Consider the incircle and A-excircle of APQ, and also the C-excircle of CPQ. Play with composition of homotheties to find several triples of collinear points.

124 The five pentagons lie inside a pentagon that is homothetic to each of them and twice as big.

125 Compose the projection maps from each line to the next to obtain a transformation of a line to itself. What kind of transformation is it?

126 Use homotheties centered at the vertices of the triangle to increase the range where the grasshopper can start in order to land close to the flower.

127 Change the scale at which you look at the turtles.

128 Scaling! If a square K has too little black, cut it into four parts.

129 Use the following result due to Isaak Moiseevich Yaglom: every convex polygon that is not a parallelogram has three sides with the property that the polygon itself lies inside the triangle formed by the lines of support of these sides.

130 There is a spiral similarity mapping the line through B_1, B_2, B_3 to the line through C_1, C_2, C_3.

131 Use Theorem 2.22.

132 Use Problem 131 and Theorem 2.23.

133 Use Theorem 2.22.

134 There are two such spiral similarities.

135 On a side of the polygon, choose a point M that is not a vertex and consider a spiral similarity s with center M, angle $\angle BAC$, and ratio AB/AC. Look at the points where the image of the polygon crosses the polygon itself.

136 Consider a spiral similarity of center H, angle $90°$, and ratio BP/BC.

137 For (a) use the Averaging Principle. For (b), think circles! Use Problem 131.

138 Use the Averaging Principle.

139 Let M be the common midpoint; use a spiral similarity of center M.

140 Use the Averaging Principle.

141 There is a spiral similarity of center M that maps $A_1A_2\ldots A_n$ to $M_1M_2\ldots M_n$.

142 Construct a point C' such that a spiral similarity of center A maps AMB to $AC'N$. Show that $C = C'$.

143 Use Problem 131.

144 Theorem 2.25 implies that there are spiral similarities that map the triangle ABC to PBR and to QRC.

145 The property is true if P and Q are the midpoints of the sides.

146 There is a spiral similarity of center M that maps O_1 to O_2 and D to C, and spiral similarities come in pairs.

147 Use the Miquel point.

148 Use the Averaging Principle.

149 Map P to Q using the composition of two spiral similarities. What transformation is this composition?

150 Can you use a spiral similarity that appears in the proof of Miquel's Theorem (Theorem 2.24)?

151 The common point of these circles is the center of the spiral similarity that maps ABC to $A_1B_1C_1$.

152 If X is the intersection point of the circumcircles of APC and AQB, then AX is symmedian and AX, BM, CN are concurrent.

153 There is a spiral similarity that maps DAE to DEB.

154 Use Theorem 2.22.

155 Let D, E, F be the feet of the altitudes from A, B, C, respectively. Then there are spiral similarities that map the triangles ADX, BEY, CFZ into one another.

These spiral similarities are paired with three others. Locate the centers of all these spiral similarities.

 156 There are spiral similarities that map the triangle NXW to BNC and the triangle YMW to BMC, and spiral similarities come in pairs. Use the new spiral similarities and their composition.

Chapter 7
Inversions

157 Use coordinates.

158 Take a Möbius transformation (or just an inversion) that maps the given circle to a line.

159 Use complex coordinates.

160 There is a hidden circle in the figure, and the intersection point is the radical center of the three circles.

161 Where does the tangency point map through the inversion with respect to ω?

162 Apply an inversion that maps two of the circles into two parallel lines.

163 Take an inversion of center A, and use the invariance of angles.

164 Map the circle of inversion to a line, and then use the Symmetry Principle.

165 Reduce the problem to an easy particular situation by making use of the invariance of the cross-ratio under circular transformations and of the Symmetry Principle.

166 Such Möbius transformations map the real axis to itself.

167 Compose with a Möbius transformation to obtain a transformation that maps lines to lines and circles to circles. Can you then use a result from a previous chapter?

168 How do the zeros of the polynomial change when (a) and (b) are applied?

169 Invert about A; what kind of quadrilateral is $AM'B'N'$?

170 Use formula (3.1). For an analytic solution, write the identity as an equality of cross-ratios.

171 Invert about the inscribed circle and use Proposition 3.3.

© The Author(s), under exclusive license to Springer Nature Switzerland AG 2022
R. Gelca et al., *Geometric Transformations*, Problem Books in Mathematics,
https://doi.org/10.1007/978-3-030-89117-6_7

172 Use an inversion of center P. For (a) use the formula (3.1) or the invariance of cross-ratios.

173 Consider an inversion of center M that maps ω into the line ST.

174 Use an inversion that transforms Ω_1 and Ω_2 into two parallel lines.

175 Invert with respect to the circumcircle and use Ceva's Theorem.

176 Take an inversion of center O and radius $\sqrt{Rr}$, where R and r are the radii of the two circles.

177 Let BE and CF be altitudes. Map, by an inversion, the line EF to the circumcircle of ABC.

178 Use an inversion of center A and radius AB.

179 An inversion can help reduce the number of circles.

180 The following can help with (a): an inversion of center M, an inversion of center B composed with reflection over B, or an inversion of center A. One of these transformations helps with (b) and (c).

181 Work backward, let B_1 be the intersection of the common interior tangent of ω and ω_1 with the line ℓ_2, and show that $B_1 = B$. For this consider an inversion of center B_1 and radius $B_1 A$.

182 Can you reduce the question to Problem 15 by using an inversion?

183 Locate the second intersection point of the circle with EM using an inversion of center E and radius $EB = EC$.

184 Part (a) follows from the fact that if ABC is a triangle that is not isosceles, I is its incenter, and D, E, F are the tangency points of the incircle to the sides BC, AC, and AB, respectively, and then the circumcircles of AID, BIE, CIF have a second intersection point besides I. Part (b) follows from the fact that if ABC is a triangle, and if M is on BC, N is on AC, and P is on AB such that AM, BN, CP are concurrent, and if moreover X is on NP, Y is on MP, and Z is on MN such that MX, NY, and PZ are concurrent, then AX, BY, and PZ are also concurrent.

185 Consider the inversion whose center is the tangency point of C_3 and ℓ and has the property that maps C into itself.

186 Use an inversion of center F.

187 Use Proposition 3.3.

188 Consider an inversion of center A. Then D' is diametrically opposite to A in the circumcircle of $AB'C'$, E' is the midpoint of the segment $B'C'$, and F' is the point where the perpendicular bisector of $B'C'$ intersects the circumcircle of $AB'C'$.

189 Power-of-a-point leads to inversion.

190 There are three circles passing through A: the circumcircles of ABL, AKC, and AKL. Turn them into lines.

191 Consider the $\sqrt{bc}$ inversion corresponding to the vertex A in the triangle ABC.

192 Consider an inversion of center B. Do you see a right triangle?

193 A circular transformation that is the composition of an inversion of center H and the reflection over H might help.

194 There are too many circles in the configuration.

195 Show first that AO is perpendicular to B_1C_1. Then use an inversion of center D and radius DA.

196 Reduce the problem to the following: two circles Ω and ω intersect at B and X and are tangent to the line ℓ at A and D, respectively. Let S and T be points on Ω such that BS is parallel to ℓ and AT is a diameter. Then the lines BD, SA, and TX intersect (at some point C).

197 Map the circle to a line.

198 Prove that the bisectors of $\angle ABP$ and $\angle ACP$ intersect at the same point on the line AP by showing that $AB/BP = AC/CP$. Write this as $\frac{BA}{CA} : \frac{BP}{CP} = 1$.

199 Does a $\sqrt{bc}$ inversion help?

200 Take an inversion with respect to a circle centered at A and orthogonal to ω, and use the fact that the polar of A with respect to ω is the line that passes through Q and through the intersection of the diagonals of $BCFE$.

201 Modify the $\sqrt{bc}$ inversion corresponding to the vertex C.

202 Use a circular transformation that is the composition of an inversion of center H and the reflection over H (i.e., an inversion of negative ratio).

203 The quadrilateral $ABCD$ will have to be harmonic. Have you seen this before?

204 An inversion that takes one circle to another also induces a homothety between the circles!

205 A, B, X, Y, I are on a circle. Invert about this circle.

206 Consider an inversion of center B, and then build an argument based on the polar with respect to a circle.

207 Use a $\sqrt{bc}$ inversion defined by the triangle ABC and the vertex B. Reduce the problem to the following: in triangle ABC, O is the circumcenter. The perpendicular ℓ to the radius OB through O passes through the orthocenter H of ABC if and only if the distance from the reflection of O over BC to this perpendicular is equal to the circumradius R of ABC.

208 Invert about a circle centered at X (to make things simple, keep the circle of center O invariant).

209 $ADBC$ is harmonic.

210 Consider the inversion of negative ratio centered at the orthocenter that maps the circumcircle to the nine-point circle.

211 Invert about a circle centered at H. Use properties of symmedians.

212 Use a $\sqrt{bc}$ inversion. Theorem 4.2 from the next chapter turns out to be useful.

213 We have written three solutions at the end of the book: one that uses the $\sqrt{bc}$ inversion corresponding to the vertex A, one that uses the $\sqrt{bc}$ inversion corresponding to the vertex C, and one that uses inversion centered at K.

214 Use the $\sqrt{bc}$ inversion corresponding to the vertex A.

215 It is easier to prove that a line is tangent to a circle.

216 Let I be the incenter of ABC. Prove that SBP and SIQ are similar, and deduce another similarity to finish the problem. There is also a symmedian in the diagram!

217 Perform the $\sqrt{bc}$ inversion ϕ_A determined by the triangle ABC and the vertex A. Use Proposition 3.41 to find the images of the B- and C-mixtilinear excircles.

218 First, prove that the point X is unique. Then work backward, showing that if $\angle BAD = \angle CAX$ then $MB = MC$. Consider the A-excenter I_a, and look at the triangle BI_aC. Think mixtilinear excircles!

219 First order of business is to take the $\sqrt{bc}$ inversion corresponding to the vertex A. Where does the circumcircle of DEF go? Consider also separately the reflection across the bisector of $\angle BAC$.

220 Invert about the circle centered at the tangency point of ω_1 and ω_2 and of radius $2\sqrt{r_1 r_2}$.

221 Reduce the problem to proving that A, T, the incenter I, and the point D_0 diametrically opposite in the incircle to D are concyclic.

Chapter 8
A Synthesis

222 Use either spiral similarities or rotations. There is a short solution of (a) using the Averaging Principle. For (b), use the spiral similarities centered at the vertices A, B, C of angles $30°$ and ratios $2\cos 30°$. Complex numbers work too.

223 For an inversion-based solution, use the invariance of the cross-ratio. There is also an analytic solution that uses the system of coordinates introduced before Problem 4.1 from the introduction.

224 Reflect the triangle SAD over the line through S that is orthogonal to PQ. Or take an inversion centered at S.

225 Show that $A_1A_2A_3$ and $M_1M_2M_3$ have parallel sides.

226 Notice that the property is true if the three triangles are part of a regular hexagon. To get to the general configuration, you can transform each triangle by a spiral similarity centered at the common point. Show that the property is preserved when changing one triangle at a time. Alternatively, the complex number solution is quite easy.

227 An inversion centered at A yields a symmetrical figure.

228 The line through E, P, F is the image of the Simson line of D with respect to ABC under a homothety of center D and ratio 2.

229 Midpoints lead to dilations; equilateral triangles lead to $60°$ rotations.

230 Cross-ratios, of course! But you can also solve this problem using just spiral similarities.

231 Use a homothety and an inversion both centered at the intersection of the common tangents of the two circles.

232 Use a transformation λ that is the composition of a spiral similarity and a reflection over a line such that $\lambda(HPQ) = ABC$.

© The Author(s), under exclusive license to Springer Nature Switzerland AG 2022   265
R. Gelca et al., *Geometric Transformations*, Problem Books in Mathematics,
https://doi.org/10.1007/978-3-030-89117-6_8

233 For the direct implication, examine the image of ℓ through the homothety of center C and ratio $1/2$. For the converse, use the rotation about the circumcenter of CKL that maps K to C.

234 Use Problem 131, the fact that spiral similarities come in pairs, and the Averaging Principle for a proof via spiral similarities. For proof by inversion, check that M, A, P, Q are concyclic by inverting with respect to a circle of center A and proving that the images of M, P, Q are collinear.

235 For (a), prove that the orthocenter lies on the line determined by the reflections of A over BC and DC, and for (b) use Steiner's Theorem proved in (a).

236 The pairs of triangles FAB and FCD respectively FAD and FBC are similar, but not directly similar. Use reflections over lines to make them directly similar in order to use homothety for (a) and spiral similarity for (b).

237 There is a $90°$ rotation that maps $P_1Q_1R_1S_1$ into $P_2Q_2R_2S_2$. Its center lies at the intersection of the bisectors of the angles formed by AI and LC, BK and FD, CE and HA, and DG and JB. What points do this angle bisectors pass through? Can you use the fact that the diagonals of $ABCD$ are orthogonal to find the points where these angle bisectors meet?

238 Start with a spiral similarity that maps the triangle ABO to AOQ' and prove that $Q' = Q$. Solutions based on isometries are also possible.

239 Let A' be the point that is diametrically opposite to A in the given circle. Prove that $A'D$ is perpendicular to BC. A spiral similarity of center C' mapping the triangle $C'CA'$ to the triangle $C'MD$ might be useful.

240 Invert about I and observe that the lines A_iA_{i+1} become equal circles.

241 Let O be the center of Γ. Use a spiral similarity that maps EAF to EOD and an inversion that maps Γ to ℓ.

242 Start with the $\sqrt{bc}$ inversion defined by the triangle QRP and the vertex Q. Theorem 1.22 might be useful for the rest of the argument.

243 Take an inversion of center R. Or, for a solution without inversion, reflect A over S.

244 Let A_1, B_1, C_1 the points where the incircle is tangent to the sides, and let M_A, M_B, M_C be the midpoints of the arcs $\overarc{BC}, \overarc{AC}, \overarc{AB}$ of the circumcircle. Then, the intersection point of AA', BB', CC' is the center of the homothety that maps $A_1B_1C_1$ to $M_AM_BM_C$.

245 One possibility is to use inversion and show that the line that joins C with the intersection of AN and Γ_2 is tangent to both Γ_1 and Γ_2, and the same is true for the line that joins D with the intersection of BN and Γ_2. Or you can use homothety to show that CD is parallel to AB and attempt a computational solution to prove that the distance from O_2 to CD is equal to the radius of Γ_2.

246 You can perform a regular inversion about C and do some computations, but the $\sqrt{bc}$ inversion can circumvent these: consider the midpoints of the arcs $\overset{\frown}{AC}$ and $\overset{\frown}{BC}$ that do not contain the other vertex, and find their images under the inversion determined by the vertex C of the triangle ABC.

247 Invert about the incircle of ABC. Problem 204 can be used to locate the inverse of the desired point.

248 Let Y be the intersection of BC and FG. Then, Y is the center of both an inversion and a homothety that exchange the circumcircles of FBD and GCE.

249 Inverting about Γ gives rise to the following question: fix a circle Γ of center G and radius r, a circle ℓ of radius $r/2$ which passes through G, and a circle Ω inside ℓ and disjoint from ℓ. The circles ω_1 and ω_2 of radii $r/2$ pass through G and through a variable point on Ω, and cross again ℓ at Y and Z. Prove that as X traces Ω, the circumcircle of XYZ is tangent to two fixed circles.

250 Recall the coordinate system introduced before Problem 4.1 from the essay part of this chapter.

251 Parametrize the triangles (up to similarity) by points in an equilateral triangle with altitude equal to π, so that the angles of a triangle are the distances from a point in the equilateral triangle to the sides. How does H act on this equilateral triangle?

252 In (a), use inscribed angles. How do O and P behave under the inversion that is part of the $\sqrt{bc}$ inversion?

253 Try to use the coordinate system introduced at the beginning of Sect. 4.2.

254 Use homotheties that map the two circumcircles to ω. Or use a $\sqrt{bc}$ inversion.

255 This problem is a showcase for spiral similarities and the Miquel point.

256 Spiral similarities and the Miquel point.

257 Take advantage of spiral similarities and homotheties that arise in the configuration. Or consider a $\sqrt{bc}$ inversion defined by the vertex A in the triangle ABN.

258 There is a spiral similarity that takes A to D, C to B, and P to Q.

259 The tangency point is the Miquel point of the quadrilateral determined by ℓ_P, ℓ_Q, AB, and MP. There is a solution based on inversion and one based on homothety.

260 Use Theorems 4.7, 4.6, and Problem 23.

261 Turn $BCEF$ into a complete quadrilateral and then consider an inversion of center D.

Part III
Solutions

Chapter 9
Isometries

1 The composition of the two reflections is either a translation or a rotation, and this transformation maps the triangle ABC to itself. It cannot be a translation because its repeated applications would move the triangle away from itself. Therefore, it is a rotation. The triangle that is invariant under some rotation is the equilateral triangle. This can be proved as follows. Let ρ be the rotation. Then $\rho(A)$ is either B or C, say $\rho(A) = B$. Then $\rho(B) = C$ and $\rho(C) = A$. It follows that $\rho^3(A) = A, \rho^3(B) = B, \rho^3(C) = C$, and since the vertices are three distinct noncollinear points, ρ^3 must be the identity map. This implies that ρ is a rotation by $120°$, and consequently A, $B = \rho(A)$, $C = \rho^2(A)$ are the vertices of an equilateral triangle. The image $A'B'C'$ of ABC through a reflection is also an equilateral triangle.

Remark This will be a recurring theme in some problems: instead of exploring the diagrams obtained after the transformations, we focus on the structure of the transformations themselves.

Case in point: while we can work on the two triangles at hand, we prefer to compose the reflections, because we already know well the structure of composition of reflections. Since we know that the composition of two reflections is a rotation or a translation, and we know what properties it must have, we can infer what kind of triangle resonates with them.

Source D. Smaranda, N. Soare, *Transformări Geometrice (Geometric Transformations)*, Ed. Academiei, Bucharest, 1988.

2 The answer is negative. Assume that such a set S exists and let O_1 and O_2 be the two centers of symmetry. The set S must be invariant with respect to the reflections σ_1 and σ_2 over the points O_1 and O_2, respectively, so it is invariant under $\sigma_2 \circ \sigma_1$. But the latter is the translation τ by $2\overrightarrow{O_1O_2}$.

The composition of a translation and a reflection is a reflection, and the simplest reflection that we can obtain this way that is not σ_1 or σ_2 is $\tau^2 \circ \sigma_1$. Indeed, this is the reflection over $O_3 = \sigma_2(O_1)$, for example, because

© The Author(s), under exclusive license to Springer Nature Switzerland AG 2022
R. Gelca et al., *Geometric Transformations*, Problem Books in Mathematics,
https://doi.org/10.1007/978-3-030-89117-6_9

$$\tau^2 \circ \sigma_1(O_3) = \tau^2 \circ \sigma_1(\sigma_2(O_1)) = \sigma_2 \circ \sigma_1 \circ \sigma_2 \circ \sigma_1^2 \circ \sigma_2(O_1)$$

$$= \sigma_2 \circ \sigma_1 \circ \sigma_2^2(O_1) = \sigma_2 \circ \sigma_1(O_1) = \sigma_2(O_1) = O_3.$$

So S has at least three centers of symmetry, O_1, O_2, O_3, and therefore it cannot have exactly two.

Remark We see that in fact the figure must have infinitely many centers of symmetry.

A good way to express the symmetries of a set is as invariance under a set of transformations. In our case, an axis of symmetry corresponds to a reflection over that axis and a center of symmetry corresponds to a reflection over the center. Then we can *compose* these transformations, giving birth to new transformations under which the set is invariant as well. In fact, these transformations form a *group* with composition as operation, and can be fully characterized. With this in mind, finding another reflection, and thus another center of symmetry, becomes an easy task.

Do you realize that there are several other options of compositions to find other centers? You only need to make sure you compose an odd number of reflections.

3 *First solution.* Let A be a point such that $f(A) \neq A$. Then $f(f(A)) = A$, which means that f maps the segment $Af(A)$ to $f(A)A$. It follows that the midpoint M of $Af(A)$ is a fixed point of f. This property is trivially true for fixed points of f, so we conclude that for every point A, the midpoint of $Af(A)$ is a fixed point of f.

We distinguish two situations. If f has exactly one fixed point, then f is just the reflection over this point.

If f has at least two fixed points, say M and N, then the entire line MN is fixed. In this case if f has a fixed point that does not lie on MN, then f is the identity map. Otherwise, if $A \notin MN$, then triangles AMN and $f(A)MN$ are congruent, so A and $f(A)$ are symmetrical with respect to line MN. Hence, f is the reflection over line MN.

Second solution. Suppose f is not the identity map. If $f(z) = rz + s$, then $f(f(z)) = z$ yields $r^2 z + rs + s = z$, for all z; thus, $r^2 = 1$ and $s(r+1) = 0$. Since f is not the identity map, this is only possible if $r = -1$, which corresponds to the reflection over a point.

If $f(z) = r\bar{z} + s$, then $f(f(z)) = z$ yields $r\bar{r}z + r\bar{s} + s = z$, which means that $r\bar{s} + s = 0$. If $s = 0$, then we are done by Theorem 1.5. Otherwise, $r = -\frac{s}{\bar{s}}$, and Theorem 1.5 shows again that f is the reflection over a line.

Remark A transformation f with the property that $f \circ f$ is the identity map is called an involution.

4 The composition $\sigma_{2j-1} \circ \sigma_{2j}$ equals the translation by $2\overrightarrow{O_{2j}O_{2j-1}}$. Thus,

$$\sigma_1 \circ \sigma_2 \circ \cdots \circ \sigma_{2n-1} \circ \sigma_{2n} = 1$$

is equivalent to the fact that the composition of the translations by $\overrightarrow{O_{2j}O_{2j-1}}$, $j = 1, 2, \ldots, n$, is the identity map. Since the composition of the translations by $\overrightarrow{v}$ and

$\vec{w}$ is the translation by $\overrightarrow{v+w}$, this is further equal to the translation by

$$\overrightarrow{O_2 O_1} + \overrightarrow{O_4 O_3} + \cdots + \overrightarrow{O_{2n} O_{2n-1}}.$$

So the composition of the reflections is equal to zero if and only if this sum of vectors is equal to zero. The sum from the statement is actually the negative of this sum, and the conclusion follows.

Remark For $n = 2$ we obtain the following characterization of parallelograms: the quadrilateral $O_1 O_2 O_3 O_4$ is a parallelogram if and only if $\sigma_1 \circ \sigma_2 \circ \sigma_3 \circ \sigma_4 = 1$ where σ_k is the reflection over O_k, $k = 1, 2, 3, 4$.

Source D. Smaranda, N. Soare, *Transformări Geometrice (Geometric Transformations)*, Ed. Academiei, Bucharest, 1988.

5 *First solution.* The solution is based on the following result:

Lemma *If $\sigma_1, \sigma_2, \sigma_3, \sigma_4$ are reflections over the points P_1, P_2, P_3, P_4 in the plane, respectively, then*

$$\sigma_1 \circ \sigma_2 \circ \sigma_3 \circ \sigma_4 = \sigma_3 \circ \sigma_4 \circ \sigma_1 \circ \sigma_2.$$

Proof The composition $\sigma_1 \circ \sigma_2$ is the translation by $2\overrightarrow{P_2 P_1}$, and the composition $\sigma_3 \circ \sigma_4$ is the translation by $2\overrightarrow{P_4 P_3}$. The conclusion of the lemma follows from the fact that translations commute.

Using the lemma, we can write

$$\sigma_1 \circ \sigma_2 \circ \sigma_3 \circ \sigma_4 \circ \sigma_5 = \sigma_1 \circ \sigma_4 \circ \sigma_5 \circ \sigma_2 \circ \sigma_3 = \sigma_5 \circ \sigma_2 \circ \sigma_1 \circ \sigma_4 \circ \sigma_3$$

$$= \sigma_5 \circ \sigma_4 \circ \sigma_3 \circ \sigma_2 \circ \sigma_1.$$

Second solution. Let σ_j be the reflection over the point of complex coordinate z_j. Then $\sigma_j(z) = 2z_j - z$, so

$$\sigma_1 \circ \sigma_2 \circ \sigma_3 \circ \sigma_4 \circ \sigma_5(z) = 2(z_1 - z_2 + z_3 - z_4 + z_5) - z$$

$$= \sigma_5 \circ \sigma_4 \circ \sigma_3 \circ \sigma_2 \circ \sigma_1(z)$$

$\square$

Remark In general, if in a composition of reflections over points two pairs of consecutive reflections are separated from each other by an even number of reflections, then the two pairs can be swapped.

The lemma and the conclusion to the problem hold true for reflections over lines as well, provided that we impose that those lines have a common point. In that case, we use the fact that two rotations about the same point commute.

Source D. Smaranda, N. Soare, *Transformări Geometrice (Geometric Transformations)*, Ed. Academiei, Bucharest, 1988.

6 We start with the computation

$$(\sigma_a \circ \sigma_b \circ \sigma_c)^2 = (\sigma_a \circ \sigma_b \circ \sigma_c) \circ (\sigma_c \circ \sigma_b \circ \sigma_a) = \sigma_a \circ \sigma_b \circ \sigma_c^2 \circ \sigma_b \circ \sigma_a$$

$$= \sigma_a \circ \sigma_b^2 \circ \sigma_a = \sigma_a^2 = 1,$$

which shows that the equation from the statement is equivalent to the fact that $\sigma_a \circ \sigma_b \circ \sigma_c$ is an involution (when composed with itself yields the identity map). And this is equivalent to the fact that it is a reflection (see Problem 3), in fact reflection over a line because it changes orientation.

To understand why if a, b, c intersect, then they satisfy the condition from the statement, note that, by Theorem 1.10, $\sigma_a \circ \sigma_b \circ \sigma_c$ is the composition of a reflection and a translation parallel to the line of reflection. But the common point of a, b, c is a fixed point for this transformation, so no translation is involved. Consequently, we are in the presence a reflection, and the condition from the statement is satisfied.

For a, b, c parallel, $\sigma_b \circ \sigma_c$ is a translation in a direction perpendicular to c. Pick a point at distance half the length of the translation vector from a. Then $\sigma_a \circ \sigma_b \circ \sigma_c$ keeps this point fixed, so again this transformation has to be a reflection.

For the converse implication, we give two proofs.

First proof. We know that $\sigma_a \circ \sigma_b \circ \sigma_c$ is the reflection over a line. Let ℓ be this line and let σ_ℓ be the reflection over ℓ. Then

$$\sigma_a \circ \sigma_b \circ \sigma_c = \sigma_\ell$$

implies

$$\sigma_b \circ \sigma_c = \sigma_a \circ \sigma_\ell.$$

If a and ℓ intersect, then $\sigma_a \circ \sigma_\ell$ is a rotation about their intersection point. But then $\sigma_b \circ \sigma_c$ is a rotation, and it must be a rotation about the intersection point of b and c. It follows that a, b, c, ℓ intersect at one point and we are done.

If a and ℓ are parallel, then $\sigma_a \circ \sigma_\ell$ is a translation by a vector perpendicular to a. But then $\sigma_b \circ \sigma_c$ is a translation by the same vector, so these two lines must be parallel to each other, and must be parallel to a and ℓ. Thus, a, b, c, ℓ are parallel lines.

Second proof. If b is parallel to both a and c, there is nothing to prove. So suppose, without loss of generality, that a and b meet at the point P. Therefore, $\sigma_a(P) = \sigma_b(P) = P$. Let $Q = \sigma_c \circ \sigma_b \circ \sigma_a(P) = \sigma_c(P)$. Thus, $Q = \sigma_a \circ \sigma_b \circ \sigma_c(P) = \sigma_a \circ \sigma_b(Q)$. But $\sigma_a \circ \sigma_b$ is a rotation with center P, and the only fixed point by a rotation is its center. Hence, $P = Q$ if and only if $\sigma_b(P) = P$. But the only points that are fixed by a reflection over a line are the points on this line, so P is on c. We conclude that P lies on $a, b,$ and c.

Remark The solution is based on understanding how reflections generate the group of isometries. We use the very important fact that the composition of two reflections is either a rotation or a translation.

7 Assume that the lines intersect at the three different points A, B, C and that the lines are $a = BC, b = AC, c = AB$. Denote the reflections by $\sigma_a, \sigma_b, \sigma_c$, correspondingly. We are to prove that the transformation $\sigma_a \circ \sigma_b \circ \sigma_c$ has no fixed points.

Note that $\sigma_a \circ \sigma_b$ is the rotation about C by $2\angle ACB$, $\sigma_b \circ \sigma_c$ is the rotation about A by $2\angle BAC$, and $\sigma_c \circ \sigma_a$ is the rotation about B by $2\angle CBA$ (the angles are directed). Therefore,

$$(\sigma_a \circ \sigma_b \circ \sigma_c)^2 = (\sigma_a \circ \sigma_b) \circ (\sigma_c \circ \sigma_a) \circ (\sigma_b \circ \sigma_c)$$

is an orientation preserving isometry that rotates by $2 \cdot 360°$, so it does not rotate at all. It must therefore be either the identity map or a translation.

The first case is excluded since the transformation does not keep A fixed. Hence, $(\sigma_a \circ \sigma_b \circ \sigma_c)^2$ is a translation by a nonzero vector. Because a translation does not have fixed points, neither does its square root "$\sigma_a \circ \sigma_b \circ \sigma_c$" have.

Remark We revisited here the idea of understanding first the square of the transformation in question. Note that if $f(x) = x$, then $f(f(x)) = x$, so if $f \circ f$ has no fixed points, then neither does f.

Source D. Smaranda, N. Soare, *Transformări Geometrice (Geometric Transformations)*, Ed. Academiei, Bucharest, 1988.

8 Multiply the relation from the statement on the left by $\sigma_a \circ \sigma_b$ and on the right by $\sigma_d \circ \sigma_c$ to obtain

$$(\sigma_a \circ \sigma_b)^2 = (\sigma_d \circ \sigma_c)^2.$$

Now we distinguish three cases:

Case 1. If a is parallel to b, then there is a vector $\overrightarrow{v}$ such that $\sigma_a \circ \sigma_b$ is the translation by $\overrightarrow{v}$. Then $(\sigma_a \circ \sigma_b)^2 = (\sigma_d \circ \sigma_c)^2$ is the translation by $2\overrightarrow{v}$. This implies that $\sigma_d \circ \sigma_c$ must be a translation (it cannot be a rotation or else its square would be a rotation as well). And the translation vector must be $\overrightarrow{v}$. Consequently, a, b, c, d are parallel, being all perpendicular to $\overrightarrow{v}$ and the distance between a and b equals the distance between c and d.

Case 2. If a and b are orthogonal, then $(\sigma_a \circ \sigma_b)^2$ is the identity map, and so $(\sigma_d \circ \sigma_c)^2$ is the identity map. Now $\sigma_d \circ \sigma_c$ is an orientation-preserving isometry with square equal to 1, so it is either the identity map or the reflection over a point. The first case is ruled out by $c \neq d$, so it is the reflection over a point. But then c and d are orthogonal. Note that if a and b are orthogonal and c and d are orthogonal, then

$$\sigma_a \circ \sigma_b \circ \sigma_c \circ \sigma_d = \sigma_b \circ \sigma_a \circ \sigma_d \circ \sigma_c = \sigma_d \circ \sigma_c \circ \sigma_b \circ \sigma_a,$$

where for the last step we have used the lemma proved in the solution to Problem 5.

Case 3. If a and b intersect at a point O and are not orthogonal, then by examining Cases 1 and 2 and noticing that we can switch the pair (a, b) with the pair (d, c), we conclude that c and d also intersect at one point and are not orthogonal.

Then $\sigma_a \circ \sigma_b$ is the rotation by twice the angle $\alpha = \angle(a, b)$, so $(\sigma_a \circ \sigma_b)^2 = (\sigma_d \circ \sigma_c)^2$ is the square of this rotation. This means that $\sigma_d \circ \sigma_c$ is a rotation around O by some angle 2β such that

$$360° \mid 4\alpha - 4\beta.$$

This divisibility condition holds only if either $\alpha = \beta$ or $\alpha = 90° + \beta$. And in this case, c and d must intersect at the same point O and they form the angle β.

The answer to the problem consists of the following three possibilities:

  (i) a, b, c, d are parallel and the distance between a and b is the same as that between d and c, measured in the same direction,
 (ii) a and b are orthogonal and c and d are orthogonal,
(iii) a, b, c, d intersect at the same point and the angle between c and d either is equal to that between a and b or exceeds this angle by $90°$.

Remark We exploit the group structure on the set of isometries. And again we decode the properties of an isometry from the properties of its square.

Source D. Smaranda, N. Soare, *Transformări Geometrice (Geometric Transformations)*, Ed. Academiei, Bucharest, 1988.

9 Note that the line ℓ is invariant under both transformations $\sigma_b \circ \sigma_{\ell_1} \circ \sigma_a$ and $\sigma_B \circ \sigma_{\ell_1} \circ \sigma_A$. The restrictions to this line of the two transformations are equal (see Fig. 9.1), and they must be reflections over a point (prove it!).

Fig. 9.1 The reflections restricted to ℓ

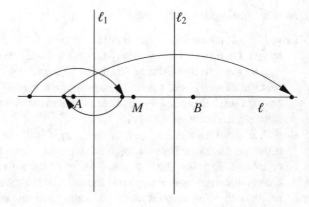

Fig. 9.2 The transformation
$\sigma_B \circ \sigma_{\ell_1} \circ \sigma_A$

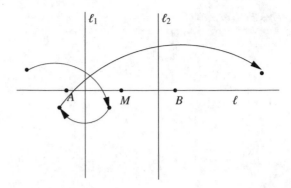

Now working back in the plane, it is not hard to see that both transformations
are reflections over lines perpendicular to ℓ (because the transformations reverse
orientation, and the line determined by a point and its image is parallel to ℓ, see
Fig. 9.2). Since the two transformations coincide on ℓ, they are reflections over the
same line, so they are equal.

Therefore, it suffices to show that σ_{ℓ_2} equals either $\sigma_b \circ \sigma_{\ell_1} \circ \sigma_a$ or $\sigma_B \circ \sigma_{\ell_1} \circ \sigma_A$.
We can check synthetically that $\sigma_{\ell_2} = \sigma_b \circ \sigma_{\ell_1} \circ \sigma_a$ because this is the same as

$$\sigma_b \circ \sigma_{\ell_1} = \sigma_{\ell_2} \circ \sigma_a.$$

If L_1 and L_2 are the intersection points of ℓ_1 and ℓ_2 with ℓ, respectively, then $\sigma_b \circ \sigma_{\ell_1}$
is translation by $2\overrightarrow{L_1 B}$ and $\sigma_{\ell_2} \circ \sigma_b$ is translation by $2\overrightarrow{AL_2}$. Since these two vectors
are equal, so are the corresponding translations, and we are done.

Or we can check analytically that $\sigma_{\ell_2} = \sigma_B \circ \sigma_{\ell_1} \circ \sigma_A$ by choosing coordinates
such that $A = \alpha$, $B = \beta$, $\alpha, \beta \in \mathbb{R}$, and ℓ_1 is given by the equation $z = \gamma + it$, where
$\gamma \in \mathbb{R}$ and t is a real parameter. Then $\sigma_A = 2\alpha - z$, $\sigma_B = 2\beta - z$, $\sigma_{\ell_1} = 2\gamma - \overline{z}$,
and

$$\sigma_B \circ \sigma_{\ell_1} \circ \sigma_A = 2(\alpha + \beta) - \gamma - \overline{z}.$$

The latter is σ_{ℓ_2}.

Remark In the particular case where $A = B = M$, if we let $m = a = b$, then we
obtain $\sigma_{\ell_2} = \sigma_M \circ \sigma_{\ell_1} \circ \sigma_M = \sigma_m \circ \sigma_{\ell_1} \circ \sigma_m$.

10 The set of vertices of the polygon is invariant under the reflection over an axis
of symmetry, and a vertex and its image determine the reflection axis (provided that
the vertex is not on the axis of symmetry). Since there are finitely many pairs of
vertices, and each reflection maps some vertex to a different vertex, there can only
be finitely many axes of symmetry. As the axes of symmetry divide the polygon into
equal parts, no two axes of symmetry can be parallel, so any two intersect. Assume
there are more than two axes of symmetry, or else the problem has no content.

Fig. 9.3 The axes of
symmetry $\ell_1, \ell_2, \ldots, \ell_n$

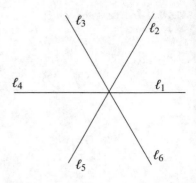

Let ℓ_1 and ℓ_2 be two axes that minimize the angle between them. Consider the reflection ℓ_3 of ℓ_1 over ℓ_2, of ℓ_2 over ℓ_3 and so on. Then $\ell_1, \ell_2, \ell_3, \cdots$ are axes of symmetry of the polygon. Note that they intersect at one point, namely, the intersection of ℓ_1 and ℓ_2. Because there are only finitely many axes of symmetry, the sequence $\ell_1, \ell_2, \ell_3, \ldots$ is periodic. The figure formed by these lines is invariant under the reflection over any of them. So they form a pencil of lines so that any two neighboring lines form equal angles (which by minimality is $\angle(\ell_1, \ell_2)$). If n is the period of our sequence, then in counterclockwise order, the sequence is $\ell_1, \ell_2, \ldots, \ell_{n+1} = l_1$ (see Fig. 9.3).

Now let ℓ be a different axis of symmetry. Let ℓ' be the translate of ℓ to the common point of $\ell_1, \ell_2, \ldots, \ell_n$. Then ℓ' lies inside the angle formed by some ℓ_j and ℓ_{j+1}. But then ℓ forms with ℓ_j an angle smaller than the angle between ℓ_1 and ℓ_2. This contradicts minimality. Hence, $\ell_1, \ell_2, \ldots, \ell_n$ are necessarily all the axes of symmetry, and the conclusion follows.

Remark An n-gon can have at most n axes of symmetry (as one vertex can be paired with itself or with any other vertex, and when it is paired with itself, then the neighboring vertices are mapped into each other). The regular n-gon has exactly n axes of symmetry, and is the only n-gon with this property.

Source M. Pimsner, S. Popa, *Probleme de geometrie elementară (Problems in elementary geometry)*, Ed. Didactică şi Pedagogică, Bucharest, 1979.

11 The solution can be followed on Fig. 9.4. Let M, N, P, Q, and R be the given midpoints, of the sides AB, BC, CD, DE, and EA, respectively. The transformation

$$\sigma_R \circ \sigma_Q \circ \sigma_P \circ \sigma_N \circ \sigma_M$$

is an isometry that rotates geometric figures by $180°$, so it must be a reflection over a point. Since it leaves A fixed, it is the reflection over A, σ_A. This can also be checked easily using complex coordinates.

For the construction, start with a point X, and then reflect it successively over M, N, P, Q, and R. The result is $X' = \sigma_A(X)$. Since X' is the reflection of X over

Fig. 9.4 The construction of
the pentagon

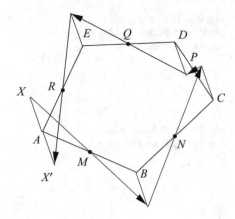

A, we can construct A as the midpoint of XX'. Then we reflect A successively over
M, N, P, and Q to obtain B, C, D, and E.

Remark The construction can be adapted to any polygon with an odd number of
sides. The midpoints do not determine uniquely the polygon if the number of sides
is even. Indeed, if $a_1, a_2, \ldots, a_m$ are the vertices and $m_j = \frac{1}{2}(a_j + a_{j+1})$, $j =
1, 2, \ldots, m$, $(a_j = a_{j+1})$ are the midpoints. The midpoint conditions

$$a_j + a_{j+1} = 2m_j, \quad j = 1, 2, \ldots, 2n + 1,$$

form a system of linear equations in the unknowns $a_1, a_2, \ldots, a_m$. This can
be solved by Cramer's rule provided that the determinant is nonzero. But the
determinant of the system is a particular case of a circulant determinant, and the
particular case of Cremona's formula shows that it is equal to

$$(1 + 1)\left(1 + e^{\frac{2\pi i}{m}}\right)\left(1 + e^{\frac{4\pi i}{m}}\right)\cdots\left(1 + e^{\frac{2(m-1)\pi i}{m}}\right).$$

This product is zero precisely when m is even, because then $1 + e^{\frac{2(m/2)\pi i}{m}} = 1 - 1 = 0$.
So the system can be solved if and only if m is odd.

Source I.M. Yaglom, *Geometric Transformations I*, (transl. by A. Shields), MAA,
1975.

12 Let $P(t)$ and $Q(t)$ be the locations of the two cars at time t, and let us orient
each line in the direction of the motion. If there exists a rotation that maps the
first line into the second, preserving the orientation and mapping $P(0)$ to $Q(0)$,
then its center O has the desired property. Indeed, in the triangles $OP(0)P(t)$
and $OQ(0)Q(t)$, $\angle OP(0)P(t) = \angle OQ(0)Q(t)$, because the angles are mapped
into each other by the rotation, $OP(0) = OQ(0)$, because of the rotation, and
$P(0)P(t) = Q(0)Q(t)$ because the cars travel with the same speed. So $OP(t) =
OQ(t)$ and we are done.

Fig. 9.5 The locus of the centers of rotations mapping one line into the other

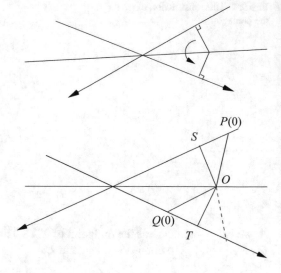

Fig. 9.6 Finding the point that is at equal distance from the cars

It remains to show that such a rotation exists. The locus of the points that are centers of rotations that map the first line into the second by preserving the orientation is one of the angle bisectors of the angles formed by the two lines. This can be seen easily on Fig. 9.5, since the center of rotation should be at equal distance from the two lines, and if a point is indeed at equal distance from the lines, then the rotation that maps the perpendiculars to the lines from the point into one another also maps the lines into one another. Note also that the rotations about the points on only one of the bisectors preserve orientations.

On the other hand, the locus of the centers of the rotations that map $P(0)$ to $Q(0)$ is the perpendicular bisector of the segment $P(0)Q(0)$. This line cannot be parallel to the angle bisector discussed above unless the two coincide. It follows that the two lines always intersect. If the intersection point is unique, let it be O. Then O is the desired center of rotation. Indeed, let us consider the rotation of center O that maps the first line into the second, preserving orientations. We claim that the image of $P(0)$ is $Q(0)$. Let S and T be the projections of O onto the first and second line, respectively. Then the right triangles $OSP(0)$ and $OTQ(0)$ are congruent (Fig. 9.6), because $OS = OT$ and $OP(0) = OQ(0)$. It follows that $\angle P(0)OS = \angle Q(0)OT$, and these angles are directed the same way (otherwise $P(0)$ reflects to $Q(0)$ over the bisector of the angle formed by the two lines and O is not unique). Hence, $\angle SOT$, which is the angle of rotation, is congruent to $\angle P(0)OQ(0)$. This combined with $OP(0) = OQ(0)$ implies that $P(0)$ is mapped to $Q(0)$ by the given rotation, as claimed. Thus, O is the center of the rotation that maps one of the car trajectories to the other. The image through the rotation of center O of $P(t)$ is $Q(t)$, and we are done.

If O is not unique, that is, if the bisector of the angle formed by the two lines coincides with the perpendicular bisector of $P(0)Q(0)$, change the time reference,

Fig. 9.7 The 90° rotation of
the square

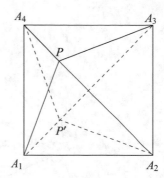

Fig. 9.8 Construction of a
triangle in which MN is a
midline

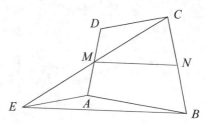

that is, work with the points $P(t_1)$ and $Q(t_1)$, where $t_1 \neq 0$. This will make O unique.

Remark There is only one point in the plane that satisfies the requirements of the problem.

Source *Kvant (Quantum)*, proposed by Igor Fedorovich Sharygin.

13 The 90° rotation about the center of the square maps A_1P, A_2P, A_3P, and A_4P to the perpendiculars from A_2, A_3, A_4, A_1, respectively (see Fig. 9.7). Hence, the intersection of the four perpendiculars is the point P' which is the image of P under the 90° rotation about the center.

Remark The 90° rotation is suggested by the construction of the perpendiculars. Try to use the same idea in other problems where perpendicular lines are present.

Source *Kvant (Quantum)*, proposed by A.N. Vilenkin.

14 Translate the segment DC to the segment EA, as shown in Fig. 9.8. Then $EACD$ is a parallelogram, so M is the midpoint of CE. In the triangle CEB, MN is a midline, so $EB = 2MN$. The condition from the statement implies that $EB = EA + AB$. But because of the triangle inequality in triangle EAB, this can only happen if E, A, B are collinear. And this happens if and only if CD and AB are parallel.

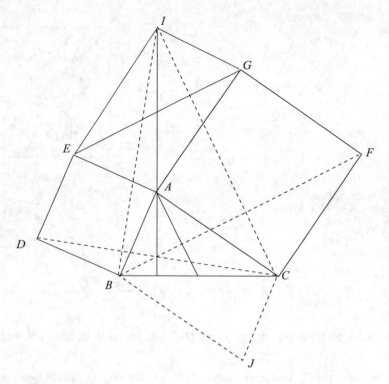

Fig. 9.9 Triangle with squares on its sides

Remark Perhaps the best inspiration for the solution is the well-known proof of the trapezoid midline formula, in which we extend the base AB and intersect it with the line CM, obtaining point E. However, doing exactly the same here is not nearly as good, because we lose the fact that CDM and EAM are congruent. So, to preserve this, we instead perform a translation that guarantees the congruence, and then see that the key to the solution lies in the triangle inequality.

The lesson here is: when you try to replicate a proof, focus on *why* that proof works rather than simply redoing it.

15 *First solution.* Complete the triangle ABC to a parallelogram $ABJC$ (see Fig. 9.9). Because AB and AE are equal and form a $90°$ angle and AC and AG are equal and form a $90°$ angle, it follows that there is a $90°$ rotation that maps the parallelogram $ABJC$ to the parallelogram $EAGI$. Hence, the segments EG and AJ, being transformed into each other by the rotation, are perpendicular and have equal lengths. Because AJ is the diagonal in the parallelogram $ABJC$, it passes through the midpoint of BC and has length equal to twice the median from A. This proves (a).

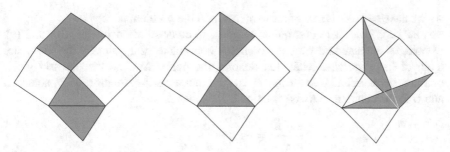

Fig. 9.10 Three 90° rotations

Additionally, the segments AI and BC are transformed into each other by the 90° rotation, which proves (b).

Moreover, $BC = AI$. The segments AC and CF are also perpendicular and of the same length. This shows that the triangle AIC is mapped to the triangle CBF by a 90° rotation. We deduce that BF is perpendicular to IC. Similarly, CD is perpendicular to IB. Thus, AI, CD, and BF are the altitudes of the triangle IBC. They intersect at the orthocenter of this triangle, which proves (c). The three rotations used for proving (a), (b), and (c) are sketched in Fig. 9.10.

(d) Denote by $d(X, \ell)$ the distance from a point X to the line ℓ. Because O is on the perpendicular bisectors of AE and AG, the distance from O to AB is $d(O, AB) = AE/2 = AB/2$, and to AC is $d(O, AC) = AG/2 = AC/2$. Thus,

$$\frac{d(O, AB)}{d(O, AC)} = \frac{AB}{AC}.$$

On the other hand, if M is a point on BC, then the fact that M is the midpoint of BC is characterized by the fact that the triangles AMB and AMC have the same area, namely, by the equation $d(M, AB) \cdot AB = d(M, AC) \cdot AC$. This equation can be written as

$$\frac{d(M, AB)}{d(M, AC)} = \frac{AC}{AB}.$$

Using similar triangles, we deduce that

$$\frac{d(X, AB)}{d(X, AC)} = \frac{AC}{AB}$$

characterizes the points that lie on the line of support of the median through A. Reflecting over the angle bisector of A, we deduce that the equation of the points that lie on the line of support of the symmedian is

$$\frac{d(O, AB)}{d(O, AC)} = \frac{AB}{AC}.$$

As we have seen, O is one of these points, and the problem is solved.

Second solution. There is a quick complex number solution for (a), (b), and (c). Denote, in the standard way, the complex coordinate of a point by the lowercase letter of the uppercase letter that denotes the point. Assume that the triangle is oriented counterclockwise. Then E is the clockwise $90°$ rotation of B about A, and G is the counterclockwise rotation of C about A, so

$$\frac{e-a}{b-a} = -i \text{ and } \frac{g-a}{c-a} = i.$$

It follows that $e = i(a-b)+a$ and $g = i(c-a)+a$, so the segment EG has length $|g-e| = |b+c-2a|$ and direction specified by the angle $\arg(g-e) = \arg(i(b+c-2a))$. On the other hand, the midpoint of BC has coordinate $\frac{b+c}{2}$, so the median has length

$$\left|\frac{b+c}{2} - a\right| = \frac{1}{2}|b+c-2a|$$

and its direction is specified by the angle

$$\arg\left(\frac{b+c}{2} - a\right) = \arg(b+c-2a).$$

Hence, the length of the median is half the length of the segment EG and the median makes with this segment a $90°$ angle, proving (a).

For (b) note that I is the reflection of A over the midpoint of EG, so it suffices to show that the segment formed by this midpoint and A is orthogonal to BC. This is the same as showing that

$$\frac{\frac{e+g}{2} - a}{c-b}$$

is imaginary. This fraction is equal to

$$\frac{\frac{i(a-b)+a+i(c-a)+a}{2} - a}{b-c} = i,$$

which is, indeed, imaginary.

For (c) we use the same observation that IA, BF, and CD are the altitudes of the triangle IBC after checking with complex coordinates that CD is orthogonal to IB and BF is orthogonal to IC. To check, for example, that BF is orthogonal to IC, we compute the complex coordinate of I using the fact that I is the reflection of A over the midpoint of EG. Hence, its coordinate is $g+e-a = i(c-b)+a$. The point F is the clockwise rotation of A about C so is $f = i(c-a)+c$. Then the line

BF has direction given by the angle $\arg(f-b) = \arg[i(c-a)+c-b]$, and the line CI has direction given by the angle $\arg[i(c-b)+a-c] = \arg[i((c-b)+i(c-a))]$ and the latter angle differs by $90°$ from the former. The conclusion follows.

Remark The idea of the problem is to prove congruence and orthogonality of segments by including them in figures that rotate into each other by $90°$, and these $90°$ rotations are suggested by the presence of the squares.

Also question (a) gives away the construction of the parallelogram $ABJC$, since constructing it is the best way to produce a segment twice as long as the median from A. Part (d) exhibits another construction of the symmedian; see also the discussion in Sect. 2.4.3.

16 It suffices to prove that the four perpendiculars transform into one another by the $90°$ rotation ρ about the center of the square. Let the parallelogram be $ABCD$ and let the square be $EFGH$, as in Fig. 9.11.

If $C' = \rho(C)$, then ρ maps the triangle CFG to the triangle $C'GH$. So $C'G$ is perpendicular to CF and hence to AD, and $C'H$ is perpendicular to CG. It follows that D is the orthocenter of triangle $C'GH$. From this we deduce that $C'D$ is perpendicular to GH. Hence, the perpendicular line from C onto FG is mapped by the $90°$ rotation ρ to the perpendicular line from D onto GH. The same holds true for the other perpendiculars from the vertices. This means that the rectangle formed by the four perpendiculars is invariant under a $90°$ rotation, so it is a square.

Remark You should compare this solution to the one given to Problem 15. Again you include the segments that you want to map into one another by $90°$ rotations into figures that are mapped as such. And again a key step is the fact that the altitudes of a triangle intersect at one point.

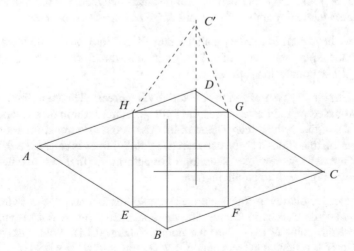

Fig. 9.11 The square inscribed in a parallelogram

Fig. 9.12 Rectangles $ABCD$
and $MNPQ$ such that
$AM = BN = CP = DQ$

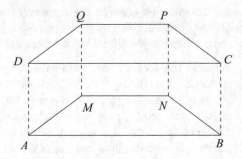

Source *Kvant (Quantum)*, proposed by Nikolai Borisovich Vassiliev.

17 Consider an isosceles trapezoid $ABNM$ that is not a rectangle, and translate
it by a vector $\vec{v}$ that is perpendicular to AB. If C, D, P, Q are the images of
B, A, N, M, respectively, then $ABCD$ and $MNPQ$ have the desired property. The
construction is shown in Fig. 9.12.

18 This is a true "geometry in motion" solution, presented in Fig. 9.13. Translate
MN along its line of support to $M'R$ and PQ along its line of support to $P'R$.
Then $M'B$, RA, $P'C$, CS, and BT are all parallel and equal to each other. The
parallelograms $ABMN$ and $ABM'R$ have the same area, because they have the
same base and the same height. The same is true for the parallelograms $ACPQ$ and
$ACP'R$.

   Now translate RA along its line of support to $R'A'$ so that R' is on BC.
Then $BR'A'T$ and $R'CSA'$ are parallelograms that partition $BCST$. Note also
that $BR'A'T$ has the same area as $ABM'R$, because they have the same base
($RA = R'A'$) and the same height (the distance between lines AR and BM').
Similarly, the parallelograms $ACP'R$ and $R'CSA'$ have the same area.

   Thus, we have partitioned the parallelogram $BCST$ into two parallelograms, one
whose area is equal to the area of $ABMN$ and one whose area is equal to the area
of $ACPQ$. The conclusion follows.

Remark This is a generalization of the Pythagorean Theorem, and so this
translation-based method also proves the Pythagorean Theorem (see Chapter 4 for
other proofs of the Pythagorean Theorem that use geometric transformations). The
transformation that slides a shape between two parallel lines is called *shearing*. This
proof of Pappus' Theorem uses shearing to transform the first two parallelograms
into a partition of the third parallelogram.

19 The metric relation in the statement should remind us of the Pythagorean
Theorem, and the trick is to arrange the three segments into a right triangle. For
that, reflect the point N over O to the point N' (Fig. 9.14). Note that the entire
triangle CNO is reflected to the triangle $AN'O$, and so $CN = AN'$.

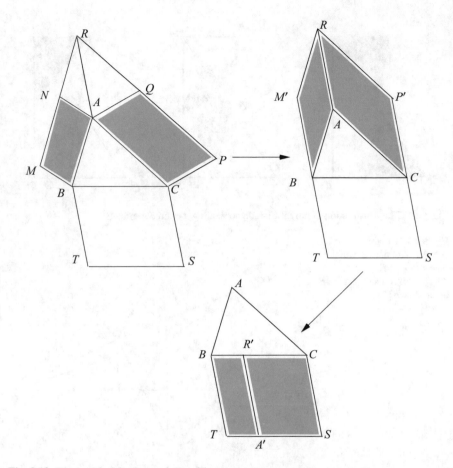

Fig. 9.13 The proof of the Pappus' Area Theorem

Moreover, because $\angle NOM = 90°$, N' is also the reflection of N over line OM, so $MN = MN'$. Thus, we have placed the three segments CN, AM, MN in the triangle AMN'.

The last observation is that

$$\angle N'AM = \angle N'AO + \angle OAM = \angle OCN + \angle OAM = 90°,$$

so the triangle AMN' is right. The Pythagorean Theorem gives $AN'^2 + AM^2 = MN'^2$, and hence $AM^2 + CN^2 = MN^2$, as desired.

Source Moscow Mathematical Olympiad, 1999.

20 Like in the first proof of Theorem 1.24, let P_A, P_B, P_C be the reflections of P over BC, CA, AB, respectively. Let also Q_A be the reflection of Q over BC

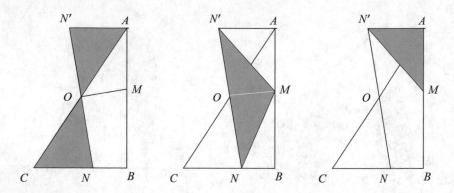

Fig. 9.14 The construction of a right triangle containing the three segments

Fig. 9.15 The reflections of
AP and AQ over the angle
bisectors of $\angle BPC$ and
$\angle BQC$ are reflections of each
other over BC

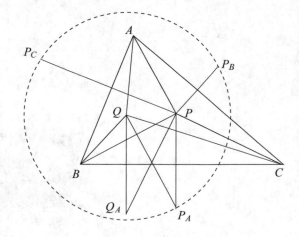

(Fig. 9.15). Then the lines QP_A and PQ_A are the reflections of each other over
BC; they are symmetric with respect to BC.

By Proposition 1.25 (i), Q is the circumcenter of $P_A P_B P_C$, so

$$\angle Q P_A P_C = 90° - \angle P_A P_B P_C,$$

In other words, $\angle(Q P_A, P_C P_A)$ and $\angle(P_B P_C, P_A P_B)$ are complementary. In the
proof of Theorem 1.24, we have seen that AQ, BQ, CQ are the perpendicular
bisectors of the segments $P_B P_C$, $P_A P_C$, $P_A P_B$, respectively, so

$$\angle(AQ, P_B P_C) = \angle(BQ, P_C P_A) = \angle(CQ, P_A P_B) = 90°.$$

Consequently,

$$\angle(BQ, QP_A) = 90° - \angle(QP_A, P_C P_A) = \angle(P_B P_C, P_A P_B) = \angle(CQ, AQ).$$

Thus, the line QP_A is the reflection of the line AQ with respect to the angle bisector of $\angle BQC$. Switching the roles of P and Q, we deduce that the line PQ_A is the reflection of the line AP with respect to the angle bisector of $\angle BPC$. So PQ_A and QP_A are the two lines from the statement, and since Q and P_A are the reflections of P and Q_A over BC, the two lines reflect into each other over BC.

Source Jean-Pierre Ehrmann.

21 We will use Theorem 1.26. Let $a_1, a_2, \ldots, a_{2n}$ be the complex coordinates of the vertices $A_1, A_2, \ldots, A_{2n}$. Translate $a_1, a_3, \ldots, a_{2n-1}$ by $v \in \mathbb{C}$. Then the area of the modified polygon is

$$\frac{1}{2}\Im[(\overline{a_1} + \overline{v})a_2 + \overline{a_2}(a_3 + v) + \cdots + (\overline{a_{2n-1}} + \overline{v})a_{2n} + \overline{a_{2n}}(a_1 + v)]$$

$$= \frac{1}{2}\Im(\overline{a_1}a_2 + \overline{a_2}a_3 + \cdots + \overline{a_{2n}}a_1) + \frac{1}{2}\Im(\overline{v}a_2 + v\overline{a_2} + \cdots + \overline{v}a_{2n} + v\overline{a_{2n}})$$

$$= \frac{1}{2}\Im(\overline{a_1}a_2 + \overline{a_2}a_3 + \overline{a_3}a_4 + \cdots + \overline{a_{2n}}a_1),$$

because the second term of the sum is a real number, so its imaginary part is zero. And this is the area of the original polygon. The problem is solved.

Source Kiril Dochev.

22 (a) The shortest solution uses complex coordinates. We place the origin at the circumcenter of the triangle ABC, and then use the property that the orthocenter's reflection over the midpoint of one side lies on the circumcircle and is diametrically opposite to the third vertex. Here is how to prove this fact (this proof can be followed on Fig. 9.16). If A' is the reflection of the orthocenter H of triangle ABC over BC, then $HBA'C$ is a parallelogram; therefore, $\angle A'BA$ is equal to $\angle(CH, AB)$, so it is right. For the same reason, $\angle A'CA$ is right, which implies that A' lies on the circumcircle of ABC diametrically opposite to A.

Fig. 9.16 Reflection of the orthocenter over the midpoint of a side

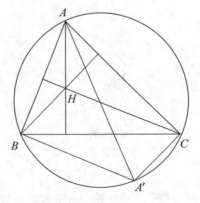

Returning to the problem, consider a coordinate system with the origin at the circumcenter of the quadrilateral $ABCD$, and let the coordinates of A, B, C, D be a, b, c, d, respectively. Then the coordinate of a' of the point A' that lies diametrically opposite to A (and so is the reflection of A over the origin) is $-a$. Because H_d is the reflection of A' over the midpoint of BC, its complex coordinate is

$$h_d = 2\frac{b+c}{2} - (-a) = a + b + c.$$

Similarly, the coordinates of H_a, H_b, H_c are

$$h_a = b + c + d, \quad h_b = a + c + d, \quad h_c = a + b + d.$$

And we can see that H_a, H_b, H_c, H_d, are the respective reflections of A, B, C, D over the centroid of $ABCD$. Indeed, the coordinate of the midpoint of AH_a is

$$\frac{a + (b+c+d)}{2} = \frac{a+b+c+d}{2},$$

which is the coordinate of the centroid, and the same is true for BH_b, CH_c, and DH_d. The problem is solved.

(b) *First solution.* Let $ABCD$ be the cyclic quadrilateral and let M, N, P, Q be the respective midpoints of AB, BC, CD, and DA (Fig. 9.17). Then $MNPQ$ is a parallelogram, and let S be its center. The perpendicular bisectors of the sides meet at the center O of the circumcircle. The perpendiculars from M, N, P, Q onto the opposite sides are reflections of the perpendicular bisectors over O. So they meet at the reflection over O of S.

Second solution. With the notation from the proof of (a), if we choose the radius of the circle to be 1, so that $a\bar{a} = b\bar{b} = c\bar{c} = d\bar{d} = 1$, then the fact that the line through the midpoint M of AB (whose coordinate is $\frac{a+b}{2}$) and the anticenter (whose coordinate is $\frac{a+b+c+d}{2}$) rotates by 90° to line CD, means that the ratio

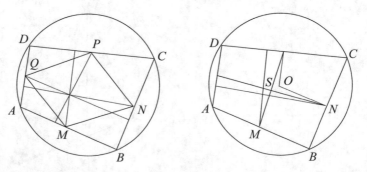

Fig. 9.17 The proof of the existence of the Mathot point

$$\frac{\dfrac{a+b+c+d}{2}-\dfrac{a+b}{2}}{c-d}$$

is a purely imaginary number. This is equivalent to

$$\frac{c+d}{c-d}$$

being an imaginary number. We compute

$$\frac{c+d}{c-d}=\frac{c+d}{c-d}\cdot\frac{\overline{c}-\overline{d}}{\overline{c}-\overline{d}}=\frac{c\overline{c}-c\overline{d}+d\overline{c}-d\overline{d}}{|c-d|^2}$$

$$=\frac{1}{|c-d|^2}(1-c\overline{d}+d\overline{c}-1)=\frac{i}{|c-d|^2}\Im(d\overline{c}),$$

which is, indeed, an imaginary number. The problem is solved.

Remark As a consequence of (a), the quadrilateral formed by the orthocenters is congruent to the original quadrilateral. The four perpendiculars from the midpoints of the sides onto the opposite sides are also known as maltitudes. Their point of intersection, which is also the circumcenter of $H_a H_b H_c H_d$, is known as the Mathot point or the anticenter of $ABCD$.

It is worth remembering the fact that if the circumcenter of a triangle lies at the origin of the coordinate system, then the coordinate of the orthocenter is the sum of the coordinates of the vertices.

23 Reflect Γ over AB to complete a circle, and let D' be the reflection of D (see Fig. 9.18). Then $\angle D'CA = \angle ACD = \angle ECB$, so E, C, D' are collinear. We have

$$\angle DEF = \frac{1}{2}\,\overset{\frown}{DE} = \angle DD'E = \angle CDD' = 90° - \angle ACD.$$

Given that the triangle FED is isosceles, we have

$$\angle EFD = 180° - 2\angle DEF = 180° - 2(90° - \angle ACD) = 2\angle ACD$$

$$= \angle ACD + \angle ECB,$$

and we are done.

Remark The solution is built on the classical trick used for proving that, for a ray of light reflected by a mirror, the angle of incidence equals the angle of reflection. This is a consequence of Fermat's principle that light travels on the fastest path (which in a homogeneous atmosphere is also the shortest), and by "reflecting" the beam into the mirror (which is the same as removing the mirror), the angle of reflection becomes opposite, and hence equal, to the angle of incidence.

Fig. 9.18 Reflection of D
over AB

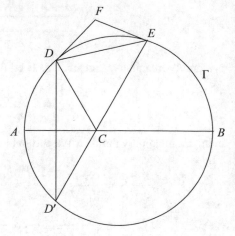

Fig. 9.19 Triangle with two
equal inscribed squares

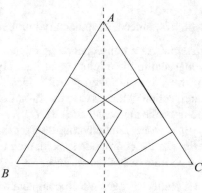

Source British Mathematical Olympiad, 2018-2019, proposed by D. Griller.

24 Let the triangle be ABC, and let us concentrate on the angle $\angle BAC$ and the two squares that have sides on AB and AC (Fig. 9.19). Because the squares are equal, the figure consisting of the squares and the angle is symmetric with respect to the bisector of the angle. So the squares are mapped into each other by the reflection with respect to this bisector. Consequently, the vertices of these squares that lie on side BC are mapped into one another by the reflection; hence, the side BC is perpendicular to the angle bisector of $\angle BAC$.

The same argument shows that the angle bisectors of $\angle ABC$ and $\angle ACB$ are perpendicular to the sides AC and AB, respectively. And because all three angle bisectors are also altitudes, the triangle is equilateral.

Fig. 9.20 Proof of the
existence of the orthopole

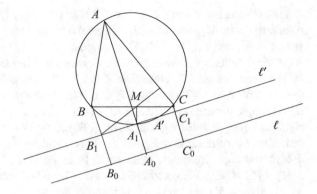

Remark The fact that two equal squares are inscribed in a triangle forces the existence of a symmetry of the triangle.

Source D. Smaranda, N. Soare, *Transformări Geometrice (Geometric Transformations)*, Ed. Academiei, Bucharest, 1988.

25 If you translate the line ℓ, the points A_0, B_0, C_0 are translated by the same vector, and so are the perpendiculars from these points to the sides of the sides of the triangle. Translation preserves concurrency, so it suffices to prove the result for a special translate ℓ' of ℓ.

If A_1, B_1, C_1 are the images of A_0, B_0, C_0 under the translation, we may enforce that A_1 be the second intersection point of AA_0 with the circumcircle of the triangle ABC. The proof of this particular case can be followed on Fig. 9.20. Let M be the projection of A_1 onto BC, and let A' be the point that is diametrically opposite to A, so that $\angle A'A_1A$ is right. Then A' must be on ℓ' (because ℓ' is orthogonal to AA_1 at A_1).

Since the quadrilateral A_1MCC_1 is cyclic (having two opposite right angles),

$$\angle(CB, C_1M) = \angle(CM, C_1M) = \angle(CA_1, C_1A_1) = \angle(CA_1, A'A_1)$$
$$= 90° - \angle(AA_1, CA_1) = 90° - \angle(AB, CB).$$

Thus, $\angle(CB, C_1M) + \angle(AB, CB) = 90°$, from where we infer that that C_1M is orthogonal to AB. Similarly, B_1M is orthogonal to AC. Hence, the desired point of concurrency for ℓ' is M, and consequently the corresponding point for ℓ is the preimage of M under the translation.

Remark The point where the three perpendiculars intersect is called the orthopole of the line ℓ with respect to the triangle ABC.

Source This proof for the existence of the orthopole was communicated to us by James Tao.

26 *First solution.* We argue on Fig. 9.21. Let $\rho_1, \rho_2, \rho_3, \rho_4$ be the (clockwise) $90°$ rotations about M_1, M_2, M_3, M_4, respectively, so that $\rho_1(A) = B$, $\rho_2(B) = C$, $\rho_3(C) = D$, and $\rho_4(D) = A$. The composition $\rho_2 \circ \rho_1$ rotates figures by $180°$, so it is the reflection over some point P. Because this reflection maps A to C, P must be the midpoint of AC. Note that P is also the midpoint of the segment joining the M_1 with $\rho_2 \circ \rho_1(M_1) = \rho_2(M_1)$, so it is the midpoint of the hypotenuse in the right triangle formed by M_2, M_1, and $\rho_2(M_1)$. Consequently, the triangle PM_1M_2 is right isosceles, with the right angle at P. Similarly, the composition $\rho_4 \circ \rho_3$ is is the reflection over the same midpoint P of AC, and for the same reason, the triangle PM_3M_4 is right isosceles, with the right angle at P. We conclude that triangle PM_1M_3 is obtained from triangle PM_2M_4 by a $90°$ rotation about P, and hence, M_1M_3 and M_2M_4 are equal and perpendicular.

Second solution. For the complex number solution, note that M_1 is the midpoint of the segment joining A to the $90°$ rotation of A about B. If the coordinates of the vertices are a, b, c, d, correspondingly, then the coordinate x of this $90°$ rotation of A satisfies

$$\frac{x - b}{a - b} = i,$$

so $x = i(a - b) + b$. Then M_1 has coordinate

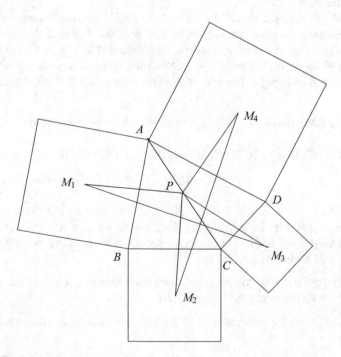

Fig. 9.21 Quadrilateral with squares constructed on its sides

$$m_1 = \frac{x+a}{2} = \frac{a+b}{2} + i\frac{a-b}{2}.$$

Similarly, the coordinates of M_2, M_3, M_4 are, respectively,

$$m_2 = \frac{b+c}{2} + i\frac{b-c}{2}, \quad m_3 = \frac{c+d}{2} + i\frac{c-d}{2}, \quad m_4 = \frac{d+a}{2} + i\frac{d-a}{2}.$$

We are left to check that $m_4 - m_2 = i(m_3 - m_1)$, namely, that the vector $\overrightarrow{M_1M_3}$ rotates by $90°$ to the vector $\overrightarrow{M_2M_4}$. And indeed

$$i(m_3 - m_1) = i\left(\frac{c+d}{2} + i\frac{c-d}{2} - \frac{a+b}{2} - i\frac{a-b}{2}\right)$$

$$= \frac{-c+d+a-b}{2} + i\frac{c+d-a-b}{2}$$

$$= \frac{d+a}{2} + i\frac{d-a}{2} - \frac{b+c}{2} - i\frac{b-c}{2} = m_4 - m_2,$$

which completes the solution.

Remark Can you find a complex number solution that runs parallel to the first solution?

It should be noted that when two consecutive vertices of the quadrilateral coincide, the problem degenerates into the following: On the sides BC, CA, AB of the triangle ABC, one constructs in the exterior squares; let M_1, M_2, M_3 be their centers, respectively. Show that AM_1 and M_2M_3 are perpendicular and of equal length. This configuration can be related to that from Problem 15. In this situation, we challenge you with the following problem, which was published by V.I. Dubrovsky in *Kvant (Quantum)*: at every vertex, the two sides of squares that are not sides of ABC determine a triangle, and in each of these triangles, consider the side that lies opposite to the corresponding vertex of ABC. Show that the perpendicular bisectors of these three sides are concurrent.

Source I.M. Yaglom, *Geometric Transformations I*, (transl. by A. Shields), MAA, 1975.

27 *First solution.* Rotate about C by $60°$ so that B is mapped to A (Fig. 9.22). Let D' be the image of D. The segment PD rotates by $120°$ to BD, and BD rotates by $60°$ to AD', so PD rotates by $180°$ to AD'. It follows that the segments DP and AD' are parallel and equal, so $DPD'A$ is a parallelogram. As E is the midpoint of the diagonal AP, it is also the midpoint of the diagonal DD'. But CDD' is an equilateral triangle; hence, DEC is the $60° - 90° - 30°$ triangle.

Second solution. We argue on Fig. 9.23. Consider the rotation about D by $120°$ that maps P to M and B to P. Then C is mapped to some point K on MP. We must have $MK = CP$, because one segment is the image of the other through the rotation, and

Fig. 9.22 Finding the angles of the triangle DEC using a $60°$ rotation

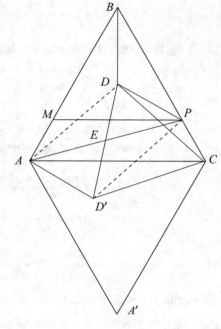

Fig. 9.23 Finding the angles of the triangle DEC using a $120°$ rotation

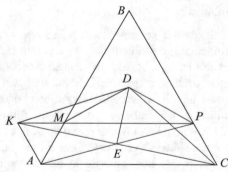

since $CP = AM$, we have $MK = AM$. It follows that triangle AMK is equilateral (being isosceles and having $\angle AMK = 60°$), so AK is parallel and equal to PC. We find that $KACP$ is a parallelogram, in which E is the midpoint of one diagonal, AP, so it is the midpoint of the other diagonal, KC. We conclude that E is the midpoint of the side KC in the isosceles triangle DKC having $\angle KDC = 120°$. Thus, $\angle DEC = 90°$, $\angle ECD = 30°$, and $\angle EDC = 60°$.

Remark When you see an equilateral triangle, you should think about either a $60°$ rotation about a vertex or a $120°$ rotation about its center.

Source *Kvant (Quantum)*, proposed by A. Kuptsov.

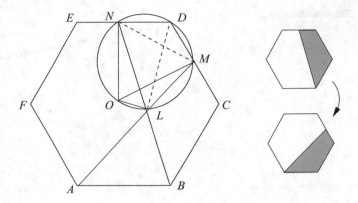

Fig. 9.24 Regular hexagon

28 The solution can be followed on Fig. 9.24. The key observation is that the quadrilateral $AMCB$ transforms into the quadrilateral $BNDC$ by a $60°$ rotation about the center O of the hexagon. By subtracting from the areas of these two quadrilaterals the area of the overlap, we obtain (a).

For (b), note first that since M rotates to N, the triangle OMN is equilateral, so $\angle OMN = 60°$. Because AM rotates to BN, $\angle NLM = 60°$, but also $\angle NOM = 60°$, and hence N, O, L, M, as well as D lie on a circle (the latter because $\angle NDM = 120°$). We deduce that $\angle OLN = \angle OMN = 60°$, and also $\angle ALO = 180° - \angle OLN - \angle NLM = 60°$.

Finally, $\angle OLD = \angle OMD = 90°$, which proves (c).

Remark Regular polygons are invariant under rotations about their centers!

Source *Kvant (Quantum)*, proposed by E. Gotman.

29 We argue on Fig. 9.25. The condition $\angle ACD = \angle BCM$ does not look very suggestive, except for the geometry expert who might think about isogonal conjugates. But the parallelogram $ABMD$ induces at least two translations. Of these, let us work with the translation that takes A to B and D to M, and let C' be the image of C. Then $\angle ACD = \angle BC'M$ and now all we need to prove is that $\angle BC'M = \angle BCM$, namely, that the quadrilateral $BC'CM$ is cyclic. But DC is mapped to MC' by the translation, so $MDCC'$ is also a parallelogram. Therefore, $\angle MC'C = \angle MDC = \angle MBC$, which proves that $BC'CM$ is indeed cyclic, and we are done.

Remark You might have already noticed that parallelograms and translations are a perfect match! In fact, a parallelogram $XYZW$ induces two translations, one that takes X to Y and W to Z and the other that takes X to W and Y to Z.

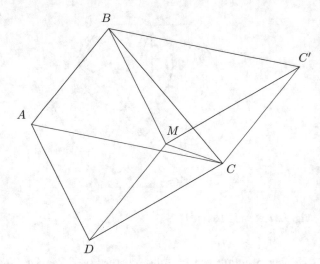

But which translation should we use for this problem? Actually, both of them work! Try performing the other translation to see what happens (in fact, you will obtain a reflection of key quadrilateral $BC'CM$).

Source Communicated by Cicero Thiago Magalhães.

30 The solution can be followed on Fig. 9.26. Let H be the orthocenter of the triangle ABC. Notice that $\angle BHC = \angle EHF = 360° - 2 \cdot 90° - \angle BAC = 180° - \angle BAC$. But, even though the angles $\angle BHC$ and $\angle BAC$ sum up to $180°$, they face the same way. How do we fix that? One idea is performing a reflection σ across BC, obtaining $\sigma(A) = A'$. Then $A'BHC$ is cyclic and its circumcircle is the reflection of the circumcircle Ω of ABC.

Is there any other symmetry between Ω and $\sigma(\Omega) = \Omega'$? Yes, there is the reflection ρ over the midpoint M of their common chord BC. Let $A'' = \rho(A)$.

Since $AD = DA'$ and $AM = MA''$, $A'A'' \parallel BC$; in particular, $\angle HA'A'' = 90°$, so HA'' is a diameter of Ω'. Since X lies on the circumcircle of AEF, which has diameter AH, $\angle HXA'' = 180° - \angle HXA = 90°$, and we conclude that X lies on Ω'.

Now an angle chasing drives the point home: since $A'BCA''$ is an inscribed trapezoid, the arcs $\overparen{BA'}$ and $\overparen{CA''}$ are equal, so, with directed angles modulo $180°$,

$$\angle ZXY = \angle BXC - \angle A''XC = \angle BHC - \angle BHA' = \angle ZHY,$$

proving that H, X, Y, Z are concyclic. Hence $\angle HZY = \angle HXY = 90°$ and so $YZ \parallel BC$, as desired.

Remark Keep in mind that the reflections of H over both the side AB and the midpoint of the side AB are on the circumcircle.

Fig. 9.26 Proof that
$YZ \parallel BC$

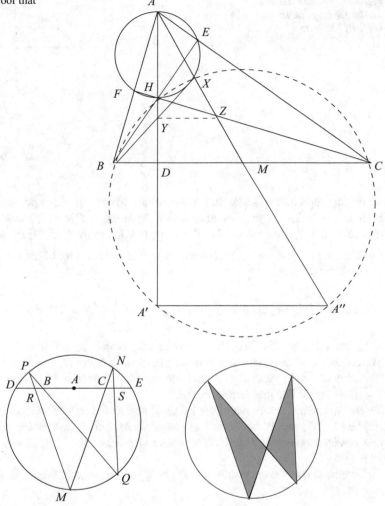

Fig. 9.27 The Butterfly Theorem

Source Cono Sur Mathematical Olympiad, 2007, proposed by Y. Lima, Brazil.

31 Look at Fig. 9.27 and recognize the shape of the butterfly, which explains the name of the theorem.

First solution. Consider the reflection over the perpendicular bisector of the chord c, and, for a point X, denote by X' its reflection. The solution can be followed on Fig. 9.28.

Fig. 9.28 Reflection over a
line for the proof of the
Butterfly Theorem

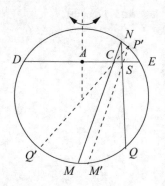

We are supposed to show that R' coincides with S. To this end, we check that
both S and R' lie on the circumcircle of the triangle CNP'. That R' is on this
circumcircle follows from the fact that the quadrilateral $NP'R'C$ is cyclic, which is
true because $\angle CNR' = \angle CP'R'$ both being half the measure of the arc $\overset{\frown}{MQ}$.

On the other hand,

$$\angle CSN = \frac{1}{2}(\overset{\frown}{ND} + \overset{\frown}{QE}) = \frac{1}{2}(\overset{\frown}{ND} + \overset{\frown}{Q'D}) = \frac{1}{2}\,\overset{\frown}{Q'N} = \angle Q'P'N = \angle CP'N.$$

This shows that $NP'SC$ is cyclic. Consequently, both R' and S lie at the intersection
of a circle and a line. The other point of intersection being C, we must have $R' = S$
(for $R' = C$ would mean $B = R$ and that can only happen if $M = Q$ in which case
$S = C$ as well). The problem is solved.

Second solution. Denote the circle by Ω. Consider the reflection over the point A,
and let Ω', M', and P' be the reflections of Ω, M, and P, respectively (Fig. 9.29).
One observation is that D reflects to E and vice versa, so D, E, as well as M' and
P' are on Ω'.

We now show that the points M', P', N, Q lie on a circle. First, using power-of-
a-point, we have

$$BM \cdot BN = BD \cdot BE = CD \cdot CE = CP \cdot CQ,$$

and using the fact that $BM = CM'$ and $CP = BP'$, we obtain

$$\frac{BN}{BP'} = \frac{CQ}{CM'}.$$

Also, since the reflection over A maps the line BN to CM' and the line BP to CQ,
$\angle NBP' = \angle M'CQ$, which together with the above metric relation, implies the
triangles BNP' and CQM' are similar. It follows that $\angle BNP' = \angle CQM'$. Using
this equality, and working with directed angles modulo π, we have

Fig. 9.29 Reflection over a
point for the proof of the
Butterfly Theorem

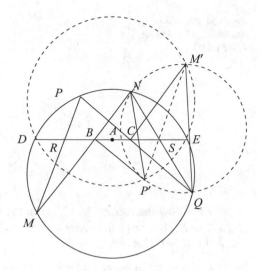

$$\angle P'M'Q = \angle(MP, M'Q) = \angle MPQ + \angle PQM'$$
$$= \angle MNQ + \angle PQM' = \angle MNP' + \angle P'NQ + \angle PQM'$$
$$= \angle P'NQ,$$

showing that M', N, P', Q are concyclic.

The lines DE, NQ, and $M'P'$ are the pairwise radical axes of Ω, Ω' and the circumcircle of M', N, P', Q, so they intersect at the radical center of the three circles. But $M'P'$ intersects DE at the symmetric of R with respect to A, while NQ and DE intersect at S. Thus, S is the reflection of R with respect to A, and the theorem is proved.

Remark The fact that A is the midpoint of both BC and DE suggests that a reflection that maps $B \mapsto C$ and $D \mapsto E$ might help, and two such reflections exist, one over a line and one over a point. Both work.

When $A = B = C$, the second solution becomes simpler and was communicated to us by Ilya Bogdanov. In that case, using once more directed angles modulo π,

$$\angle M'NQ = \angle MNQ = \angle MPQ = \angle QP'M',$$

where the last equality follows from $M'P' \| MP$, whence M', N, P', Q are concyclic, and the solution ends as above.

32 Let P be the intersection of CF with the parallel through A to BC (Fig. 9.30). Then the triangles AFP and BFC are similar, so $AP/BC = AF/BF$. Also the triangles AFE and BFD are similar; thus, $AE/BD = AF/BF$, that is, $AE/BC = AF/BF$. It follows that $AP = AE$.

Fig. 9.30 Additional
construction for finding
$\angle(BE, CF)$

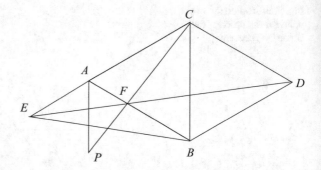

Now consider the rotation by $60°$ about A that maps B to C. This rotation also maps E to P, so it maps the segment BE to CP. It follows that the angle between BE and $CF = CP$ is $60°$.

Remark Note that in order to prove that the segment BE rotates by $60°$ to the segment CP, we have placed them in the triangles ABE and ACP that rotate into each other.

Can you write an analytical solution?

Source A. Myller, *Geometrie Analitică (Analytical Geometry)*, 3rd ed., Ed. Didactică şi Pedagogică, Bucharest, 1972.

33 For easy reference, draw the triangle ABC so that it is oriented counterclockwise. Let ρ_A and ρ_B be the $60°$ counterclockwise rotations about and A and B, respectively. Then $\rho_B(A_1) = C$ and $\rho_A(C) = B_1$ (see Fig. 9.31). So

$$\rho_A \circ \rho_B(A_1) = B_1.$$

The composition of the two rotations is a rotation itself, by $120°$. The center of this rotation is its fixed point. Note that $\rho_B(M)$ is the reflection M' of M over AB and $\rho_A(M') = M$, and hence $\rho_A \circ \rho_B(M) = M$. So M is this fixed point. We conclude that A_1 is mapped to B_1 by a $120°$ rotation about M, and so $MA_1 = MB_1$ and $\angle A_1 M B_1 = 120°$, as desired.

Remark In order to prove that a triangle is isosceles, we show that there is a rotation centered at one vertex that maps the second vertex to the third.

Source I.M. Yaglom, *Geometric Transformations I*, (transl. by A. Shields), MAA, 1975.

34 First note that two lines with the desired property divide the polygon $\mathcal{P}$ into four parts, with the parts inside opposite angles having the same area. Reflect $\mathcal{P}$ over O to $\mathcal{P}'$ (see Fig. 9.32). The boundaries of $\mathcal{P}$ and $\mathcal{P}'$ intersect inside each angle formed by two lines that divide the area in half, or else the part of $\mathcal{P}$ inside this angle would be either larger or smaller than the corresponding part of $\mathcal{P}'$, and the

Fig. 9.31 Equilateral triangles constructed on the sides of a triangle

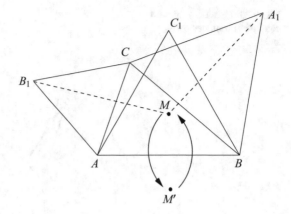

Fig. 9.32 Lines that bisect the area of a polygon

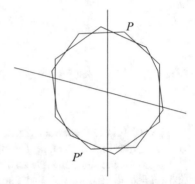

opposite equality would hold in the opposite angle. Of course this cannot happen because then the part of $\mathcal{P}$ in one region would be strictly larger than the part of $\mathcal{P}$ in the other.

If there are k lines, there are $2k$ angles formed by them, hence $2k$ intersection points of the two polygons. However, one side of $\mathcal{P}$ cannot contain more than two intersection points, or else $\mathcal{P}'$ would not be convex. Therefore, $2k \leq 2n$, that is, $k \leq n$, as desired.

Remark By sliding a horizontal line continuously, we can always find a situation where it bisects the area of the polygon. Same with a vertical line. So there is a point in the interior where at least two lines that bisect the area exist. The reflection of the polygon over this point intersects the original polygon in at least four points.

Source All-Union Mathematical Olympiad, 1973, solution from S. Savchev, T. Andreescu, *Mathematical Miniatures*, MAA, 2003.

35 We assume that the quadrilateral is oriented counterclockwise and that all rotations in this solution are counterclockwise as well. We will prove first that $AN = CM$, by showing that they map into each other by a $120°$ rotation. The argument can be followed on Fig. 9.33.

Fig. 9.33 Quadrilateral with
equilateral triangles on its
sides

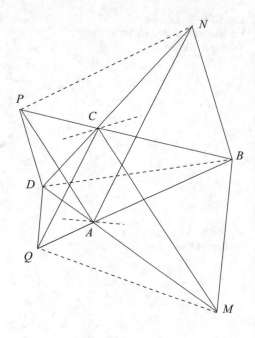

The 60° rotation ρ_1 about A maps M to B, and the 60° rotation ρ_2 about C maps
B to N. Note that if $S = \rho_1(C)$, then ACS is equilateral and $\rho_2(S) = A$. Hence,
$\rho_2 \circ \rho_1$ maps the segment AN to the segment CM. This composition is an isometry
that rotates every figure by 120°, and we can conclude that it is a 120° rotation. And
so $AN = CM$ and the two segments make an angle of 120°.

Similarly, $\rho_2 \circ \rho_1(AP) = CQ$ so $AP = CQ$ and they form a 120° angle. We
deduce that the angle between AP and CQ is equal to the angle between AN and
CM, both angles being equal to 120°.

On the other hand, BD maps to PN by the reflection over the exterior angle
bisector of $\angle BCD$, and BD maps to QM by the reflection over the exterior angle
bisector of $\angle BAD$. Hence, $QM = BD = PN$. We deduce that triangles APN and
CQM are congruent having the corresponding sides equal. Consequently, $\angle PAN = \angle QCM$. So the angle between CQ and CM equals the angle between AP and
AM, which combined with the fact that the angle between AN and CM equals the
angle between AP and CQ implying that the four lines AP, AN, CM, CQ form a
parallelogram, and the conclusion follows.

Remark The problem is based on the properties of the Fermat-Torricelli point. This
point can be defined for triangles whose angles are less than 120°, and is the point
in the interior of the triangle with the property that the sum of the distances to
the vertices is minimal. The Fermat-Torricelli point is obtained by constructing
equilateral triangles on the sides of the triangle and then taking the common
intersection point of the circumcircles of these triangles. It can also be obtained

as the common point of the lines that join each vertex of the original triangle with the exterior vertex of the equilateral triangle that lies opposite to it.

In the case of our problem, both triangles ABC and ADC have angles less than $120°$ (or else the sum of the angles of the quadrilateral would exceed $360°$). The intersection F of AN and CM is the Fermat-Torricelli point of the triangle ABC. Let ACT be the third equilateral triangle that defines the Fermat-Torricelli point F. An easy angle chasing using the fact that $FATC$ is cyclic shows that BT passes through F, and the above argument based on rotations shows that $BT = AN = CM$ and that they form $60°$ angles. Note also that by applying Ptolemy's Theorem in the quadrilateral $FATC$ $(FA \cdot CT + FC \cdot AT = AC \cdot FT)$ implies $FA + FC = FT$, so $FA + FB + FC = BT = AN = CM$. Note that if F' is another point in the interior, then $F'A + F'C \geq F'B$, by Ptolemy's Theorem $(F'A \cdot CT + F'C \cdot AT \geq AC \cdot F'T)$ in the quadrilateral $F'ATC$, which is not necessarily cyclic, and $TF' + F'B \geq TB$ by the triangle inequality, so $F'A + F'B + F'C > FA + FB + FC$ (you cannot have equality in both of the above inequalities unless $F = F'$).

36 The condition $AB = BC + AE$ is not very geometry-friendly unless we make an additional construction. This construction is motivated by a number of factors: that M is the midpoint of CE, the condition from the statement, and the fact that BC is parallel to AE. Because of these considerations, we take the reflection over M (Fig. 9.34). Let this reflection map A to A', B to B', and D to D'. Notice that C is mapped to E. Now the condition $BC \parallel AE$ implies that $C \in A'B$ and $E \in AB'$, and we obtain the parallelogram $ABA'B'$. Furthermore, since $AB = BC + AE = BC + CA' = BA'$, the parallelogram $ABA'B'$ is actually a rhombus.

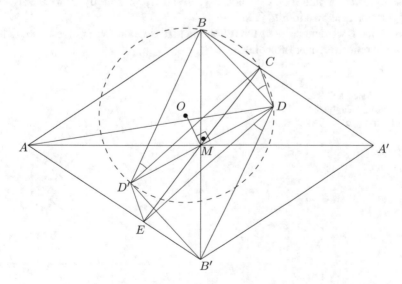

Fig. 9.34 The reflection of $ABCD$ over M

The condition $\angle DMO = 90°$ means that OM is the perpendicular bisector of DD', which implies that D' lies on the circumcircle of BCD. Keeping in mind the cyclic quadrilateral $BCDD'$ and the reflection, we obtain

$$\angle BDC = \angle BD'C = \angle B'DE,$$

so

$$\angle BDB' = \angle BDE + \angle EDB' = \angle BDE + \angle BDC = \angle CDE$$
$$= \angle ABC = 180° - \angle BAB'.$$

This means that $BDB'A$ is a cyclic quadrilateral itself. Combining this with the fact that $ABA'B'$ is a rhombus, we obtain

$$\angle BDA = \angle BB'A = \frac{1}{2}\angle ABC = \frac{1}{2}\angle CDE,$$

and we are done.

Remark Whenever the conditions of a problem include a sum of two segments, it is good to think about an additional condition that creates a segment equal to the sum.

Source Short list of the International Mathematical Olympiad, 2010, proposed by N. Serdyuk, Ukraine.

37 *First solution.* Let the perpendiculars to the sides BC, CA, AB at M and M_1, N and N_1, and P and P_1 be m and m_1, n and n_1, p and p_1, respectively. The configuration is shown in Fig. 9.35.

For a line l, we denote by σ_l the reflection across l, and for a point X, we denote by σ_X the reflection over the point X.

Fig. 9.35 Concurrent lines that are orthogonal to the sides of a triangle

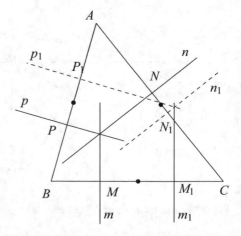

By the result proved in Problem 6, three nonparallel lines l_1, l_2, l_3 intersect at one point if and only if $\sigma_{l_1} \circ \sigma_{l_2} \circ \sigma_{l_3} = \sigma_{l_3} \circ \sigma_{l_2} \circ \sigma_{l_1}$, that is, if and only if

$$(\sigma_{l_1} \circ \sigma_{l_2} \circ \sigma_{l_3})^2 = 1.$$

Using this fact, we see that we have to show that

$$(\sigma_m \circ \sigma_n \circ \sigma_p)^2 = 1$$

is equivalent to

$$(\sigma_{m_1} \circ \sigma_{n_1} \circ \sigma_{p_1})^2 = 1.$$

Of course we only have to check the direct implication (by symmetry). We can write

$$\sigma_{m_1} = \sigma_B \circ \sigma_m \circ \sigma_C, \quad \sigma_{n_1} = \sigma_C \circ \sigma_n \circ \sigma_A, \quad \sigma_{p_1} = \sigma_A \circ \sigma_p \circ \sigma_B.$$

Then

$$(\sigma_{m_1} \circ \sigma_{n_1} \circ \sigma_{p_1})^2$$
$$= \sigma_B \circ \sigma_m \circ \sigma_C^2 \circ \sigma_n \circ \sigma_A^2 \circ \sigma_p \circ \sigma_B^2 \circ \sigma_m \circ \sigma_C^2 \circ \sigma_n \circ \sigma_A^2 \circ \sigma_p \circ \sigma_B$$
$$= \sigma_B \circ (\sigma_m \circ \sigma_n \circ \sigma_p)^2 \circ \sigma_B = \sigma_B^2 = 1.$$

The problem is solved.

Second solution. With the notation from the first solution, the reflections of m, n, and p over the circumcenter O of triangle ABC are m_1, n_1, and p_1, respectively. So if the lines m, n, p intersect at one point, then the lines m_1, n_1, p_1 intersect at the reflection of this point over O and vice versa.

Remark The first solution relies on algebraic manipulations in the group of isometries.

The second solution relies on the following observation: if two lines ℓ and ℓ' are the reflections of each other over the point O, then they are reflections over any point that lies on the parallel through O to ℓ and ℓ'.

Source I.M. Yaglom, *Geometric Transformations I*, (transl. by A. Shields), MAA, 1975.

38 The sums of angles from the statement can be computed, because the sum of the angles of a hexagon is $720°$, so

$$\angle A + \angle B + \angle C = \angle A_1 + \angle B_1 + \angle C_1 = 360°.$$

The hexagon is the juxtaposition of the triangles ABC, A_1BC, AB_1C, and ABC_1, so the problem actually requires us to show that the area of ABC is equal to the

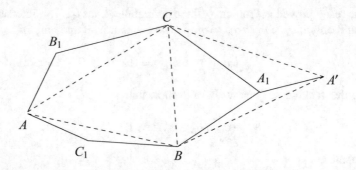

Fig. 9.36 Dissection of the triangle ABC into the triangles ABC_1, AB_1C, A_1BC

sum of the areas of the triangles A_1BC, AB_1C, and ABC_1. As we are focused on isometries, a first idea that should come to mind is to dissect the triangle ABC into those three triangles. This is possible, indeed, and we will prove it with the aid of Fig. 9.36.

We are given that $B_1C = A_1C$, so there is a rotation about C that maps B_1 to A_1. Let A' be the image of A through this rotation. We compute

$$\angle A'A_1B = 360° - \angle A'A_1C - \angle BA_1C = 360° - \angle AB_1C - \angle BA_1C$$
$$= 360° - \angle B_1 - \angle A_1 = \angle C_1 = \angle AC_1B.$$

So we have $AC_1 = AB_1 = A'A_1$, $BA_1 = BC_1$, and $\angle AC_1B = \angle A'A_1B$ (for the equality of segments, we have used the hypothesis of the problem). It follows that the triangles AC_1B and $A'A_1B$ are congruent. We deduce that the triangle $A'BC$ can be dissected into triangles congruent to A_1BC, AB_1C, and ABC_1.

Let us compare the triangles ABC and $A'BC$. They share the side BC, while $CA = CA'$ as one rotates into the other. But also $AB = A'B$ because of the congruence of triangles AC_1B and $A'A_1B$. So the triangles ABC and $A'BC$ are congruent, having respectively equal sides. Therefore, the triangle ABC can be dissected into triangles congruent to A_1BC, AB_1C, and ABC_1, and the problem is solved.

Remark This "cut and paste" technique will appear in some other problems.

Source Moscow Mathematical Olympiad, 1997.

39 Translate the triangles A_1BC_2, A_2B_1C along their respective equilateral triangle sides until C_1 coincides with C_2 and B_1 coincides with B_2. Now complete the figure to a triangle (see Fig. 9.37). The result is an equilateral triangle.

Next, let α, β, and $60°$ be the measures of the angles around the vertex P, defined in Fig. 9.38. We have $\alpha + \beta + 60° = 180°$, so in the upper triangle, the remaining angle is β, and in the bottom rightmost triangle, the remaining angle is α. It follows

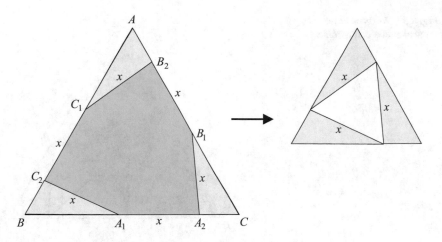

Fig. 9.37 Translating the small triangles to form an equilateral triangle

Fig. 9.38 Analysis of the small equilateral triangle

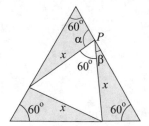

that the three shaded triangles are congruent. So what we have here are the translates of the triangles AB_2C_1, A_1BC_2, A_2B_1C.

Back to the original diagram, we now deduce that the triangles $A_1A_2B_1$, $B_1B_2C_1$, and $C_1C_2A_1$ are congruent (by side-angle-side), so the triangle $A_1B_1C_1$ is equilateral.

But then the equalities $C_1B_1 = C_1A_1$ and $A_2B_1 = A_2A_1$ mean that the line C_1A_2 is the perpendicular bisector of A_1B_1 (see Fig. 9.39). Analogously, the line A_1B_2 is the perpendicular bisector of B_1C_1 and the line B_1C_2 is the perpendicular bisector of C_1A_1. Therefore, the lines A_1B_2, B_1C_2, and C_1A_2 are concurrent at the circumcenter of the triangle $A_1B_1C_1$.

Remark The main motivation behind the "cut and paste" idea is that we know much more about triangles than hexagons; so we prefer working on a diagram with only triangles! In this problem, the "cut and paste" translates into... translations!

The translations work well because they preserve many of the angles, so most of the angle chasing that we have done can be carried back to the original diagram. Finally, the many equal segments lead to perpendicular bisectors and to yet another equilateral triangle, and this is what finishes the problem.

Fig. 9.39 Analysis of the original equilateral triangle

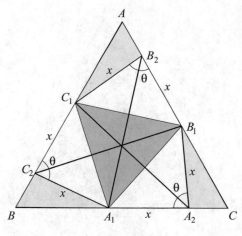

Fig. 9.40 A rotation about I maps DBF to GCE

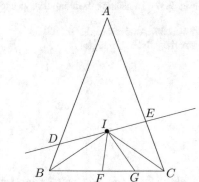

Source International Mathematical Olympiad, 2005, proposed by Bogdan Enescu, Romania.

40 Because $BD = CG$, $BF = CE$, and $\angle DBF = \angle ABC = \angle ACB = \angle GCE$, the triangles DBF and GCE are congruent and have the same orientation (see Fig. 9.40).

Since $BI = CI$, there is a rotation of center I and angle $\alpha = \angle BIC$ that takes B to C and, consequently, the triangle DBF to GCE. So $\angle DIG = \angle FIE = \alpha$. Since α does not depend on r, $\angle FIG = \angle DIG + \angle FIE - 180° = 2\alpha - 180°$, which also does not depend on r.

Remark With some angle chasing (try it!), you can prove that $\angle FIG = \angle BAC$.

Source Cono Sur Mathematical Olympiad, 2005.

41 *First solution.* For a point M in the plane, denote by f_M the clockwise $120°$ rotation about M. Then $P_1 = f_A(P_0)$, $P_2 = f_B(P_1)$, $P_3 = f_C(P_2)$, etc., as you can see in Fig. 9.41. The problem states that $(f_C \circ f_B \circ f_A)^{662}(P_0) = P_0$. But

Fig. 9.41 The path of the man

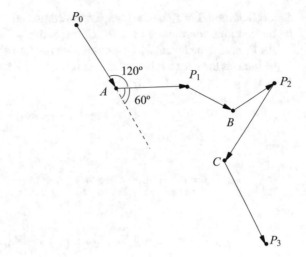

Fig. 9.42 C' is a fixed point of $f_{C'} \circ f_B \circ f_A$.

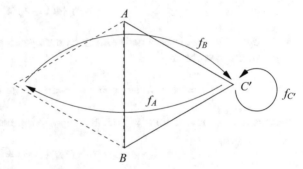

the transformation $g = f_C \circ f_B \circ f_A$ is an orientation-preserving isometry, so by Theorem 1.9, it is either a rotation, a translation, or the identity map. It rotates figures by $120° + 120° + 120° = 360°$, so it is not a rotation. It cannot be a translation because g^{662} is a translation as well, which cannot have P_0 as a fixed point. So it must be the identity map.

It follows that $f_C = (f_B \circ f_A)^{-1}$. Now construct a point C' such that ABC' is equilateral and oriented counterclockwise. Then, by the same argument, $f_{C'} \circ f_B \circ f_A$ is either a translation or the identity map. Figure 9.42 shows that $f_{C'} \circ f_B \circ f_A(C') = C'$, so this map is not a translation; it is the identity map. But then

$$f_{C'} \circ f_B \circ f_A = f_C \circ f_B \circ f_A,$$

so $f_C = f_{C'}$, and hence $C = C'$. The problem is solved.

Second solution. Let z_k be the complex coordinate of P_k, $k \geq 0$, and let w_1, w_2, w_3 be the complex coordinates of A, B, C, respectively. Set also $\epsilon = e^{4\pi i/3}$.

As P_1 is obtained from P_0 by a clockwise rotation about A of angle $2\pi/3$, which is the same as the counterclockwise rotation by $4\pi/3$, we have $z_1 - w_1 = \epsilon(z_0 - w_1)$. Hence

$$z_1 = \epsilon z_0 + (1 - \epsilon)w_1.$$

Then

$$z_2 = \epsilon z_1 + (1 - \epsilon)w_2 = \epsilon^2 z_0 + (1 - \epsilon)(\epsilon w_1 + w_2)$$

and

$$z_3 = \epsilon z_2 + (1 - \epsilon)w_3 = \epsilon^3 z_0 + (1 - \epsilon)(\epsilon^2 w_1 + \epsilon w_2 + w_3) =$$
$$z_0 + (1 - \epsilon)(\epsilon^2 w_1 + \epsilon w_2 + w_3)$$

We recognize that z_3 is obtained from z_0 by a translation by $(1 - \epsilon)(\epsilon^2 w_1 + \epsilon w_2 + w_3)$. Iterating we obtain

$$z_{3n} = z_0 + n(1 - \epsilon)(\epsilon^2 w_1 + \epsilon w_2 + w_3), \quad n \geq 1.$$

In particular this is true for $n = 662$. But $z_{1986} = z_0$; hence,

$$n(1 - \epsilon)(\epsilon^2 w_1 + \epsilon w_2 + w_3) = 0.$$

The only possibility is that $\epsilon^2 w_1 + \epsilon w_2 + w_3 = 0$. And this is the same as

$$w_3 + e^{\frac{2\pi i}{3}} w_1 + e^{\frac{4\pi i}{3}} w_2 = 0,$$

which, by Lemma 1.27, implies that w_1, w_2, w_3 are the coordinates of the vertices of an equilateral triangle.

Source International Mathematical Olympiad, 1986, proposed by China, first solution from D. Djukić, V. Janković, I. Matić, N. Petrović, *The IMO Compendium*, Springer, 2006, second solution from T. Andreescu, D. Andrica, *Complex numbers from A to ... Z*, Birkhäuser, Second Ed. 2014.

42 Arguing on Fig. 9.43, we begin with some angle chasing in which we take advantage of the fact that C_1 and C_2 have equal radii:

$$\angle ADC = \angle ADB = \frac{\overset{\frown}{APB}}{2} = \frac{\overset{\frown}{AEB}}{2} = \angle ACB = \angle ACD.$$

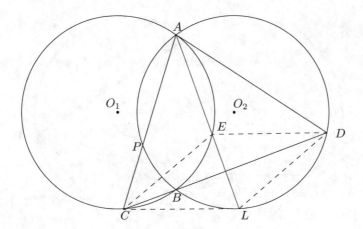

Fig. 9.43 Angle chasing in C_1 and C_2

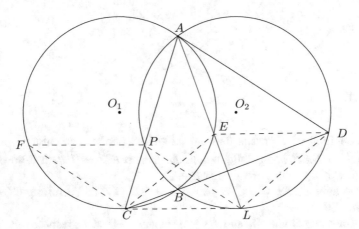

Fig. 9.44 Analysis of the translation τ

So the triangle ACD is isosceles which implies that the line AL is also the perpendicular bisector of the segment CD, thus $AL \perp CD$. Also, the equal angles $\angle BAE$ and $\angle BAL$ are respectively inscribed in C_1 and C_2, so $BE = BL$. Moreover, since $EL \perp CD$, BD is the perpendicular bisector of EL. This proves that $CLDE$ is a rhombus and, in particular, it is a parallelogram. As a consequence of this, there is a translation τ that takes E to D and C to L.

Let us take a closer look at τ examining Fig. 9.44. Because τ takes the chord CE to the chord LD and because the circles C_1 and C_2 have equal radii, τ also takes C_1 to C_2, and hence it takes the center O_1 of C_1 to the center O_2 of C_2. Then the distance between any two points Y and $\tau(Y)$ is equal to $O_1 O_2$.

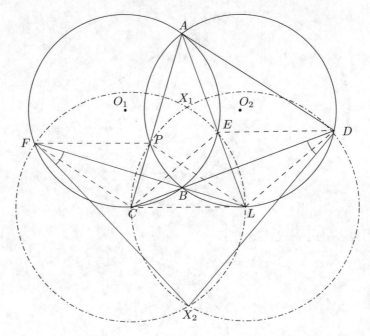

Fig. 9.45 The construction of X

The definition of F means that the midpoints of the segments PE and DF coincide, so $PDEF$ is a parallelogram, and hence $\overrightarrow{ED} = \overrightarrow{FP}$. It follows that τ takes F to $P \in C_2$, which proves that $F \in C_1$.

We also have $\angle PAL = \angle CAE$, so $PL = CE = CL$. Therefore, $CLPF$ is a rhombus.

Now draw two circles S_1 and S_2 with centers C and L, respectively, and the same radius $r = O_1 O_2$, as shown in Fig. 9.45. Then L and F lie on S_1 and C and D lie on S_2. We claim that any of the two intersection points of S_1 and S_2 can be chosen as point X (note that the circles intersect because the distance $CL = r$ between their centers is smaller than the sum of the radii, which is $2r$, and greater than the difference, which is zero). Indeed, the point X satisfies the desired property because $XL = XC = CL = r$, so CLX is an equilateral triangle. Then $\angle XFL$ is inscribed in S_1 and is equal to $\angle XCL/2 = 30°$ and $\angle XDC$ is inscribed in S_2 and is equal to $\angle XLC/2 = 30°$. The problem is solved.

Remark The first idea of the solution is that circles with equal radii are good for angle chasing, because of the symmetry in arcs: not only angles inscribed in the same arc are congruent, but angles inscribed in arcs of the *same length* are congruent. Watch how effective this idea is in the way we have proved that triangle ACD is isosceles and that $BE = BL$.

Also, two congruent circles can be mapped into each other by all three isometries: we can reflect across AB, rotate about A or B, or translate by the vector $\overrightarrow{O_1 O_2}$.

The parallelograms $CLDE$ and $FCLP$, which share side CL, suggest that the translation is the best option here. Indeed, the translation is crucial to prove that F lies on C_1. It also spreads the length O_1O_2 across the diagram, causing both rhombi to pop up.

Once we discover that these parallelograms are rhombi, the equality $FC = CL = LD$ gives us a good reason to construct the circles S_1 and S_2: the centers and any of the two intersection points are vertices of an equilateral triangle (aha, this is where the 30° came from!), and both D and F are in at least one of the circles (inscribed angles!).

Source Iberoamerican Mathematical Olympiad, 2009, proposed by A. Aguilar.

43 *First solution.* Let ABC be the triangle; reflect it successively over AC, BC, AB, AC, BC, as shown in Fig. 9.46 (we have labeled the sides a, b, c so that you can read the figure with ease). A careful examination of the configuration finishes the proof.

Let us explain. First, it should be noted that each pair of sides of the orthic triangle that meet at a vertex form with the side of the triangle that passes through that vertex equal angles (the well-known fact that the altitudes of ABC are the angle bisectors of the orthic triangle is a consequence of this). Moreover, this property characterizes the orthic triangle. Indeed, assume that some triangle UVW inscribed in the triangle ABC (with $U \in BC, V \in AC$, and $W \in AB$) has the property that

$$\angle BUW = \angle CUV, \quad \angle CVU = \angle AVW, \quad \angle AWV = \angle BWU.$$

Set $\angle BUW = \angle CUV = u, \angle CVU = \angle AVW = v, \angle AWV = \angle BWU = w$. Then in the triangles AVW, BUW, CUV, we can write

$$v + w = 180° - A$$
$$u + w = 180° - B$$
$$u + v = 180° - C.$$

Adding we obtain $u + v + w = 180°$, so $u = A, v = B, c = C$, showing that the sides of UVW and those of the orthic triangle are parallel. This forces the two triangles to coincide. We say that the orthic triangle forms a billiard path, since if the original triangle were a billiard table and a ball were shot from one of the vertices of the orthic triangle toward another, then the ball would reflect off the sides of the table and follow a periodic trajectory along the sides of the orthic triangle.

In Fig. 9.46, we have drawn the orthic triangle DEF and some other inscribed triangle UVW. Because DEF is a billiard path, the successive reflection of its sides yields the straight line FP from the figure. On the other hand, the successive reflections of the sides of UVW yield the broken line from W to Q as seen in the same figure. The length of FP is twice the perimeter of DEF, and the length of the broken line is twice the perimeter of UVW (because each side of the triangle

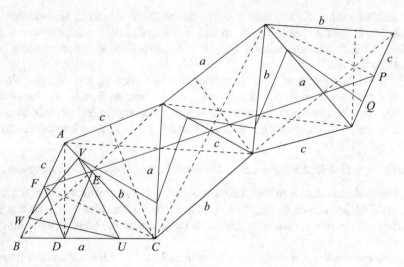

Fig. 9.46 A proof without words: Hermann Schwarz's proof of Fagnano's Theorem

appears twice in the sum). By the triangle inequality, the broken line is strictly longer than the segment WQ. But the latter is equal to FP because $FPQW$ is a parallelogram. We infer that the perimeter of DEF, which is half of the length of FP and hence of WQ, is strictly less than the perimeter of UVW. We conclude that the orthic triangle has minimal perimeter among all triangles inscribed in a given acute triangle.

Second solution. Orient the triangle counterclockwise, and let z_1, z_2, z_3 be the complex coordinates of the vertices A, B, C, respectively. The function $S : [0, 1] \times [0, 1] \times [0, 1] \to \mathbb{R}$,

$$S(s, t, u) = |z_1 + s(z_2 - z_1) - z_2 - t(z_3 - z_2)|$$
$$+ |z_2 + t(z_3 - z_2) - z_1 - u(z_1 - z_3)| + |z_3 + u(z_1 - z_3) - z_3 - s(z_1 - z_3)|$$

computes the perimeter of a triangle whose vertices have coordinates $z_1 + s(z_2 - z_3)$, $z_2 + t(z_3 - z_1)$, and $z_3 + u(z_1 - z_3)$, which are the coordinates of some arbitrary points on AB, BC, CA, respectively. Note that the cube $[0, 1] \times [0, 1] \times [0, 1]$ is closed and bounded in $\mathbb{R}^3$, and S is continuous. Hence, S has a minimum in this cube, showing that a triangle with minimal perimeter exists. We prove that the minimum is attained for the orthic triangle.

The minimum corresponds to a point that lies either in the interior or on the boundary of the cube. Let us rule out the second possibility, in which case at least one of the vertices of the inscribed triangle lies at a vertex of the original triangle, say at A. Since another vertex of the inscribed triangle must be on BC, the perimeter of this inscribed triangle must be at least twice the altitude from A. Let us show that the altitudes of an acute triangle are greater than the semiperimeter of the orthic triangle.

For this, let D, E, F be the feet of the altitudes from A, B, C, and let the lengths of BC, CA, AB be a, b, c, respectively. We want to show that $DE + EF + FE$ is less than twice the altitude from A (Fig. 9.47). Because $AE = c \cos A$, by applying the Law of Sines in the triangle AEF, we obtain

$$EF = \frac{c}{\sin C} \sin A \cos A = 2R \sin A \cos A,$$

where R is the circumradius. Similarly, $DF = 2R \sin B \cos B$ and $DE = 2R \sin C \cos C$. Also, in the triangle ABD, $AD = c \sin B = 2R \sin C \sin B$. The inequality $DE + EF + FE < 2AD$ reduces to

$$\sin A \cos A + \sin B \cos B + \sin C \cos C < 2 \sin B \sin C.$$

Rewrite this as

$$\sin 2A + \sin 2B + \sin 2C < 4 \sin B \sin C,$$

and then notice that since the triangle is acute, $\pi/2 < B + C < \pi$, so

$$\sin 2A = -\sin 2(B + C) = -\sin 2B \cos 2C - \sin 2C \cos 2B.$$

Thus, we have to show that

$$\sin 2B(1 - \cos 2C) + \sin 2C(1 - \cos 2B) < 4 \sin B \sin C,$$

or

$$\sin B \cos B \sin^2 C + \sin C \cos C \sin^2 B < \sin B \sin C.$$

Dividing by $\sin B \sin C$, we transform this into the equivalent inequality $\sin C \cos B + \sin B \cos C < 1$, which is obvious because the left-hand side is $\sin(B + C)$ (and $B + C > \pi/2$).

Fig. 9.47 The orthic triangle

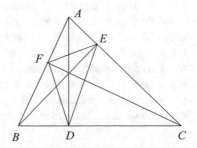

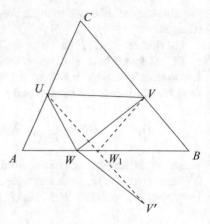

Fig. 9.48 Proof of the minimum forms a billiard path

So the minimum of the function S is not attained on the boundary of the cube, namely, for a degenerate triangle. Could the minimum be a triangle UVW ($U \in BC, V \in AC, W \in AB$) different from the orthic triangle? Notice that the minimum should be a billiard path, meaning that

$$\angle BUW = \angle CUV, \quad \angle CVU = \angle AVW, \quad \angle AWV = \angle BWU.$$

Indeed, if one of these inequalities does not hold, say the first, then reflect V across A to V' (Fig. 9.48), and let W_1 be the intersection of the segments AB and UV' (they intersect because the triangle is acute). Then the triangle inequality in UWV' implies

$$UW_1 + W_1V = UW_1 + W_1V' = UV' < WV' + WU = WV + WU,$$

so the triangle UVW_1 has a smaller perimeter, showing that UVW is not a minimum. But, as we have seen above, the orthic triangle is the only triangle that forms a billiard path, so it has to be the minimum.

Remark Note that both arguments fail for obtuse triangles, and indeed, in such a triangle, the shortest altitude is less than the semiperimeter of the orthic triangle. Do you see why the proofs fail?

In the second argument, we have used a method for finding the minimum which is referred to as Sturm's Principle: given a function f that has a minimum on a certain domain, if there is a point x_0 such that every point $x \neq x_0$ is not a minimum for f, then x_0 is the minimum of f. It is very important that you check first that f has a minimum on the given domain, for example, here we could not just work with the interior of the cube, since it is not true that a continuous function on the interior of the cube has a minimum. This property is, however, true for the closed cube, as the Heine-Borel Theorem states that a set in $\mathbb{R}^n$ is compact if and only if is closed and bounded, and a function defined on a compact set has a both a maximum and a minimum.

Source The first proof of Fagnano's Theorem was found by Hermann Schwarz (and was communicated to us by Cosmin Pohoață).

44 We argue on Fig. 9.49. Let Q be the isogonal conjugate of P with respect to the triangle AED. Then, by using the definition of the isogonal conjugate as well as inscribed angles in ω and Ω, we obtain

$$\angle(QA, AD) = \angle(EA, AP) = \angle(EB, BA)$$

and

$$\angle(QD, DA) = \angle(ED, DP) = \angle(EC, CD).$$

Note that PE tangent to Ω is equivalent to $\angle(PE, DE) = \angle(EC, DC)$, which, via the isogonal conjugate, translates to $\angle(AE, QE) = \angle(EC, DC)$, and the latter is equivalent to $QE \parallel CD$.

So the problem reduces to showing that QE is parallel to CD. To prove this, consider the degenerate triangle whose sides are the segment BC and the rays $|BA$ and $|CD$, and let R be the isogonal conjugate of E with respect to this triangle. Then

Fig. 9.49 Isogonal conjugates of P and E with respect to $\triangle AED$ and $\triangle BC\infty$

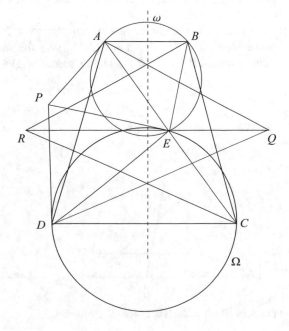

$$\angle(RB, BC) = \angle(AB, BE) = \angle(PA, AE) = \angle(DA, AQ),$$

where the first equality follows from the fact that R and E are isogonal conjugates, the third follows from the fact that P and Q are isogonal conjugates, and the second equality follows from inscribed angles in circle ω. So $\angle RBC = \angle DAQ$. Similarly, $\angle RCB = \angle ADQ$, which implies that R and Q coincide under a reflection over the common perpendicular bisector of AB and CD (which is the symmetry axis of the trapezoid). Since E is on the line that is parallel to BA and CD and at equal distance from the two, so is its isogonal conjugate R. But then the reflection Q of R over the perpendicular bisector of AB is also on this line, so EQ is parallel to CD, and the problem is solved.

Remark The reader should observe that key step of the solution is not the use of an explicit isometry, but a construction based on isometries: the construction of the isogonal conjugate.

Source Romanian Master of Mathematics, 2019, proposed by Jakob Jurij Snoj.

45 Denote the lengths of the sides BC, AC, AB of the triangle, respectively, by a, b, c. We reason on Fig. 9.50.

Let $AB'C'$ be the reflection of the triangle ABC over the angle bisector of $\angle BAC$. Construct the parallelograms $APNB'$, $AC'MP$, $B'NMC'$. Now we are almost in the setting of the Pappus' Area Theorem (see Problem 18), except that two of the parallelograms are constructed internally. Nevertheless, it is still true that the sum of the areas of the first two parallelograms is equal to the third. This can be shown as follows.

Translate AP along its line of support to $A'P'$ where $A' \in B'C'$. Then the parallelograms $APNB'$ and $A'P'NB'$ have the same area because they have the same base ($AP = A'P'$) and the same height. Similarly, the parallelograms

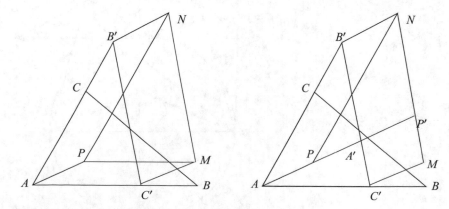

Fig. 9.50 Proof of the Erdős-Mordell Theorem

$AC'MP$ and $A'C'MP'$ have the same area. But the sum of the areas of $A'P'NB'$ and $A'C'MP'$ is the area of $B'NMC'$, which proves our claim.

Returning the problem, note that the area of $APNB'$ is $d_b \cdot c$ and the area of $AC'MP$ is $d_c \cdot b$, while the parallelogram $B'NMC'$ has sides BC and AP and so its area is less than or equal to $BC \cdot AP$. We thus have the inequality

$$d_b \cdot c + d_c \cdot b \le a \cdot AP.$$

Dividing by a, we obtain

$$\frac{c}{a} d_b + \frac{b}{a} d_c \le AP.$$

Similar arguments yield

$$\frac{a}{b} d_c + \frac{c}{b} d_a \le BP \text{ and } \frac{b}{c} d_a + \frac{a}{c} d_b \le CP.$$

Adding the three inequalities, we obtain

$$\left(\frac{c}{b} + \frac{b}{c}\right) d_a + \left(\frac{a}{c} + \frac{c}{a}\right) d_b + \left(\frac{b}{a} + \frac{a}{b}\right) d_c \le AP + BP + CP.$$

Using the fact that for $x > 0$, $x + \frac{1}{x} \ge 2$, we obtain that the left-hand side is greater than or equal to $2(d_a + d_b + d_c)$, and we are done.

Remark The key step of the proof is an inequality for areas proved by shearing, exactly like in the case of the Pappus' Area Theorem.

46 The broken tool makes constructions very elaborate. We have to divide the work into small steps; the steps are presented in the guise of two lemmas.

Lemma 1 *Given the points A, B, and C in the plane and a compass with fixed opening, one can construct a point D such that ABCD is a parallelogram.*

Proof We are supposed to translate the point A by the vector $\overrightarrow{BC}$. Let a be the width of the opening of the compass. The construction is very simple in the particular case where $AB = BC = a$. Indeed, if we construct the two circles of radius a centered at A and C, one of their intersections is B and the other is the desired point D. Our intention is to reduce the general case to this particular one.

Let us show how to translate A by the arbitrary vector $\overrightarrow{BC}$ when $AB = a$. While moving toward C, by using the compass, we can construct a sequence of points $P_1, P_2, \ldots, P_n$ such that $BP_1 = P_1P_2 = \cdots = P_{n-2}P_{n-1} = a$ a and P_{n-1} is within a from C. Intersect the circles of radii a centered at C and P_{n-1} to obtain P_n such that $P_{n-1}P_n = P_nC = a$. This is illustrated in Fig. 9.51.

The point D is obtained from A by a translation of vector $\overrightarrow{BP_1}$, followed by a translation of vector $\overrightarrow{P_1P_2}$, then $\overrightarrow{P_2P_3}$, \ldots, and finally $\overrightarrow{P_nC}$.

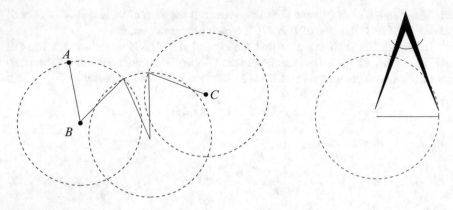

Fig. 9.51 Reduction of the translation of a segment to the simplest case

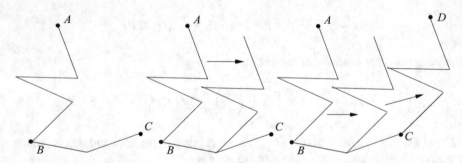

Fig. 9.52 Construction of D such that $ABCD$ is a parallelogram

If AB has arbitrary length, construct $Q_1, Q_2, \ldots Q_n$ such that $BQ_1 = Q_1Q_2 = \cdots = Q_nA = a$. Translate Q_1 by $\overrightarrow{BC}$ to R_1, so that BQ_1R_1C is a parallelogram, then Q_2 to R_2 so that $Q_1Q_2R_2R_1$ is a parallelogram, and so on. The translate of A will be the desired point D. This entire process is illustrated in Fig. 9.52.   □

Lemma 2 *Given the points A and B in the plane and a compass with fixed opening, one can construct a point C such that the triangle ABC is equilateral.*

Proof Let the width of the opening of the compass be a. We start with an observation. Given two points M, N in the plane, let N' be the rotation of N around M by α. Let also τ and ρ be the translation by $\overrightarrow{NN'}$ and the rotation around N' by α; then $\rho \circ \tau$ is some rotation by α (Theorem 1.9). If $M' = \tau(M)$, then $MM'N'N$ is a parallelogram. In this parallelogram $MN' = MN = M'N'$ and $\angle M'N'M = \angle N'MN = \alpha$, because N' is the rotation of N about M by α. Consequently, $\rho \circ \tau(M) = \rho(M') = M$; thus, M is the center of the rotation. Therefore, $\rho \circ \tau$ is the rotation around M by α. This means that to construct the image of a point P through the rotation around M by α, you first translate P by $\overrightarrow{NN'}$ and then rotate around N' by α.

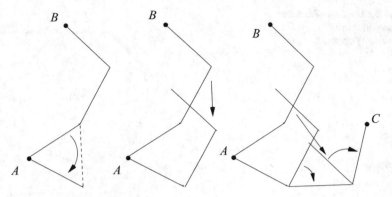

Fig. 9.53 Construction of the equilateral triangle

Fig. 9.54 Construction of D
such that $AB \perp CD$

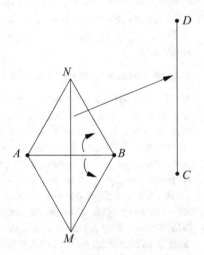

Our goal is to rotate B around A by $60°$. Choose the points $P_0 = A$, P_1, ...,
$P_n = B$ such that $P_1 P_2 = P_2 P_3 = \cdots = P_{n-1} P_n = a$. One can easily construct
an equilateral triangle $P_0 P_1 Q_1$, with Q_1 being the rotation of P_1 around $P_0 = A$
by $60°$. Indeed, you just intersect the circles of radius a centered at P_0 and P_1.
Then translate the broken line $P_1 P_2 \ldots P_n$ to a broken line that starts at Q_1. By the
observation made at the beginning of the proof, C is the rotate of the end of this
broken line around Q_1 by $60°$. We have reduced the problem from a broken line
with n segments to a broken line with $n - 1$ segments. Repeating we produce in the
end the point C. These steps are shown schematically in Fig. 9.53.   □

Now we solve the actual problem. The construction can be followed on Fig. 9.54.
Using Lemma 2 construct M and N such that AMB and ANB are equilateral. Then
MN is perpendicular to AB. Next translate MN to CD using Lemma 1. The line
CD is parallel to MN so it is perpendicular to AB, and we are done.

Fig. 9.55 The case where the
contact points are
diametrically opposite

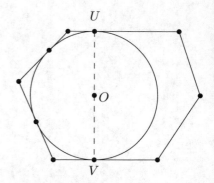

Remark In other words, D is obtained as the image of B under a composition of a
$60°$ rotation about A followed by a translation by $\overrightarrow{AC}$.

Note that with the rusty compass, you cannot construct the $90°$ rotation of B
about A.

Source Lemma 1 was given at the Bundeswettbewerb Mathematik in 1977.

47 Let us first understand what "making it impossible to move $\mathcal{P}$" means, in the
language of isometries. Let $P_1, P_2, \ldots, P_k$ be the points where the nails have been
placed. Then the impossibility of moving the polygon $\mathcal{P}$ means that any small
movement f forces $f(\mathcal{P})$ to contain at least one of the points P_i. Here "small
movement" means any orientation-preserving isometry with small parameter (small
vector in the case of a translation, small angle in the case of a rotation).

To gain insight, we examine some particular cases. It is clear that we need three
nails for a triangle. What about a square? Were there less than four nails, one of the
sides would have no nails at all, and we could translate the square in the direction
normal to this side, so we need at least four nails. And four nails at the midpoints of
the sides fix the square.

The idea that solves the general case is the use of *inscribed circles*. That is, to
consider a maximal circle ω contained in $\mathcal{P}$ and investigate the points at which ω
touches the sides of $\mathcal{P}$; these points will be the candidates for nails. With this in
mind, let $\mathcal{H}$ be the convex hull of these tangency points.

First consider the case where two vertices of $\mathcal{H}$, say, U and V, are such that
UV is a diameter of ω, as in Fig. 9.55. Place two nails at U and V. The sides that
contain U and V are parallel (both perpendicular to UV), so the only movement
these nails allow is a translation perpendicular to UV. Indeed, any small translation
not perpendicular to UV would make $f(\mathcal{P})$ cover either U or V: just consider the
component parallel to UV. Also, any small clockwise rotation with center to the left
of $\overrightarrow{UV}$ would make $f(\mathcal{P})$ cover V and so on. Two additional nails, one to the left of
$\mathcal{P}$ and one to the right of $\mathcal{P}$, prevent $\mathcal{P}$ from moving to the left or to the right.

What if this is not the case? Let O be the center of ω. Arguing by contradiction,
suppose that O is not inside $\mathcal{H}$. Let PQ be the side that separates O from $\mathcal{H}$ and

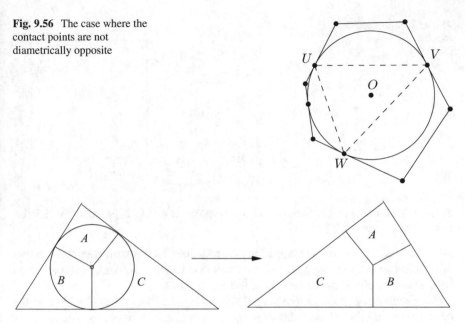

Fig. 9.56 The case where the contact points are not diametrically opposite

Fig. 9.57 Cutting the cake

let the tangents to ω at P and Q meet at K. A homothety (contraction) with center K and ratio slightly larger than 1 keeps ω inside $\mathcal{P}$, a contradiction. Hence, O is inside $\mathcal{H}$. Dissect $\mathcal{H}$ into triangles, and let UVW be the triangle that contains O, as shown in Fig. 9.56. Placing nails at U, V, and W we can see that they fix the triangle formed by the tangents to ω at these points (any small movement but a rotation with center O makes ω cover one of these points, and a small rotation with center O makes the polygon $\mathcal{P}$ cover all three points). We conclude that three nails fix the polygon $\mathcal{P}$.

Therefore, the answer to the problem is 4.

Remark Here we think of geometric transformations as "geometry in motion."

Source 10th Sharygin Geometry Olympiad, 2014, proposed by N. Beluhov and S. Gerdgikov, Russia.

48 Cut the cake along the perpendiculars from the incenter to the sides into pieces A, B, C arranged in counterclockwise order. Now place these slices in clockwise order around the incenter, as shown in Fig. 9.57. The result is the mirror image of the triangle.

Remark Recall the Wallace-Bolyai-Gerwien Theorem which says that the problem has solution if we do not limit the number of slices. The problem is made interesting by requiring the use of exactly three slices.

Fig. 9.58 Invariance under
rotation

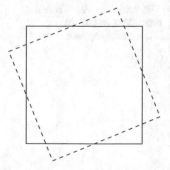

Source V.G. Boltyanskii, *Hilbert's Third Problem* (transl.R. Silverman), V.H. Winston, Washington, DC, 1978.

49 The blue square can be mapped to the red square by a rotation about the center of the blue square that brings it to the correct orientation, followed by a translation by the vector defined by the centers of the two squares.

We check first that the sum of the red sides is equal to the sum of blue sides in the case where only the rotation about the center is performed. In our figures, we draw the red square with continuous line and the blue square with dotted line. Examining Fig. 9.58, we see that four equal right triangles of the red square are left outside of the octagon. They are similar (because of equal angles) to the four triangles of the blue square that are left outside of the octagon. But since the squares have equal areas, the areas of the triangles cut out from the red square are equal to the areas of the triangles cut out from the blue square. So all eight triangles are equal, and hence they have equal hypotenuses. Thus, the sum of the four red hypotenuses equals the sum of the four blue hypotenuses, proving the equality of the sums of the sides of each color in this case.

Next we will show that the sum of the red sides and the sum of the blue sides of the octagon are both invariant when a translation is performed. In fact we only have to check the invariance of the sum of the blue sides, because the two squares are translated one with respect to the other, so the situation is symmetric.

Observe that the translation can be written as the composition of two translations, each in the direction of a pair of opposite sides of the red square (horizontal and vertical in Fig. 9.59). So it suffices to check invariance under translations parallel to the sides (see Fig. 9.59).

It is important that after each translation, the two squares still overlap to form an octagon. And because a pair of parallel lines crossed by another pair of parallel lines determine equal segments (because the segments form parallelograms), we see that the segments that are lost on one side are gained on the other (examine the pairs of boldface parallel segments in Fig. 9.59!). Hence, the sum of the segments of one color is preserved under translations, and the problem is solved.

Fig. 9.59 Invariance under
translation

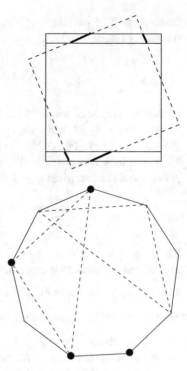

Fig. 9.60 A red triangle
congruent to a black triangle

Remark Here we think about geometric transformations as "geometry in motion."
We first check an easy case, then transform the easy case into the general case by
a translation, and check invariance of the sum of red (respectively blue) segments
under the translation. To this end, we decompose an arbitrary translation into simpler
translations and check invariance under the simpler translations. We use the fact that
the group of translations is generated by horizontal and vertical translations (with
the "horizontal" and "vertical" chosen at our discretion).

   An idea worth remembering: to check invariance under a group of transformations, it suffices to check invariance under the group generators.

Source *Kvant (Quantum)*, proposed by Vyacheslav Viktorovich Proizvolov.

50 An example of a coloring, with the black vertices indicated by black dots, is
shown in Fig. 9.60.

   Consider the rotations ρ_k, $k = 1, 2, \ldots, n - 1$, of the polygon by the angles
$2\pi k/n$. Let a_i be the number of red vertices that are mapped to black vertices by
the rotation ρ_i. To solve the problem, it suffices to show that there is i such that
$a_i \geq \lfloor p/2 \rfloor + 1$. Because when performing the $n - 1$ rotations each red vertex
overlaps with exactly $n - p$ black vertices (because that is how many black vertices
there are), we must have

$$a_1 + a_2 + \cdots + a_n = p(n - p).$$

But $n - p \geq p$ implies $n \geq 2p$, so $n > 2p - 1$. A little algebra shows that this inequality implies

$$\frac{p(n - p)}{n - 1} > \frac{p}{2}.$$

Thus, the average of the a_i's is larger than $p/2$, proving that some a_i is greater than or equal to $\lfloor p/2 \rfloor + 1$. But wait, we are not yet done! We must also show that we have indeed a polygon, and not just a segment. We do have a polygon if $\lfloor p/2 \rfloor + 1 \geq 3$. However, it is also possible that $\lfloor p/2 \rfloor + 1 < 3$, which happens precisely when $p = 3$. In that case, when $n \geq 8$ we automatically have

$$\frac{3(n - 3)}{n - 1} > 2.$$

The cases $n = 6$, $p = 3$, and $n = 7$, $p = 3$ should be examined case-by-case (there are only seven cases to check and they are left to the reader).

Remark Note the use of the Pigeonhole Principle.

Source Romanian Mathematical Olympiad, 1995.

51 (a) We denote the set by $\mathcal{P}$ and its area by S. Place the square so that one of its corners is at the origin and two sides are aligned along the positive x- and y-axes. Let $\mathcal{P}_1$ and $\mathcal{P}_2$ be the translates of $\mathcal{P}$ by two vectors of length 0.001, the first of which pointing in the positive direction of the x-axis, and the other making a $60°$ angle with the first, as sketched in Fig. 9.61. Then $\mathcal{P}$, $\mathcal{P}_1$, and $\mathcal{P}_2$ are mutually disjoint, while $\mathcal{P}_1$ and $\mathcal{P}_2$ are also mapped into each other by a translation whose vector has length .001 (observe the equilateral triangle formed by the translation vectors). All three surfaces lie inside a square of side 1.001; hence, $3S < 1.001^2$. We obtain $S < 0.335$.
(b) We have to improve the estimate from (a), and we do this using the four vectors from Fig. 9.61. The vectors $\vec{v_3}$ and $\vec{v_4}$ make an angle of $2 \arcsin \frac{1}{2\sqrt{3}}$ and have length $0.001 \times \sqrt{3}$, with $\vec{v_3}$ pointing in the positive direction of the x-axis. Let $\mathcal{P}_3$ and $\mathcal{P}_4$

Fig. 9.61 The set $\mathcal{P}$, made of a triangle, quadrilateral, and pentagon, and its translates

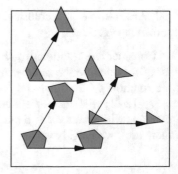

Fig. 9.62 The four vectors
by which we translate the set
$\mathcal{P}$

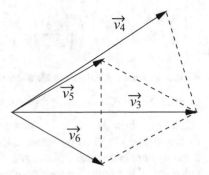

be the translates of $\mathcal{P}$ by these vectors. The two vectors form a triangle with sides
$.001 \times \sqrt{3}, 0.001 \times \sqrt{3}, 0.001$, so $\mathcal{P}_4$ can be obtained from $\mathcal{P}_3$ by a translation by a
vector of length 0.001. It follows that $\mathcal{P}_3$ and $\mathcal{P}_4$ do not intersect (Fig. 9.62).

Of $\mathcal{P}_3$ and $\mathcal{P}_4$, choose the one whose overlap with $\mathcal{P}$ has the smaller area; let
this be $\mathcal{P}_3$. Then the area of the overlap of $\mathcal{P}$ and $\mathcal{P}_3$ is at most $S/2$, so the total
area covered by $\mathcal{P}$ and $\mathcal{P}_3$ is at least $3S/2$. Now consider the translates $\mathcal{P}_5$ and $\mathcal{P}_6$
of $\mathcal{P}$ in the directions of the vectors $\vec{v_5}$ and $\vec{v_6}$ which have lengths equal to 0.001
and make angles of $30°$ with $\vec{v_3}$. Then $\mathcal{P}_5$ and $\mathcal{P}_6$ are also mapped into each other
by a translation whose vector has length 0.001 (notice the equilateral triangle). So
$\mathcal{P}, \mathcal{P}_5, \mathcal{P}_6$ do not intersect each other. Moreover, both $\mathcal{P}_5$ and $\mathcal{P}_6$ translate to $\mathcal{P}_3$ by
vectors of length 0.001, so they do not intersect $\mathcal{P}_3$. It follows that $\mathcal{P}, \mathcal{P}_3, \mathcal{P}_5, \mathcal{P}_6$
altogether cover a total area that is at least $S + S/2 + S + S = 7S/2$. These sets lie
inside a square of side-length $1 + 0.001 \times \sqrt{3} < 1.0018$. Therefore,

$$S < \frac{2}{7}(1.0018)^2 < 0.287,$$

and the inequality is proved.

Remark The given condition means that every translation $\mathcal{P}'$ of $\mathcal{P}$ by a vector of
length 0.001 does not intersect $\mathcal{P}$. So, in (a), we use three copies of $\mathcal{P}$ each of which
is a translation by a vector of length 0.001 of the others. Of course, the best we can
do in the plane is using an equilateral triangle.

To improve that, in (b) we try to add a new translation, forming two equilateral
triangles (hence, the angle $2 \arcsin \frac{1}{2\sqrt{3}}$). But with that, we introduce some overlaps
between translates. This is not desirable, since we do not have much control of the
overlaps. Having two nonoverlapping options $\mathcal{P}_3$ and $\mathcal{P}_4$ gives us some control, as
$\mathcal{P}$ would intersect the least with one of them.

A natural question is what the optimal set $\mathcal{P}$ is. We can construct an example
to obtain a lower bound by tessellating the plane with copies of a regular hexagon
H with side-length $0.001/2 - \epsilon$ (and thus diameter $0.001 - 2\epsilon$) surrounded by
strips of width $0.001/2 + \epsilon$ (assuring that points from different hexagons are at least
$0.001 + 2\epsilon$ apart) and taking $\mathcal{P}$ to be the union of the hexagons H in the tessellation
within the square. This gives us an area arbitrarily close to

$$\left(\frac{\sqrt{3}/2}{1+\sqrt{3}/2}\right)^2 = 21 - 12\sqrt{3} > 0.215.$$

So the optimal set $\mathcal{P}$ has area between 0.215 and 0.287.

Source *Kvant (Quantum)*, proposed by G.V. Rozenblium.

52 Let us show that there is a monochromatic set (i.e., a set colored by one color) that is mapped to itself by some reflection over a point. Arguing by contradiction, assume there is no such infinite set. Then every reflection over a point changes the colors of all but finitely points (maybe none); call these points "exceptional."

Now consider two arbitrary reflections, σ_1 and σ_2, over the points O_1 and O_2, respectively. Consider two lines ℓ_1 and ℓ_2 parallel to O_1O_2 such that $\sigma_1(\ell_1) = \sigma_2(\ell_1) = \ell_2$ and such that all the exceptional points of σ_1 and σ_2 are between ℓ_1 and ℓ_2, as shown in Fig. 9.63.

The composition $\sigma_1 \circ \sigma_2$ is a translation by $2\overrightarrow{O_1O_2}$, and it transforms every point in the exterior of the band bounded by ℓ_1 and ℓ_2 into a point of the same color. If X is such a point, then all points Y with $\overrightarrow{XY} = 2k\overrightarrow{O_1O_2}$, with $k \in \mathbb{Z}$, are of the same color, because they are obtained by reflecting X and even number of times. But then the set of Y's is infinite and monochromatic and has a center of symmetry (which can be any of the Y's). Hence, our assumption was false, and so there exists an infinite, monochromatic set with a center of symmetry.

Remark Can you adapt this argument to show that in the same conditions there is a monochromatic set having an axis of symmetry?

Source Moscow Mathematical Olympiad, 1996-1997, proposed by V. Protasov.

53 Let the disks be D_i, $1 \le i \le n$, and let D be their union. We argue on Fig. 9.64.

Construct first a tessellation of the plane by congruent regular hexagons H_j, $j \ge 1$, chosen so that the distance between the centers of two neighboring hexagons is slightly greater than 4, say $4 + 2\epsilon$. Let τ_j be the translation mapping H_1 to H_j, $j \ge 1$.

Fig. 9.63 The region between ℓ_1 and ℓ_2 containing all exceptional points

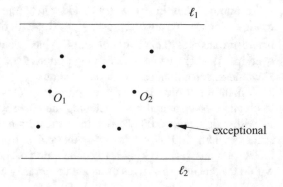

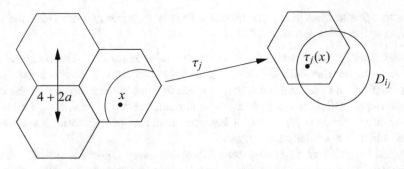

Fig. 9.64 How to pick the disks using the tessellation of the plane by hexagons

Given a point x in the plane, if its translates $\tau_j(x)$ and $\tau_k(x)$ belong to disks D_{i_j} and D_{i_k} from the family, then D_{i_j} and D_{i_k} must be disjoint due to the fact that the distance between $\tau_j(x)$ and $\tau_k(x)$ is greater than 4, while the disks D_j have diameters equal to 2. Thus, for a point x, we could choose all disks to which the translates of x belong, and these disks would then be pairwise disjoint. Our aim is to pick an x that maximizes the number of such disks, and hence their total area.

To this end, let $M_j = D \cap H_j$ and let $F_j = \tau_j^{-1}(M_j)$. The F_j's, $j \geq 1$, lie all inside H_1 and so there should be a lot of overlap (unless there are very few disks in the family, in which case the problem is trivial). Let $n(x)$ be the number of sets F_j that cover point x and let $n = \max_x n(x)$. Then

$$S = \sum_j \text{Area}[M_j] = \sum_j \text{Area}[F_j] \leq n\text{Area}(H) = 2n\sqrt{3}(2 + \epsilon)^2.$$

It follows that

$$n \geq \frac{S}{2\sqrt{3}(2 + \epsilon)^2}.$$

If we choose an x that is covered by n sets F_j, then the disks that contain translates of x have total area equal to $n\pi$, and the above inequality implies

$$n\pi \geq \frac{S\pi}{2\sqrt{3}(2 + a)^2}.$$

All we need to do is choose a sufficiently small so that $\pi/2\sqrt{3}(2 + a)^2$ is greater than 2/9. This is possible because, when $a = 0$,

$$\frac{\pi}{8\sqrt{3}} > \frac{2}{9}.$$

Note that this last inequality can be proved easily by squaring both sides and then observing that $\pi^2 > 9.6 > 256/27$.

Remark The key idea is to look at the problem "modulo H_1." The points P and Q are "congruent modulo H_1" when $Q = \tau_i(P)$ for some i. In this sense, since disks that contain different translates of x are disjoint, we transform a problem of nonoverlapping disks into a problem of maximum overlap of translates; we reduce the problem "modulo H_1." Now it becomes a matter of computing overlapping areas, which is essentially averaging.

A related problem is *Minkowski's Theorem*: every convex set in $\mathbb{R}^n$ that is symmetric with respect to the origin and has volume greater than 2^n contains a point of integer coordinates other than the origin. Try to prove it!

Source Short list of the International Mathematical Olympiad, 1981, proposed by Yugoslavia.

54 Assume that the sum of the lengths of the chords is greater than or equal to $k\pi$. Then the sum of the lengths of the arcs subtended by these chords is greater than $k\pi$. The key idea is to add to the picture the reflections of these arcs over the center of the circle. The sum of the lengths of all arcs is now greater than $2k\pi$, and so there exists a point covered by at least $k + 1$ arcs. The diameter through that point intersects at least $k + 1$ chords, contradicting our assumption. Hence the conclusion.

Remark Adding the reflections of the chords over the center of the circle was essential for being able to apply the Pigeonhole Principle. Note that a diameter intersects an arc if and only if it intersects the reflection of the arc over the center of the circle (see Fig. 9.65).

Source *Kvant (Quantum)*, proposed by A.T. Kolotov.

55 Call the people who are not enemies friends. Choose an arbitrary assignment of places in which two neighbors, A and B, are enemies. Assume that A sits to the right of B. A has at least n friends; choose n of them: $A_1, A_2, \ldots, A_n$. Each of $A_1, A_2, \ldots, A_n$ has one person sitting to its left; denote these people by $B_1, B_2, \ldots, B_n$ (some B_k's can overlap with A_k's). A certain B_k is necessarily a friend of B (because B has at most $n - 1$ enemies).

Fig. 9.65 A diameter intersects an arc if and only if it intersects its reflection

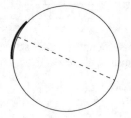 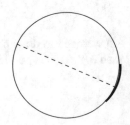

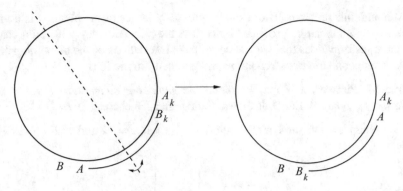

Fig. 9.66 The reflection of the arc $A\widehat{B}_k$

In this case, take the arc $A\widehat{B}_k$ of the round table that lies inside the arc $B\widehat{A}_k$, and reflect it, so that the order of people lying in that arc is reversed (see Fig. 9.66). Now B_k sits next to B and A sits next to A_k, so we have reduced the number of enemy pairs by 1. Because there are only finitely many enemy pairs, this move cannot be repeated forever. So after repeating this move sufficiently many times, we arrive at an arrangement where all neighbors are friends.

56 Consider the isometries of the cylinder that are the compositions of a translation in the direction of the axis of the cylinder and a rotation of the cylinder around its axis. Imagine instead that the figure is fixed and the points move on the cylinder, all rigidly linked to each other. Let P be one of the n points; when another point traces S, P itself traces a figure congruent to S. So after all points traced S, P alone traced a surface F of area strictly less than n.

On the other hand, if we rotate P around the cylinder or translate it back and forth by $\frac{n}{4\pi r}$, we trace a surface of area exactly equal to n. Choose on this surface a point P' that does not lie in F, and consider the isometry that maps P to P'. The fact that P' is not in F means that at this moment none of the points lies in S. This transformation, therefore, satisfies the required condition.

Remark All the tracing described in the first paragraph describes the *forbidden positions* of S, using P as a reference; in fact, by tracing S with each point, we are doing exactly what we do *not* want S to do. This is useful since it greatly organizes and simplifies the reasoning; after all, we are trying to find a possible position for S, and using one of the points can be useful.

Since we only want to prove that the transformation is possible, we do not actually need to find it explicitly. So we resort to areas, again using P as a reference. Because the surface we obtain has area greater than the forbidden positions, we find a suitable position for S. This is very similar to applying the Pigeonhole Principle, with area; notice that this principle is useful to prove that something exists, but not to find an explicit example.

Although this problem is not exactly about isometries of the plane, it is almost, since the cylinder can be produced by rolling the plane the way you roll a carpet, and the transformations that we use come from translations of the plane. Again we think about geometric transformations as "geometry in motion."

Source M. Pimsner, S. Popa, *Probleme de geometrie elementară (Problems in elementary geometry)*, Ed. Didactică şi Pedagogică, Bucharest, 1979.

57 We claim that the number t_n is equal to $k^2 - k$ if $n = 2k$ and k^2 if $n = 2k + 1$. In fact

$$t_n = t_{n-1} + \left\lfloor \frac{n-1}{2} \right\rfloor .$$

Let us first check that

$$t_n \geq t_{n-1} + \left\lfloor \frac{n-1}{2} \right\rfloor .$$

To this end, place the guests at the vertices of a regular n-gon $A_1 A_2 \ldots A_n$. The final configuration is obtained from the initial configuration by an orientation-reversing isometry; from Theorem 1.10, we deduce that it is the reflection of the initial configuration over some axis ℓ. Without loss of generality, we may assume that A_n is one of the vertices at greatest distance from ℓ, and let A_j be its image under the reflection over ℓ. Between A_n and A_j, there are $k = \lfloor \frac{n-1}{2} \rfloor$ sides of the polygon, so to bring A_n to the location of A_j, at least k seat exchanges are required, and these all happen between the guest A_n and some other guest. But the guests $A_1, A_2, \ldots, A_{n-1}$ need themselves t_{n-1} swaps to reverse their own order. This proves the inequality.

To show that equality is actually achieved, start by exchanging the locations of A_n and A_{n-k} by swapping successively A_n with $A_{n-1}, A_{n-2}, \ldots, A_{n-k}$ and then swapping successively A_{n-k} with $A_{n-1}, A_{n-2}, \ldots, A_{n-k+1}$. Using the same reflection axis but ignoring A_n and A_{n-k}, perform the same moves on the remaining polygon (with vertices labeled in the same order). Notice that the moves do not require crossing the locations of A_n and A_{n-k}. Repeat inductively until all vertices are in the desired position.

An easy induction shows that

$$\sum_{j=1}^{n} \left\lfloor \frac{j-1}{2} \right\rfloor$$

is equal to $k^2 - k$ if $n = 2k$ and to k^2 if $n = 2k + 1$.

Remark Another way to prove the inequality is to observe that in order for A_n and A_j to switch places, they need at least $k + (k - 1)$ jumps, while the other guests

Fig. 9.67 Partitioning of the triangle S_i

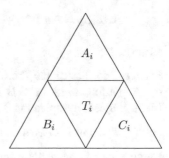

must realize the same reflection but this time with fewer guests. Thus, we obtain an inequality of the form $t_n \geq t_{n-2} + 2k - 1$, which yields the same conclusion.

Source *Kvant (Quantum)*, proposed by V.B. Alexeev.

58 Partition each triangle S_i into four congruent triangular regions, A_i, B_i, C_i, and T_i, $i = 1, 2, \ldots, n$, arranged as in Fig. 9.67. Define $A = (A_1 \cup A_2 \cup \ldots \cup A_n) \setminus T$, and define B and C similarly. Notice that $S = A \cup B \cup C \cup T$. We will prove that $\text{area}(A) \leq \text{area}(T)$.

Let ℓ_i be the line that contains the common segment of A_i and T_i. We can suppose, without loss of generality, that the horizontal lines ℓ_i are ordered from top to bottom, dividing the plane into $n + 1$ strips.

Partition A into sets $A(1)$, $A(2)$, $\ldots$, $A(n)$ according to the strip (or A_i) they are in. More precisely, $A(1) = A_1$ and $A(i)$ is the set of points from A_i that are not in any of the triangles A_1, A_2, $\ldots$, A_{i-1}.

Let $T(i)$ be the image of $A(i)$ under the reflection over ℓ_i. It is clear that $T(i) \subseteq T_i$ and that $\text{Area}[T(i)] = \text{Area}[A(i)]$. We now prove that $T(i)$ and $T(j)$ are disjoint whenever $i \neq j$. Suppose without loss of generality that $i < j$ and, for the sake of contradiction, suppose that there is a common point X of $T(i)$ and $T(j)$. Let X_i and X_j be the reflections of X over ℓ_i and ℓ_j, respectively. Notice that X, X_i, and X_j are collinear, because they belong to a line perpendicular to all lines ℓ_k. From the definitions of $T(i)$ and $T(j)$, we have that $X_i \in A(i)$ and $X_j \in A(j)$. Since $i < j$, X_j is not in A_i.

Also, X is below ℓ_j and so X_j lies between X and X_i. Notice that A_i and T_i cover the segment XX_i (since $X \in T_i$ and $X_i \in A_i$), so X_j is in T_i. But $X_j \in A(j) \subseteq A$, and so, by its definition, X_j cannot be in T_i, a contradiction.

Therefore, $T(1)$, $T(2)$, $\ldots$, $T(n)$ are disjoint, and hence

$$\text{Area}[A] = \text{Area}[A(1)] + \text{Area}[A(2)] + \cdots + \text{Area}[A(n)]$$

$$= \text{Area}[T(1)] + \text{Area}[T(2)] + \cdots + \text{Area}[T(n)]$$

$$= \text{Area}[T(1) \cup T(2) \cup \ldots \cup T(n)] \leq \text{Area}[T].$$

One proves similarly that $\text{Area}[B] \leq \text{Area}[T]$ and $\text{Area}[C] \leq \text{Area}[T]$. We obtain

$$\text{Area}[S] = \text{Area}[T] + \text{Area}[A] + \text{Area}[B] + \text{Area}[C] \leq 4\text{Area}[T],$$

as desired.

Source Shortlist of the Iberoamerican Mathematical Olympiad, 2010, proposed by Argentina, used in a Brazilian Team Selection Test for the International Mathematical Olympiad in 2011.

59 (a) The proof is based on the following result:

Lemma *Let Γ be a polygonal line in the plane with endpoints A and B, and consider a number $\alpha \in (0, 1)$. Then among all segments with endpoints in Γ that are parallel to AB, there is either one of length αAB or one of length $(1 - \alpha)AB$.*

Proof (a) Let us choose the x-axis to be parallel to AB and let us assume that $AB = 1$. For $\beta > 0$ we denote by τ_β the translation of the plane to the right by β units. The problem reduces to showing that Γ cannot be disjoint from both $\tau_\alpha \Gamma$ and $\tau_{1-\alpha} \Gamma$. Rephrasing, we have to show that $\tau_\alpha \Gamma$ cannot be disjoint from both Γ and $\tau_1 \Gamma$. Suppose it were. We argue with the aid of Fig. 9.68.

Choose $p, q \in \tau_\alpha \Gamma$ having respectively maximal and minimal y-coordinates. Let L^+ be a vertical ray extending upward from p and let L^- be a vertical ray extending downward from q. Let L be the (infinite) polygonal line consisting of L^+, L^- and the polygonal line that is part of $\tau_\alpha \Gamma$ and runs between p and q, i.e.,

$$L = L^+ \cup (pq) \cup L^-.$$

Then L separates the plane into two regions. Moreover, Γ lies to the left of L, while $\tau_1 \Gamma$ lies to the right. But $\Gamma \cap \tau_1 \Gamma \neq \emptyset$, since B is obviously in this intersection. We reached the contradiction that shows that our assumption was false. This proves the lemma.   □

Fig. 9.68 Proof of the Chord
Theorem

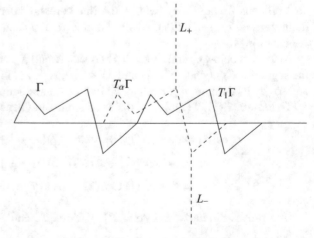

Returning to the original problem, we conclude that there is a parallel segment of length $\frac{1}{n}AB$ or one of length $\frac{n-1}{n}AB$.

If the second case holds true, then by repeating the argument for the segment $\frac{n-1}{n}AB$ and the number $\frac{1}{n-1}$, we conclude that there is either a segment of length

$$\frac{1}{n-1} \cdot \frac{n-1}{n}AB = \frac{1}{n}AB,$$

or one of length

$$\frac{(n-1)-1}{n-1} \cdot \frac{n-1}{n}AB = \frac{n-2}{n}AB.$$

If we continue the argument, we eventually reach a segment equal to $\frac{1}{n}AB$ and parallel to AB.

(b) Let us show that for every $\alpha \in (0, 1)$ not of the form $\frac{1}{n}$ with n an integer, there is a polygonal line Γ joining points $A \neq B$ such that there is no segment parallel to AB, with endpoints on Γ, and equal to αAB.

Choose n such that $\alpha \in (\frac{1}{n+1}, \frac{1}{n})$. Consider the segment AB, and consider points $P_0 = A, P_1, P_2, \ldots, P_n = B$ on AB such that for all $j = 0, 1, \ldots, n-1$, the segment $P_j P_{j+1}$ has length $\frac{1}{n}AB$ and points $Q_0 = A, Q_1, \ldots, Q_n = B$ such that for all $j = 0, 1, \ldots, n$, the segment $Q_j Q_{j+1}$ has length $\frac{1}{n+1}$. Now draw a family of parallel lines $\ell_0, \ell_1, \ldots, \ell_n$ through $P_0, P_1 \ldots, P_n$, respectively, and another family of parallel lines $L_1, L_2, \ldots, L_n$ through $Q_1, Q_2, \ldots, Q_{n-1}$, respectively, so that the two families are not parallel to each other. Let C_j be the intersection of ℓ_{j-1} and L_j, and let D_j be the intersection of ℓ_j and L_j, $j = 1, 2, \ldots, n$. Then $AC_1 D_1 C_2 D_2 \ldots D_n B$ is a polygonal line with the desired property (prove it!). The example for $n = 2$ is shown in Fig. 9.69.

Remark The easiest way to prove facts about parallel segments is usually by performing translations of those segments, and that is exactly what we do here. It is natural to try to relate the translation by αAB and its "complementary" translation by $(1 - \alpha)AB$; this is the subject of the lemma. Also, the key idea in the lemma is shifting Γ αAB units to the left and $(1 - \alpha)AB$ units to the right, so we force a translation by AB, which naturally takes A to B.

Finally, the proof of (a) is based on the lemma as well as on the identity

$$\frac{1}{n} = \frac{1}{2} \cdot \frac{2}{3} \cdot \frac{3}{4} \cdots \frac{n-2}{n-1} \cdot \frac{n-1}{n}.$$

Fig. 9.69 Counterexample for the second part of the Chord Theorem

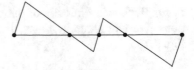

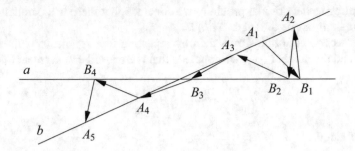

Fig. 9.70 A possible trajectory of the flea

Fig. 9.71 The vectors of the
jumps determine the
trajectory

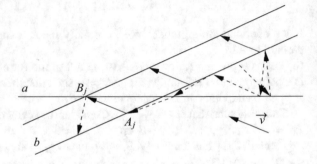

Source This result is well-known (see, e.g., D. Rolfsen, *Knots and Links*, AMS
Chelsea Publ. 2003) and is true for any curve, not just polygonal lines.

60 Figure 9.70 illustrates a possible trajectory of the flea.

Looking at the sequence of vectors that describe the jumps, $\overrightarrow{A_1 B_1}$, $\overrightarrow{B_1 A_2}$, $\overrightarrow{A_2 B_2}$,
..., we realize that by knowing this sequence, we can recover the path of the flea.
For if we know the vector $\overrightarrow{v} = \overrightarrow{A_j B_j}$, then the point B_j can only be the intersection
of the line b with the translate of the line a by $\overrightarrow{v}$, and the point A_j is the preimage
of B_j under this translation (see Fig. 9.71).

Thus, for the purpose of bookkeeping, we can translate all the vectors to originate
at some point O, and we let these translates be, in order, $\overrightarrow{OC_1}$, $\overrightarrow{OD_1}$, $\overrightarrow{OC_2}$, $\overrightarrow{OD_2}$,
... This procedure is shown in Fig. 9.72. We can say something about the positions
of these vectors.

Lemma *Assume that the line a transforms into the line b by a counterclockwise
rotation of angle α. Then, for all j, the angle between $\overrightarrow{OC_j}$ and $\overrightarrow{OC_{j+1}}$ and the
angle between $\overrightarrow{OD_j}$ and $\overrightarrow{OD_{j+1}}$ are both equal to -2α.*

Proof The main difficulty of the proof is that there are many possible configurations. We will argue in a way that covers all cases simultaneously, reasoning on
Fig. 9.73.

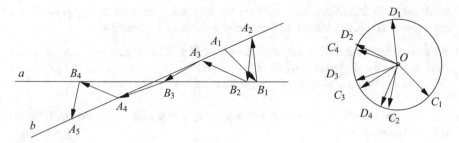

Fig. 9.72 The construction of the vectors $OC_j, OD_j, j \geq 1$

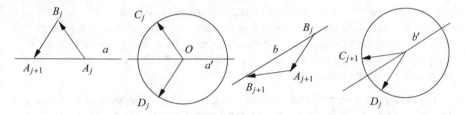

Fig. 9.73 Proof of the lemma

Let a' and b' be the translates of a and b that pass through O. Then because the triangle $B_j A_j A_{j+1}$ is isosceles, $\overrightarrow{OC_j}$ and $\overrightarrow{OD_j}$ are symmetric with respect to the line a'. Similarly, $\overrightarrow{OD_j}$ and $\overrightarrow{OC_{j+1}}$ are symmetric with respect to the line b'. Hence, C_j is mapped to C_{j+1} by the composition of two reflections over lines. This composition is a *clockwise* rotation by an angle that is twice the angle between the two lines. The lemma is proved. □

Returning to the problem, the path of the flea is periodic if and only if the sequences $C_1, C_2, \ldots, C_n, \ldots$ and $D_1, D_2, \ldots, D_n, \ldots$ are both periodic. By the lemma, this happens if and only if there is n such that $2n\alpha$ is a multiple of $360°$. The problem is solved.

Remark The construction that associates to the trajectory of the flea the points $C_1, D_1, C_2, D_2, \ldots$ on the circle is a particular example of the spherical image of a curve. In general, if $t \mapsto \gamma(t)$ is a smooth curve, its spherical image is obtained by normalizing all velocity vectors $\gamma'(t)$ to have length 1 (i.e., by considering the vectors $\gamma'(t)/\|\gamma'(t)\|$), and then mapping these vectors to have the same origin. The result is a curve on the unit sphere. The usefulness of the construction comes from the fact that the length of the spherical image equals the total curvature of the curve.

Source *Kvant (Quantum)*, proposed by Nikolai Borisovich Vassiliev.

61 The second player has a strategy. We call an *almost-circuit* a polygonal line that is missing one segment in order to become closed. Here is the strategy of the second player: choose one diagonal, and then for whatever segment the first player marks, mark the reflection of that segment over the diagonal. Continue until you notice the presence of an almost-circuit, and then close up the polygonal line.

Remark An example of a game is shown in Fig. 9.74, where the second player reflects over the diagonal marked at the beginning. On an $n \times n$ table with n odd, the second player can also use the reflection with respect to the center.

Source This is the first half of a problem published in *Kvant (Quantum)* by I. Vetrov and A. Kogan

62 We can assume that p and q are coprime, otherwise shrink the size of the chessboard by the greatest common divisor of p and q as the knight is confined to a lattice of square-size equal to this greatest common divisor.

Consider the Klein 4-group, which is the group of symmetries of a (nonsquare) rectangle discussed in Sect. 1.1.5, with e the identity, a and b the reflections over the perpendicular bisectors of the sides, and c the reflection over the center of the rectangle. Color the chessboard as in Fig. 9.75.

If p and q are both odd, then at each jump, the color of the location of the knight is multiplied by c. Thus, after n jumps, the knight is on a square colored by c^n. The initial square was colored by e, and the equality $c^n = e$ is only possible if n is even.

If one of p and q is even and the other is odd, then at each jump, the color of the square is multiplied by either a or b. After n jumps, the color will be $a^k b^{n-k}$, for some k. The equality $a^k b^{n-k} = e$ implies $a^k = b^{n-k}$ so both k and $n-k$ have to be even. So n itself has to be even, and we are done.

Remark This problem shows how to distinguish configurations using *invariants*. We distinguish squares where the knight can land from those where it cannot by their "color" (which "color" is an element of the Klein 4-group).

Source Bundeswettbewerbe Mathematik.

63 In both cases, the first player has a winning strategy, and it works on an $m \times n$ chocolate bar with $m = 2k$, $k \in \mathbb{N}$.

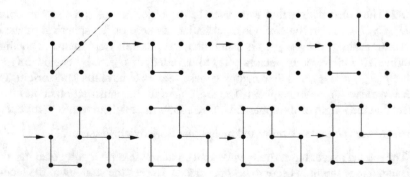

Fig. 9.74 An example of a game

Fig. 9.75 Coloring of the chessboard by elements of the Klein 4-group.

c	b	c	b	c	b
a	e	a	e	a	e
c	b	c	b	c	b
a	e	a	e	a	e
c	b	c	b	c	b
a	e	a	e	a	e

Fig. 9.76 In each of the two figures, the shaded rectangles have the same number of a's, b's, c's

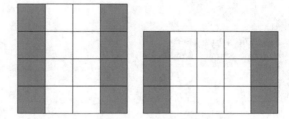

(a) The winning strategy for the first player is to first cut the bar into two equal parts of size $k \times n$, and then play *symmetrically*.

(b) The first player can still play symmetrically until the second produces a $1 \times \ell$ bar, $\ell > 1$, and then cut out a 1×1 square.

Remark Note the similarity with Problem 61. The "symmetrization strategy" is commonly used in games.

Source *Kvant (Quantum)*, proposed by S.V. Fomin.

64 We assume that such a coloring exists, and try to understand its structure. Working in some region away from the boundary, we can reason under the assumption that the board is infinite. We start with two key observations, illustrated in Fig. 9.76:

(1) If two 1×4 rectangles coincide by a horizontal or vertical translation by 3 units in the direction of their shorter side, then they must contain an equal number of a's, b's, and c's.

(2) If two 1×3 rectangles coincide by a horizontal or vertical translation by 4 units in the direction of their shorter side, then they must contain an equal number of a's, b's, and c's.

    This is because in both cases, the rectangles can be added to the same rectangle to create a 3×4 rectangle. The solution consists of a repeated application of these "translations" in order to determine the possible locations of the a, b, c's. We state, then prove, a series of claims.

Claim 1 In every 1×3 rectangle, there is at most one a.

    For simplicity, we argue inside the rectangle $[0, 4] \times [0, 5]$ shown in Fig. 9.77, with $[0, 3] \times [0, 1]$ being the rectangle in question (by flipping the figure, we can adapt this argument to vertical rectangles as well as those close to the upper side

Fig. 9.77 Why a 1×3
rectangle contains at most
one a

Fig. 9.78 Why a 1×4
rectangle contains at most
one a

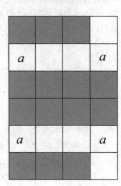

of the 100×100 square). If the 1×3 rectangle has two or more a's, then after we complete the rectangle to $[0, 4] \times [0, 1]$ and use the first observation, we deduce that the rectangle $[0, 4] \times [3, 4]$ contains at least two a's. By using the second observation, we deduce that the rectangle $[0, 3] \times [4, 5]$ also contains at least two a's, and so the rectangle $[0, 4] \times [3, 5]$ contains at least 4 a's, impossible.

Claim 2 In every 1×4 rectangle, there is at most one a.

To prove this, we argue on the rectangle $[0, 4] \times [0, 6]$ shown in Fig. 9.78. If in the rectangle $[0, 4] \times [1, 2]$ there are more than one a's, then there are exactly two, by Claim 1 applied to $[0, 3] \times [1, 2]$ and $[1, 4] \times [1, 2]$, and they must be in the squares $[0, 1] \times [1, 2]$ and $[3, 4] \times [1, 2]$. Then the rectangle $[0, 4] \times [4, 5]$ has the same property with the a's in the corresponding locations. But then the rectangle $[0, 4] \times [2, 4]$ contains exactly one a, so the rectangles $[0, 3] \times [0, 1]$ and $[0, 3] \times [5, 6]$ have an a each. But then the same is true for their translates by 3 units $[0, 3] \times [3, 4]$ and $[0, 3] \times [2, 3]$. This forces $[0, 4] \times [1, 4]$ to have four a's.

Claim 3 In every 1×4 rectangle, there is exactly one a.

We have seen that there is at most one a. Arguing on the same $[0, 4] \times [0, 6]$ rectangle, we see that if $[0, 4] \times [1, 2]$ and $[0, 4] \times [4, 5]$ have no a's, then $[0, 4] \times [2, 3]$ and $[0, 4] \times [3, 4]$ have together three a's, which is impossible, since each has at most one a.

Fig. 9.79 Why a 1×4 rectangle contains at most two b's

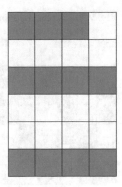

Claim 4 In every 1×4 rectangle, there are at most two b's.

In the rectangle $[0, 4] \times [0, 6]$ (Fig. 9.79), if $[0, 4] \times [0, 1]$ contains three b's, then so does $[0, 4] \times [3, 4]$. The second "translation" property forces the rectangle $[0, 3] \times [5, 6]$ to have at least two b's, so $[0, 4] \times [3, 6]$ has at least 5 b's, which is not allowed.

Claim 5 In every 1×4 rectangle, there are at most two c's.

We argue on $[0, 4] \times [0, 5]$. If in $[0, 4] \times [0, 1]$ there are three or more c's, then this is also true for $[0, 4] \times [3, 4]$. Moreover, there are at least two c's in $[0, 3] \times [4, 5]$. In $[0, 4] \times [2, 3]$, there is at least one c because there is at most one a and two b's. So the rectangle $[0, 4] \times [1, 4]$ has at least six c's, a contradiction.

Claim 6 In every 1×3 rectangle, there is at most one b.

Indeed, if in the rectangle $[0, 3] \times [0, 1]$ there are two or more b's, then so are in the rectangles $[0, 4] \times [3, 4]$ and $[0, 3] \times [4, 5]$. But then the rectangle $[2, 5] \times [0, 4]$ has five b's, because by Claims 3 and 5, the rectangle $[0, 4] \times [2, 3]$ has at most two c's and one a, so it must contain a b as well.

Claim 7 In every 1×3 rectangle, there is exactly one b.

If the rectangle $[0, 3] \times [0, 1]$ has no b, then in each of $[0, 3] \times [k, k + 1]$, $k = 1, 2, 3$ there is at most one b by Claim 6, so $[0, 3] \times [0, 4]$ has at most three b's, which is not allowed.

With these observations at hand, let us analyze a possible configuration. There must be many squares containing an a, so let us start with one of these somewhere in the middle of the table. Either to the left or to the right of it, there is a b, because every 1×3 rectangle contains exactly one b. Let this be to the right (Fig. 9.80). Then the next two on that row must be c's, so we have the configuration $abcc$. Immediately after the second c should follow an a by Claim 3 and at the same time a b by Claim 7. This is impossible! Hence, there is no way we can produce the desired coloring.

Remark The condition from the statement allows us to see patterns, structures that repeat periodically. The translation invariant pattern "equally spaced 1×3 and $1 \times$

Fig. 9.80 Explanation for
why such a configuration
does not exist

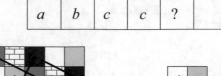

a	b	c	c	?	

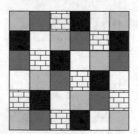 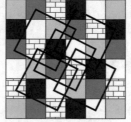 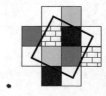

Fig. 9.81 Coloring by five colors

4 rectangles that contain an equal number of a, b, c's" completely identifies the
configuration, but leads to contradictory constraints.

Source *Kvant (Quantum)*, proposed by V.E. Lapitski, solution by Yu.I. Yonin.

65 Let us consider a coloring as described in (a), such as the one with five colors
depicted on the left of Fig. 9.81.

Each color forms a square lattice, and the square lattices can be translated to
overlap with one another. Reasoning on Fig. 9.81, we take the two orthogonal
vectors, $\vec{v}$ and $\vec{w}$, that determine any of these lattices, and define an equivalence
relation on the points of the plane, declaring two points to be equivalent if one is
translated into the other by some vector of the form $r\vec{v} + s\vec{w}$, with r and s integers.
The equivalence relation identifies points of the same color, and all squares of the
same color are identified with one another. Moreover, the equivalence classes are
parametrized by the points in one square S formed by $\vec{v}$ and $\vec{w}$, meaning that every
point in the plane is identified with one and only one point in S. Well, not quite, the
points on opposite sides of this square are identified under the translations by $\vec{v}$ and
$\vec{w}$. But if we glue the opposite sides, so as to obtain a torus[1] (Fig. 9.82), then the
points of the torus are in one-to-one correspondence with the equivalence classes.
So the torus parametrizes the equivalence classes.

This torus contains precisely one unit square for each color, so its area is n, the
number of colors. Consequently the side-length of the square S is $\sqrt{n}$. Examining
the original lattice, we see that the side of S can be placed in a right triangle with
integer sides, by centering the vertices of S at the centers of the colored squares. In
this configuration, let us say that there are m units on the vertical and k units on the
horizontal. By the Pythagorean Theorem,

$$n = m^2 + k^2.$$

---

[1] A torus is the surface that bounds a donut.

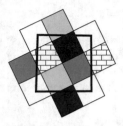

Fig. 9.82 The torus that parametrizes the equivalence classes

Fig. 9.83 Tessellation of the plane by hexagons of three and four colors

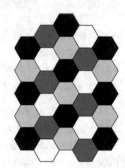

This gives a necessary condition for n: to be the sum of two perfect squares.

Conversely, let us show that every positive integer n of the form $m^2 + k^2$ has an associated coloring. Color one unit square by the first color, and then move m units to the right and k units up. Color this square by the same color. Then complete a square lattice containing these two squares. Next, choose an uncolored unit square, color it by the second color, and repeat the above construction. Repeating with each of $n = m^2 + k^2$ colors, we obtain the desired coloring.

Part (b) is similar to (a) in the sense that it uses translations parallel to the sides of the equilateral triangular lattice. In this case, the role of the square S is played by a rhombus R with a $60°$ angle, which again parametrizes the equivalence relations, namely, each point in the plane is parametrized by one and only one point on this rhombus, provided that again we identify the opposite sides, to obtain again a torus (you have to stretch the torus, and your imagination, to see this happening). Let the distance between the center of two neighboring hexagons be 1, so that the area of one hexagon is $\sqrt{3}/2$. The torus contains n hexagons; thus, its area is $n\sqrt{3}/2$. If the side of the rhombus is a, its area is $a^2\sqrt{3}/2$; thus, $a^2 = n$.

Examining the configurations from Fig. 9.83, we notice that to move one hexagon of the tessellation to another, you have to translate it some m "steps" vertically and k "steps" up to the left or right at an angle of $60°$ with the horizontal, with m, k nonnegative integers. The length of the total translation vector, which is a, is, therefore, the third side in a triangle whose other two sides have lengths m and k and make an angle of $120°$. Using the Law of Cosines, we obtain that

$$n = a^2 = m^2 + mk + k^2.$$

This is a necessary condition for the coloring to exist, but it is also sufficient, because this reasoning also discloses the algorithm for the coloring: color one hexagon by some color, go m steps vertically and k steps to the left and up by 60°, and then color the hexagon where you land by the same color. Complete the lattice, and then repeat for every other color.

Remark What if we allow lattices of different sizes? And if we do not impose that two lattices have parallel sides?

Source *Kvant (Quantum)*, proposed by Andrey Nikolaevich Kolmogorov.

66 The situation is described in Fig. 9.84. We will label the trees counterclockwise by the numbers from 0 to $n - 1$ (these are the residue classes modulo n). Let the trees be equally spaced around the unit circle in the complex plane, so that the 0th tree is at 1. Then the kth tree is at the kth root of unity: $e^{\frac{2\pi i k}{n}}$ (Fig. 9.85).

Denote by ρ the counterclockwise rotation by $\frac{2\pi i}{n}$. The flight of a bird from the kth tree to the mth tree corresponds to a rotation by $\frac{2\pi i (m-k)}{n}$; it is realized by ρ^{m-k}.

Assume that at the beginning all birds are on the tree number 0. To bring one bird on the tree number k, we must act upon it by ρ^k. Now for every configuration of birds, let us consider the composition, in the group of rotations about the origin, of the rotations that have brought these birds from the tree number 0 to this configuration. For example, for the initial configuration from our problem, this product is equal to

$$\rho^0 \circ \rho^1 \circ \rho^2 \circ \cdots \circ \rho^{n-1} = \rho^{\frac{n(n-1)}{2}}.$$

In general, this product is

Fig. 9.84 Birds on trees

Fig. 9.85 Modeling the flight
of birds using rotations

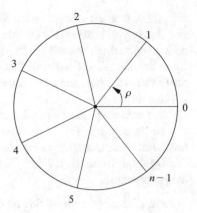

$$\left(\rho^0\right)^{N_0} \circ \left(\rho^1\right)^{N_1} \circ \cdots \circ \left(\rho^{n-1}\right)^{N_{n-1}} = \rho^{\sum\limits_{k=0}^{n-1} k N_k},$$

where N_k is the number of birds on kth tree.

Note that the product of a counterclockwise rotation and a clockwise rotation by $\frac{2\pi i}{n}$ is just $\rho \circ \rho^{-1}$, which is the identity map. So the move described in the statement, where two birds change their location, the first by an action of ρ and the second by ρ^{-1}, does not change the product. So this product is an *invariant*, which does not change under the move.

When the birds accumulate on the kth tree, the product is

$$\rho^{kn} = \left(\rho^n\right)^k = 1^k = 1,$$

where 1 denotes the identity map. So the birds can accumulate on a single tree only if

$$\rho^{\frac{n(n-1)}{2}} = 1.$$

This is equivalent to $\frac{n(n-1)}{2} \equiv 0(\text{mod } n)$, which happens if and only if n is odd. Thus, the birds cannot accumulate on the same tree when n is even.

And indeed, when n is odd, that is, $n = 2m + 1$, the birds can accumulate on the 0th tree in the following way: for each $k = 1, 2, \ldots, m$, the birds on the kth and $m - k$th tree fly toward each other until that arrive on the 0th tree.

Remark We can ask a more general question, given two configurations of birds on trees, when can they be transformed into each other by the move specified in the statement? Clearly, for this to happen, the above invariant should have the same value for the two configurations. We will show that this is also a sufficient condition. Let the value of the invariant be ρ^r, with r one of the numbers $0, 1, 2, \ldots, n$. We will show that any configuration having this invariant can be transformed into the one that has all birds on the 0th tree except for one which is on the rth tree. Indeed, pick a bird, which we call the "flyer." Pair each bird with the flyer, and let the two fly in opposite directions until that bird lands on the 0th tree. After all birds are on the 0th tree, the flyer must be on some tree, and given that the invariant has the value ρ^r, this must be the rth tree.

We are in the presence of what is called a *complete invariant*: two configurations can be transformed into each other if and only if they have the same value of the invariant. In mathematics complete invariants are rare, but they are very desirable.

You should note the similarity with the problem about balls in a box discussed in the introduction.

Source *Kvant (Quantum)*, this problem was given at the university entrance exam at the Moscow State University.

67 Translate A to the left by $k = \max A$ (i.e., transform $A \mapsto A - \{k\}$) and B to the left by $\ell = \min B$ ($B \mapsto B - \{\ell\}$) so as to obtain the sets A' and B' with $\max A' = \min B' = 0$. All sums in $A + B$ are translated by $k + \ell$, so $|A' + B'| = |A + B|$, and, of course, $|A| = |A'|$ and $|B| = |B'|$. And we can argue on $A' + B'$ instead, which is significantly easier! This is because all elements of A' and B' are in $A' + B'$, as $0 \in A'$ and $0 \in B'$, and also $A' \cap B' = \{0\}$. We therefore have $|A' + B'| \geq |A' \cup B'| = |A'| + |B'| - 1$, proving the inequality.

For the equality case, let $t \neq 0$ be the element from $A' \cup B'$ of minimal absolute value; suppose without loss of generality that $t \in B$, so that $t > 0$. Notice that because we are in the equality case, $A' + B' = A' \cup B'$. Consider the translation $A' \mapsto A' + \{t\}$. Since $\max(A' \setminus \{0\}) + t < 0$, $A' + \{t\}$ consists of $A' \cup \{t\}$ from which we exclude $\min A'$. Thus, A' consists of the elements of an arithmetic progression with common difference t. Now that we have established that the maximal nonzero element of A' is $-t$. One can also deduce that the elements of B' are in arithmetic progression as well, by noticing that $B' + \{-t\}$ is $B' \cup \{-t\}$ from which we exclude $\max B'$. Reversing the translation, we have the answer $|A + B| = |A| + |B| - 1$ if and only if the elements of each of sets A and B are in arithmetic progressions with same common difference (the arithmetic progressions do not need to overlap, though).

Remark Translating A and B does not alter the cardinality of $A + B$, so we can shift them in a convenient manner, so as to have $A \cap B = \{0\}$.

68 Consider the translates $A' = A + \{-a\}$ and $B' = B + \{b\}$. Then condition (ii) is the same as $A \cup B \subseteq A' \cup B'$. But we have the following inequality for cardinalities: $|A' \cup B'| \leq |A'| + |B'| = |A| + |B|$, and, since A and B are disjoint, $|A| + |B| = |A \cup B|$. Thus,

$$|A \cup B| \leq |A' \cup B'| \leq |A'| + |B'| = |A| + |B| = |A \cup B|.$$

This means that $|A \cup B| = |A' \cup B'|$ and that A' and B' are disjoint. It follows that $A \cup B = A' \cup B'$. So we have four sets A, B, A', B' so that $A \cap B = A' \cap B' = \emptyset$

and $A \cup B = A' \cup B'$. If $s(X)$ is the sum of the elements of X, then summing over $A \cup B$ and $A' \cup B'$, which are equal, we have

$$s(A) + s(B) = s(A') + s(B').$$

This is equivalent to

$$s(A) + s(B) = s(A + \{-a\}) + s(B + \{b\}),$$

and this latter equality is equivalent to

$$s(A) + s(B) = s(A) - a|A| + s(B) + b|B|.$$

This yields $a|A| = b|B|$, and we are done.

Remark The trick of the solution was to compare the elements of A and B and their union to the elements of their translates $A + \{-a\}$ and $B + \{b\}$ and their union.

Source Asian Pacific Mathematical Olympiad, 2013.

69 The number c has the property that no integer $-k$ can be found in the translates by c of any of the intervals $[a_i, b_i]$. The inequality from the hypothesis of the problem can be written as

$$\sum_{i=1}^{n} (b_i - a_i) < 1,$$

meaning that the sum of the lengths of the intervals $[a_i, b_i]$ is less than 1.

So the problem asks us to show that the union S of finitely many nondegenerate intervals with total length less than 1 may be translated so as not to intersect the set $\mathbb{Z}$ of integer numbers. For $k \in \mathbb{Z}$, consider the sets $S_k = S \cap [k, k+1)$ and their translates $S_k - \{k\} \subset [0, 1)$. The total length of the union $\cup_{k \in \mathbb{Z}} (S_k - \{k\})$ is at most equal to the total length of S (due to overlaps, the total length of the first might be smaller). So the length of $\cup_k (S_k - \{k\})$ is less than 1. If c is a number in $[0, 1)$ that does not lie in this union, then $S + c$ does not intersect $\mathbb{Z}$.

Remark The proof generalizes to n dimensions to show that a union of several solid bodies of total volume less than 1 can be translated so that it does not contain any point of integer coordinates.

Source Romanian Team Selection Test for the International Mathematical Olympiad, 2008.

70 We will prove that $k = 2$. Let f be a permutation, which we want to write as the product of two involutions. We know that f can be written as the composition of disjoint cycles, so let

$$f = g_1 \circ g_2 \circ \cdots \circ g_r,$$

where g_j are the disjoint cycles, $1 \leq j \leq r$.

Identify a cycle of length m with the rotation ρ of the regular m-gon $A_0 A_1 \ldots A_{m-1}$ by $\frac{2\pi}{m}$ about its center. Then this rotation is the composition of two reflections: first the reflection over the perpendicular bisector of $A_0 A_1$, followed by the reflection over the perpendicular bisector of $A_0 A_2$. The two reflections are themselves permutations (as they map the polygon to itself), but what is more important is that they are involutions.

Thus, we can write $g_j = \sigma_j \circ \sigma_j'$, where σ_j and σ_j' are involutions. Setting

$$\sigma = \sigma_1 \circ \sigma_2 \circ \cdots \sigma_r, \quad \sigma' = \sigma_1' \circ \sigma_2' \cdots \sigma_r',$$

we have that σ and σ' are involutions (because they are products of involutions on disjoint sets) and $f = \sigma \circ \sigma'$.

Remark This is an application of Theorem 1.8, but it is important to notice that the given rotation can be written as a composition of two reflections in many ways. But we need to choose reflections that map the regular polygon to itself, because only these correspond to permutations.

71 *First solution.* Let us assume that this is possible. Denote by A the set that contains 1. The other sets are obtained from A by translations by positive integers. We group the integers that define translations in a set B, to which set we add the element 0. Then every positive integer can be represented uniquely as $a + b$ with $a \in A$ and $b \in B$.

Conversely if A, B are infinite sets of nonnegative integers such that $1 \in A$ and $0 \in B$ and if every integer is represented uniquely as $a + b$ with $a \in A$ and $b \in B$, then the translates of A by elements of B determine a partition of positive integers into sets congruent to each other.

To construct an example, let us change slightly the problem by working with nonnegative integers instead of positive integers. To this end, we translate to the left by 1 unit, so A must now contain the number 0. We then define A to be the set of nonnegative integers whose binary expansion has 1s only in odd positions, including the number 0, and let B be the set of nonnegative integers whose binary expansion has 1s only in even positions, including 0. The sets A and B have the desired property, so the answer to the question is affirmative.

Second solution. Quite analogous to the above construction, we can start with 1 and then construct an infinite array as follows: at the first step, write 2 to the right of 1, 3 below 1 and 4 to the right of 3. In general at the nth step, translate the $2^{n-1} \times 2^{n-1}$ array obtained so far to the right, down, and diagonally as to obtain a $2^n \times 2^n$ array, and then add 4^n to the elements in the $2^{n-1} \times 2^{n-1}$ block on the left, 2×4^n to the elements in the $2^{n-1} \times 2^{n-1}$ block below, and 3×4^n to the elements in the $2^{n-1} \times 2^{n-1}$ block diagonally opposite to the original block. The second step is shown below:

$$
\begin{array}{cc}
\begin{array}{cc} 1\ 2 \\ 3\ 4 \end{array}
\ \to\
\begin{array}{cccc}
1 & 2 & 5 & 6 \\
3 & 4 & 7 & 8 \\
9 & 10 & 13 & 14 \\
11 & 12 & 15 & 16
\end{array}
\end{array}.
$$

The columns of the array obtained at the end of the process give the required partition. The rows of the array give another partition.

Remark The second solution becomes even more transparent if you write the numbers in binary form.

Source *Kvant (Quantum)*, proposed by A. Fedorov, solutions by A. Fedorov and S. Slosman.

72 We begin the solution by describing a combinatorial device that encodes both the key-string and the resulting string of characters.

Each string of key presses corresponds to a rooted planar tree with letters labeling the non-root vertices. The algorithm is as follows. First, add to the plane a coordinate system so that we can talk about up-down and left-right. Given a key string, start at the root, which will be an unlabeled vertex and serves now as the *current vertex*. When pressing a key α on the keyboard, draw a vertex directly underneath the root, connected to the root by an edge, and then label this vertex by the letter α. This now becomes the current vertex and pressing a key would produce a new vertex underneath this one. Pressing the key $\leftarrow$ moves us up one vertex on the tree, thus moving the current vertex one step up. From this moment on, the pressing of a literal key would produce a vertex below the current vertex, connected to it by an edge and positioned to the right of all other vertices connected by an edge to the current vertex, and this new vertex becomes the current vertex. An illustration of the graph produced by the instances of key pressings

$$
a, \quad ab, \quad ab \leftarrow cd, \quad ab \leftarrow cd \leftarrow e, \quad ab \leftarrow cd \leftarrow e \leftarrow\leftarrow f,
$$

which gives rise to the word *faecdb* is shown in Fig. 9.86.

Given such a tree, one constructs the associated key-string by doing a *left-to-right depth-first search*. This means that you start at the root, go down all the way on the leftmost branch, writing the vertices in the order they are encountered from left to right in the string (in our case ab). Then backtrack until reaching a vertex from which more branches bifurcate downward, where at each upward step, an $\leftarrow$ is written to the right of the string (in our case $ab \leftarrow$). Then travel on the new leftmost branch downward, writing to the letters encountered in the order from left to right (in our case $ab \leftarrow cd$). Continue until all vertices are exhausted. The backtracking can be easily realized by drawing a contour around the tree and then traveling counterclockwise around this contour starting at the root (see Fig. 9.87) while skipping all vertices that were already recorded.

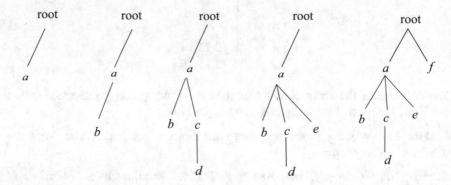

Fig. 9.86 The encoding of a sequence of key strokes

Fig. 9.87 Contour around a
tree

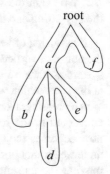

Lemma *The string produced by a key-string is obtained from the corresponding
rooted tree via a right-to-left depth-first search (meaning that we travel on the same
contour but this time counterclockwise).*

Proof We induct on the number of vertices of the tree (i.e., the number of letters in
the key-string). The base case is trivial, so let us prove the induction step.

Let us assume that the property holds true for all smaller trees and let us prove
it for the tree in question. Deleting the root, we obtain a family of sub-trees
$A_1, A_2, \ldots, A_k$ rooted at the "children" of the root (where the numbering is from
left to right). If $k = 1$, then both the string and the key-string start with the letter
at the unique "child" of the root, and we can apply the inductive hypothesis to the
sub-tree A_1.

If $k > 1$, the path that defines both the left-to-right and the right-to-left depth-
first searches decomposes into the paths corresponding to the trees $A_1, A_2, \ldots, A_k$,
and, after surrounding each of the A_j's, the cursor is again at the beginning in the
case of the key-string, or at the end, in the case of the string. So the concatenation of
paths corresponds to the concatenation of both the key-strings and the strings, with
the caveat that at the beginning of each of the subpaths, we have to add the letter
corresponding to its root as well (both to the key-string and to the corresponding
string). This completes the induction, and the lemma is proved.   □

We conclude that to every pair of words (A, B) with B reachable from A, we can associate a tree T such that

- the left-to-right depth-first search starting with the root yields A, and
- the right-to-left depth-first search yields B.

The key element that solves the problem, and which is the reason why this problem is included in a book on geometric transformations, is the reflective symmetry between the left-to-right and right-to-left depth-first searches. So if B is reachable from A and the tree for the pair (A, B) is T, and if we let T' be the reflection of T (over the y-axis), then T' corresponds to the pair (B, A) showing that A is reachable from B. Problem solved.

Remark Depth-first searches are widely used in combinatorics, for example, this is the strategy for solving a maze.

Source United States of America Team Selection Test for the International Mathematical Olympiad, 2014, proposed by Linus Hamilton, solution by contestant James Tao.

73 Look at all possible colorings of the vertices of a regular p-gon by n colors (see Fig. 9.88). This set has n^p elements.

We call two colorings equivalent if they can be obtained from each other by a rotation of the polygon. Except for monochromatic colorings, each coloring is equivalent to exactly $p - 1$ others. It is here where we use the fact that p is prime, because if a rotation maps a coloring into itself, then all rotations map the coloring into itself (this is because any nontrivial rotation of the regular p-gon generates the group of all rotations).

Thus, if a coloring overlaps with itself after a nontrivial rotation, then by repeatedly applying that rotation, we find that coloring coincides with all its rotations and that can only happen if it is monochromatic. Therefore, the colorings that are not monochromatic can be partitioned into equivalence classes of p elements, which implies that their total number, which is $n^p - n$, must be divisible by p.

Fig. 9.88 A coloring of the vertices of the regular heptagon by three colors

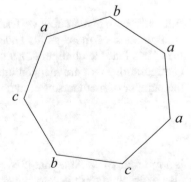

Fig. 9.89 A configuration for
$p = 5$, $a = 12$ and $b = 8$

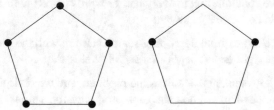

Remark Note the similarity with the proof of Wilson's Theorem. It is important to observe that the proof fails for p not prime, because there exist non-monochromatic colorings that are invariant under some nontrivial rotations (which are they?).

Source This proof to the Fermat's Little Theorem was given by Julius Petersen.

74 We start by writing $a = a_0 + \alpha_1 p$ and $b = b_0 + \beta_1 p$ with $0 \le a_0, b_0 < p$ and α_1, β_1 some nonnegative integers that are not necessarily less than p. Consider the vertices of α_1 regular p-gons and, in addition to that, a_0 points that are not among these vertices. Altogether we have $a_0 + \alpha_1 p$ points, of which we mark $b_0 + \beta_1 p$ points. An example is shown in Fig. 9.89, with the marked points specified by the larger dots.

We allow the α_1 polygons to rotate independently about their centers so that each polygon is mapped into itself, but we keep the other a_0 points fixed. Let us count the number of configurations with $b_0 + \beta_1 p$ marked points that are invariant under these transformations. The only way a configuration is invariant is if on any regular polygon either all vertices are marked or no vertex is marked (see the solution to the previous problem). This can only happen if we mark all vertices of β_1 polygons, and then choose b_0 of the remaining a_0 points. So of all $\binom{a}{b}$ configurations, there are $\binom{\alpha_1}{\beta_1}\binom{a_0}{b_0}$ that are fixed by the transformations.

It follows that the number of configurations that are not fixed is

$$\binom{a}{b} - \binom{a_0}{b_0}\binom{\alpha_1}{\beta_1}.$$

But when on a given circle we have marked some but not all of the points, then the configuration of those points and its $p - 1$ rotates are all distinct configurations. It follows that the configurations that are not fixed can be grouped into families of configurations that are mapped into one another by the given transformations, and the number of members of each family is a multiple of p. Hence,

$$\binom{a}{b} - \binom{a_0}{b_0}\binom{\alpha_1}{\beta_1}$$

is a multiple of p. Note that if $b_0 > a_0$, then there are no fixed configurations, and also the second term is zero because there is no way you can choose b_0 objects out of a_0.

Now we can prove by induction on k that if $a = a_0 + a_1 p + a_2 p^2 + \cdots + a_{k-1} p^{k-1} + \alpha_k p^k$, $b = b_0 + b_1 p + b_2 p^2 + \cdots + b_{k-1} p^{k-1} + \beta_k p^k$ with $0 \leq a_j, b_j < p$, $j = 0, 1, \ldots, k-1$ and $\alpha_k, \beta_k \geq 0$, then

$$\binom{a}{b} \equiv \binom{a_0}{b_0}\binom{a_1}{b_1} \cdots \binom{\alpha_k}{\beta_k} \pmod{p}.$$

The base case for $k = 1$ was proved above. For the induction step, assume the property is true for k and let us prove it for $k + 1$. Write $a = a_0 + a_1 p + a_2 p^2 + \cdots + a_{k-1} p^{k-1} + \alpha_k p^k$, $b = b_0 + b_1 p + b_2 p^2 + \cdots + b_{k-1} p^{k-1} + \beta_k p^k$ with $0 \leq a_j, b_j < p$, $j = 0, 1, \ldots, k-1$, and write $\alpha_k = \alpha_{k+1} p + a_k$, $\beta_k = \beta_{k+1} p + b_k$. Then, by using the induction hypothesis and the base case, we have

$$\binom{a}{b} \equiv \binom{a_0}{b_0}\binom{a_1}{b_1} \cdots \binom{\alpha_k}{\beta_k} \equiv \binom{a_0}{b_0}\binom{a_1}{b_1} \cdots \binom{a_k}{b_k}\binom{\alpha_{k+1}}{\beta_{k+1}} \pmod{p},$$

and the theorem is proved.

Remark The proof can be done in one step by arranging the $a_0 + a_1 p + a_2 p^2 + \cdots + a_k p^k$ points as follows: first put the a_0 points somewhere. Then consider the vertices of a_1 regular p-gons. At each of the vertices of a_2 regular p-gons, place (formally) one tiny regular p-gon. In this configuration, we have rotations that move the tiny regular polygons by the rotations of the regular polygon that has them as vertices, and rotations of the tiny regular polygons themselves, and we look at configurations that are invariant under all these transformations. For a_3 we iterate the process with regular polygons placed at the vertices of regular polygons that are placed at the vertices of regular polygons, and so on.

75 We will concentrate on the points of integer coordinates in the plane and among those we mark the ones that have both coordinates nonnegative. For every positive integer n, we let d_n be the line of equation

$$ax + by = n.$$

Next, we mark the lines that contain at least one marked point. As an illustration, Fig. 9.90 shows a particular case with marked lines shown in solid line and the unmarked lines shown in dotted line.

Now we focus on two families of isometries of the plane that map the lines d_n into one another:

- translations by (p, q): $(x, y) \mapsto (x + p, y + q)$ and
- reflections over the points $(p/2, q/2)$: $(x, y) \mapsto (p - x, q - y)$,

where p and q range among integers.

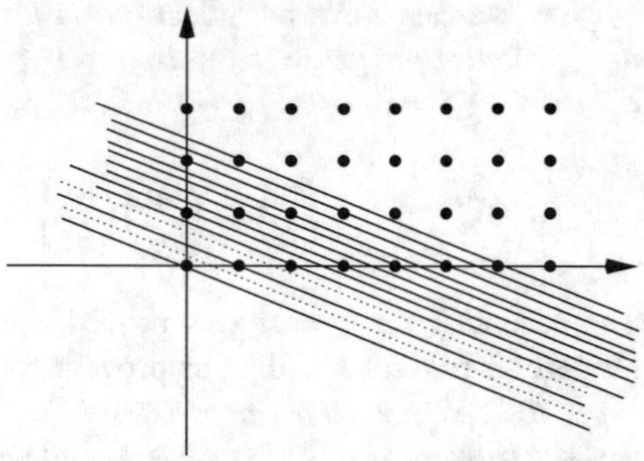

Fig. 9.90 The case of the
family of equations
$2x + 5y = n$

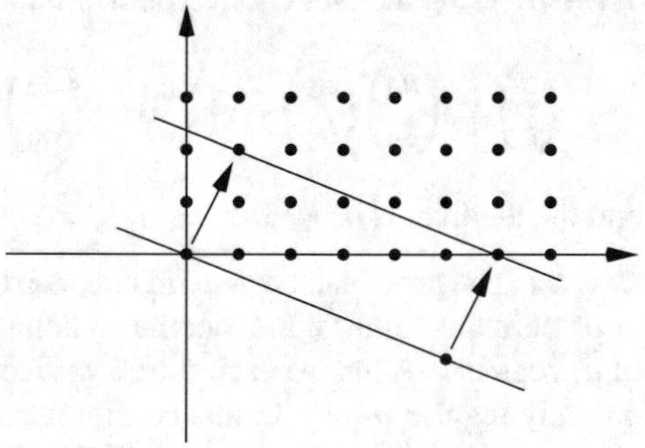

Fig. 9.91 The translation of
$2x + 5y = 0$ by the vector
$(1, 2)$

Lemma *If (x_0, y_0) is a point of integer coordinates on the line d_n, then the points of integer coordinates on d_n that are closest to it are $(x_0 - b, y_0 + a)$ and $(x_0 + b, y_0 - a)$.*

Proof Consider the line d_0, which passes through $(0, 0)$. Let $(-b_1, a_1)$ be the integer point on this line that is the closest to $(0, 0)$ such that $a_1, b_1 > 0$. Translating by the vector (x_0, y_0), the segment from $(0, 0)$ to $(-b_1, a_1)$ on d_0 becomes the segment from (x_0, y_0) to $(x_0 - b_1, y_0 + a_1)$ on d_n (see Fig. 9.91). Then $(x_0 - b_1, y_0 + a_1)$ must be the closest to (x_0, y_0) in the upward direction (or else we can translate the closest point backward to the line d_0 and contradict the minimality of $(-b_1, a_1)$).

Similarly, by translating by the vector $(x_0 + b_1, y_0 - a_1)$, the segment from $(0, 0)$ to $(-b_1, a_1)$ becomes the segment from $(x_0 + b_1, y_0 - a_1)$ to (x_0, y_0). And again we must have minimality for a downward segment starting at (x_0, y_0). We have reduced the problem to the line d_0 and the point $(0, 0)$. From here we deduce that the points of integer coordinates on any of the lines d_n are equally spaced, with the distance between consecutive points independent of n. In other words, the distance between two consecutive integer points on d_n is a_1 units on the y-axis and b_1 units on the x-axis.

Since $(-b, a)$ is on d_0, it follows that $b = tb_1$ and $a = ta_1$. But a and b are coprime, so $t = 1$, that is, $a_1 = a$, $b_1 = b$. The lemma is proved.   □

Fig. 9.92 The slice
$0 \leq x \leq 4$

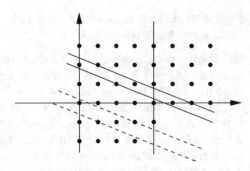

Returning to the proof of the theorem, from the lemma, it follows that each d_n
passes through exactly one point in the slice

$$S = \{(x, y) \mid 0 \leq x \leq b - 1\}.$$

If the line is marked, then necessarily a marked point on this line must be in this
slice. So we have a bijection between lines and their associated points in S, and
a line is marked if its associated point is marked. The reader can follow this on
Fig. 9.92.

The reflection σ over $\left(\frac{b-1}{2}, -\frac{1}{2}\right)$:

$$(x, y) \mapsto (b - 1 - x, -1 - y)$$

maps the slice S to itself, and it maps the marked points in this slice to unmarked
points and vice versa. Moreover, because

$$a(b - 1 - x) + b(-1 - y) = ab - a - b - ax - by,$$

we have

$$\sigma(d_n) = d_{ab-a-b-n}.$$

It follows that of the lines d_n and $d_{ab-a-b-n}$, exactly one is marked. This proves
that for every n, exactly one of the numbers n and $ab - a - b - n$ can be represented
as $ax + by$ with x, y nonnegative integers.

Finally, it is clear that the lowest marked line is d_0, and so the highest unmarked
line must be d_{ab-a-b}. This proves that $ab - a - b$ is the largest positive integer that
cannot be represented as $ax + by$.

Remark The geometric proof might suggest why solving the same problem for more
variables is considerably more difficult, as repeating the argument in a space of three
or more dimensions is impossible.

Source This proof to Sylvester's Theorem was published by Nikolai Borisovich Vassiliev in *Kvant (Quantum)*.

76 We prove the inequality by induction on the cardinality $|A|$ of the set A. The base case $|A| = 1$ is obvious. Now let us assume that the inequality holds for all pairs (A, B) with $|A| < n$, and let us prove it when $|A| = n$.

Place the residue classes modulo p at the vertices of a regular polygon. We want to rotate A until it intersects B, but in such a way that A does not lie entirely inside B. Assuming that this is not possible, then the rotates of A are either disjoint of B or lie inside B. Because the rotates of A cover the circle, some will be disjoint from B and others will lie entirely inside B.

Because A has at least two elements, we can find x, y in B that are rotates of elements in A. Let ρ be the rotation that maps x to y. Then $\rho(x) = y$ is in B, so $\rho(y) = \rho^2(x)$ must be in B too (because there is some rotation that maps two elements of A to y and $\rho(y)$). Inductively $\rho^n(x) \in B$ for all n, and since

$$\{\rho^n(x) \mid n > 0\} = \mathbb{Z}_p,$$

it follows that $B = \mathbb{Z}_p$. In this case, $A + B = \mathbb{Z}_p$, and the inequality is satisfied trivially.

Now let us assume that we are in the other case, where a rotation of A intersects B but does not lie entirely in B. Then we replace the pair (A, B) by $(A', B') = (A \cap B, A \cup B)$. In the process, we have transferred the elements of $A \backslash (A \cap B)$ from A to B. So $|A| + |B| = |A'| + |B'|$. But

$$A' + B' = [A + (A \cap B)] \cup [B + (A \cap B)]$$

and this lies inside $A + B$. Thus, $|A' + B'| \leq |A + B|$. Now we use the induction hypothesis to conclude that

$$|A + B| \geq |A' + B'| \geq \min(|A'| + |B'| - 1, p) = \min(|A| + |B| - 1, p).$$

Remark Compare the Cauchy-Davenport Theorem to Problem 67.

77 We will construct g and h that fulfill the desired equality $f = g + h$ in such a way that g is an odd function (having the graph symmetric with respect to the origin) and h is a function whose graph is symmetric with respect to the point $(1, 0)$.

Let g be any odd function on the interval $[-1, 1]$ for which $g(1) = f(1)$. Define $h(x) = f(x) - g(x)$, $x \in [-1, 1]$. Now proceed inductively as follows (Fig. 9.93). For $n \geq 1$, let $h(x) = -h(2 - x)$ and $g(x) = f(x) - h(x)$ for $x \in (2n - 1, 2n + 1]$, and then extend these functions such that $g(x) = -g(-x)$ and $h(x) = f(x) - g(x)$ for $x \in [-2n - 1, -2n + 1)$. By construction g and h satisfy the required condition, and we are done.

Remark Certainly, the centers of symmetry of the graphs of g and h have to be distinct, or else the graph of $f = g + h$ would have the same center of symmetry.

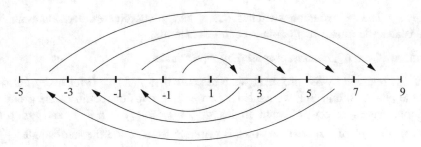

Fig. 9.93 The inductive procedure for constructing f and g

Fig. 9.94 Graph of a
function that is invariant
under a 90° rotation about the
origin

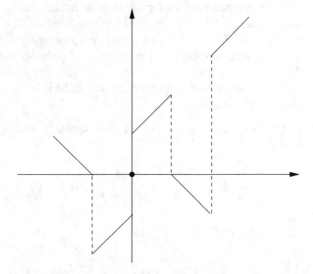

Once we decide that the centers of symmetry are distinct, then the alternating inductive construction is dictated by the reflective property of the graphs.

Source *Kvant (Quantum).*

78 The function $f : \mathbb{R} \to \mathbb{R}$, $f(0) = 0$ and

$$f(x) = \frac{x}{|x|} - (-1)^{\lfloor -|x| \rfloor} x, \quad x \neq 0,$$

where $\lfloor \cdot \rfloor$ is the greatest integer function, satisfies the required property. Because the algebraic expression is rather difficult to visualize, we have drawn the graph of the function in Fig. 9.94.

Note that because the graph is invariant under a 90° rotation about the origin, it is also invariant under the square of this transformation, which is the 180° rotation about the origin. This means that the graph is invariant under the reflection over the origin, so the function f is odd. Hence, $f(0) = 0$.

Remark The 90° rotation switches the x and y coordinates. Playing with the translates and rotates of $f(x) = x$ becomes natural.

Source *Kvant (Quantum)*, proposed by A. Karinskii.

79 As required by the problem, we will show that $f(A) = 0$ for every point A in the plane. To this end, we consider two finite groups of rotations: the group G of rotations ρ_k of center A and angles $\frac{2k\pi}{n}$, $k = 0, 1, \ldots, n-1$, and group G' of rotations ρ'_j of center some point B different from A and the same angles $\frac{2j\pi}{n}$, $j = 0, 1, \ldots, n-1$. Define $A_{kj} = \rho_k(\rho'_j(A))$, $k, j = 0, 1, \ldots, n-1$. Note that $A_{k0} = A$ for all k.

There are two types of regular n-gons that arise in the configuration: the regular n-gon obtained by acting on A with the elements of G' and the rotates of this n-gon by elements of G (shown with continuous line in Fig. 9.95) and the regular n-gons that arise from making G act on some $\rho'_j(A)$ (shown with dotted line in Fig. 9.95). The key observation is that the sum of the values of f at the vertices of each of these regular n-gons is zero, and because A is fixed by G, it will follow that $f(A) = 0$. Here are the details:

For each k, $A_{k0}A_{k1} \ldots A_{k,n-1}$ is a regular n-gon, so $\sum_{j=0}^{n-1} f(A_{kj}) = 0$, and hence

$$\sum_{k=0}^{n-1}\sum_{j=0}^{n-1} f(A_{kj}) = 0.$$

This can be rewritten as

$$nf(A) + \sum_{k=0}^{n-1}\sum_{j=1}^{n-1} f(A_{kj}) = 0.$$

We can change the order of summation:

Fig. 9.95 Regular n-gons that arise after rotations

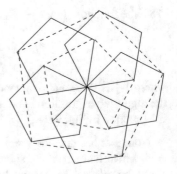

$$nf(A) + \sum_{k=0}^{n-1}\sum_{j=1}^{n-1} f(A_{kj}) = nf(A) + \sum_{j=1}^{n-1}\sum_{k=0}^{n-1} f(A_{kj}).$$

Now observe that for every j,

$$A_{0j}A_{1j}\ldots A_{n-1,j} = \rho_0(\rho_j'(A))\rho_1(\rho_j'(A))\ldots \rho_{n-1}(\rho_j'(A))$$

is a regular n-gon, so $\sum_{k=0}^{n-1} f(A_{kj}) = 0$, and then

$$\sum_{j=1}^{n-1}\sum_{k=0}^{n-1} f(A_{kj}) = 0.$$

We deduce that $nf(A) = 0$, so $f(A) = 0$. Since A was arbitrary, f is identically equal to zero.

Remark Rephrasing in complex coordinates, the problem states that if

$$\sum_{j=0}^{n-1} f\left(a + be^{\frac{2\pi i}{n} j}\right) = 0,$$

then $f(z) = 0$ for all $z \in \mathbb{C}$.

Source Romanian Team Selection Test for the International Mathematical Olympiad, 1996, proposed by Geffry Barad.

80 Assume first that the axes of symmetry are parallel, and let m be their common slope (which is either a nonzero number or infinite when the axes are vertical). Then the graph of f is invariant under the composition of the two reflections, which is a translation in the direction perpendicular to the axes of symmetry of length twice the distance between the lines (see Fig. 9.96). So the graph is invariant under the translation in the direction of a line of slope $-1/m$. Because f is continuous, $f(x) - (-1/m)x$ must be bounded, so

$$\lim_{x\to\infty} \frac{f(x)}{x} = \lim_{x\to\infty} \frac{(-1/m)x}{x} = -\frac{1}{m}.$$

If the two axes of symmetry are not parallel, then the composition of the reflections over the two axes is a rotation ρ by twice the angle between them. Let us first consider the case where the axes are not orthogonal. Then the angle α of rotation is different from $180°$. Let us consider the group

$$G = \{\rho^k \mid k \in \mathbb{Z}\}$$

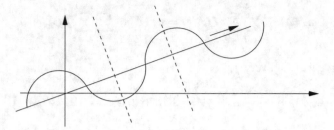

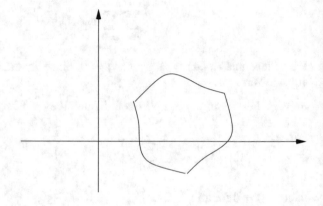

Fig. 9.96 A graph that is translation invariant

Fig. 9.97 A graph that is
rotation invariant

generated by ρ. It is either infinite or cyclic, depending on whether α is rational or irrational. Consider a point p on the graph of f. Then $\rho^k(p)$ must be on the graph of f for every k.

If G is infinite, then $\rho^k(p)$ is dense in a circle, and since f is continuous, this would mean that the graph of f contains a circle, which is absurd.

If G is cyclic, say of order n, then $\rho^k(p)$, $k = 0, 1, \ldots, n$, is a regular n-gon. Because f is continuous, the graph of f contains an arc of a curve from $p = \rho^0(p)$ to some $\rho^k(p)$ that contains no other vertices of the n-gon. Then the rotates of this arc are also on the graph (Fig. 9.97). Consequently, the graph contains a cycle, which again is impossible.

We are left with the situation where the two symmetry axes are orthogonal. If there is a point p on the graph of f that does not belong to either of the axes, then by reflecting it over the symmetry axes and over the intersection point of the symmetry axes, we obtain three more points that are on the graph of f. On the graph of f, there is an arc that connects p to one of these three points such that neither of the remaining two points is on the arc. If we take this arc together with its reflections over the symmetry axes and over their intersection point, we produce a cycle on the graph of f, which is impossible.

So the only possibility is that the points on the graph of f lie on the union of the two symmetry axes. The graph is either V-shaped (which is ruled out by the symmetry requirement) or it is a line. In the latter case, f is a linear function, and hence it satisfies the desired property.

Remark In Problem 78, we have seen that the graph of a function can be invariant under the group of rotations by $0°, 90°, 180°, 270°$, but if we impose continuity, an example can no longer be constructed, as explained above.

The fact that a subgroup of the rotation group about a point is either finite and cyclic or it is given by a set of rotation angles that are dense in $[0, 2\pi)$ is a direct consequence of Kronecker's Theorem, which was proved in the solution to the example from the introduction.

Source Răzvan Gelca.

81 With the substitution $x = \cos t$, the equation from the statement reads

$$P(\sqrt{2}\cos t) = P\left(\sqrt{2}\cos\left(t - \frac{\pi}{4}\right)\right).$$

After a new substitution $z + \frac{1}{z} = 2\cos t$, we turn $P(\sqrt{2}\cos t)$ into a Laurent polynomial in z (i.e., a polynomial in z and z^{-1}),

$$Q(z) = P\left(\frac{1}{\sqrt{2}}\left(z + \frac{1}{z}\right)\right).$$

The functional equation implies that $Q(z)$ is invariant under the group generated by the 45° rotation $z \mapsto e^{\frac{\pi i}{4}}z$, namely, that

$$Q(z) = Q\left(e^{\frac{\pi i}{4}}z\right), \quad k = 0, 1, \dots, 7.$$

And we have the following result.

Lemma *Let $p, q \in \mathbb{Z}$, $q \neq 0$, $\gcd(p, q) = 1$. If the Laurent polynomial function*

$$R(z) = a_{-m}\frac{1}{z^m} + \cdots + a_0 + a_1 z + \cdots + a_n z^n$$

is invariant under rotations of z by the angle $\frac{p\pi}{q}$, then R is a Laurent polynomial in z^{2q}.

Proof This means that $R(e^{\frac{p\pi i}{q}}z) = R(z)$, so the coefficients of R do not change when the variable is multiplied by $e^{\frac{p\pi i}{q}}$. Since the coefficient of z^k is multiplied by $e^{\frac{kp\pi i}{q}}$ under this transformation is applied, k must be a multiple of $2q$. □

Applying the Lemma for $p = 1$ and $q = 4$, we obtain that Q is a Laurent polynomial in z^8. Note also that Q is invariant under the transformation $z \mapsto 1/z$, consequently only terms in z^{8k}, $k \in \mathbb{Z}$, appear and the coefficient of z^{8k} equals that of z^{-8k}. In other words,

$$P\left(\frac{1}{\sqrt{2}}\left(z+\frac{1}{z}\right)\right)$$

is a polynomial in $z^8 + \frac{1}{z^8}$. But we know that

$$z^8 + \frac{1}{z^8} = T_8\left(z+\frac{1}{z}\right),$$

where $T_8(y) = y^2(y^2 - 2)^2(y^2 - 4) + 2$ is a rescaling of the eighth Chebyshev polynomial, and since $y = 2x$, we obtain that

$$P\left(\frac{1}{\sqrt{2}}x\right) = P_1(T_8(2x)),$$

for some polynomial P_1. Consequently, $P(x) = P_1(T_8(2\sqrt{2}x))$.

This means that $P(x)$ is a polynomial in $8x^2(8x^2 - 1)^2(8x^2 - 2) + 2$. So the answer to the problem is

$$P(x) = \sum_{j=1}^{m} a_j[8x^2(8x^2 - 1)^2(8x^2 - 2) + 2]^j$$

for some nonnegative integer m and real coefficients a_j, $j = 0, 1, \ldots, m$.

Remark How does the problem change if we want the variable to be invariant under rotations by $\frac{\pi}{3}$? What is the corresponding functional equation and what is the answer to the problem?

Source United States of America Team Preselection Test for the International Mathematical Olympiad, 2014, proposed by Răzvan Gelca.

82 The answer to (a) is negative, a counterexample being the greatest integer function $f(x) = \lfloor x \rfloor$.

For (b) the answer is positive. We begin by showing that if $|z - w| = \sqrt{3}$, then $|f(z) - f(w)| = \sqrt{3}$. For this, consider two arbitrary points z and w in $\mathbb{C}$ such that $|z - w| = \sqrt{3}$, and construct the configuration consisting of two equilateral triangles zz_1z_2 and z_1z_2w of side 1 sharing the side z_1z_2. Then, because of the hypothesis, $f(z)f(z_1)f(z_2)$ and $f(z_1)f(z_2)f(w)$ are also equilateral triangles of side 1, which either are distinct and share one side or are the same triangle and then $f(w) = f(z)$. Let us rule out the second possibility.

Assume therefore that $f(z) = f(w)$, and consider another configuration of equilateral triangles of side 1, $zz_1'z_2'$ and $z_1'z_2'w'$ such that $|w - w'| = 1$, as shown in Fig. 9.98. Then as before, there are two possible situations: $f(w') = f(z)$, or $f(z)f(z_1')f(z_2')$ and $f(z_1')f(z_2')f(w')$ are distinct equilateral triangles sharing the side $f(z_1')f(z_2')$. But we should also have

Fig. 9.98 Equilateral triangles zz_1z_2, wz_1z_2, $zz_1'z_2'$, and $wz_1'z_2'$

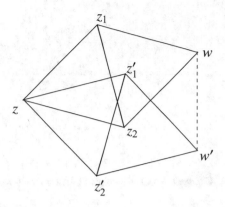

$$|f(z) - f(w')| = |f(w) - f(w')| = 1,$$

and this does cannot happen in either situation. Hence, $f(z) \neq f(w)$, and therefore $|f(z) - f(w)| = \sqrt{3}$, as claimed.

From here we deduce that f maps the vertices of any tessellation of the plane by equilateral triangles of side 1 into the vertices of a tessellation of the same type. Additionally, we deduce that if $|z - w| = \sqrt{3}$, then $|f(z) - f(w)| = \sqrt{3}$, and a similar argument then shows that f maps the vertices of any tessellation of the plane by equilateral triangles of side-length $\sqrt{3}$ is mapped into a tessellation of the same type. Hence, $|f(z) - f(w)| = |z - w|$ whenever $|z - w| = n$ or $|z - w| = n\sqrt{3}$ for any positive integer n.

Now let us concentrate on a line l that passes through some fixed point z_0, and on this line, let us consider the sets S and T of points at distance equal to an integer, respectively an integer multiple of $\sqrt{3}$, from z_0. We will prove that $f(S \cup T)$ lies on a line, and this proof can be followed on Fig. 9.99. Because f preserves distances in S and it also preserves distances in T (so it is an isometry when restricted to either of these sets), $f(S) = \rho \circ \tau(S)$ and $f(T) = \rho' \circ \tau(T)$ where τ is the translation from z_0 to $f(z_0)$ and ρ, ρ' are rotations about $f(z_0)$.

The numbers of the form $m + n\sqrt{3}$ form a dense set in $\mathbb{R}$. If $\rho \neq \rho'$, then by using this density property, we can find $z \in S$, $w \in T$ with $|z - w| < 1$ and $|f(z) - f(w)|$ arbitrarily large. But then there is $u \in \mathbb{C}$ with $|z - u| = |z - w| = 1$, and so

$$|f(z) - f(w)| < |f(z) - f(u)| + |f(u) - f(w)| = 2,$$

a contradiction. Hence, $\rho = \rho'$; thus, $f|(S \cup T)$ is an isometry. The end result of this discussion is that if z and w are such that $|z - w| = |m + n\sqrt{3}|$ with $m, n \in \mathbb{Z}$, then $|f(z) - f(w)| = |z - w|$. It is important that, by Kronecker's Theorem, there are arbitrarily small distances of this form.

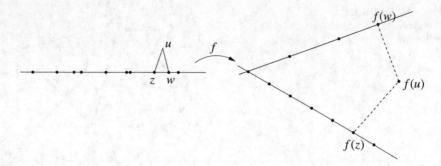

Fig. 9.99 Proof that points of the form $m + n\sqrt{3}$ on a line are mapped to a line

Now let z, w be two arbitrary points. We will show that for every $\epsilon > 0$,

$$|z - w| - 3\epsilon < |f(z) - f(w)| < |z - w| + \epsilon.$$

Choose points $z_1, z_2, \ldots, z_n$ such that

$$|z - z_1|, |z_1 - z_2|, \ldots, |z_n - w| \in \{m + n\sqrt{3} \mid m, n \in \mathbb{Z}\}$$

and

$$|z - z_1| + |z_1 - z_2| + \cdots + |z_{n-1} - z_n| + |z_n - w| < |z - w| + \epsilon.$$

Basically we use a broken line of segments of nice lengths to approximate the segment joining z and w. Then, by using the triangle inequality, we obtain

$$|f(z) - f(w)| \leq |f(z) - f(z_1)| + |f(z_1) - f(z_2)| + \cdots + |f(z_n) - f(w)|$$
$$= |z - z_1| + |z_1 - z_2| + \cdots + |z_{n-1} - z_n| + |z_n - w| < |z - w| + \epsilon.$$

This proves the inequality on the right.

For the other inequality, choose $w_1, w_2, \ldots, w_m$ on the line zw such that

$$|z - w_1|, |w_1 - w_2|, \ldots, |w_{n-1} - w_n| \in \{m + n\sqrt{3} \mid m, n \in \mathbb{Z}\},$$

and such that

$$|w - w_1| > |w - z_2| > \cdots > |w - w_n| \text{ and } |w - w_n| < \epsilon.$$

Then by the triangle inequality,

$$|f(z) - f(w)| > |f(z) - f(w_n)| - |f(w_n) - f(w)|$$
$$= |z - w_n| - |f(w_n) - f(w)|.$$

But we have shown that $|f(w_n) - f(w)| \le |w_n - w| + \epsilon$ (the inequality on the right), so the right-hand side is greater than

$$|z - w_n| - |w_n - w| - \epsilon > |z - w_n| - 3\epsilon,$$

which proves the inequality on the left. Since ϵ was arbitrary, by letting $\epsilon \to 0$, we obtain

$$|f(z) - f(w)| = |z - w| \text{ for all } z, w \in \mathbb{C},$$

as desired.

Source *Kvant (Quantum)*, proposed by A. Tyshka.

Chapter 10
Homotheties and Spiral Similarities

83 *First solution.* Let us ask a different question: how many transformations from the group of translations and homotheties map a circle ω_1 to a circle ω_2? If $f_1 \neq f_2$ are two such transformations, then $f_2^{-1} \circ f_1$ is a translation or a homothety that maps ω_1 to itself. It cannot be a translation, so it is a homothety. The center of the homothety must be the center of the circle, since the center is mapped into itself, so it is the fixed point of the homothety. As the radius of the circle equals the radius of the image, the homothety ratio must be -1 (it cannot be 1 because we ruled out the case where $f_2^{-1} \circ f_1$ is the identity map). Then $f_2^{-1} \circ f_1 = \sigma$, where σ is the reflection over the center of the first circle. It follows that $f_2 = f_1 \circ \sigma$. So we can have at most two such transformations, f_1 and $f_1 \circ \sigma$.

Case 1. The original circles have different radii. Let us show that there is a homothety mapping ω_1 to ω_2. Consider a circle ω that is exterior to both. Take the internal tangents of ω_1 and ω, and let A be there intersection (Fig. 10.1). Then there is an inverse homothety h_1 of center A that maps ω_1 to ω. Now take the internal tangents of ω_2 and ω, and let B be their intersection. Again there is an inverse homothety h_2 of center B that maps ω to ω_2. Then $h_2 \circ h_1$ is a homothety that maps ω_1 to ω_2. And $h_2 \circ h_1 \circ \sigma$ is the other homothety (which is not a translation because the circles have different radii).

Case 2. The original circles have equal radii. One such transformation is the translation by the vector with endpoints the centers of the circles. In this case there is at most one homothety. Perform the construction from Case 1. If the result is a translation τ, take $\tau \circ \sigma$, which now is necessarily a homothety.

Second solution. Let the circles be $|z - c_1| = R_1$ and $|z - c_2| = R_2$. Then a homothety $f(z) = kz + (1 - k)a$ must map c_1 to c_2, so $kc_1 + (1 - k)a = c_2$. We deduce that

$$a = \frac{c_2 - kc_1}{1 - k}.$$

R. Gelca et al., *Geometric Transformations*, Problem Books in Mathematics, https://doi.org/10.1007/978-3-030-89117-6_10

Fig. 10.1 Finding the
homothety that maps a circle
into another

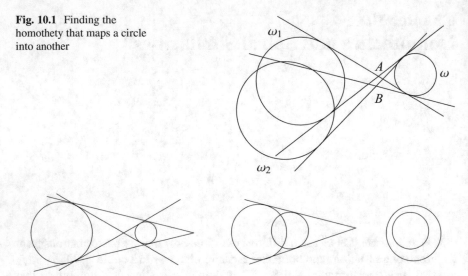

Fig. 10.2 The center of homothety mapping one circle into another

But we know that $|k| = R_2/R_1$. There are two situations $k = R_2/R_1$ or $k = -R_2/R_1$. We have therefore two homotheties if $R_2 \neq R_1$ and one homothety if $R_2 = R_1$.

Remark A first observation: a function f maps a γ_1 into γ_2 if when substituting the variable in the equation of γ_2 by f one obtains the equation for γ_1. So if γ_1 is given by $g_1(z) = 0$ and γ_2 is given by $g_2(z) = 0$, then γ_1 is mapped to γ_2 if and only if $g_2(f(z)) = 0$ is the same curve as $g_1(z) = 0$. In our situation, if the circles are $|z - c_1| = R_1$ and $|z - c_2| = R_2$, then the homothety $f(z) = kz + (1 - k)a$ maps the first circle into $|kz + (1 - k)a - c_1| = R_1$, and this should be the equation of the second circle.

By examining Fig. 10.2, we see that if the circles are exterior, then the centers of homothety are the intersection of the external tangents and the intersection of the internal tangents. If the circles intersect, one of the centers of homothety is at the intersection of the external tangents, and the other is interior to both circles. If one of the circles is inside the other, both centers of homothety are inside both circles, with the special case where the circles are concentric, in which case the centers of homothety coincide and are the common center of the circles.

For two circles of different radii, when two homotheties exist, the centers of these homotheties are collinear with the centers of the circles and form with them a harmonic ratio. This can be shown easily using complex coordinates: the centers of the homotheties be

$$c_3 = \frac{c_2 - kc_1}{1 - k}, \quad c_4 = \frac{c_2 + kc_1}{1 + k}.$$

Then

$$\frac{c_3 - c_1}{c_3 - c_2} : \frac{c_4 - c_1}{c_4 - c_2} = \frac{\dfrac{c_2 - kc_1 - c_1 + kc_1}{1 - k}}{\dfrac{c_2 - kc_1 - c_2 + kc_2}{1 - k}} : \frac{\dfrac{c_2 + kc_1 - c_1 - kc_1}{1 + k}}{\dfrac{c_2 + kc_1 - c_2 - kc_2}{1 + k}}$$

$$= \frac{c_2 - c_1}{k(c_2 - c_1)} : \frac{c_2 - c_1}{k(c_1 - c_2)} = -1.$$

84 *First solution.* We prove directly the general case of polygons. Let $A_1 A_2 \ldots A_n$ and $A_1' A_2' \ldots A_n'$ be two polygons such that $A_j A_k$ is parallel to $A_j' A_k'$ for every j, k. Consider a translation τ that maps A_1 to A_1'. Then $\tau(A_2)$ is on the line $A_1' A_2'$. Consider a homothety h_0 of center A_1' that maps $\tau(A_2)$ to A_2'. Then $h = h_0 \circ \tau$ is a homothety that maps A_1, A_2 to A_1', A_2', respectively. For $j \neq 1, 2$, $A_1 A_j || A_1' A_j'$, and $A_2 A_j || A_2' A_j'$. Because h preserves parallelism, $A_1' h(A_j) || A_1' A_j$ and $A_2' h(A_j) || A_2 A_j'$. This can only happen if $h(A_j) = A_j$. So h maps the first polygon into the second.

Second solution. (a) Let ABC and $A'B'C'$ be the two triangles, with $AB || A'B'$, $AC || A'C'$, and $BC || B'C'$. Having parallel sides, they have equal angles, and so they are similar. At least two of the lines AA', BB', and CC' must be distinct, say AA' and BB'. They are not parallel because the triangles are not congruent. Assume also that CC' does not coincide with AA'.

Arguing on Fig. 10.3, let O be the intersection of AA' and OC. Then from similarity

$$\frac{A'C''}{AC} = \frac{OA'}{OA} = \frac{A'B'}{AB} = \frac{A'C'}{AC}.$$

Hence $C' = C''$. It follows that AA', BB', CC' are concurrent and

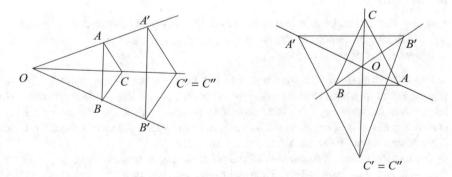

Fig. 10.3 Triangles with parallel sides are homothetic

$$\frac{OA'}{OA} = \frac{OB'}{OB} = \frac{OC'}{OC}.$$

There are two cases: if A is between O and A', then AB separates A' from O, so it must separate B' from O; thus B is between O and B'. Similarly, C is between O and C'. Thus there is a direct homothety with center A and ratio $\frac{OA'}{OA}$ that maps ABC to $A'B'C'$. If A is between O and A', there is an inverse homothety with ratio the negative of this quotient.

For (b) proceed by induction on the number of sides, with the base case being the triangle. Let $A_1 A_2 \ldots A_n$ and $A'_1 A'_2 \ldots A'_n$ be two polygons such that $A_j A_k$ is parallel to $A'_j A'_k$ for every j, k. We may assume (why?) that $A_1, A'_1, A_{n-1}, A'_{n-1}$ are not collinear. By the induction hypothesis and the base case $A_1 A_2 \ldots A_{n-1}$ and $A'_1 A'_2 \ldots A'_{n-1}$ are homothetic and $A_{n-1} A_n A_1$ and $A'_{n-1} A'_n A'_1$ are homothetic. The center of each homothety is the intersection O of $A_1 A'_1$ and $A_{n-1} A'_{n-1}$, and the homothety ratio is $\overline{OA_1}/\overline{OA'_1}$ (we work with directed segments to take into account a possible inverse homothety). So they are the same homothety and we are done.

Remark The homothety is unique. If the polygons are congruent, then they are mapped into each other by either a translation or an inverse homothety of ratio -1. Part (a) is a particular case of Desargues' Theorem (Problem 115).

85 It is possible for two such homotheties to exist, for example, the square with vertices $1, i, -1, -i$ is mapped to the square with vertices $2, 2i, -2, -2i$ by both the homothety of center 0 and ratio 2 and the homothety of center 0 and ratio -2.

But three homotheties cannot exist, for if h_1, h_2, h_3 are distinct homotheties of ratios k_1, k_2, k_3 that map two polygons into each other, then $|k_1| = |k_2| = |k_3|$ because they are equal to the similarity ratios of the polygons. So two of the ratios must be equal, say k_1 and k_2. Then $\tau = h_2^{-1} \circ h_1$ is an isometry obtained by composing two homotheties, so it is a translation. But this translation maps the first polygon into itself, which is impossible, since a repeated application of this translation would render the polygon unbounded. Hence we cannot have more than two homotheties.

Remark Two homotheties might exist only if the polygons have centers of symmetry, which happens when they have an even number of sides and the opposite sides are parallel and of equal length.

86 (a) Let h_{12}, h_{13}, and h_{23} be the direct homotheties of the pairs (ω_1, ω_2), (ω_1, ω_3), and (ω_2, ω_3), respectively. From Problem 83, we know that they exist and are unique. Because $h_{12} \circ h_{23}$ is a homothety that maps ω_1 to ω_3, $h_{13} = h_{23} \circ h_{12}$. But then the center of h_{13} is on the line determined by the centers of h_{12} and h_{23}, as it has been shown in Theorem 2.6, and we are done.

(b) Let the inverse homotheties of the pairs (ω_1, ω_2), (ω_1, ω_3), and (ω_2, ω_3) be h'_{12}, h'_{13}, and h'_{23}, respectively. Then, since the composition of either two inverse homotheties or two direct homotheties is a direct homothety, and the composition of an inverse homothety with a direct homothety is an inverse homothety, we have

$$h_{12} \circ h_{23} = h_{13}, \quad h_{23} \circ h'_{12} = h'_{13}, \quad h'_{23} \circ h_{12} = h'_{13}, \quad h'_{23} \circ h'_{12} = h_{23}.$$

Here again we used the uniqueness of the homothety that was proved in Problem 83. By the same Theorem 2.6, the centers of the homotheties in each equality are collinear, and they determine the four lines.

Remark If the circles are pairwise exterior, the centers of direct homotheties are the intersection points of the external common tangents, and the centers of inverse homotheties are the intersection points of the internal common tangents. Monge's Theorem implies that the three points of intersection of common external tangents are collinear, while d'Alembert's Theorem implies that the six intersection points of the common tangents lie on four lines (Fig. 10.4).

87 Consider the homothety of center A that maps ω_2 to ω_1, and let M', N' be the images of M, N, respectively (Fig. 10.5). Then the line $M'N'$ is the image of the line MN, so it is parallel to it. It follows that in the circle ω_1, the arcs $\overset{\frown}{PM'}$ and $\overset{\frown}{QN'}$ are equal. We deduce that the angles $\angle PAM'$ and $\angle QAN'$ are equal, and these are the same as the angles from the statement.

Remark If $M = N$, that is, if the line is actually tangent to ω_2, then the line AM is the angle bisector of $\angle PAQ$. This property can be rephrased as follows: if ABC is a triangle, and if a circle ω is tangent to BC at D and to the circumcircle at A, then AD is the angle bisector of $\angle BAC$.

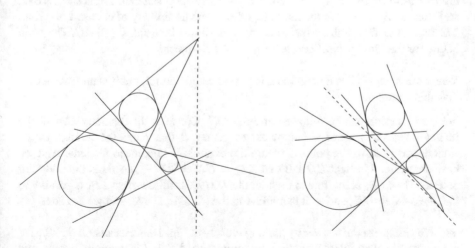

Fig. 10.4 The Theorems of Monge and d'Alembert

Fig. 10.5 ω_2 is mapped into
ω_1 by a homothety

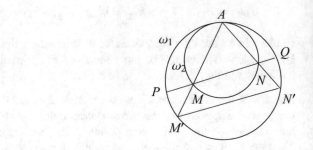

Fig. 10.6 The projections of
a vertex of a triangle onto
angle bisectors are collinear

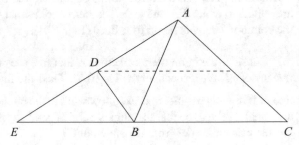

Source V.A. Gusev, V.N. Litvinenko, A.G. Mordkovich, *Praktikum po Resheniyu
Matematicheskih Zadach, Geometria (Practice for Solving Mathematics Problems,
Geometry)*, Prokveshcheniye, Moscow, 1985.

88 Let D be the projection of A onto one of the bisectors of $\angle B$, say the exterior
bisector as in Fig. 10.6. The reflection E of A over BD satisfies $\angle EBD = \angle ABD$,
so E lies on BC. As D is the midpoint of AE, the homothety of center A and ratio
$1/2$ maps E to D. It follows that D is on the midline of the sides AB, AC. The same
is true for the other projections, so they are all collinear.

Remark An angle chasing argument is possible and natural, but it is far less elegant
than this solution.

89 The solution can be followed on Fig. 10.7. Consider the diameter DM of the
incircle. We know that the homothety of center B that maps the incircle to the
excircle maps M to the point K where the excircle is tangent to the side. And we
have determined in Sect. 2.2.1 II that $AD = CK = \frac{a+b-c}{2}$, so B_1 is the midpoint
of DK. Thus IB_1 is midline in the triangle DMK. It follows that IB_1 is parallel to
the line $MK = BK$, and so it is midline in the triangle DBK, and we are done.

90 We can guess the answer by looking closely at Fig. 10.8. Because AB and MN
are diameters, they intersect at their midpoints, so $ANBM$ is a parallelogram, and
so BN is parallel to AM. Thus the triangle AKP is mapped to the triangle BKM
by an inverse homothety of center K and ratio KB/KA. Consequently the locus is
the circle that is the image of Ω through this homothety, shown with a dotted line in
the figure.

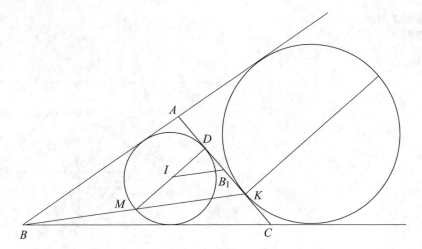

Fig. 10.7 Proof that $B_1 I$ passes through the midpoint of BD

Fig. 10.8 Finding the locus of P

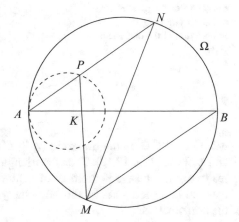

91 The solution can be followed on Fig. 10.9. The triangles $A'B'C'$ and $A''B''C''$ are homothetic, and the triangles ABC and $A'B'C'$ are homothetic, too. So these three triangles are pairwise homothetic, and their centers of homothety are collinear (Theorem 2.6). Since the center of the (inverse) homothety that maps ABC to $A'B'C'$ is the centroid G, and the center of the (direct) homothety that maps $A'B'C'$ to $A''B''C''$ is the circumcenter O, it follows that the center of the (inverse) homothety that maps ABC to $A''B''C''$, which is the point where the lines AA'', BB'', CC'' intersect, is on Euler's line OG.

Remark The locus of the points of intersection of AA'', BB'', CC'', when A'', B'', C'' vary on those rays, is the segment that connects the circumcenter with the orthocenter.

Fig. 10.9 Proof of Boutin's
Theorem

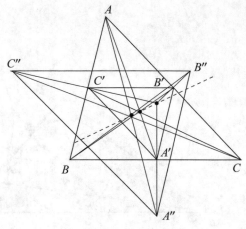

Fig. 10.10 In a trapezoid, the
line determined by the
midpoints of the parallel sides
passes through the
intersection of the diagonals

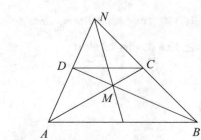

92 Let $ABCD$ be the trapezoid $(AB||CD)$ and let M be the intersection of AC
and BD (See Fig. 10.10). Then the triangles MAB and MCD are homothetic with
respect to an inverse homothety of center M. The midpoints of AB and CD are
mapped into each other by this homothety, so they are collinear with the center of
the homothety M.

Extend the sides AC and BD until they intersect at some point N. The triangles
NAB and NDC are directly homothetic with respect to a homothety of center N.
So N and the midpoints of AB and CD are collinear. It follows that NM is median
in the triangle NAB, and since by the hypothesis M is at equal distance from AD
and BC, NM is also the angle bisector of $\angle ANB$. It follows that the triangle NAB
is isosceles, and consequently the trapezoid $ABCD$ is isosceles, too.

Remark The property that in a quadrilateral the line joining the midpoints of
two opposite sides passes through the intersection of the diagonals characterizes
trapezoids (prove it!).

93 *First solution.* Let $a_1, a_2, \ldots, a_n$ be the complex coordinates of the vertices.
Then the centroid G of the polygon has the coordinate $g = \frac{a_1+a_2+\cdots+a_n}{n}$, while the
centroids G_j have the coordinates

$$\frac{a_1 + a_2 + \cdots + a_n - a_j}{n-1} = \frac{ng - a_j}{n-1} = -\frac{1}{n-1}a_j$$

$$+ \left(1 + \frac{1}{n-1}\right)g, \ j = 1, 2, \ldots n.$$

Hence G_j is the image of A_j through the homothety of center G and ratio $-\frac{1}{n-1}$.

Second solution. Let $a_1, a_2, \ldots, a_n$ be the complex coordinates of the vertices. Since $G_jG_k = \frac{1}{n-1}(a_k - a_j)$, the two polygons have parallel sides and diagonals. In view of Problem 84, they are homothetic.

Remark Note that for $n = 3$ we obtain the well-known fact that a triangle is homothetic to the triangle formed by the midpoints of the sides under a homothety of ratio $1/2$ whose center is the centroid.

94 Because the sides of the polygons are parallel, and the polygons are similar, the diagonals must be parallel, too (prove it!). By Problem 84 the polygons are homothetic. Since one polygon lies inside the other, the center of homothety lies inside both.

Let u and v be the distances from the center O of homothety to two sides of the original polygon. Then the distances from O to the corresponding sides of the new polygon are $u + 1$ and $v + 1$. The fact that the homothety maps one pair of sides to the other implies that

$$\frac{u}{u+1} = \frac{v}{v+1},$$

hence $u = v$. Thus all sides of the original polygon are at the same distance from O, showing that there is a circle of center O that is tangent to all sides.

95 *First solution.* Let X, Y, Z, W be the intersections of the parallels through A and B, B and C, C and D, and D and A, as shown in Fig. 10.11. A first observation is that $MN\|AC\|XY$, $NP\|BD\|YZ$, $PQ\|AC\|ZW$, and $QM\|BD\|WX$, so $MNPQ$ is a parallelogram, and the parallelograms $MNPQ$ and $XYZW$ have parallel sides. It is a good guess that they are homothetic.

To see why $MNPQ$ and $XYZW$ are homothetic, note that $AXBI$ is a parallelogram, so M is not only the midpoint of AB but also the midpoint of IX. Similarly, N, P, Q are the midpoints of IY, IZ, IW, respectively. Thus $MNPQ$ is mapped into $XYZW$ by the homothety of center I and ratio 2. The centers J and K of the two parallelograms are mapped by this homothety into each other, so they must be collinear with the center I of the homothety.

Second solution. Let the complex coordinates of A, B, C, D be a, b, c, d, respectively. The equations of the two diagonals are

$$(\bar{a} - \bar{c})z - (a - c)\bar{z} + a\bar{c} - c\bar{a} = 0$$
$$(\bar{b} - \bar{d})z - (b - d)\bar{z} + \bar{b}d - b\bar{d} = 0.$$

Fig. 10.11 Two homothetic
parallelograms

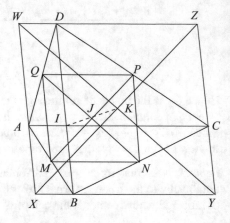

The coordinate z_1 of their intersection point satisfies both equations. The lines
parallel to AC and BD through B and A, respectively, have the equations

$$z - b = \frac{a - c}{\bar{a} - \bar{c}}(\bar{z} - \bar{b})$$

$$z - a = \frac{b - d}{\bar{b} - \bar{d}}(\bar{z} - \bar{a}).$$

The coordinate of their intersection point X is the unique solution z_2 to the system
of equations

$$(\bar{a} - \bar{c})z - (a - c)\bar{z} - \bar{a}b + \bar{c}b + a\bar{b} - c\bar{b} = 0$$
$$(\bar{b} - \bar{d})z - (b - d)\bar{z} - a\bar{b} + a\bar{d} + b\bar{a} - d\bar{a} = 0.$$

Then $z_1 + z_2$ is the unique solution to

$$(\bar{a} - \bar{c})z - (a - c)\bar{z} + a\bar{c} - c\bar{a} - \bar{a}b + \bar{c}b + a\bar{b} - c\bar{b} = 0$$
$$(\bar{b} - \bar{d})z - (b - d)\bar{z} + \bar{b}d - b\bar{d} - a\bar{b} + a\bar{d} + b\bar{a} - d\bar{a} = 0.$$

This system can be rewritten as

$$(\bar{a} - \bar{c})z - (a - c)\bar{z} + (a - c)(\bar{a} + \bar{b}) - (\bar{a} - \bar{c})(a + b) = 0$$
$$(\bar{b} - \bar{d})z - (b - d)\bar{z} + (b - d)(\bar{a} + \bar{b}) - (\bar{b} - \bar{d})(a + b) = 0,$$

and we observe that it has the unique solution $z_1 + z_2 = a + b$. This shows that M,
having the coordinate equal to $\frac{a+b}{2}$, is the midpoint of IX. Similarly N, P, Q are the
midpoints of IY, IZ, IW, respectively; thus $MNPQ$ and $XYZW$ are mapped into

Fig. 10.12   A configuration
in which the nine-point
circles of $A_1B_1C_1$ and
$A_2B_2C_2$ are exterior

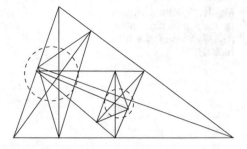

each other by a homothety of center I. Then J is mapped to K by this homothety,
and thus I, J, K are collinear.

Remark Drawing the correct figure discloses immediately that the two parallelograms have parallel sides and diagonals and then, in view of Problem 84, they are
homothetic.

96   This configuration is possible, as Fig. 10.12 demonstrates, but difficult to draw.
The argument is rather simple: Because ABC and $A'B'C'$ are homothetic with
center of homothety the centroid G and ratio $-1/2$, the same is true for $A_1B_1C_1$
and $A_2B_2C_2$. Hence the two nine-point circles are homothetic, with G the center
of homothety. Because the homothety has negative ratio, the interior tangents of the
two circles intersect at the center of homothety, which is G. And G is on the Euler
line of the triangle ABC.

Remark Certainly the two nine-point circles are homothetic regardless of their
relative position, but the interior tangents exist only if the circles are exterior.

Source   C. Mihalescu, *Geometria Elementelor Remarcabile (The Geometry of the
Remarkable Elements)*, Ed. Tehnică, Bucharest, 1957.

97   *First solution.* Examining the particular case of a square, we guess that the
homothety is centered at the centroid of the quadrilateral and has ratio equal to -3.
And indeed, if M and N are the midpoints of AB and CD, respectively, and if G is
the centroid of the quadrilateral, then G is the midpoint of MN, so the ratio between
the distances from G to AB and the reflection of AB over N is $1/3$ (Fig. 10.13).
Hence the reflection of AB over N coincides with the image of AB through the
homothety of center G and ratio -3. The same is true for the other four sides, and
the problem is solved.

Second solution. In coordinates, a and b reflect to $c+d-a$ and $c+d-b$ over
the midpoint $\frac{c+d}{2}$ of the segment with endpoints c and d. The line through $c+d-a$
and $c+d-b$ has the parametric form $c+d-a+t(a-b)$, $t \in \mathbb{R}$. Similarly, the line
through a and d reflects to the line whose parametric form is $b+c-a+s(a-d)$,
$s \in \mathbb{R}$. The two lines intersect when $t = s = -1$, and the intersection point is
$b+c+d-2a$. So the vertices of the quadrilateral determined by the four reflected
lines are $a' = b+c+d-2a$, $b' = a+c+d-2b$, $c' = a+b+d-2c$,

Fig. 10.13 Reflection over
the midpoint is the same as
homothety centered at the
centroid

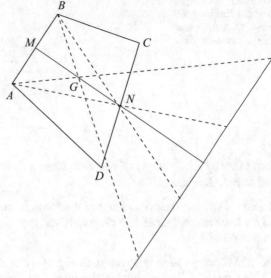

Fig. 10.14 The points
$A_0, B_0, C_0,$ and G

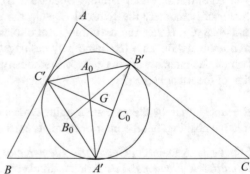

$d' = a + b + c - 2d,$ and

$$\frac{3}{4}a + \frac{1}{4}a' = \frac{3}{4}b + \frac{1}{4}b' = \frac{3}{4}c + \frac{1}{4}c' = \frac{3}{4}d + \frac{1}{4}d' = \frac{a+b+c+d}{4},$$

showing that the quadrilateral $a'b'c'd'$ is obtained from $abcd$ through a homothety
of center $\frac{a+b+c+d}{4}$ and ratio -3.

Source *Mathematical Reflections*, proposed by Francisco Javier García Capitán and
Juan Bosco Romero Márquez.

98 Let the midpoints of $B'C', C'A', A'B'$ be $A_0, B_0, C_0,$ respectively, and let G
be the centroid of the triangle $A'B'C'$, as shown in Fig. 10.14. Denote the perpendiculars to the sides BC, CA, AB through A_0, B_0, C_0 by ℓ_A, ℓ_B, ℓ_C, respectively.

Fig. 10.15 Four circles in a triangle

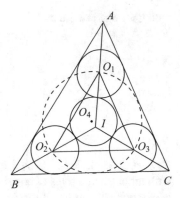

The homothety of center G and ratio -2 takes A_0, B_0, C_0 to A', B', C', respectively. It also carries ℓ_A, ℓ_B, ℓ_C to the lines $\ell'_A, \ell'_B, \ell'_C$ through A', B', C', that are perpendicular to the sides. These three lines meet at the incenter. It follows that ℓ_A, ℓ_B, ℓ_C meet at the preimage of the incenter through the homothety.

Remark A configuration that uses just the points A', B', C' is easier to control. The use of the homothety that maps the triangle formed by the midpoints to the triangle itself is then automatic.

Source Romanian Team Selection Test for the International Mathematical Olympiad, 1986, solution from S. Savchev, T. Andreescu, *Mathematical Miniatures*, MAA, 2003.

99 Let O_1, O_2, O_3, O_4 be the centers of $\alpha, \beta, \gamma, \delta$, as shown in Fig. 10.15.

Then the triangles ABC and $O_1 O_2 O_3$ have parallel sides, so they are homothetic. Because O_1 is on the angle bisector of $\angle A$, O_2 is on the angle bisector of $\angle B$, and O_3 is on the angle bisector of $\angle C$, and because of parallel sides, the triangles ABC and $O_1 O_2 O_3$ have the same incenter. Since the homothety maps the incenter of one triangle to the incenter of the other, the center of homothety is the common incenter of the two triangles. The center O_4 of δ is also the circumcenter of $O_1 O_2 O_3$, because the points O_1, O_2, O_3 are at distance 2ρ from O_4. Hence O_4 is on the line connecting the center of homothety with the circumcenter of ABC, and this is the line connecting the incenter to the circumcenter of ABC. This solves (a).

For (b), let R, r be the circumradius and the inradius of the triangle ABC. The circumradius of $O_1 O_2 O_3$ is 2ρ, and so the homothety ratio is $2\rho/R$. But the homothety ratio can also be computed using the distance from the center of homothety, I, to the sides of $O_1 O_2 O_3$ and ABC as follows: the distance from I to AB is r, and the distance from I to $O_1 O_2$ is $r - \rho$; thus the homothety ratio is $(r - \rho)/r$. We thus have the equality

$$\frac{r - \rho}{r} = \frac{2\rho}{R},$$

which we can rewrite as

$$\frac{1}{\rho} = \frac{2}{R} + \frac{1}{r}.$$

Hence

$$\rho = \frac{Rr}{R + 2r}.$$

Remark The circles $\alpha, \beta, \gamma, \delta$ exist for every triangle ABC. Indeed, we can start with three small circles of equal radii tangent to pairs of sides, consider a circle that is exterior tangent to all three, and vary them continuously until the circle that is tangent to all three has radius equal to the radii of the three circles.

Even if δ does not have the same radius as α, β, γ, the argument still works to show that its center is on OI.

You can draw a parallel with Problem 94. Here the triangle $O_1O_2O_3$ is obtained by translating inward the sides of ABC by the same length. The two triangles have parallel sides, so they are homothetic by Problem 84. The center of homothety is necessarily the incenter.

Source Part (a) appeared in *Kvant (Quantum)*, proposed by V. Yagubiants.

100 We argue on Fig. 10.16. Let the quadrilateral be $ABCD$; let M, N, P, Q be the midpoints of AB, BC, CD, DA, respectively; and let I, J, H, K be the projections of the intersection T of the diagonals onto AB, BC, CD, DA, respectively.

We want to determine first the center of the circle on which the eight points should lie. Switch to complex coordinates, and let the coordinate axes be the diagonals AC and BD so that A, B, C, D have coordinates a, bi, c, di, respectively, with $a, b, c, d \in \mathbb{R}$. Then the coordinates of M, N, P, Q are $m = \frac{a+bi}{2}, n = \frac{c+bi}{2}, p = \frac{c+di}{2}, q = \frac{a+di}{2}$, respectively. Note that $(m - n)/(p - n) = i(a -$

Fig. 10.16 Proof that M, N, P, Q, I, J, H, K lie on a circle

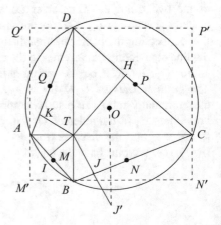

$c)/(b - d) \in i\mathbb{R}$, so $\angle MNP = 90°$, and the same is true for the other angles of the quadrilateral $MNPQ$; thus this quadrilateral is a rectangle. This can also be deduced from the fact that MN, NP, PQ, QM are midlines in the triangles ABC, BCD, DCA, ADB, so they are parallel to the bases. Therefore $MNPQ$ is cyclic, and its circumcenter is the midpoint of MP, which has coordinate $\frac{a+c}{4} + \frac{b+d}{4}i$. But the center of the circumcircle of $ABCD$ has the coordinate $\frac{a+c}{2} + \frac{b+d}{2}i$, because it lies at the intersection of the perpendicular bisectors of AC and BD, and these have equations $y = \frac{a+c}{2}$ and $x = \frac{b+d}{2}i$.

The homothety of center T and ratio 2 maps therefore M, N, P, Q, respectively, to some points M', N', P', Q' on a circle centered at O. We are left with showing that this homothety maps I, J, H, K to four points on the circumcircle of $M'N'P'Q'$. Let J' be the image of J, which is therefore the reflection of T over J. Then J' is the reflection of N' about the perpendicular bisector of BC (because the triangles $BJ'C$ and $CN'B$ transform into each other by this reflection). The perpendicular bisector passes through O, and hence $OJ' = ON'$. It follows that J' is on the circumcircle of $M'N'P'Q'$, and the same is true for I', H' and K'. This completes the proof.

Remark Of course we could have completed the solution without the use of the homothety, by computing the complex coordinates of I, J, H, K (as projections onto the sides), but that is more work.

101 We solve first the case where the circles ω_b and ω_c are not equal; for simplicity ω_b is smaller than ω_c. Let K be the intersection point of the exterior tangents ℓ_1 and ℓ_2 (Fig. 10.17). Consider the homothety of center K that takes ω_b to ω_c. The image A' of A through this homothety is diametrically opposite to A in the circle ω_c.

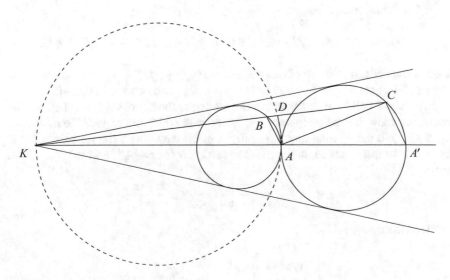

Fig. 10.17 Finding the locus of D

We therefore have $\angle ACA' = 90°$, and so $A'C$ is parallel to AB. It follows that C is the image of B through the homothety, and so K is on the line BC. We deduce that D lies on the circle ω of diameter AK. But D is also inside the angle formed by ℓ_1 and ℓ_2, and it is not hard to see that any point on the arc of ω that lies between ℓ_1 and ℓ_2 belongs to the locus. Hence the locus is this arc.

If ω_a and ω_b are equal, then the locus of D is the segment on the common tangent through A that is determined by the two common exterior tangents.

Remark A brute-force coordinate computation discloses immediately that $B \mapsto C$ by the direct homothety that maps ω_b to ω_c. Indeed, the equations of the two circles can be written as $|z - b| = b$ and $|z + c| = c$, for $b, c > 0$. Switching to polar coordinates ($z = re^{i\theta}$) turns these into $r = b \cos\theta$ and $r = -c \cos\theta$, and if B has coordinate $b \cos\theta(\cos\theta + i \sin\theta) = b(\cos^2\theta + i \sin\theta \cos\theta)$, then C has coordinate

$$-c\cos\left(\theta + \frac{\pi}{2}\right)e^{i(\theta + \frac{\pi}{2})} = c\sin\theta(i\cos\theta - \sin\theta) = c(\cos^2\theta + i\sin\theta\cos\theta) - c,$$

thus they are mapped into each other by the map $z \mapsto -\frac{c}{b}z + c$, which is a direct homothety. Since the homothety maps A to the point diametrically opposite on ω_c, the center of homothety is on the line of centers. Moreover, every point on one circle must be mapped to a point on the other and vice versa, so the homothety must be the one that maps the circles into each other.

102 We fix B and C and constrain A to lie above the line BC. Then A varies on an arc of measure $2\pi - 2\alpha$ with endpoints B and C. The centroid G of ABC lies on the image of that arc under a homothety of center the midpoint M of BC and ratio $1/3$. Call this arc, which is the locus of G, γ_G.

If I is the incenter, then

$$\angle BIC = \pi - \angle ABC/2 - \angle ACB/2 = \pi/2 + \angle BAC/2 = \pi/2 + \alpha/2.$$

So I is constrained to an arc of measure $2\pi - 2(\pi/2 + \alpha/2) = \pi - \alpha$ with endpoints B and C. Denote this arc by γ_I. The arcs γ_G and γ_I can be seen in Fig. 10.18.

We claim that the arc γ_I lies above the arc γ_G, as the figure suggests. This will be true if we show that the midpoint of γ_I is above the midpoint of γ_G. The midpoints coincide with the incenter and the centroid in the case where the triangle ABC is isosceles. In that situation, in the triangle IBC, $\angle BIM = \frac{\pi + \alpha}{4}$, and so

$$IM = \frac{BC}{2}\cot\frac{\pi + \alpha}{4} = \frac{1}{2}\cot\frac{\pi + \alpha}{4},$$

while in the triangle ABC,

$$GM = \frac{1}{3}AM = \frac{1}{6}\cot\frac{\alpha}{2}.$$

Fig. 10.18 How to minimize
the length of IG

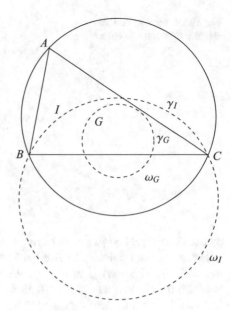

We have to show that $IM > GM$, which amounts to showing that

$$\frac{1}{2} \cot \frac{\pi + \alpha}{4} > \frac{1}{6} \cot \frac{\alpha}{2}.$$

Denoting $\alpha/4 = \theta$, we have to show that

$$\tan \left(\frac{\pi}{4} + \theta \right) < 3 \tan 2\theta, \text{ for } \frac{\pi}{12} < \theta < \frac{\pi}{4}.$$

Using the addition and the double angle formulas for the tangent, we transform this inequality into

$$\frac{1 + \tan \theta}{1 - \tan \theta} < 6 \frac{\tan \theta}{1 - \tan^2 \theta}.$$

The denominators are positive, so this is further equivalent to

$$\tan^2 \theta - 4 \tan \theta + 1 < 0.$$

Since $I = G$ for the equilateral triangle, this quadratic polynomial has a zero when $\theta = \pi/12$, or better said, when $\tan \theta = \tan \pi/12$. The product of the roots is 1, so the other root of the quadratic is greater than $1 = \tan \pi/4$. We conclude that when $\theta \in (\pi/12, \pi/4)$, the quadratic function assumes negative values, which proves the inequality.

Fig. 10.19 How to find the distance between two circles

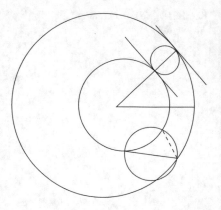

So the arcs γ_G and γ_I do not intersect, and we want to solve the easier problem of determining the minimal distance between them. Complete the arcs to two circles ω_G and ω_I. Then ω_G is inside ω_I. We want to determine the minimal distance between the circles and use Fig. 10.19 for our reasoning. Let $M \in \omega_G$ and $N \in \omega_I$ for which the minimum is attained. The circle of diameter MN is tangent to both ω_G and ω_I, for any other point of intersection with one of the two circles would determine with either M or N a shorter distance. Then the tangents at M and N to ω_G and ω_I are parallel, so $N = h(M)$, where h is the direct homothety that maps γ_G to γ_I. But if O is the center of this homothety and k is its ratio, then

$$\frac{MO + MN}{MO} = \frac{NO}{MO} = k,$$

so for MN to be minimal, MO has to be minimal. This means that the minimum is attained when M and N are on the line of centers. Rephrasing in terms of our problem, the minimum is attained for the isosceles triangle, and this minimum is

$$\frac{1}{2} \cot \frac{\pi + \alpha}{4} - \frac{1}{6} \cot \frac{\alpha}{2}.$$

The problem is solved.

Remark We were not required to find the minimal distance between the arcs γ_G and γ_I; as I and G do not vary independently on the two arcs, they both depend on one parameter, the position of A on the arc $\overset{\frown}{BAC}$. However, the configuration that minimizes the distance between the arcs, and hence between the two circles, does correspond to a triangle: the isosceles triangle. It is also important to point out that because $\angle BIC > \angle BAC$, it is the configuration formed by the incenter and the circumcenter of the isosceles triangle that yields the minimum (and not the configuration of the other two points on the line of centers of ω_I and ω_G).

Source Vietnamese Mathematical Olympiad, 1996.

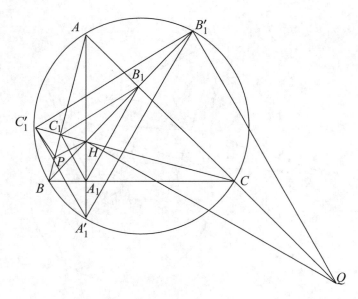

Fig. 10.20 The homothety with center H and ratio 2

103 Consider the homothety of center H and ratio 2, and let A'_1, B'_1, and C'_1 be the images of A_1, B_1, and C_1, respectively (Fig. 10.20). Then on the one hand A'_1, B'_1, C'_1 lie on the circumcircle of ABC, and on the other hand $A'_1 H, B'_1 H, C'_1 H$ are the angle bisectors of triangle $A'_1 B'_1 C'_1$ (since they are the images through the homothety of the angle bisectors $A_1 H, B_1 H, C_1 H$ of triangle $A_1 B_1 C_1$).

Because B'_1 is the reflection of H over AC, and C'_1 is the reflection of H over AB, it follows that $\angle H B'_1 Q = \angle Q H B'_1$, so

$$\angle C'_1 B'_1 Q = H B'_1 Q + \angle C'_1 B'_1 H = \angle Q H B_1 + \angle C_1 B_1 H$$
$$= \angle Q H B_1 + \angle H B_1 A_1 = 90°.$$

Similarly $\angle B'_1 C'_1 P = 90°$. Because of the same reflections $AB'_1 = AH = AC'_1$, so the perpendicular bisector of $B'_1 C'_1$ passes through A. But since PC'_1 and QB'_1 are parallel to this perpendicular bisector, it also passes through the midpoint of PQ (by Thales' Theorem). So the line that passes through A and the midpoint of PQ is the perpendicular bisector of $B'_1 C'_1$; hence it is perpendicular to $B_1 C_1$, as desired.

Remark An observation that leads to a different solution is that while H is the incenter of triangle $A_1 B_1 C_1$, A is its excenter.

Source The Danube Mathematical Competition, Romania, 2018, solution by Mircea Fianu.

104 Figure 10.21 shows a possible configuration, on which the subsequent construction is also marked. For the solution, let G be the centroid of the triangle ABC,

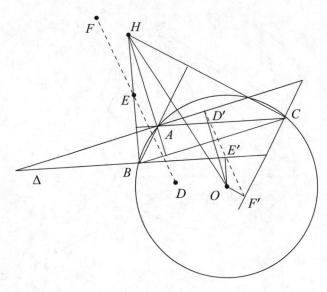

Fig. 10.21 The images of D, E, H through h

and consider the homothety h of center G and ratio $-1/2$. Then $h(H) = O$. With the standard notation: $X' = h(X)$, A', B', C' are the midpoints of BC, CA, AB, respectively. Furthermore, D' is the reflection of A' across $B'C'$. This means that the line AD' is the image of the line BC through h, so it is parallel to BC. Similarly BE' is parallel to AC, and CF' is parallel to AB, and so the lines AD', BE', CF' form a triangle Δ whose sides are parallel to those of triangle ABC and pass through the vertices of this triangle.

We should also note that OA' is perpendicular to BC and thus O, A', D' are collinear and OD' is perpendicular to AD'. Consequently D', E', F' are the projections of O onto the sides of the triangle Δ. The fact that they are collinear (which is the same as D, E, F being collinear), is equivalent, by the Simson-Wallace Theorem (Theorem 3.34), to the fact that O lies on the circumcircle of Δ. But note that Δ is the preimage of ABC through h, so the circumcenter of Δ is $H = h^{-1}(O)$ and its circumradius is $2R$. So the last condition is equivalent to $OH = 2R$, and we are done.

Remark The observation that A is the orthocenter of HBC and that the circumcircle of HBC is the reflection of the circumcircle of ABC over BC gives a clue on how to construct such a configuration.

Source Short list, 39th International Mathematical Olympiad, 1998, proposed by France, solution from D. Djukić, V. Janković, I. Matić, N. Petrović, *The IMO Compendium, A Collection of Problems Suggested for the International Mathematical Olympiads: 1959-2004*, Springer, 2006.

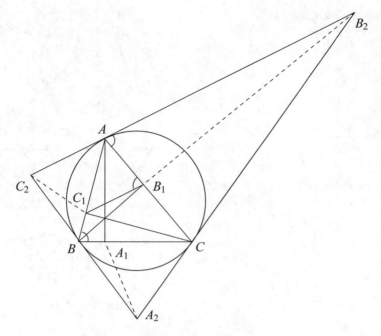

Fig. 10.22 The triangles $A_1B_1C_1$ and $A_2B_2C_2$ are homothetic

105 The configuration is shown in Fig. 10.22. Note that because BCB_1C_1 is cyclic, $\angle AB_1C_1 = \angle ABC$. Also, $\angle ABC = \overset{\frown}{AC}/2 = \angle B_2AC$. It follows that $\angle AB_1C_1 = \angle B_2AC$, so B_1C_1 and B_2C_2 are parallel. Similarly A_1C_1 is parallel to A_2C_2 and A_1B_1 is parallel to A_2B_2. Therefore the triangles $A_1B_1C_1$ and $A_2B_2C_2$ have parallel sides, and hence there is a homothety h such that $h(A_1) = A_2$, $h(B_1) = B_2$, and $h(C_1) = C_2$.

The incenter of $A_2B_2C_2$ is O, the circumcenter of ABC, while the incenter of $A_1B_1C_1$ is the orthocenter H of ABC. Then $h(H) = O$. Since the line that connects a point with its image through h passes through the center of h, it follows that A_1A_2, B_1B_2, C_1C_2, and HO are concurrent. But OH is the Euler line of the triangle ABC, and so the center of homothety, which is the intersection of A_1A_2, B_1B_2, C_1C_2, is on this line.

Source C. Mihalescu, *Geometria Elementelor Remarcabile (The Geometry of the Remarkable Elements)*, Ed. Tehnică, Bucharest, 1957.

106 Consider a homothety of center A and ratio $1/2$, and let B', C', and D' be the images of B, C, and D, respectively (see Fig. 10.23). Then B' and C' are the midpoints of the sides AB and AC, respectively. The triangles $NB'C'$ and $D'B'C'$ are both isosceles ($NB' = NC'$ and $D'B' = D'C'$), thus either $N = D'$ or the line ND' is the perpendicular bisector of $B'C'$.

Fig. 10.23 The homothety of center A and ratio $1/2$

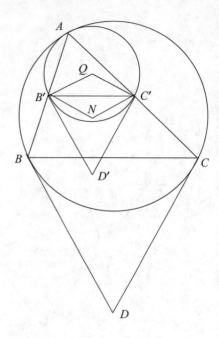

Now A, D, D' are collinear, so A, D, N are collinear if and only if A, D', N are collinear. This is impossible if $N \neq D'$, for in this case A would have to be on the perpendicular bisector of $B'C'$, and hence the triangle $AB'C'$ would be isosceles, and so would be the triangle ABC, contradicting the hypothesis of the problem. Thus A, D, N are collinear if and only if $D' = N$.

But N is the reflection over $B'C'$ of the circumcenter Q of $AB'C'$, so $\angle B'NC' = \angle B'QC' = 2\angle A$. Also, because $D'B'$ and $D'C'$ are tangent to the circumcircle of triangle $AB'C'$, $\angle B'D'C' = 180° - 2\angle A$. So A, D, N are collinear if and only if $2\angle A = 180° - 2\angle A$, that is if and only if $\angle A = 45°$, and we are done.

Source Brazilian Mathematical Olympiad, 2015.

107 Using inscribed angles in the two circles (Fig. 10.24), we obtain

$$\angle EFC = \angle ECA = \angle EDB.$$

It follows that the lines CF and AB are parallel.

Let G be the intersection of BK and CF, and let G' be the intersection of AE and CF. Now we use the fact that D is the midpoint of AB. The direct homothety that maps the triangle KAB to KCG maps D to F, and so F is the midpoint of CG. The inverse homothety that maps the triangle EAB to $EG'C$ maps D to F, so F is the midpoint of CG'. It follows that $G = G'$, and the problem is solved.

Source St. Petersburg Mathematical Olympiad, 1998.

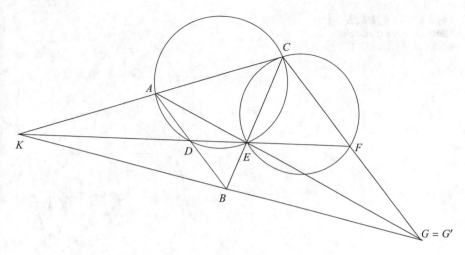

Fig. 10.24 Proof that AE, BK, DF are concurrent

108 Let us first notice that the circumcircles of the triangles ABC and DBC are congruent, being symmetric with respect to the line BC. Let R be their common radius. Then, using the well-known metric relation that computes the distance from the vertices to the orthocenter, we have

$$AH_1 = 2R \cos \angle A = 2R \cos \angle D = DH_2.$$

Since AH_1 and DH_2 are also parallel, it follows that AH_1DH_2 is a parallelogram, so the segments AD and H_1H_2 intersect at their midpoint P, as shown in Fig. 10.25.

On the other hand, if P' is the intersection of H_1H_2 and EF, by using the Law of Sines in the triangles $H_1P'E$ and $H_2P'F$, we obtain

$$\frac{H_1P'}{H_2P'} = \frac{H_1E \sin \angle BEF}{H_2F \sin \angle CFE} = \frac{AH_1 \cos \angle AH_1E \sin \angle BCF}{DH_2 \cos \angle DH_2F \sin \angle CBE}$$

$$= \frac{2R \cos \angle A \cos \angle ACB \cos \angle CBD}{2R \cos \angle D \cos \angle CBD \cos \angle ACB} = 1.$$

So $P = P'$, proving that lines AD, H_1H_2, and EF intersect at P.

Since the nine-point circle is the image of the circumcircle by the homothety centered at the orthocenter and of ratio $1/2$, and since the orthocenters H_1 and H_2 are on the circumcircles of DBC and ABC, we obtain that P, being both the image of H_2 under the homothety of center H_1 and ratio $1/2$ and the image of H_1 under the homothety of center H_2 and ratio $1/2$, is on both nine-point circles.

Remark The information that the common point of the three lines should be on the nine-point circle of the triangle ABC gives away the information that this point is

Fig. 10.25 AD, EF, H_1H_2
intersect on the nine-point
circles of ABC and BCD

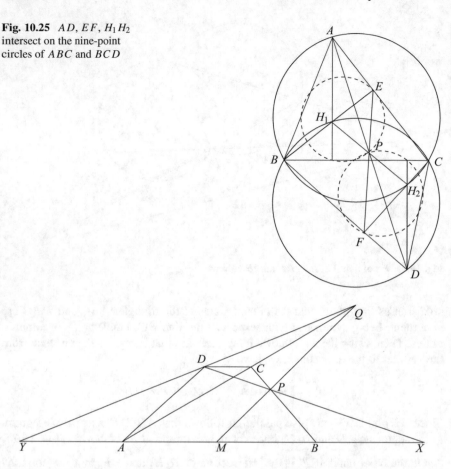

Fig. 10.26 M is the midpoint of XY

the image of H_2 (who is on the circumcircle) under the homothety of center H_1 and
ratio $1/2$, so it tells us that the common point is the midpoint of the segment H_1H_2.

Source Romanian Team Selection Test for the Junior Balkan Mathematical
Olympiad, 2013.

109 The situation is described in Fig. 10.26. Consider the direct homothety h_Q of
center Q that maps A to C and the inverse homothety h_P of center P that maps C
to B. By Theorem 2.6, the composition $h_Q \circ h_P$ is a homothety whose center is on
PQ. Also $h_Q \circ h_P$ maps A to B, and so its center is on AB. Hence $h_Q \circ h_P$ is a
homothety with center $AB \cap PQ$; this center is therefore M. But M is the midpoint
of AB, so the homothety $h_Q \circ h_P$ has ratio -1; it is the reflection over M.

We also have that $h_Q(Y) = D$ and $h_P(D) = X$, so $h_Q \circ h_P(Y) = X$. Thus the
reflection over M maps Y to X, showing that M is the midpoint of XY.

Remark The transformation $h_Q \circ h_P$ cannot be a translation because the ratios of the two homotheties have opposite signs, so they cannot be one the reciprocal of the other.

Source A. Engel, *Problem-Solving Strategies*, Springer, 1998.

110 Assume that the triangle PQR has been constructed (Fig. 10.27). Let S be the tangency point of ℓ and ω, T the reflection of S over M, and U the point diametrically opposite to S in ω. The line PU intersects ℓ at the point where the excircle is tangent to RQ, because, as we have explained in the introduction, the homothety that maps the incircle to the excircle maps U to this point. But, as we know, this point is T, the reflection of S over M. It follows that P lies on the open ray of the line UT that is opposite to $|UT$.

Conversely, for every P on this ray, we can construct the triangle PQR, and then T must be the tangency point of the excircle, and so T, U, P are collinear, and S is the tangency point of the incircle with ℓ. The problem is solved.

Source International Mathematical Olympiad, 1992, proposed by France.

111 The point Z is on the radical axis of the two circles, so power-of-a-point gives $ZA \cdot ZC = ZB \cdot ZD$ and we can set

$$k = \frac{ZC}{ZD} = \frac{ZB}{ZA}.$$

Fig. 10.27 The triangle
PQR

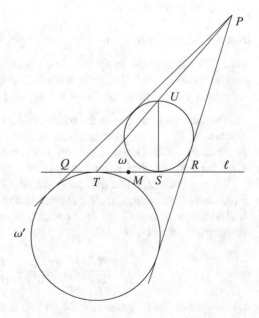

Fig. 10.28 The images of
BN and CM through the
homothety h

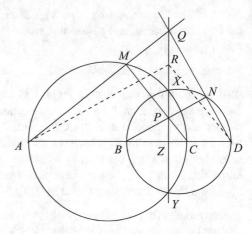

Consider the homothety h of center Z and ratio k. Then $h(B) = A$, so $h(BN)$ is a line through A that is parallel to BN and is therefore perpendicular to DN; similarly $h(CM)$ is a line through D that is perpendicular to AM (Fig. 10.28).

Let Q be the intersection of AM and DN and let R be the intersection of $h(BN)$ and $h(CM)$. Then Q is the orthocenter of the triangle RAD, and so RQ is perpendicular to AD. On the other hand since P is the intersection of BN and CM, $R = h(P)$, so Z, P, R are collinear. This means that R is on XY and since RQ is perpendicular to AD, so is Q. Therefore the lines AM, DN, and XY intersect at Q, and the problem is solved.

Source International Mathematical Olympiad, 1995, proposed by Bulgaria.

112 The solution can be followed on Fig. 10.29. Let O and G be the circumcenter and the centroid of the triangle ABC, respectively. The points O, G, O_9, H determine Euler's line, and $O_9 O = O_9 H = 3 O_9 G$.

The nine-point circles of the triangles ABC, BHC, CHA, and AHB coincide, because A, B, C are the orthocenters of BHC, CHA, AHB, respectively. Thus O_9 is the center of the nine-point circle in each of these triangles. So AO_9, BO_9, and CO_9 are the respective Euler lines of these three triangles.

The segment HA' is a median in the triangle BHC, so its intersection with the Euler line of this triangle is the centroid of BHC. Hence A_1 is the centroid of BHC, and similarly B_1 and C_1 are the centroids of the triangles CHA respectively AHB, respectively. We obtain

$$\frac{HA_1}{HA'} = \frac{HB_1}{HB'} = \frac{HC_1}{HC'} = \frac{2}{3}.$$

Consequently the triangles $A'B'C'$ and $A_1B_1C_1$ are homothetic, with the center of homothety being H and the ratio 2/3.

Fig. 10.29 The pairwise
homothetic triangles
$ABC, A'B'C', A_1B_1C_1$

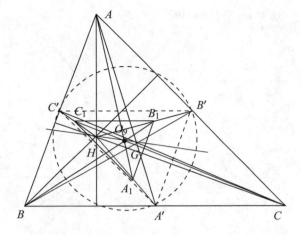

Fig. 10.30 Square inside a
circle

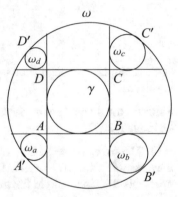

Also, the triangles ABC and $A'B'C'$ are homothetic, with the center of homothety being G and the ratio $-1/2$. By Theorem 2.6, the triangles $A_1B_1C_1$ and ABC are homothetic, with ratio $-1/3$ and the center of homothety on GH, dividing this segment in the ratio $1/3$. But the point with this property is O_9, the center of the nine-point circle of ABC. And because the Euler line of the triangle ABC passes through the center of homothety, it is mapped into itself by the homothety, so it is the Euler line of triangle $A_1B_1C_1$, as desired.

Source N. Grunbaum, *Gazeta Matematică (Mathematics Gazette, Bucharest)*, solution by E. Drăgănescu and B.M. Barbalatt, published in C. Mihalescu, *Geometria Elementelor Remarcabile (The Geometry of the Remarkable Elements)*, Ed. Tehnică, Bucharest, 1957.

113 The configuration is shown in Fig. 10.30. Let γ be the incircle of the square $ABCD$. Consider the inverse homothety h_a of center A that maps ω_a to γ and the direct homothety h'_a of center A' that maps ω to ω_a. Then $h_a \circ h'_a$ is the unique inverse homothety that maps ω to γ. Let T be its center. By Theorem 2.6 T is on

Fig. 10.31 The homothety of
center A

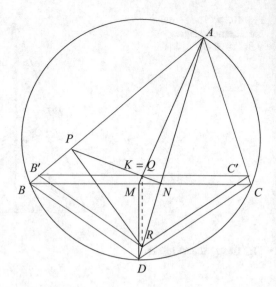

AA'. A similar argument shows that T is on BB', CC', and DD' and the problem is solved.

Source Romanian Team Selection Test for the International Mathematical Olympiad, 2003.

114 The case $AB = AC$ is obvious. Otherwise, without loss of generality, we may assume that $AB > AC$. Let D be the point where the angle bisector AN intersects the circumcircle for the second time. Then D is the midpoint of the arc $\overarc{BC}$, so $DB = DC$ and DM is the perpendicular bisector of BC. Consider now the homothety of center A that maps D to R (Fig. 10.31). Let $B' \in AB$ and $C' \in AC$ be the images of B and C. Then $B'C'$ is parallel to BC, and, because $DB = DC$, we have that $RB' = RC'$.

Let K be the point where $B'C'$ intersects PN. Then

$$\angle RB'K = \angle DBC = \angle DAC = \angle DAB = 90° - \angle ARP = \angle RPK.$$

So the points R, B', P, K are concyclic. It follows that $\angle B'KR = \angle B'PR = 90°$ and, consequently, K is the image of M through the homothety. This means that $K \in AM$, and therefore $K = Q$. We deduce that RQ is the perpendicular bisector of $B'C'$, so it is perpendicular to $B'C'$ and hence to BC, as desired.

Source Asian Pacific Mathematical Olympiad, 2000, solution by Bobby Poon.

115 Let M be the intersection of A_1A_2, B_1B_2, C_1C_2; let also P, Q, R be the intersections of the pairs (B_1C_1, B_2C_2), (A_1C_1, A_2C_2), and (A_1B_1, A_2B_2), respectively, as in Fig. 10.32. We apply repeatedly the trick behind the proof of Menelaus' Theorem (Sect. 2.2.1). We introduce six homotheties:

Fig. 10.32 Desargues'
Theorem

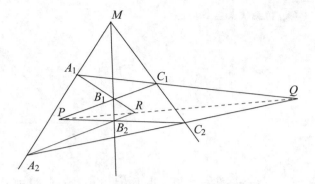

h_a of center A_1 that maps A_2 to M, h_b of center B_1 that maps B_2 to M,

h_c of center C_1 that maps C_2 to M, h_p of center P that maps B_2 to C_2,

h_q of center Q that maps C_2 to A_2, h_r of center R that maps A_2 to B_2.

Arguing like in the proof of Menelaus' Theorem for the triangle MA_2B_2 and the line of the points A_1, B_1, R, the triangle MA_2C_2 and the line of the points A_1, C_1, Q, and the triangle MB_2C_2 and the line of the points B_1, C_1, P, we obtain

$$h_b \circ h_r \circ h_a^{-1} = 1, \, h_a \circ h_q \circ h_c^{-1} = 1, \, h_c \circ h_p \circ h_b^{-1} = 1.$$

Composing we obtain

$$h_b \circ h_r \circ h_a^{-1} \circ h_a \circ h_q \circ h_c^{-1} \circ h_c \circ h_p \circ h_b^{-1} = h_b \circ h_r \circ h_q \circ h_p \circ h_b^{-1} = 1.$$

Consequently $h_r \circ h_q \circ h_p = 1$. So $h_r^{-1} = h_q \circ h_p$, and by Theorem 2.6, the centers of these homotheties, P, Q, R, are collinear.

Once the direct implication is proved, the converse is immediate. Let M be the intersection of A_1A_2 and B_1B_2, and let C_1' be the intersection of A_1C_1 and MA_2. Let also P' be the intersection of B_1C_1' and B_2C_2. Then P, Q, R are collinear by hypothesis, and P', Q, R are collinear by the direct implication. So both P and P' are at the intersection of RQ and B_2C_2. This can only happen if $P = P'$ and we are done.

Remark Of course, we could have invoked Menelaus' in the proof and do the algebra. We leave it to the reader to solve the cases where one or more of the points M, P, Q, R are at infinity.

116 Assume that AA_1, BB_1, and CC_1 intersect at M (Fig. 10.33). Consider the homotheties

Fig. 10.33 Ceva's Theorem

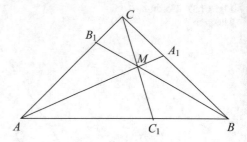

h_a of center A that maps B_1 to C, h_{a_1} of center A_1 that maps C to B,

h_{c_1} of center C_1 that maps B to A, and h_c of center C that maps A to B_1.

The composition $h_{a_1} \circ h_a$ maps B_1 to B and is not a translation because the image of A lies on AA_1, which intersects BB_1. So $h_{a_1} \circ h_a$ is a homothety, and its center lies on the line of centers AA_1 (by Theorem 2.6), but also on the segment BB_1 because B_1 is the image of B. Hence the center of $h_{a_1} \circ h_a$ is M. A similar argument shows that $h_c \circ h_{c_1}$ is a homothety of center M. But $h_c \circ h_{c_1} \circ h_{a_1} \circ h_a(B_1) = B_1$, so this composition of homotheties has two fixed points, B_1 and M. It must therefore be the identity map.

Thus the product of the ratios of the four homotheties is 1, meaning that

$$\frac{b_1 - c}{a - c} \cdot \frac{a - c_1}{b - c_1} \cdot \frac{b - a_1}{c - a_1} \cdot \frac{c - a}{b_1 - a} = 1,$$

which is equivalent to the identity from the statement.

For the converse, let M be the intersection of BB_1 and CC_1, and let A_1' be the intersection of AM and BC. Then A_1' and A_1 are the centers of homotheties of equal ratios that map B to C, so they coincide.

Remark Of course, this is just a translate in the language of homotheties of the proof of Ceva's Theorem by Menelaus' Theorem.

117 By just examining a carefully drawn figure (see Fig. 10.34), we discover some collinearities: K, M and P; K, N, and Q; and M, L, and N. To prove that the first two triples consist indeed of collinear points, consider the direct homothety that maps ω to Ω. Because homothety preserves tangencies, the image of AB is a line that is tangent to Ω at the image of M. But this tangent is also parallel to AB, so it meets Ω at the midpoint P of $\overarc{AB}$, and hence P is the image of M. Similarly, Q is the image of N, which proves the first two collinearities.

For the third collinearity, we use inscribed angles. From the fact that the quadrilaterals $BPML$ and $BQNL$ are cyclic, we obtain that the angles $\angle MLB$ and $\angle MPB$, as well as the angles $\angle NLB$ and $\angle NQB$, are supplementary. But in the quadrilateral $KPBQ$, $\angle MPB = \angle KPB$, and $\angle NQB = \angle KQB$ are supple-

Fig. 10.34 The points in the triples (K, M, P), (K, N, Q), (M, L, N) are collinear

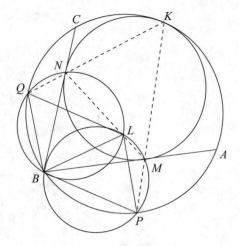

mentary, so $\angle MLB$ and $\angle NLB$ are supplementary, proving the third collinearity. Using this, the fact that the triangle BMN is isosceles (because BM and BN are the tangents from B to ω), and the fact that $BLMP$ and $BLNQ$ are cyclic, we obtain

$$\angle BPL = \angle BML = \angle BNL = \angle BQL.$$

Also

$$\angle LPB = \angle LMB = \angle NMB = \frac{1}{2}\,\overset{\frown}{MN} = \angle MKN = \angle PKQ$$

$$= 180° - \angle PBQ.$$

We conclude that the quadrilateral $BPLQ$ has two opposite equal angles and two adjacent supplementary angles, so it is a parallelogram.

Source Russian Mathematical Olympiad, 2000.

118 *First solution.* We argue on Fig. 10.35 and denote by $(I), (J), (E), (F)$ the circles specified in the statement of the problem with centers I, J, E, F, respectively. Let X be the intersection point of IJ and BC. Then X is the center of the homothety that maps the circle (I) to the circle (J), and A is the center of the homothety that maps the circle (I) to the circle (E). It follows that the center Y of the homothety that maps (J) to (E) is on AX, by the Monge-d'Alembert Theorem (Problem 86).

For the same reason, the centers of the direct homotheties that map (E) to (F), (E) to (J), and (J) to (F) are collinear. Let Z be the center of the first, while the centers of the other two are A and Y. It follows that Z is on AY. But Z is also the intersection of the exterior tangent BC of the circles (E) and (F) with the line of

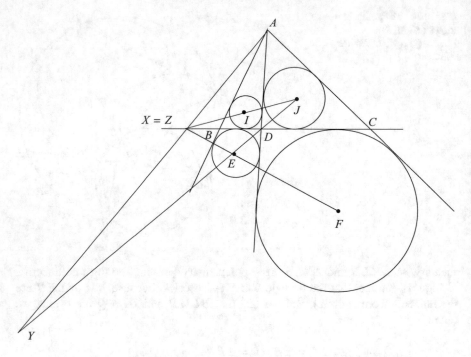

Fig. 10.35 Four circles connected by homotheties

the centers EF. We already know that the line $AY = AX$ intersects BC at X, so $Z = X$. Thus X is the intersection point of IJ, EF, and BC, and we are done.

Second solution. This problem can be solved using the fact that the triangles BIE and CJF are perspective. Note that IE and JF pass through A, being the angle bisectors of $\angle BAD$ and $\angle DAC$, respectively. Also, BI and CJ are bisectors of the interior angles at B and C, respectively, so they intersect at the incenter of ABC, and BE and CF are angle bisectors of the exterior angles at B and C, respectively, so they intersect at the center of the excircle of triangle ABC tangent to BC. The two centers are collinear with A, since they lie on the angle bisector of $\angle BAC$. So the intersection points of (IE, JF), (BE, CF), and (BI, CJ) are collinear. Desargues' Theorem (Problem 115) implies that the triangles BIE and CJF are perspective, as claimed, so IJ, BC, EF intersect at one point.

Remark As a by-product of this problem, we obtain that two excircles (E) and (F) are congruent if and only if the two incircles (I) and (J) are congruent.

119 The solution can be followed on Fig. 10.36. Let O_1, O_2, O_3 be the centers of ω_1, ω_2, ω_3, respectively. The key idea is that the intersection point of the common tangents of two exterior circles is the center of the direct homothety that maps the circles into each other. If A is the intersection of the common tangents of ω_1 and

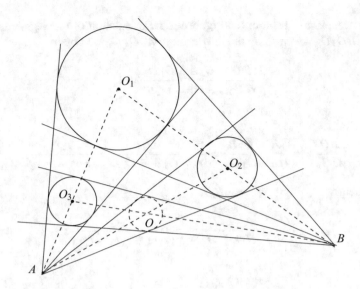

Fig. 10.36 Two pairs of common tangents for a circumscribable quadrilateral

ω_3, and B is the intersection of the common tangents of ω_1 and ω_2, then $A \in O_1O_3$ and $B \in O_1O_2$, and moreover, the ratio of the homothety that maps ω_2 to ω_1 is

$$\frac{r_1}{r_2} = \frac{BO_1}{BO_2},$$

and the ratio of the homothety that maps ω_3 to ω_1 is

$$\frac{r_1}{r_3} = \frac{AO_1}{AO_3}.$$

The center of the incircle of the quadrilateral formed by the tangents from B to ω_3 and from A to ω_2 should be at the intersection of AO_2 and BO_3, and we denote this intersection by O. We have to show that there is a circle ω centered at O that is mapped to ω_2 by a homothety of center A and to ω_3 by a homothety of center B. This amounts to showing that there is a number $r > 0$ such that

$$\frac{r}{r_2} = \frac{AO}{AO_2} \quad \text{and} \quad \frac{r}{r_3} = \frac{BO}{BO_3}.$$

And this reduces to showing that

$$r_2 \frac{AO}{AO_2} = r_3 \frac{BO}{BO_3}.$$

Writing Menelaus' Theorem (and ignoring the orientation of the segments) in the triangle BO_1O_3 traversed by the line through A, O, O_2 we obtain

$$\frac{BO}{OO_3} \cdot \frac{O_3A}{AO_1} \cdot \frac{O_1O_2}{O_2B} = 1,$$

hence

$$\frac{BO}{OO_3} = \frac{AO_1}{O_3A} \cdot \frac{O_2B}{O_1O_2} = \frac{r_1}{r_3} \cdot \frac{O_2B}{O_1B - O_2B} = \frac{r_1r_2}{r_3r_1 - r_3r_2}.$$

Consequently

$$\frac{BO}{BO_3} = \frac{BO}{BO + OO_3} = \frac{r_1r_2}{r_1r_2 + r_1r_3 - r_2r_3}.$$

Similarly

$$\frac{AO}{AO_2} = \frac{r_1r_3}{r_1r_2 + r_1r_3 - r_2r_3}.$$

We deduce that

$$r_2\frac{AO}{AO_2} = r_3\frac{BO}{BO_3} = \frac{r_1r_2r_3}{r_1r_2 + r_1r_3 - r_2r_3},$$

and the value of this expression is equal to r, the radius of the circle that is inscribed in the quadrilateral.

Remark Note that ω_2 and ω_3 are mapped into each other by a homothety of center A followed by a homothety of center B, but also by a homothety of center B followed by a homothety of center A. By the Monge-d'Alembert Theorem, the intersection of the common exterior tangents of ω and ω_1 and the intersection of the common exterior tangents of ω_2 and ω_3 lie both on the line AB.

Source All Soviet Union Mathematical Olympiad, 1984, proposed by L.P. Kuptsov.

120 (a) Consider the homothety of center A that maps D and E to M and N, respectively (Fig. 10.37). The square $BCDE$ is mapped to a square of side MN, whose other two vertices are on AB and AC. This square must be $MNPQ$.
(b) The three Lucas circles are shown in Fig. 10.38. For simplicity let $a = BC, b = AC, c = AB$. The homothety specified above maps the circumcircle of ABC to Ω_a. The radius R_a and of Ω_a and the radius R of the circumcircle satisfy $R_a/R = PQ/a$. On the other hand, using the extended Law of Sines, we obtain

$$\sin B = \frac{QM}{QB} = \frac{PQ}{c - AQ} = \frac{PQ}{c - 2R_a \sin C},$$

Fig. 10.37 Proof that
$MNPQ$ and $BCDE$ are
similar.

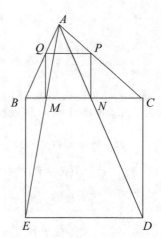

Fig. 10.38 Lucas circles

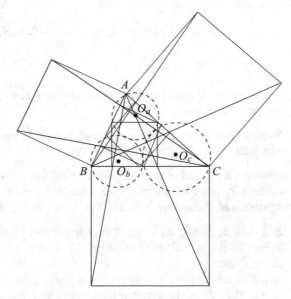

and also $\sin B = \frac{b}{2R}$ and $\sin C = \frac{c}{2R}$. An easy computation yields

$$R_a = \frac{Rbc}{bc + 2aR}$$

If O_a and O_b are the centers of the circles corresponding to the vertices A and B, respectively, then

$$OO_a = R - R_a = \frac{2aR^2}{bc + 2aR} = \frac{2aR}{bc}R_a$$

and

$$OO_b = R - R_b = \frac{2bR^2}{ac + 2bR} = \frac{2bR}{ac} R_b,$$

where R_b is the radius of the Lucas circle at B. Because $\angle O_a O O_b = \angle AOB = 2\angle C$, using the Laws of Cosines and Sines, we obtain

$$
\begin{aligned}
O_a O_b^2 &= O O_a^2 + O O_b^2 - 2 O O_a \cdot O O_b \cos 2C \\
&= (R - R_a)^2 + (R - R_b)^2 - 2(R - R_a)(R - R_b)(1 - 2\sin^2 C) \\
&= [(R - R_a - (R - R_b)]^2 + 4(R - R_a)(R - R_b)\sin^2 C \\
&= (R_a - R_b)^2 + 4\frac{2aR}{bc} R_a \cdot \frac{2bR}{ac} R_b \sin^2 C \\
&= (R_a - R_b)^2 + 4 R_a R_b \frac{4R^2 \sin^2 C}{c^2} = (R_a - R_b)^2 + 4 R_a R_b \\
&= (R_a + R_b)^2,
\end{aligned}
$$

hence Ω_a and Ω_b are tangent. And Ω_c is tangent to these two circles for a similar reason. The problem is solved.

Remark The Lucas circles and the circumcircle form a family of four circles in which any one is tangent to the other three.

Source This proof that the Lucas circles are pairwise tangent has appeared in Antreas P. Hatzipolakis and Paul Yiu, *The Lucas circles of a triangle*, in *The American Mathematical Monthly*, Vol. 108, No. 5, 2001.

121 Let $A_1 A_2$ and $A_3 A_4$ meet at Q and let $A_1 A_4$ and $A_2 A_3$ meet at P, as shown in Fig. 10.39. We will assume that P, T_1, T_3 are collinear and prove that Q, T_2, T_4 are collinear.

Construct the circle ω that is tangent to $A_1 A_2$ at T_1 and is also tangent to the line $A_3 A_4$. Then T_1 is the center of the inverse homothety that maps ω to ω_1, and P is the center of the direct homothety that maps ω_1 to ω_3. By the Monge-d'Alembert Theorem, proved in Problem 86, the center of the inverse homothety that maps ω to ω_3 is on PT_1. Furthermore, this center lies on the common tangent $A_3 A_4$ of these two circles. But PT_1 intersects $A_3 A_4$ at T_3, because this is our assumption. Thus T_3 is the center of the inverse homothety that maps ω to ω_3, and hence ω is tangent to $A_3 A_4$ at T_3. Thus $QT_1 = QT_3$, being the common tangents from Q to ω.

Using the fact that $QT_1 = QT_3$ and the fact that the two tangents from Q to ω_2 are equal, we obtain that the common internal tangents of the pair of circles ω_1 and ω_2 are equal to the common internal tangents of the pair of circles ω_3 and ω_2. Consequently, T_2 is the midpoint of the segment determined by the tangency points of $A_2 A_3$ to ω_1 and ω_3. Similarly, T_4 is the midpoint of the segment determined by

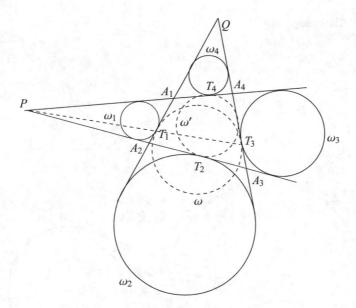

Fig. 10.39 Proof that P, T_1, T_3 collinear implies Q, T_2, T_4 collinear

the tangency points of $A_1 A_4$ to ω_1 and ω_3. Using this, and the fact that the common exterior tangents from P to ω_1 are equal and the common exterior tangents from P to ω_3 are equal, we deduce that $P T_2 = P T_4$. So there is a circle ω' tangent to $P T_2$ and $P T_4$ at T_2 and T_4, respectively. Applying this time the Monge-d'Alembert Theorem in reverse to the circles ω', ω_2, ω_4, we deduce that Q, T_2, T_4 are collinear.

Remark Note the similarity between this problem and Problem 113.

Source Romanian Master of Mathematics, 2010, proposed by Pavel Kozhevnikov, solution from T. Andreescu, C. Pohoață, *110 Geometry Problems for the International Mathematical Olympiad*, XYZ Press, 2014.

122 The argument repeats the pattern of the solution to Problem 121: metric relations that arise from equal tangents combined with applications of the Monge-d'Alembert Theorem. The four lines AB, BC, CD, and DA form a complete quadrilateral, and the circle ω is tangent to the four lines. This suggests that our quadrilateral might have properties analogous to those of a quadrilateral that admits an inscribed circle. It is a well-known fact that a quadrilateral admits an inscribed circle if and only if the sum of one pair of opposite sides is equal to the sum of the other pair of opposite sides, a result known as Pitot's Theorem. The analogous property for the complete quadrilateral in this problem states that

$$AB + AD = CB + CD.$$

Fig. 10.40 The analogue of
Pitot's Theorem

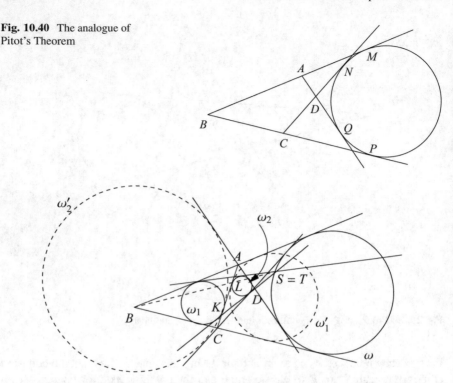

Fig. 10.41 Proof that the common exterior tangents intersect on ω

To prove the direct implication, which we need for this problem, let us denote
by M, N, P, Q the points where AB, BC, CD, DA are tangent to ω, respectively
(Fig. 10.40). Using the fact that the two tangents from a point to a circle are equal,
we can write

$$AB + AD = BM - AM + AQ - DQ = BP - AQ + AQ - DN = BP - DN$$
$$= BP - CP + CP - DN = BP - CP + CN - DN = BC + CD,$$

and the claim is proved.

We now continue the solution with the aid of Fig. 10.41. There is a surprising
consequence of the above Pitot-like relation: if K and L are the points where AC is
tangent to the circles ω_1 and ω_2, respectively, then

$$AL = \frac{AB + AC - BC}{2} = \frac{CD - AD + AC}{2} = CK.$$

Here we have used the computations from Sect. 2.2.1 II. From the observations
made in Sect. 2.2.1 II, we deduce that L is also the point of tangency of the B-

excircle ω_1' of the triangle ABC to the side AC and K is also the point of tangency of the D-excircle ω_2' of the triangle ACD to the side AC.

Once these facts have been established, we turn to geometric transformations and employ several homotheties.

Let S be the intersection of the common exterior tangents of ω_1 and ω_2. The homothety of center S that takes ω_2 into ω_1 is the composition of the homothety of center L that takes ω_2 in ω_1' and the homothety of center B that takes ω_1' to ω_1. By Theorem 2.6, S, L, B are collinear. Writing the homothety of center S that maps ω_2 to ω_1 as the composition of the homothety of center D that maps ω_2 to ω_2' and the homothety of center K that maps ω_2' to ω_1, we deduce that S, D, K are collinear. It follows that S is the intersection of the lines BL and DK.

Next, let K' be the point that is diametrically opposite to K in ω_1 and L' the point that is diametrically opposite to L in ω_2. Let T be the image of K' through the direct homothety that maps ω_1 to ω. Then B, K', T are collinear. Note that the tangents at K', L', T to the respective circles are parallel to AC, which means that T is the image of L' through the homothety of center D that maps ω_2 to ω. Consequently D, L', T are collinear. But B, K', L are also collinear because K' is mapped to L by the homothety that maps ω_1 to ω_1'. Thus B, K', L, S, T are collinear. Similarly D, K, L', S, T are collinear. It follows that $S = T$, being the unique point of intersection of these two lines. So the two tangents intersect on ω, as desired.

Remark Dan Schwarz has pointed out to us that the common interior tangents of ω_1 and ω_2 intersect at the intersection point U of the diagonals AC and BD. Obviously U is on AC, because AC is one of these common tangents. On the other hand, U is the center of the inverse homothety that maps ω_1 to ω_2, which you can write as the composition of the homothety of center B that maps ω_1 to ω, and the homothety of center D that maps ω to ω_2. Hence U, B, D are collinear.

The restriction $AB \neq BC$ from the statement is superfluous; the result holds true in the equality case as well.

Source International Mathematical Olympiad, 2008.

123 This problem is similar to the previous. We argue on Fig. 10.42, in which γ is the incircle of the triangle CPQ, Γ is the incircle of the triangle APQ, and Ω is the A-excircle of the triangle APQ.

Let γ touch PQ at U and let Ω touch PQ at V. If ω touches AB, BC, CD, and DA, respectively, at K, L, M, and N, then, by repeating the argument from the previous problem,

$$PU = \frac{CP + PQ - CQ}{2} = \frac{CP + CM + PQ - CQ - CL}{2}$$
$$= \frac{PM + PQ - QL}{2}$$

Fig. 10.42 An extension of the 2008 International Mathematical Olympiad problem

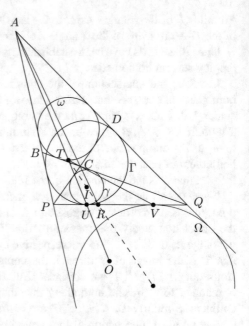

$$= \frac{PK + PQ - QN}{2} = \frac{PK + AK + PQ - QN - AN}{2}$$
$$= \frac{AP + PQ - AQ}{2},$$

so Γ also touches PQ at U. Since $PU = QV$, the C-excircle of CPQ also touches PQ at V. In some sense, APQ and CPQ are sister triangles!

Now let us take advantage of the several circles and tangencies in order to apply a series of homotheties. Place the figure such that the line PQ is horizontal, so the concepts of points being up/top or down/bottom make sense. The point U is the bottom point of both γ and Γ, and T is the bottom point of ω. Since there is a homothety with center A that takes ω to Γ, A, T, and U are collinear.

We are particularly interested in the center T' of the direct homothety that takes γ to Ω. From the observations made in Sect. 2.2.1 II applied in the triangle CPQ, if U' is the point diametrically opposite to U in γ, then, since V is also the tangency point of the C-excircle of CPQ in PQ, it follows that the points C, U', and V are collinear. Because U' and V are top points in γ and Ω, T' lies on CV as well. On the other hand, the inverse homothety that takes ω to γ has center C, so the bottom point V from ω is taken to the top point U'; that is, T lies on CV, too.

Now we add the circle ω back into the equation and consider the direct homotheties between γ, Ω, and ω. Their centers are T' (ω to γ), A (ω to Γ), and O (γ to ω). These three points are also collinear (Monge's Theorem proved in Problem 86), meaning that T' lies on AO. If instead we consider ω, γ, and Γ, the corresponding centers are O, U (γ to Γ), and A (ω to Γ). Then U, A, and O are also collinear for the same reason, and the point U also lies on AO. This means that T' lies on the line AU, which also passes through T.

We has just proved that AU and CV intersect at both T and T'. Let us rule out the case where T lies on AC, for then the configuration is symmetric with respect to AC, making the problem immediate. Thus we can conclude that $T = T'$.

Finally, we play with the circles ω, γ, and Ω once more, but now using inverse homotheties. The center of the inverse homothety that takes γ to ω is C, and the center of the direct homothety that takes ω to Γ is A, so the center of the inverse homothety that takes γ to Γ lies on the line AC (d'Alembert's Theorem; see Problem 86). It also lies on the common tangent PQ, so it is the intersection of AC and PQ, which is R.

Since R and T are the centers of the inverse and direct homotheties that take γ to Γ, both circle centers, one of which is the incenter I of CPQ, lie on the line RT, and we are done.

Source Brazilian Mathematical Olympiad, 2017, proposed by Géza Kós.

124 Place the original pentagon so that A_1 is at the origin of a system of complex coordinates, and let a_2, a_3, a_4, a_5 be the respective coordinates of the other vertices. Let A'_j be the translate of A_j by $\overrightarrow{A_1 A_j}$, $j = 2, 3, 4, 5$. These points have the coordinates $2a_2, 2a_3, 2a_4, 2a_5$; thus the pentagon $P = A_1 A'_2 A'_3 A'_4 A'_5$ is the image of $P_1 = A_1 A_2 A_3 A_4 A_5$ through a homothety of center A_1 and ratio 2. Because the center of the homothety is on the perimeter of P_1 and P_1 is convex, P_1 lies inside of P. Similar homotheties exist for P and each of P_2, P_3, P_4, P_5 (see Fig. 10.43), so P_1, P_2, P_3, P_4, P_5 lie all inside P. Since the similarity ratio between P and any of these polygons is 2, the ratio of their area is 4. But the sum of the areas of P_1, P_2, P_3, P_4, P_5 is 5 times the area of any one of them. So they must overlap inside P, and the problem is solved.

Fig. 10.43 A pentagon and its translates

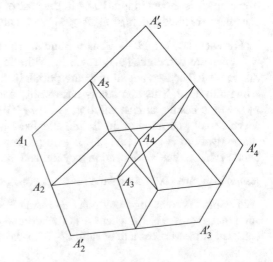

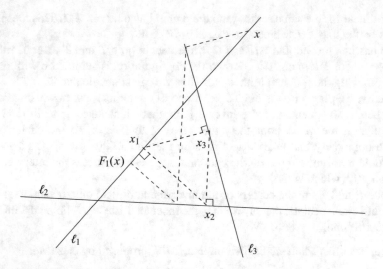

Fig. 10.44 The definition of F_1 and the points x_1, x_2, x_3 for the case of three lines

Remark Compare this to Problems 51, 53, 56. The idea is always the same: map your surface by isometries into some given region, then use the area of the region as an upper bound for the sum of the areas of the images of your surface to either bound the original area from above or to discover overlaps.

125 We study what the points x_k are supposed to be. Identify each ℓ_k with $\mathbb{R}$, and let $f_k : \ell_{k+1} \to \ell_k$ be the orthogonal projection, $1 \le k \le n$. Define

$$F_1 = f_1 \circ f_2 \circ \cdots \circ f_n, \quad F_k = f_k \circ \cdots \circ f_n \circ f_1 \circ \cdots \circ f_{k-2} \circ f_{k-1}, \quad k > 1.$$

The map F_1 is shown in Fig. 10.44 in the case of three lines. The points x_k from the statement are fixed points of $F_k, k \ge 1$. How many fixed points do these maps have?

For each k, $F_k : \mathbb{R} \to \mathbb{R}$ is a nonconstant linear transformation (because no two lines are perpendicular), hence is of the form $F_k(x) = a_k x + b_k$. Note also that $a_k < 1$ (because not all lines are parallel). It follows that F_k is a homothety of a line into itself. This homothety has exactly one fixed point because the equation $F_k(x) = x$ has the unique solution $x_k = b_k/(1 - a_k)$. But do these points fit into a cycle? All you have to do is find the fixed point x_1 of F_1 and then project it successively to $\ell_n, \ell_{n-1}, \ldots, \ell_2$. If you project the last point to ℓ_1, you must get back x_1; hence you are indeed in a cycle, and the problem is solved.

Remark Figure 10.44 shows the points x_1, x_2, x_3 in the case of three lines.

126 Note first that for every point P inside the triangle, there is a vertex, say A, such that $PA \le 9h/10$, where h is the altitude of the triangle. Indeed, the circles centered at the vertices and of radii $2h/3$ cover the interior, and $2/3 < 9/10$.

Fig. 10.45 The strategy of the grasshopper

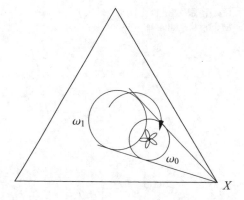

Let the flower be at C_0 and consider the disk ω_0 of center C_0 and some small radius ϵ. Let X be a vertex whose distance to the flower is less than $9h/10$. Consider the homothety of center X and ratio $10/9$, and let ω_1 the image of ω_0. The center C_1 of ω_1 is still inside the triangle, and if the grasshopper is inside ω_1, then when it jumps toward X, it lands inside ω_0.

Replace ω_0 by ω_1 and C_0 by C_1 and repeat the process to obtain the disk ω_2 centered at C_2. Inductively, obtain a sequence of disks $\omega_1, \omega_2, \omega_3, \ldots$ whose centers are inside the triangle and whose radii grow in an unbounded fashion. For some n, the nth disk will cover the triangle and will therefore contain the grasshopper inside. Now the grasshopper can jump

$$\omega_n \to \omega_{n-1} \to \cdots \to \omega_1 \to \omega_0,$$

landing within distance ϵ from the flower. The last step of grasshopper's strategy is illustrated in Fig. 10.45.

Remark It is worth mentioning that nobody solved the problem during the contest; the information that a homothety-based solution exists really helps.

Source 4th Mathematical Olympiad of the Former Soviet Countries, 2006.

127 In a system of complex coordinates, we let the positions at the start and the velocities of the turtles be $a_i, b_i \in \mathbb{C}$, respectively, $i = 1, 2, \ldots n$, where $|b_i| = r$ for all i. At time t, the homothety centered at the origin and with ratio $\frac{1}{rt}$ maps the ith turtle to the point $\frac{a_i}{rt} + \frac{b_i}{r}$, which can be made arbitrarily close to the point $\frac{b_i}{r}$ on the unit circle by choosing t sufficiently large. The points $\frac{b_i}{r}, i = 1, 2, \ldots, n$, being on the unit circle, are the vertices of a convex polygon. If t is sufficiently large, the images of the turtles are very close to these points, and so they are still the vertices of a convex polygon. The turtles themselves will be at the vertices of the inverse image of this polygon through the homothety.

Remark In this problem we change the scale, we look from afar, and the turtles seem to start at the same point. The problem is not true if the turtles have different

Fig. 10.46 Strategy for finding the squares

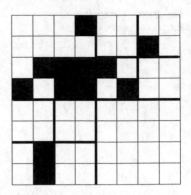

velocities, and homothety can be used to construct a counterexample. (Do you see how?)

Source *Kvant (Quantum)*, proposed by V. Prasolov.

128 If we compare the numbers $1/5$ and $4/5$, we notice the factor 4, which suggests immediately the idea of scaling: if a square K has too little black, make it four times smaller.

Take n sufficiently large so that all black squares lie in some $2^n \times 2^n$ square and such that their area is at most $1/5$ of the area of this square. Cut this square into four equal squares. The black region of each of these squares is at most $4/5$ of its area. Those squares whose black region is more than $1/5$ of their surface are retained, those that have no black squares are discarded, and for the others we repeat the process. After finitely many steps, one obtains the desired family of squares. This strategy is illustrated in Fig. 10.46, where the bold lines are the cuts and the squares that contain no black are discarded.

Remark The problem can be rephrased in space. What should the numbers $1/5$ and $4/5$ be replaced with?

Source *Kvant (Quantum)*, proposed by G.A. Rosenblium.

129 The solution uses the following result:

Lemma (I.M. Yaglom) *Every convex polygon that is not a parallelogram has three sides such that the polygon itself lies inside the triangle formed by the lines of support of these sides.*

Proof Let $\mathcal{P}$ be the polygon. If $\mathcal{P}$ is a triangle, there is nothing to prove. Otherwise $\mathcal{P}$ has two nonparallel sides that do not share a vertex. Extend these sides until they intersect, thus producing a polygon $\mathcal{P}'$ with fewer sides than P (see Fig. 10.47).

We keep repeating this procedure until we are stuck. This can only happen if we obtain a triangle, in which case we are done, or if we obtain a parallelogram. In the latter situation we go one step back and adjust our procedure so as not to

Fig. 10.47 Proof of
Yaglom's Lemma

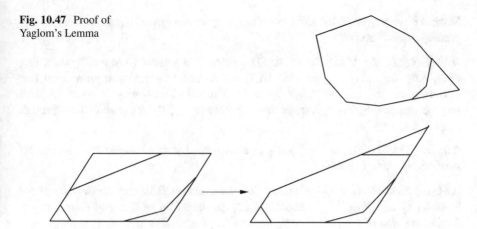

Fig. 10.48 Proof of Yaglom's Lemma

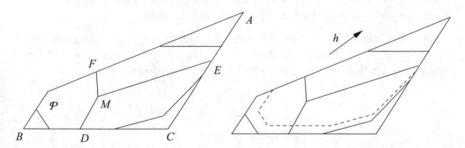

Fig. 10.49 Proof of the Gelfand-Markus Theorem

obtain a parallelogram, using a different pair of sides (see Fig. 10.48). This proves
the lemma.   □

Returning to the problem, consider three sides of P whose lines of support form
a triangle ABC containing P inside. Choose in the interiors of the three sides the
points D, E, F, so that $D \in BC$, $E \in AC$, $F \in AB$. Let also M be a point inside
P. The segments MD, ME, MF dissect the polygon into three polygons P_1, P_2,
respectively, P_3 (Fig. 10.49).

A homothety of center A and ratio slightly less than 1 transforms P into a
polygon still covering P_1. Analogous homotheties with centers B and C transform
P into polygons that cover P_2, respectively, P_3. Hence the three images of P
through these homotheties cover P entirely.

In the case of a parallelogram, the four vertices must be in different homothetic
copies of the parallelogram, so the number is at least 4. And, of course, 4 copies
suffice. The problem is solved.

Remark The result is true for general convex regions in the plane, as any such
region can be approximated by convex polygons. Israel Gohberg and Alexander

Markus have proposed the same problem in space. In space, what numbers should correspond to 3 and 4?

130 Because the triangles are similar, there is a positive number k such that $AC_1/AB_1 = AC_2/AB_2 = AC_3/AB_3 = k$, and a directed angle α such that $\angle B_1AC_1 = \angle B_2AC_2 = \angle B_3AC_3 = \alpha$. The spiral similarity of center A with angle α and scaling factor k maps the line through B_1, B_2, B_3 into the line through C_1, C_2, C_3.

Remark The triangles are mapped into each other by spiral similarities, and spiral similarities come in pairs.

131 (a) *First solution.* Consider $X_1, X_2 \in \omega$, and let P be the intersection of the lines X_1X_1' and X_2X_2'. By Theorem 2.22, the center M of the spiral similarity is the intersection of the circumcircles of PX_1X_2 and $PX_1'X_2'$. Thus P is the second intersection point of the circumcircles of MX_1X_2 and $MX_1'X_2'$, and this is the second intersection point of ω and $s(\omega)$. If follows that the lines XX' pass through this intersection point. The situation is shown in Fig. 10.50.

Second solution. Let ω be the unit circle, and let M have coordinate equal to 1. Let s have ratio r and angle α, so that $s(z) = w(z - 1) + 1$, where $w = re^{i\alpha}$. For a point $\epsilon \in \omega$, the line determined by ϵ and $s(\epsilon)$ has the equation

$$\begin{vmatrix} 1 & 1 & 1 \\ z & \epsilon & w(\epsilon - 1) + 1 \\ \overline{z} & \overline{\epsilon} & \overline{w}(\overline{\epsilon} - 1) + 1 \end{vmatrix} = 0,$$

which is the same as

$$(z - \epsilon)(\overline{w} - 1)(\overline{\epsilon} - 1) - (\overline{z} - \overline{\epsilon})(w - 1)(\epsilon - 1) = 0.$$

Let $u = (w - 1)/(\overline{w} - 1)$, which is a number of absolute value 1. Then, using the fact that $\epsilon\overline{\epsilon} = 1$, we can write the equation of the line as

Fig. 10.50 The intersection point of the pencil of lines determined by points and their images

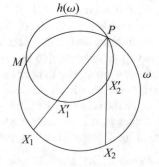

$$(\bar{\epsilon} - 1)z + (\epsilon - 1) = u(\epsilon - 1)\bar{z} + u(\bar{\epsilon} - 1).$$

Now we can write this equation for another point on the unit circle and solve the system in z and $\bar{z}$ to find the intersection point of the two lines and then notice that this intersection point does not depend on the two chosen points, or simply notice that u itself satisfies the equation of the line, and therefore u is the common intersection point of all such lines. Since u has absolute value 1, it is on the unit circle, that is, on ω. But by exchanging the roles of ω and $s(\omega)$, that is, by considering s^{-1} instead of s, we deduce that u must be also on $s(\omega)$, so u is the second intersection point of these circles.

Remark This shows how to construct the image of a point $X \in \omega$ under the spiral similarity: intersect the line AX with $s(\omega)$.

132 Look at Fig. 10.51. Consider a spiral similarity $s(B)$ of center X that maps the circle through B; call it ω to the circle through C; call it ω'. Because AB is tangent to $\omega' = s(\omega)$, $s(B) = A$ and $s(A) = C$, as we are in the limiting case of the result that was proved in Problem 131. So we are in the situation described in Sect. 2.4.3, and we can invoke Theorem 2.23 to conclude that X lies on the symmedian from A in the triangle ABC.

133 *First solution.* It is natural to start by locating the centers of possible spiral similarities (see Fig. 10.52). By Theorem 2.22, the center O of such a spiral similarity is on the circumcircle of PA_1A_2. And moreover, it should satisfy $OA_2/OA_1 = k$; hence it belongs to the locus of the points X with the property that $XA_2/XA_1 = k$, and this is the Apollonian circle in question. There are two intersection points of the two circles, so there are at most two such spiral similarities (because the spiral similarity is completely determined by the center and the image of a point).

On the other hand, if O is one of the intersection points, then using inscribed angles, we obtain $\angle(OA_1, OA_2) = \angle(PA_1, PA_2) = \angle(\ell_1, \ell_2)$, and once we know that it maps A_1 to A_2 it must map ℓ_1 to ℓ_2. Hence there are exactly two such spiral similarities.

Second solution. The above geometric facts can be read immediately in the analytic equation of the spiral similarity of center b, ratio k, and angle α:

Fig. 10.51 Spiral similarities
and symmedians

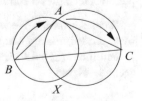

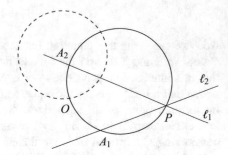

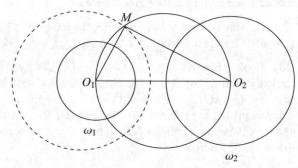

$$\frac{s(a_1) - b}{a_1 - b} = ke^{i\alpha},$$

From here we infer that the angle of the spiral similarity should be the angle between
the two lines, but that is also the angle formed by the line through b and $s(a_1) = a_2$
with the line through b and a_1, which implies that b, a_1, a_2, p are concyclic. On the
other hand, taking absolute values, we obtain $|s(a_1) - b|/|a_1 - b| = |a_2 - b|/|a_1 - b| = k$, which shows that b is on the said Appolonian circle.

134 *First solution.* Let the centers and radii of ω_1 and ω_2 be O_1, r_1 and O_2, r_2,
respectively, and let the spiral similarity be of center M and scaling factor k
(Fig. 10.53). Then $r_2/r_1 = k$, so the scaling factor is completely determined, and
since O_1 is mapped to O_2, the center M of the spiral similarity is on the Apollonian
circle $MO_2/MO_1 = k$. Also, because $\angle O_2MO_1 = 90°$, M is on the circle of
diameter O_1O_2. The two circles intersect at two points, and each of the intersection
points determines a spiral similarity.

Second solution. We want to map the circle $|z - a_1| = r_1$ to the circle $|z - a_2| = r_2$
by a spiral similarity whose equation is $s(z) - m = \pm ki(z - m)$. The spiral similarity
is therefore $s(z) = \pm kiz - (\pm kiz - 1)$, which we substitute in the equation of the
second circle.

$$|\pm kiz - (\pm kiz - 1)m - a_2| = r_2 \iff \left| z - \left(1 \mp \frac{1}{ki} \right) m - a_2 \right| = \frac{r_2}{k}.$$

The result should be the equation of the first circle. For that we should have $r_2/k = r_1$, meaning that $k = r_2/r_1$ and $\left(1 \mp \frac{1}{ki}\right)m - a_2 = a_1$, which then yields one solution for each choice of signs. Hence there are exactly two spiral similarities.

Remark To emphasize the idea of the second solution, for a curve $F_1(z) = 0$ to be mapped to a curve $F_2(z) = 0$ by s, the equations $F_2(s(z)) = 0$ and $F_1(z) = 0$ should describe the same set of points.

135 Let $\mathcal{P}$ be the polygon. On a side of $\mathcal{P}$, choose a point M that is not a vertex and consider a spiral similarity s with center M, angle $\angle BAC$, and ratio AB/AC. Then $\mathcal{P}$ and $s(\mathcal{P})$ cross at M, and so they must cross at one more point (at least). Let P be the second crossing point of $\mathcal{P}$ and $s(\mathcal{P})$, and let $N = s^{-1}(\mathcal{P})$. Then N is on $\mathcal{P}$. We have $\angle NMP = \angle BAC$, and $MN/MP = AB/AC$, so the triangles MNP and ABC are similar. Done.

136 *First solution.* Consider the spiral similarity of center H, angle $90°$, and ratio BP/BC, which maps the triangle BHC to PHB (see Fig. 10.54; the angle of the spiral similarity in the figure is oriented clockwise). This spiral similarity maps C to B, and therefore it maps the line CD to a line through B that is perpendicular to CD, and this line is BC. Since $BQ/CD = BP/BC$, it follows that D is mapped to Q, and therefore $\angle DHQ = 90°$, as desired.

Second solution. For a coordinate solution place the origin at B, and let A, C, D have coordinates $i, 1, 1 + i$, respectively. Let $0 < t < 1$ so that $P = it, Q = t$. The line CP has the equation

$$\begin{vmatrix} 1 & 1 & 1 \\ 1 & it & z \\ 1 & -it & \overline{z} \end{vmatrix} = 0 \iff (t - i)z + (t + i)\overline{z} - 2t = 0.$$

The projection of the origin B onto PC is the point H whose coordinate is

$$h = \frac{(t - i) \cdot 0 - (t + i) \cdot 0 + 2t}{2(t - i)} = \frac{t}{t - i}.$$

Fig. 10.54 $\angle DHQ = 90°$

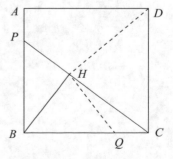

The vector $\overrightarrow{HD}$ has the complex number form

$$1 + i - h = 1 + i - \frac{t}{t-i} = \frac{it - i + 1}{t-i},$$

while the vector $\overrightarrow{HQ}$ has the complex number form

$$t - h = t - \frac{t}{t-i} = \frac{t^2 - it - t}{t-i} = (-it)\frac{it - i + 1}{t-i}.$$

We recognize that the second vector is the first multiplied by $-it$; thus Q is the image of D through the spiral similarity of center H, clockwise angle $90°$, and ratio $t = BP/BC$.

137 (a) The pentagons can be mapped into each other by a spiral similarity or a translation. We may therefore invoke the Averaging Principle (Theorem 2.18) to conclude that the midpoints of the given segments form a pentagon similar to each of the two pentagons and hence regular. An instance of this result is shown in Fig. 10.55.

(b) Consider the spiral similarity s of center A_1 that maps the circumcircle of $A_1A_2A_3A_4A_5$ to the circumcircle of $A_1A_2'A_3'A_4'A_5'$. Then $s(A_j) = A_j'$, $j = 2, 3, 4, 5$, and the conclusion follows from Problem 131.

Remark So the Averaging Principle is true for all transformations of the form $f(z) = rz + s$, $r, s \in \mathbb{C}$, including translations.

138 Examining Fig. 10.56 we observe that the triangle ABC is mapped into the triangle AOD by a spiral similarity of center A. The points M, N, A are the midpoints of the segments BO, CD, and the degenerate segment AA. Applying the

Fig. 10.55 The Averaging Principle applied to regular pentagons

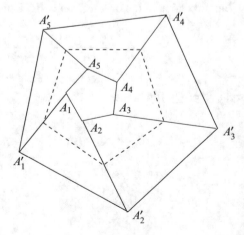

Fig. 10.56 A spiral
similarity hidden in a square

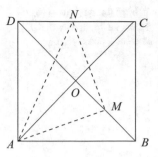

Fig. 10.57 Equilateral
triangles with two sides
sharing the midpoint

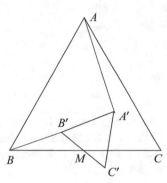

Averaging Principle (Theorem 2.18), we obtain that the triangle AMN is similar to
each of the triangles ABC and AOD, so it is an isosceles right triangle.

139 Let M be the common midpoint of BC and $B'C'$ (Fig. 10.57). Consider the
spiral similarity s of center M, angle $90°$, and ratio $\sqrt{3}$ that maps B to A. Then
$s(B') = A'$ and consequently $s(BB') = AA'$. We thus obtain that $AA'/BB' = \sqrt{3}$,
and the two segments form a $90°$ angle.

140 *First solution.* Let A_3, B_3, C_3 be the midpoints of the segments OA, OB, OC,
respectively. Then ABC is mapped to $A_3B_3C_3$ by a homothety h of center O and
ratio $1/2$. On the other hand, the triangles ABC and $A_1B_1C_1$ are directly similar, so
by Proposition 2.21, there is a spiral similarity s that maps $A_1B_1C_1$ to ABC. Then
$h \circ s$ is a spiral similarity that maps $A_1B_1C_1$ to $A_3B_3C_3$. The points A_3, B_3, C_3
divide the segments A_1A_2, B_1B_2, and C_1C_2 in half (as they are the centers of the
three parallelograms), so we can apply the Averaging Principle (Theorem 2.18) to
the triangles $A_1B_1C_1$ and $A_3B_3C_3$ and the points A_2, B_2, C_2 in order to deduce
that there is a spiral similarity that maps $A_1B_1C_1$ to $A_2B_2C_2$. Therefore $A_2B_2C_2$
is similar to $A_1B_1C_1$, and so it is similar to ABC. The situation is described in
Fig. 10.58, with the three triangles that appear when using the Averaging Principle
drawn with dotted line.

Second solution. We can be bolder and apply the general form of the Averaging
Principle (Theorem 2.19) to the similar triangles ABC, $A_1B_1C_1$ and the complex

Fig. 10.58 The Averaging
Principle applied to $A_1B_1C_1$,
$A_3B_3C_3$, and $A_2B_2C_2$

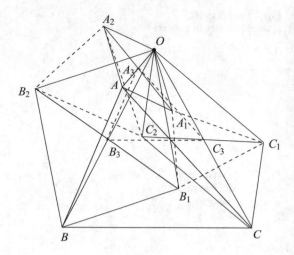

coefficients 1 and -1. These coefficients define the points A_2, B_2, C_2 if the origin
of the coordinate system is chosen at O, because $a_1 + a_2 = a + 0$, $b_1 + b_2 = b + 0$,
and $c_1 + c_2 = c + 0$. Hence $A_2B_2C_2$ is similar to each of the triangles ABC and
$A_1B_1C_1$.

141 Taking a closer look at the triangle MA_jM_j, we see that M_j is the image
of A_j under the spiral similarity s of center M, angle $\frac{\alpha}{2} - 90°$, and ratio $2\sin\frac{\alpha}{2}$.
Consequently $M_1M_2 \ldots M_n$ is the image of $A_1A_2 \ldots A_n$ by the spiral similarity s,
and so its area is $4S\sin^2\frac{\alpha}{2}$.

Remark Note the spiral similarity hidden in the rotation!

Source Short list of the International Mathematical Olympiad, 1989, proposed by
Mongolia.

142 The argument can be followed on Fig. 10.59. Using inscribed angles and the
fact that AM and AN are tangents, we obtain that

$$\angle MAB = \angle BNA \text{ and } \angle AMB = \angle BAN.$$

Consider a spiral similarity of center A that maps AMB to $AC'N$. Then the angle
between the lines $C'N$ and MB is the angle of the spiral similarity, which is
$\angle BAN = \angle AMB$. Hence the lines AM and $C'N$ form the same angle with MB, so
they are parallel.

On the other hand, as spiral similarities come in pairs (Theorem 2.25), there is a
spiral similarity that maps the triangle AMC' to ABN. The angle between MC' and
BN is the angle of the spiral similarity, so it is equal to $\angle MAB = \angle BNA$. Thus
MC' and AN make the same angle with BN, so they are parallel. It follows that
$AMC'N$ is a parallelogram, showing that $C = C'$.

Fig. 10.59 Proof that
$\angle APQ = \angle ANC$

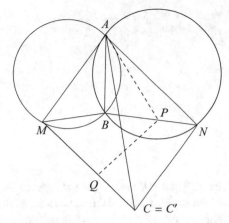

Fig. 10.60 The closed path
of the grasshopper

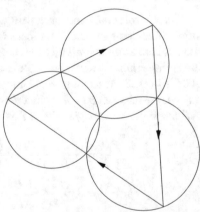

We deduce that AMB and ACN are mapped into each other by a spiral similarity, so we can invoke the Averaging Principle (Theorem 2.18) to conclude that there is a spiral similarity that maps ACN to AQP. Consequently $\angle ANC = \angle APQ$, and we are done.

143 A situation with three circles is shown in Fig. 10.60. Let s_i be the spiral similarity of center O that maps ω_i to ω_{i+1}, ($\omega_{n+1} = \omega_1$). Problem 131 shows that $s_i(X_i) = X_{i+1}$. Also $s_n \circ s_{n-1} \circ \cdots \circ s_1$ is a spiral similarity that maps ω_1 into itself. As the center of this transformation lies on ω_1, this transformation can map ω_1 into itself only if it is the identity transformation. Which implies that $X_{n+1} = X_1$, and the problem is solved.

144 *First solution.* The configuration is shown in Fig. 10.61. Let $\alpha_1 = \angle PBA = \angle RBC = \angle QAC$ and $\alpha_2 = \angle PAB = \angle RCB = \angle QCA$ and let also $k_1 = PB/AB = RB/BC = QA/CA$, $k_2 = PA/AB = RC/BC = QC/AC$. The spiral similarity of center B, angle α_1, and ratio k_1 maps A to P and C to R. The spiral similarity of the same angle and ratio but of center A maps C to Q and keeps A

Fig. 10.61 $APRQ$ is a
parallelogram

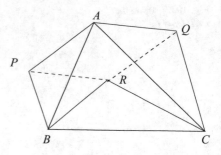

fixed. Thus PR and AQ are images of the same segment through spiral similarities
of the same angle and the same ratio, so they are parallel and have equal length. This
shows that $APRQ$ is a parallelogram.

Second solution. The argument is an application of Theorem 2.25: "spiral
similarities come in pairs." As there is a spiral similarity that maps APB to RBC,
there is also a spiral similarity that maps PBR to ABC. Its angle is $-\alpha_1$. Also there
is a spiral similarity that maps BRC to AQC, so there is a spiral similarity that maps
ABC to QRC. Its angle is α_2. Composing we find that there is a spiral similarity of
angle $\alpha_2 - \alpha_1 = \angle(BP, AP)$ that maps the triangle PBR to QRC. Hence QR and
AP make the same angle with BP, so they are parallel. Moreover,

$$\frac{PB}{QR} = \frac{PB}{AB} \cdot \frac{AB}{QR} = \frac{PB}{AB} \cdot \frac{AC}{QC} = \frac{PB}{AB} \cdot \frac{AC}{AQ} \cdot \frac{AQ}{QC}$$

$$= k_1 \cdot \frac{1}{k_1} \cdot \frac{PB}{AP} = \frac{PB}{AP}.$$

Hence $QR = AP$, showing that $APRQ$ is a parallelogram.

Third solution. We use coordinates, with the usual convention that the lower case
letter is coordinate of the upper case point. With the same notation as in the first
solution, P is obtained from A by a spiral similarity of center B, ratio k_1, and angle
α_1, so

$$\frac{p - b}{a - b} = k_1 e^{i\alpha_1}.$$

Therefore $p = b + k_1 e^{i\alpha_1}(a - b)$.

The same spiral similarity maps C to R, while a spiral similarity of center A and
same angle and ratio maps C to Q. Hence

$$r = b + k_1 e^{i\alpha_1}(c - b), \quad q = a + k_1 e^{i\alpha_1}(c - a).$$

Then

$$\frac{a+r}{2} = \frac{a+b}{2} + \frac{k_1}{2}e^{i\alpha_1}(c-b) = \frac{p+q}{2},$$

showing that PQ and AR have the same midpoint; hence $APRQ$ is a parallelogram.

Remark The original version of the problem was phrased with three isosceles triangles. This more general version discloses the trick, as directly similar triangles lead to spiral similarities.

Source Romanian Mathematical Olympiad, 2001, the particular case where the three similar triangles are isosceles was on the short list of the International Mathematical Olympiad in 1983, proposed by Belgium.

145 Denote by K, L, M the midpoints of the sides BC, CA, AB, respectively. First, we verify the case where $P = M$ and $Q = L$ (note that in this case $AMOL$ is cyclic because it has two opposite right angles). Because OK, OL, OM are orthogonal to BC, CA, AB, respectively, they are also orthogonal to LM, MK, KL, respectively. It follows that O is the orthocenter of KLM, and consequently K is the orthocenter of OML, which solves the problem in this case.

In the general case, shown in Fig. 10.62, we have

$$\angle APO = 180° - \angle AQO = \angle OQC,$$

and so $\angle OPM = \angle OQL$, showing that the right triangles OPM and OQN are directly similar. It follows that there is a spiral similarity s_1 of center O such that $s_1(OPM) = OQN$. And since spiral similarities come in pairs (Theorem 2.25), there is a spiral similarity s_2 of center O such that $s(OML) = OPQ$.

Let $H = s_2(K)$. As O is the orthocenter of KLM, O is also the orthocenter of HPQ, so H is the orthocenter of POQ. And as $s_2(OMK) = OPH$ and spiral similarities come in pairs, there is a spiral similarity that maps OMP to OKH. Hence $\angle OKH = \angle OMP = 90°$. This implies that the line KH is perpendicular to OK, so it must coincide with the line BC. Therefore H is on BC, and the problem is solved.

Fig. 10.62 The orthocenter of POQ is on BC

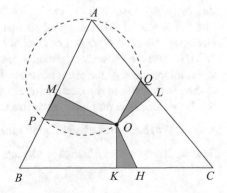

Fig. 10.63 O_1O_2 passes through N

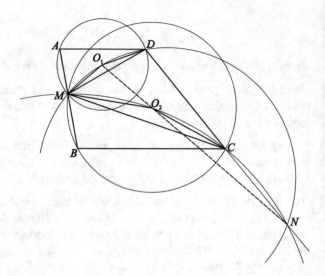

Source All-Russian Mathematical Olympiad, 2011, solution from T. Andreescu, M. Rolínek, J. Tkadlec, *Geometry Problems from the AwesomeMath Year-Round Program*, XYZ Press, 2013.

146 The solution can be followed on Fig. 10.63. By angle chasing in the circumcircles of MAD and MBC and noticing that the points A and O_1 are on the same side of the line MD and B and O_2 are separated by the line MC, we can write

$$\angle MO_1D = 2\angle A = 2(180° - \angle B) = \angle MO_2C.$$

It follows that the isosceles triangles MO_1D and MO_2C are similar and oriented the same way. Hence there is a spiral similarity of center M that maps these triangles into each other. In other words, there is a spiral similarity of center M that maps O_1 to O_2 and D to C.

By Theorem 2.25 (spiral similarities come in pairs), there is a spiral similarity that maps O_1 to D and O_2 to C. Now we can apply Theorem 2.22 to determine that the center of this last spiral similarity is the second intersection point of the circumcircles of $N'O_1D$ and $N'O_2C$, where N' is the intersection of the lines O_1O_2 and CD. But we know that the center of this spiral similarity is the point M, so M is at the intersection of the circumcircles of $N'O_1D$ and $N'O_2C$. Consequently N' is at the intersection of the circumcircles of MO_1D and MO_2C, and hence $N' = N$. It follows that N is on O_1O_2, as desired.

Remark We have proved additionally that N is on the line CD.

Source International Zhautykov Olympiad, 2013.

Fig. 10.64 Finding the
common center of the spiral
similarities.

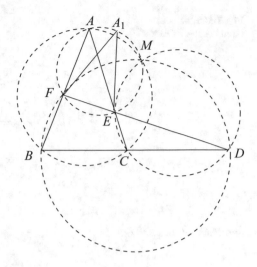

147 We argue on Fig. 10.64. The points A, B, C, D, E, F form a complete
quadrilateral. We will show that center of the spiral similarities in question is the
Miquel point of this complete quadrilateral.

By Theorem 2.22, the center of the spiral similarity that maps the segment EF
to the segment BC is the intersection point of the circumcircles of DFB and DEC.
This is the Miquel point M of the complete quadrilateral. Because the triangles
A_1FE and ABC are similar, M is also the center of the spiral similarity that maps
the two triangles into each other. As the composition of two spiral similarities with
the same center is a spiral similarity with the same center, M is the center of the
spiral similarities that map the four triangles into one another.

Source C. Mihalescu, *Geometria Elementelor Remarcabile (The Geometry of the
Remarkable Elements)*, Ed. Tehnică, Bucharest, 1957.

148 The situation is described in Fig. 10.65. By the result proved in Problem 131,
the spiral similarity of center S that maps ω_1 to ω_2 maps A to C and B to D. It also
maps the circumcenter O_1 of ω_1 to the circumcenter O_2 of ω_2. By the Averaging
Principle (Theorem 2.18), there is a spiral similarity of center S that maps the
configuration determined by the points S, A, B, O_1 to the configuration determined
by the points S, M, N, O. It follows that O is the circumcenter of SMN. But P is
the reflection of S over O_1O_2, and since $O \in O_1O_2$, it follows that $SO = PO$.
Hence P is on the circumcircle of SMN. We conclude that O is the circumcenter
of MNP, as desired.

Source T. Andreescu, M. Rolínek, J. Tkadlec, *Geometry Problems from the
AwesomeMath Year-Round Program*, XYZ Press, 2013.

Fig. 10.65 O is the
circumcenter of MNP

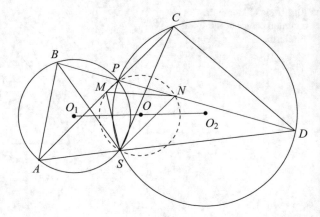

Fig. 10.66 P rotates to Q
about R

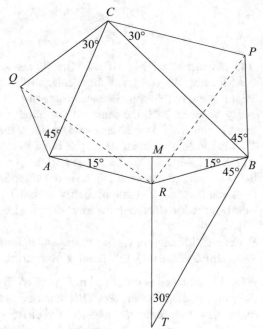

149 *First solution.* The point P is transformed into the point Q by a spiral
similarity of center B, angle $45°$, and ratio BC/BP, which maps P to C, followed
by a spiral similarity of center A, angle $45°$, and ratio AQ/AC, which maps C to
Q (see Fig. 10.66). But $AQ/AC = BP/BC$, because the triangles ACQ and BCP
are similar. So the product of the ratios of the two spiral similarities is 1, and hence
their composition is a rotation. The angle of this rotation is $45° + 45° = 90°$.

    If we show that the center of rotation is R, then we are done. For this we have to
show that R is fixed by the composition of the two spiral similarities.

Let T be the image of R through the first spiral similarity, that is, $\angle RBT = 45°$ and $BT/BR = BC/BP$. We deduce that the triangles BRT and BPC are similar, so $\angle BRT = \angle BPC = 180° - 75°$. Denoting by M the midpoint of AB, RM is the perpendicular bisector of AB, so $\angle BRM = 75°$. Consequently T, R, M are collinear, and therefore T is on the perpendicular bisector of AB. But then using the symmetry of the figure consisting of A, B, R, and T with respect to the perpendicular bisector of AB, we deduce that the second spiral similarity maps T back to R. The problem is solved.

Second solution. Here is the analytical solution, maybe not as elegant as the synthetic one. With the standard convention, we will use lower case letters to denote the coordinates of the points designated by upper case letters. We place the origin of the coordinate system at R, so that $r = 0$. Then $a = e^{\frac{5\pi i}{6}} b$.

The point Q is obtained from C by a spiral similarity of center A, angle $45°$, and ratio $AQ/AC = BP/BC = \sin\frac{\pi}{6}/\sin\frac{7\pi}{12}$. So

$$\frac{q - a}{c - a} = \frac{\sin\frac{\pi}{6}}{\sin\frac{7\pi}{12}} e^{\frac{\pi i}{4}}.$$

We obtain

$$q = \frac{\sin\frac{\pi}{6}}{\sin\frac{7\pi}{12}} e^{\frac{\pi i}{4}} c + \left(1 - \frac{\sin\frac{\pi}{6}}{\sin\frac{7\pi}{12}} e^{\frac{\pi i}{4}}\right) a.$$

Similarly

$$p = \frac{\sin\frac{\pi}{6}}{\sin\frac{7\pi}{12}} e^{-\frac{\pi i}{4}} c + \left(1 - \frac{\sin\frac{\pi}{6}}{\sin\frac{7\pi}{12}} e^{-\frac{\pi i}{4}}\right) b.$$

We must show that $q = ip = e^{\frac{\pi i}{2}} p$. Note that the coefficient of c in q does have this property. On the other hand, using the fact that $a = e^{\frac{5\pi i}{6}} b$, we are left to show that

$$\left(\sin\frac{7\pi}{12} - \sin\frac{\pi}{6} e^{\frac{\pi i}{4}}\right) e^{\frac{5\pi i}{6}} = \left(\sin\frac{7\pi}{12} - \sin\frac{\pi}{6} e^{-\frac{\pi i}{4}}\right) i.$$

The solution to the problem ends with a computation based on the formulas

$$\sin\frac{7\pi}{12} = \sin\left(\frac{\pi}{3} + \frac{\pi}{4}\right) = \sin\frac{\pi}{3}\cos\frac{\pi}{4} + \cos\frac{\pi}{3}\sin\frac{\pi}{4} = \frac{\sqrt{6}+\sqrt{2}}{4}$$

$$e^{\pm\frac{\pi i}{4}} = \frac{\sqrt{2}}{2} \pm \frac{\sqrt{2}}{2}i, \quad e^{\frac{5\pi i}{6}} = -\frac{\sqrt{3}}{2} + \frac{1}{2}i.$$

Source 17th International Mathematical Olympiad, 1975, proposed by the Netherlands.

Fig. 10.67 $D \in BC$ if and
only if O is on the
circumcircle of AEF

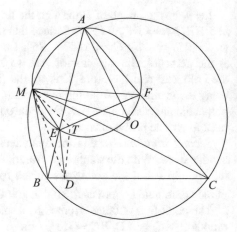

150 Both the direct and the converse implication can be followed on Fig. 10.67. The point M is the Miquel point of the complete quadrilateral determined by the lines AC, AB, BC, and EF. As seen in the proof of Miquel's Theorem (Theorem 2.24), there is a spiral similarity s of center M that maps the segment EF to the segment BC.

To prove the direct implication, assume $D \in BC$. Then $T = s^{-1}(D)$ is on EF. The triangles MTD and MEB are similar (automatic similarity). But EF is the perpendicular bisector of the line segment MD, so the triangle MTD is isosceles. It follows that the triangle MEB is isosceles as well, so $\angle EMB = \angle EBM$. The Exterior Angle Theorem in this triangle then gives

$$\angle AEM = 2\angle ABM = \angle AOM.$$

This shows that O is on the circumcircle of the triangle AEF.

For the converse we use the same spiral similarity. Define T in the same way, but now we do not know that $T \in EF$. But we know that O is on the circumcircle of AEF and from this we obtain

$$2\angle ABM = \angle AOM = \angle AEM = \angle ABM + \angle BME.$$

So the triangle EMB is isosceles. This implies that the triangle TMD (which is similar to EMB by automatic similarity) is also isosceles, and so T belongs to the perpendicular bisector of MD, which is the line EF. But then $D = s(T) \in s(EF) = BC$, and we are done.

Remark The result remains valid if the circumcircles of the triangles AEF and ABC are tangent. In this case $M = A$, and the solution is somewhat simpler: We consider the homothety centered at A that transforms the circumcircle of AEF into the circumcircle of ABC. This homothety takes EF into BC. Then $D \in BC$ is equivalent to the fact that EF is the midline of the triangle ABC, which is equivalent to $AEOF$ being cyclic.

Source Romanian Team Selection Test for the Junior Balkan Mathematical Olympiad, 2018, solution by contestant Ioana Popescu.

151 The reader can examine this configuration of six intersecting circles in Fig. 10.68. The triangles ABC and $A_1B_1C_1$ are oriented the same way because the order of the sides of ABC dictates the order of the vertices of $A_1B_1C_1$. So there is a spiral similarity that takes the triangle $A_1B_1C_1$ to the triangle ABC; let O be its center. The segment AB is mapped to the segment A_1B_1, so O is also the center of the spiral similarity that maps the segment AA_1 to BB_1, as Theorem 2.25 shows that spiral similarities come in pairs. This latter spiral similarity is centered at the second intersection point of the circumcircles of ABC_2 and $A_1B_1C_2$ (by Theorem 2.22). So O lies at the intersection of the circumcircles of ABC_2 and $A_1B_1C_2$, the same argument for the other pairs of circles.

152 Let ω_B and ω_C be the circumcircles of the triangles ACP and ABQ, respectively, and let $X \neq A$ be their second intersection point. The angle conditions from the statement imply that AB is tangent to ω_B and AC is tangent to ω_C. Problem 132 shows that X is on the symmedian from A of the triangle ABC.

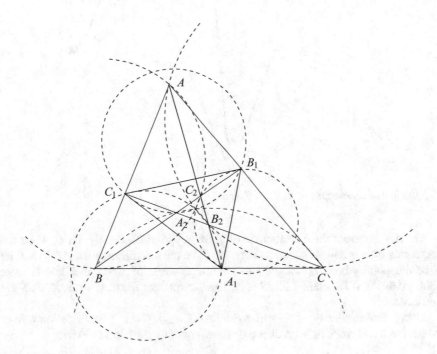

Fig. 10.68 Six circles intersecting at one point

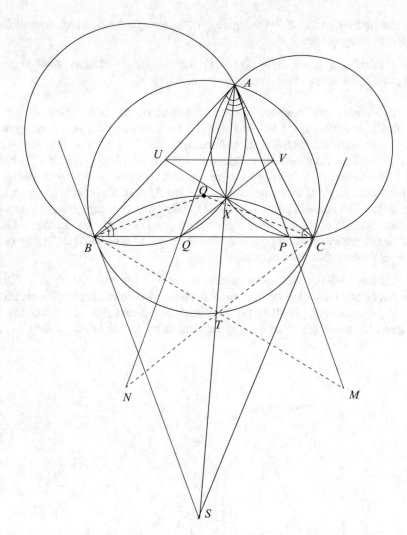

Fig. 10.69 The construction of X, T, U, V, and S

Let AX intersect the circumcircle of ABC again at T. Finally, let U, V be the midpoints of AB and AC, respectively, let O be the circumcenter of ABC, and let S be the point where the tangents to the circumcircle of ABC at B and C meet (Fig. 10.69). By Theorem 1.22, S is on the symmedian from A, so A, X, T, S are collinear.

Using the similarity of the triangles ABX and CAX (which follows from tangencies) and working with directed angles modulo $180°$, we can write

$$\angle BXC = \angle BXA + \angle AXC = \angle XBA + \angle BAX + \angle XAC + \angle ACX$$

$$= \angle XAC + \angle BAX + \angle XAC + \angle BAX = 2\angle BAC = \angle BOC,$$

so B, X, O, C are concyclic. Since $\angle OBS = \angle OCS = 90°$, S is also on this circle, so $\angle OXS = \angle OBS = 90°$. Then $\angle AXO = \angle OXT = 90°$, and since $AO = OT$, we have $AX = XT$.

Next observe that, using the tangencies of ω_B and ω_C to AB and AC, respectively, imply that the triangles AQC and BPA are both inverse similar to the triangle ABC and hence are directly similar to each other. But the similarity of the triangles ABX and CAX implies that X is the center of the spiral similarity that maps the segment BA to the segment AC. And the similarity of triangles AQC and BPA implies that the same spiral similarity maps P to Q. Furthermore, it maps U to V since they are the midpoints of sides BA and AC.

We conclude that X is the center of a spiral similarity mapping UP to VQ and thus of the one mapping UV to PQ (spiral similarities come in pairs). But UV is parallel to BC, and hence to PQ, so the triangles XUV and XPQ must actually be homothetic. It follows that U, X, P are collinear, and the same is true about Q, X, V. Dilating by a factor of 2 about A gives that B, T, M and, respectively, C, T, N are collinear. So BM and CN meet at T, which is on the circumcircle, as desired.

Source 55th International Mathematical Olympiad, 2014, proposed by Georgia, solution by contestant Sammy Luo, United States of America.

153 Because the triangle DBC is isosceles, the Exterior Angle Theorem implies that $\angle ADB = 2\angle C$. So $\angle ADE = \angle ADB/2 = \angle C$, and therefore $DE\|BC$. We deduce that $\angle DEB = \angle DAE$, both being equal to $180° - \angle EBC$. It follows that the triangles DAE and DEB are directly similar (having two pairs of equal angles), so they are mapped one into the other by a spiral similarity of center D. If we denote by I and J the respective incenters of these triangles, as shown in Fig. 10.70, then the spiral similarity maps I to J.

Fig. 10.70 A spiral similarity that maps DAE to DEB

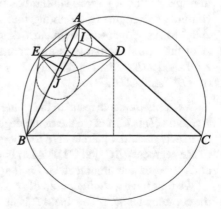

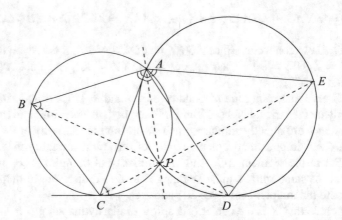

Fig. 10.71 ABC, ACD, ADE map into each other by spiral similarities

But as the spiral similarity maps the triangle DAI to DEJ and spiral similarities come in pairs (Theorem 2.25), another spiral similarity will map the triangle DAE to the triangle DIJ. From this we infer that

$$\angle(IJ, AE) = \angle(DI, DA) = \angle IDA,$$

and this is further equal to $\angle C/2 = 90° - \angle AEB/2$ (because $AEBC$ is cyclic). Thus $\angle(IJ, AE)$ is the complement of the angle formed by AE with the angle bisector of $\angle AEB/2$. And this implies that IJ is orthogonal to the angle bisector of $\angle AEB$, as desired.

Source Romanian Team Selection Test for the International Mathematical Olympiad, 2011.

154 The triangles ABC, ACD, ADE are directly similar, so they are mapped into one another by spiral similarities centered at A. Reasoning on Fig. 10.71, we continue the argument by noting that the spiral similarity that maps the triangle ABC to ADE maps the segment BC to DE, and so, by Theorem 2.22, the center A of this spiral similarity is the second intersection point of the circumcircles of BCP and DEP. Said differently, P is the second intersection of the circumcircles of ABC and ADE.

We apply the same result to the triangles ABC and ACD. Here the intersection of BC and CD is C, and we are in the degenerate situation of Sect. 2.4.3. The line CD intersects the circumcircle of ABC once at C and a second time at the intersection of the segments BC and CD, which is C again. Hence CD is tangent to this circle at C. By a similar argument, CD is tangent to the circumcircle of ADE.

We thus have a configuration of two circles that intersect at A and P with CD as the common tangent. The line AP is the radical axis of the two circles, and it passes through the point on CD that has equal powers with respect to the circles. And this point is the midpoint of CD, as desired.

Remark If you do not like the limiting argument in the degenerate case, you can also prove that CD is tangent to the two circles by an angle chasing argument.

Source Short list of the International Mathematical Olympiad, 2006, proposed by the United States of America.

155 Let D, E, F be the feet of the altitudes from A, B, C, respectively (Fig. 10.72).

There is a spiral similarity mapping the triangle CFZ to the triangle ADX. Using Theorem 2.22 we deduce that the center of this spiral similarity is the second intersection point of the circumcircles of BXZ and BDF, and by the same argument, because spiral similarities come in pairs (Theorem 2.25), this is also the second intersection point of the circumcircles of HDF and HAC. Hence the circumcircle of XBZ passes through the intersection point M of the circumcircles of $BFHD$ and HAC (Fig. 10.73) different from H.

But M is also on BG. Indeed, if B' is the point in the circumcircle of AHC that is diametrically opposite to H, then $AB' \perp CH$ and $CB' \perp AH$, showing that $AB' \parallel BC$ and $CB' \parallel AB$. Thus $BAB'C$ is a parallelogram and so BB' passes through the midpoint of AC; it is the median. Once we know this fact, then we can consider the spiral similarity of center H that maps the circumcircle of $BFHD$ to the

Fig. 10.72 Construction of P and H

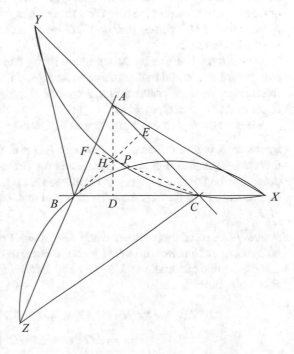

Fig. 10.73 Construction of
M

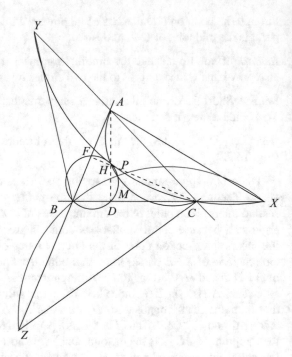

circumcircle of AHC and notice that it must map B to B'. But then B, H, B' are collinear, by Problem 131. So M is on the median from B.

Note that HM is perpendicular to BG, so M is on the circle of diameter HG that we are interested in.

On the other hand, using the spiral similarity that maps the triangle ADX to the triangle BEY, we obtain that the circumcircle of CXY passes through the second intersection point N of the circumcircles of HAB and $CEHD$. And N is on the circle of diameter HG, as well.

We can now rephrase our problem as follows:

Problem It is given a circle of diameter HG and the points B and C outside of this circle. The lines BG and CG intersect the circle at M and N, respectively. The point X is on the line BC (say C is between B and X). Prove that second point of intersection, P, of the circumcircles of XBM and XCN is on the circle of diameter HG.

Solution We solve this problem using Fig. 10.74. Let R be the second intersection point of the circumcircle of XBM with the circle of diameter HG. Then $\angle XRM = \angle MBC$ (cyclic quadrilateral $XBMR$) and $\angle MRN = \angle BGC$ (cyclic quadrilateral $RMNG$). Thus

$$\angle XRN + \angle XCN = \angle XRM + \angle MRN + \angle BCG$$

$$= \angle GBC + \angle BGC + \angle BCG = 180°.$$

Fig. 10.74 The modified
problem

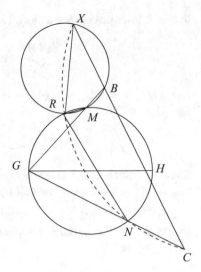

Fig. 10.75 There is a spiral
similarity taking X to W and
B to C and a spiral similarity
taking B to C and W to Y

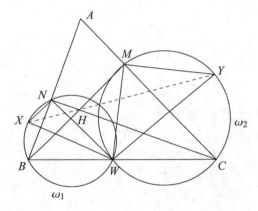

Hence $XCNR$ is cyclic, and therefore $R = P$ and the problem is solved.   □

Source Brazilian Mathematical Olympiad, 2015.

156 Because WX is a diameter (Fig. 10.75), $\angle XNW = 90° = \angle BNC$. Also,
$\angle NXW = \angle NBW = \angle NBC$. It follows that the triangles NXW and NBC are
similar in the same orientation, so they are mapped one into the other by a spiral
similarity with center N. But spiral similarities come in pairs (Theorem 2.25); hence
there exists a spiral similarity s_N with center at N, rotation angle $90°$, and ratio
$\frac{NB}{NC} = \cot \angle B$ that takes X to W and B to C. Analogously, there is a spiral similarity
s_M with center at M, rotation angle $90°$, and ratio $\frac{BM}{MC} = \tan \angle C$ that takes B to C
and W to Y.

If we compose s_N and s_M, we obtain another spiral similarity, $s = s_M \circ s_N$, such
that $s(X) = s_M(W) = Y$. This transformation has rotation angle $90° + 90° = 180°$,

that is, it is an inverse homothety. Let us find its center. Notice that

$$\frac{NH}{NA} = \cot \angle AHN = \cot \angle B$$

and $\angle HNA = 90°$, so $s_N(H) = A$. We can deduce in the same fashion that $s_M(A) = H$. So

$$s(H) = s_M \circ s_N(H) = s_M(A) = H.$$

The only fixed point of a homothety is its center, so the center of s is H. This means that H, X, and $s(X) = Y$ are collinear.

Source International Mathematical Olympiad, 2013, proposed by Warut Suk-sompong and Potcharapol Suteparuk, Thailand, solution by contestant Aleddine Sabbagh, Tunisia.

Chapter 11
Inversions

157 (a) Placing the center of inversion at the origin of the coordinate system, the inversions have the equations $\chi_1 = R_1^2/\overline{z_1}$ and $\chi_2 = R_2^2/\overline{z_2}$, so their composition has the equation $\chi_2 \circ \chi_1(z) = (R_2^2/R_1^2)z$, which is the homothety of center O and radius R_2^2/R_1^2.

(b) and (c) Let the inversion be $\chi(z) = R^2/\overline{z}$ and the homothety $h(z) = kz$. Then $h \circ \chi = kR^2/\overline{z}$ which is the inversion of ratio kR^2 with the same center and $\chi \circ h = (R^2/k)/\overline{z}$ which is the inversion of ration R^2/k with the same center.

Remark The composition of two inversions of the same center is not commutative, that is, $\chi_2 \circ \chi_1 \neq \chi_1 \circ \chi_2$. Neither is the composition of an inversion and a homothety with the same center commutative.

158 Take a Möbius transformation (or just an inversion, or a circular transformation) that maps the circle to a line. Using the Symmetry Principle (Theorem 3.15) and the fact that inversion preserves angles, we reduce the problem to showing that two points are reflections of each other over a line if and only if every circle or line passing through the two points is orthogonal to the line. But a line is orthogonal to a circle if and only if it passes through its center, and so the line in question must be the locus of the centers of the circles that pass through the two points, and this locus is the perpendicular bisector of the segment determined by the two points (see Fig. 11.1).

Remark There is a less sophisticated way to see why if A and B reflect to each other over a circle ω, then any circle ω' passing through A and B is orthogonal to ω. As one of A, B is inside ω and the other one is outside, ω' intersects ω twice. Those points are invariant under the inversion, and A and B are mapped into each other, so ω' must be invariant under the inversion. It is therefore orthogonal to ω.

It might be worth remembering that a circle that passes through two points that correspond under an inversion is invariant under that inversion.

© The Author(s), under exclusive license to Springer Nature Switzerland AG 2022

R. Gelca et al., *Geometric Transformations*, Problem Books in Mathematics, https://doi.org/10.1007/978-3-030-89117-6_11

Fig. 11.1 Line orthogonal to circles passing through two points

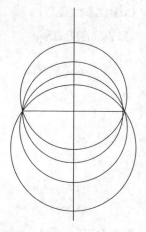

159 We can solve this problem easily using coordinates. Let a, b, p be the complex coordinates of A, B, P, respectively, and let u be the complex coordinate of the center of inversion, to be determined, and let R be the radius of inversion. Then the images of the three points have complex coordinates

$$\frac{R^2}{a-u}+u, \quad \frac{R^2}{b-u}+u, \quad \frac{R^2}{p-u}+u.$$

The condition that the third is the midpoint of the segment determined by the first two is equivalent to the equality

$$\frac{1}{p-u}=\frac{\dfrac{1}{a-u}+\dfrac{1}{b-u}}{2}.$$

Solving for u we obtain

$$u=\frac{2ab-(a+b)p}{a+b-2p}.$$

Since $p \neq \frac{a+b}{2}$, as P is not the midpoint of AB, this last formula makes sense, and it yields the complex coordinate of the center of inversion.

Remark The center of the inversion must be on the line through a, b, p, and you can read this in the formula by choosing the coordinates a, b, p to be real (i.e., the line through a and b is the x-axis), and then u is real, too. If we relax the requirement by replacing inversion with Möbius transformation, then the problem becomes trivial since Möbius transformations are parametrized by the images of three given points.

160 Let O be the center of inversion. By Theorem 3.2, $\angle OAB = \angle OB'A'$, hence $ABB'A'$ is a cyclic quadrilateral; let ω'' be its circumcircle. Then AB is the radical

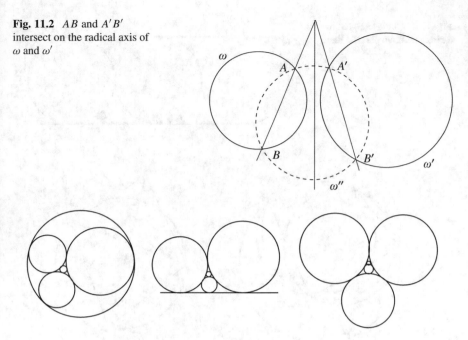

Fig. 11.2 AB and $A'B'$ intersect on the radical axis of ω and ω'

Fig. 11.3 Circles tangent to three given pairwise tangent circles

axis of ω and ω'', while $A'B'$ is the radical axis of ω' and ω''. The two radical axes intersect at the radical center of the three circles, which lies also on the radical axis of ω and ω'. The situation is described in Fig. 11.2.

161 Being orthogonal to ω, the two circles are invariant under inversion with respect to ω. As inversion preserves tangencies, it maps the point of tangency of the two circles to another point of tangency of the two circles. But there is only one point of tangency, so this point has to be mapped into itself, proving that it lies on the circle of inversion.

162 The three cases described in the statement can be seen in Fig. 11.3.

Let O be the tangency point of ω_1 and ω_2. Consider an inversion of center O. This maps ω_1 and ω_2 to two parallel lines ω'_1 and ω'_2, with the image ω'_3 of ω_3 being a circle tangent to the two lines. Any circle tangent to ω_1, ω_2, and ω_3 becomes a circle tangent to the two lines and to ω'_3. There are exactly two such circles, ω'_4 and ω'_5, one on each side of ω'_3, as shown in Fig. 11.4. There are three possible cases:

Case I. O is inside one of ω'_4 or ω'_5.
Case II. O is on one of ω'_4 and ω'_5.
Case III. O is in the exterior of both ω'_4 and ω'_5.

In cases I and III, the inverses of ω'_4 and ω'_5 are circles. If one of ω'_4 or ω'_5 contains O inside, then ω_1, ω_2, ω_3 are interior tangent to its inverse and are exterior tangent

Fig. 11.4 Circles tangent to
a circle and two parallel lines

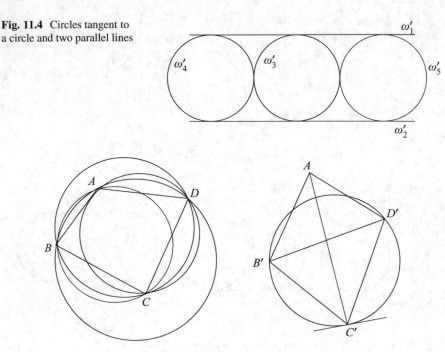

Fig. 11.5 Simplification of the verification of the equality of angles using inversion

to the inverse of the other circle. If neither of the circles ω'_4 and ω'_5 contains O inside, then ω_4 and ω_5 are exterior tangent to $\omega_1, \omega_2, \omega_3$. Thus I corresponds to (i) and III to (iii).

Finally, if O is on one of the circles ω'_4 or ω'_3, then the preimage of that circle is a line that is a common tangent of $\omega_1, \omega_2, \omega_3$. Therefore II corresponds to (ii), and we are done.

Remark A corollary of what we have just proved is that the largest number of circles in the plane that are pairwise tangent with the tangency points being distinct is 4. Can you reformulate this problem in space?

163 Invert about a circle centered at A. With the usual convention for the notation, the circumcircles of ABC, ABD, and ACD are mapped to the lines $B'C'$, $B'D'$, and $C'D'$, respectively. And the circumcircle of BCD is mapped to the circumcircle of $B'C'D'$. Because inversion preserves angles (Theorem 3.21), we are left with showing that the angle between the lines $B'C'$ and $B'D'$ is equal to the angle between the line $C'D'$ and the circumcircle of $B'C'D'$ (Fig. 11.5). But both angles measure half of the arc $\overset{\frown}{C'D'}$, and we are done.

Remark This is a recurring theme: use a circular transformation to simplify the figure as much as possible without altering the hypothesis and the conclusion, and

Fig. 11.6 Proof that ω is orthogonal to Ω

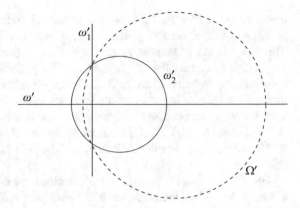

then solve the problem for the simpler configuration. In Fig. 11.5 we have placed side-by-side the noninverted and the inverted figure, so that you can contrast the two configurations and understand the advantage of this method.

164 Let ω be a circle that is orthogonal to both ω_1 and ω_2, and let Ω be the circle of inversion. Take an inversion centered at one of the intersection points of ω and ω_1. The images of these two circles are the orthogonal lines ω' and ω_1', and the image ω_2' of ω_2 is orthogonal to ω', a configuration which is presented in Fig. 11.6. Note that ω_2 cannot pass through the center of inversion, so ω_2' is a circle.

By the Symmetry Principle, ω_1' and ω_2' are reflections of each other over the image Ω' of Ω. By Theorem 3.4 the center of Ω' is on ω'. Thus Ω' is orthogonal to ω', and consequently Ω is orthogonal to ω.

Remark If instead ω just cuts ω_1 and ω_2 at equal angles, the conclusion might not be true.

165 *First solution.* All three parts of the problem, which comprise a famous discovery of Apollonius, are easy to check when $k = 1$, for in this case we recover the characterization of the perpendicular bisector of the segment AB as the locus of points P satisfying $PA = PB$.

The general situation is a consequence the Symmetry Principle (Theorem 3.15) and the invariance of the cross-ratio (Proposition 3.17). In detail: start with A and B and some point P such that $PA/PB = k$. We want to understand the locus of the points Q characterized by $QA/QB = PA/PB$, namely, for which

$$\frac{QA}{QB} : \frac{PA}{PB} = .1$$

As we have seen in Sect. 3.1.3, the cross-ratio of the points P, Q, A, B is invariant under Möbius, even circular transformations. Now a Möbius transformation is determined by specifying the images of three points, so let us choose such a transformation that maps A, B, P to A', B', P', respectively, such that P' is the midpoint of $A'B'$. The locus of the points Q such that $QA/QB = PA/PB$ is mapped to the locus of points Q' such that $Q'A'/Q'B' = P'A'/P'B' = 1$. This locus is the perpendicular bisector of $A'B'$ (the particular case discussed in the beginning). Note that A' and B' are the reflections of each other over this perpendicular bisector. Then $\Gamma_k(A, B)$ is the preimage of the perpendicular bisector through the Möbius transformation, so it has to be a circle or a line. But for $k \neq 1$ it cannot be a line because it intersects the line AB twice (once inside the segment AB and once outside). So it is a circle. Moreover, A, B are symmetric with respect to this circle, by the Symmetry Principle (Theorem 3.15). This proves both (a) and (b).

For (c), note that by the same considerations, we can start by assuming that Γ is a line, and then it is the perpendicular bisector of AB. And so it is also $\Gamma_1(A, B)$, and we are done.

Second solution. Here is a proof of (a) using coordinates. Without loss of generality, we may assume $0 < k < 1$ and then choose coordinates such that the origin is at B. Let a be the complex coordinate of A. Then the locus is given by the equation

$$|z - a| = k|z|,$$

which can be written as

$$(z - a)(\bar{z} - \bar{a}) = k^2 z\bar{z}.$$

Transform this into

$$(1 - k^2)z\bar{z} + a\bar{z} + z\bar{a} + a\bar{a} = 0.$$

Now set $1 - k^2 = t^2$, and "complete the square" to obtain

$$\left(tz - \frac{a}{t}\right)\left(t\bar{z} - \frac{\bar{a}}{t}\right) = \left(\frac{1}{t^2} - 1\right)a\bar{a}.$$

Divide both sides by t^2 and set $r = \sqrt{\left(\frac{1}{t^4} - \frac{1}{t^2}\right)a\bar{a}}$ to obtain the equation of a circle

$$\left|z - \frac{a}{t^2}\right|^2 = r^2.$$

Can you verify (b) and (c) using coordinates?

Remark The key fact used in solving this problem is that a Möbius transformation is determined by the images of three points. This allows us to map our configuration into one where P is the midpoint of AB without changing the statement. The problem is now easy, and the property that Möbius transformations map lines or circles into lines or circles solves the general case.

The circles defined in this problem are called the *circles of Apollonius* after Apollonius of Perga, who, according to Pappus and other sources, has introduced them for the first time. It is important to notice that the Apollonian circles of the points A and B are orthogonal to any circle passing through A and B (Problem 158). So we are in the presence of two mutually orthogonal families of circles.

166 Such a Möbius transformation

$$\phi(z) = \frac{az + b}{cz + d}$$

must map the real axis to itself. Let us first work in the case $d \neq 0$. For $t \in \mathbb{R}$,

$$\phi(t) = \frac{az + b}{cz + d} = \frac{(at + b)(\bar{c}t + \bar{d})}{(ct + d)(\bar{c}t + \bar{d})} = \frac{a\bar{c}t^2 + (a\bar{d} + b\bar{c})t + b\bar{d}}{|ct + d|^2} \in \mathbb{R}$$

Hence $a\bar{c}t^2 + (a\bar{d} + b\bar{c})t + b\bar{d} \in \mathbb{R}$ for all $t \in \mathbb{R}$. Setting $t = 0$ we deduce that $b\bar{d} \in \mathbb{R}$. But then $[a\bar{d}t + (a\bar{d} + b\bar{c})]t \in \mathbb{R}$ for all $t \in \mathbb{R}$, so $a\bar{c}t + (a\bar{d} + b\bar{c}) \in \mathbb{R}$. Again setting $t = 0$ we obtain $a\bar{d} + b\bar{c}$, and finally $a\bar{c} \in \mathbb{R}$. Since $a\bar{c}, b\bar{d} \in \mathbb{R}$, there are $s, t \in \mathbb{R}$ such that $a = sc$ and $b = td$. But then

$$a\bar{d} + b\bar{c} = sc\bar{d} + tc\bar{d} = sc\bar{d} + \overline{tc\bar{d}} \in \mathbb{R}.$$

Set $w = c\bar{d}$. Then $sw + t\bar{w}$ should be real. This can happen if either w is real or if $s = t$. But if $s = t$, then ϕ is constant, which is not allowed. So $c\bar{d} \in \mathbb{R}$, meaning that $c = ud$, $u \in \mathbb{R}$. Normalizing so that $d \in \mathbb{R}$, we see that every such Möbius transformation is of the form

$$\phi(z) = \frac{az + b}{cz + d}, \quad a, b, c, d \in \mathbb{R}.$$

The case where $d = 0$ can be treated similarly, and the same conclusion is reached. We thus conclude that these are precisely the Möbius transformations that map the real axis to itself. But do they all map the upper half-plane to itself?

The upper half-plane is mapped to either the upper half-plane or the lower half-plane $\{z \mid \operatorname{Im} z < 0\}$. It is mapped to the upper half-plane when the imaginary part of $\phi(i)$ is positive. We have

$$\phi(i) = \frac{ai + b}{ci + d} = \frac{(ai + b)(-ci + d)}{(ci + d)(-ci + d)} = \frac{ac + bd + (ad - bc)i}{|c|^2 + |d|^2}.$$

The imaginary part of this number is positive when $ad - bc > 0$. Thus the Möbius transformations that map the upper half-plane to itself are

$$z \mapsto \frac{az+b}{cz+d}, \quad a,b,c,d \in \mathbb{R}, \quad ad-bc > 0.$$

Remark The Möbius transformations that map the upper half-plane to itself are the real projective special linear group $PSL(2, \mathbb{R})$, that is, the group of real fractional linear transformations. In the Poincaré model of the hyperbolic plane (i.e., the plane of the Lobachevskian geometry in which given a point and a line not containing it, there are infinitely many lines that pass through the point and are parallel to the line), the elements of $PSL(2, \mathbb{R})$ are the orientation preserving isometries.

167 Certainly, circular transformations have this property.

Conversely, let f be such a transformation. Choose a Möbius transformation ϕ such that $\phi(f(\infty)) = \infty$. Then

$$g : \mathbb{C} \cup \{\infty\} \to \mathbb{C} \cup \{\infty\}, \quad g = \phi \circ f$$

maps lines to lines and circles to circles. By Theorem 2.28 $g(z) = az + b$ or $g(z) = a\bar{z} + b$, for some $a, b \in \mathbb{C}$, $a \neq 0$. Then $f = \phi^{-1} \circ g$ is a circular transformation. Hence the conclusion.

Remark Let us embed the complex plane into the three-dimensional space by $z = x_1 + ix_2 \mapsto (x_1, x_2, 0)$, and then take the inversion of center $N = (0, 0, 1)$ and radius 1. This inversion maps a point P to P' such that $P' \in |NP$ and $NP \cdot NP' = 1$. The complex plane together with the point at infinity is mapped onto the sphere Σ of center $(0, 0, 1/2)$ and radius $1/2$ (note that N is the image of ∞). The lines and circles of $\mathbb{C} \cup \{\infty\}$ are in one-to-one correspondence with the circles on Σ. This is why we call $\mathbb{C} \cup \{\infty\}$ the Riemann sphere, because it can be actually realized as a sphere.

The inversion maps the plane to the sphere and hence also the sphere to the plane. This map from the sphere to the plane is called stereographic projection.

The result we have just obtained allows us to describe all transformations of the sphere to itself that map circles to circles. These correspond, via the stereographic projection, to transformations of the plane that map every line or circle to a line or a circle, that is, to the circular transformations.

168 (a) If the polynomial has the complex zeros x_1, x_2, x_3, then the first operation yields a polynomial whose zeros are $\frac{1}{x_1}, \frac{1}{x_2}, \frac{1}{x_3}$, and the second yields a polynomial whose zeros are $x_1 - t, x_2 - t, x_3 - t$. Clearly one has to take into account the case where 0 is a zero of the polynomial, as well as that where the polynomial has degree less than 3. For that we have to work on the Riemann sphere, thus including the point at infinity. The successive application of the two operations amounts to changing the zeros of the polynomial by a Möbius transformation. The first polynomial is $P_1(x) = x^3 + x^2 - 2x = x(x-1)(x+2)$, which has three distinct zeros, while the second is $P_2(x) = x^3 - 3x - 2 = (x+1)^2(x-2)$, which has only two distinct zeros. And no Möbius transformation maps a set with three elements to a set with two elements. Thus the answer to the question is negative.

(b) Two zeros of the polynomial $x^3 - 3x - 3$ are nonreal, while all zeros of $P_1(x)$ are real. The two operations map polynomials with real zeros into polynomials with real zeros, so the answer is still negative.

(c) The polynomial $x^3 - 3x - 1$ has three distinct real zeros. Adding the third operation to the picture reduces the problem to one about the possibility of mapping the sets of zeros of the polynomials into each other. The operations (i) and (ii) correspond to the inversion and translation on the real axis, and we will now show that by composing them we can also obtain homothety on the real axis. Let α be a positive number, and let $\beta = 1/\sqrt{\alpha}$. Consider the sequence of transformations

$$z \to z + \beta \mapsto \frac{1}{z + \beta} \mapsto \frac{1}{z + \beta} - \frac{1}{\beta} = \frac{-z}{\beta z + \beta^2}$$

$$\mapsto -\frac{\beta z + \beta^2}{z} = -\beta - \frac{\beta^2}{z} \mapsto -\frac{\beta^2}{z} \mapsto -\frac{z}{\beta^2} = -\alpha z.$$

Of course by applying this twice (and working with $\sqrt{\alpha}$ instead of α), we can write the map $z \mapsto \alpha z$ as the composition of the complex inversion and translations. Thus, using Theorem 3.13, we conclude that every linear fractional transformation with real coefficients can be written as a composition of maps of the form $z \mapsto \frac{1}{z}$ and $z \mapsto z + t, t \in \mathbb{R}$.

Now let x_1, x_2, x_3 be the zeros of $x^3 - 3x - 1$. Consider the linear fractional transformation that maps them into $2, 0, 1$, respectively. Write this transformation as a composition of the inversions and translations, and apply the corresponding sequence of operations (i) and (ii) to $P_1(x)$. In the end, you have obtained a polynomial that is a real multiple of $x^3 - 3x - 1$. The third operation allows us to transform these two polynomials into each other, and we are done.

Remark Working with complex numbers and considering for $\alpha \neq 0$ a complex number β such that $\beta^2 = -\frac{1}{\alpha}$, by performing the sequence of transformations in the solution to (c), we can strengthen the statement of Theorem 3.13: every Möbius transformation can be written as the composition of translations and the complex inversion.

Source Part (a) of the problem was given at a test at the Mathematical Olympiad Summer Program, being proposed by Răzvan Gelca.

169 Consider an inversion of center A (Fig. 11.7). The circles ω_1 and ω_2 become the lines $B'M'$ and $B'N'$, respectively, where B', M', N' are the images of B, M, N. The lines AM and AN are mapped into themselves; the fact that they form $0°$ degree angles with the circles means that the opposite sides in $AM'B'N'$ form $0°$ angles, so this quadrilateral is a parallelogram. Also, because B is the midpoint of AC, C' is the midpoint of AB', so C' is the center of the parallelogram. Hence C' is on $M'N'$, that is, M', C', N' are collinear. But then Theorem 3.4 implies that A, M, C, N are concyclic, and we are done.

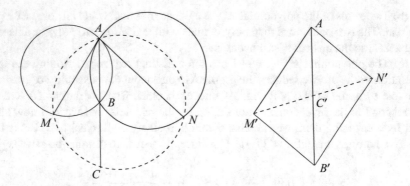

Fig. 11.7 *AMCN* is cyclic

Fig. 11.8 Proof of Steiner's
identity

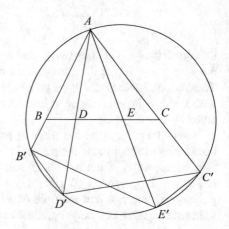

170 *First solution.* Consider an inversion with center A and radius R and let
B', C', D', E' be the inverses of B, C, D, E, respectively (Fig. 11.8). Then by
Theorem 3.4, A, B', C', D', E' are concyclic. And since $\angle D'AB' = \angle DAB = \angle EAC = \angle E'AC'$, it follows that $B'D' = E'C'$ and $B'E' = D'C'$. The
formula (3.1) turns these equalities into

$$\frac{R^2}{AB \cdot AD} BD = \frac{R^2}{AE \cdot AC} EC \quad \text{and} \quad \frac{R^2}{AB \cdot AE} BE = \frac{R^2}{AD \cdot AC} DC.$$

Dividing the two equalities yields

$$\frac{AE}{AD} \cdot \frac{BD}{BE} = \frac{AD}{AE} \cdot \frac{EC}{DC},$$

which is equivalent to the formula from the statement.

Second solution. The coordinate solution is a variation of the synthetic one and is based on the rewriting of Steiner's identity as

$$\frac{DB}{DC} : \frac{AB}{AC} = \frac{AB}{AC} : \frac{EB}{EC}.$$

We denote the coordinates of the points by lowercase letters and let $\angle DAB = \angle CAE = \alpha/2$ and $\angle EAD = \beta/2$. The problem reduces to proving the equality of cross-ratios

$$(d, a, b, c) = \overline{(a, e, b, c)}.$$

(The complex conjugate is motivated by the fact that DB/DC and EB/EC are real, while the argument of AB/AC is the negative of the argument of AC/AB.) Consider an inversion of center A and choose the complex coordinates so that the image of the line BC is a circle centered at the origin. Then the images of b, d, e, c are $b', d' = b'e^{\alpha i}, e' = b'e^{(\alpha+\beta)i}, c' = b'e^{(2\alpha+\beta)i}$, respectively, because the arcs $\overparen{B'D'}$ and $\overparen{E'C'}$ have measure α and the arc $\overparen{D'C'}$ has measure β. Since inversion transforms a cross-ratio to its complex conjugate, the problem reduces to showing that

$$(d', \infty, b', c') = \overline{(\infty, e', b', c')},$$

and this is equivalent to

$$\frac{d' - b'}{d' - c'} = \frac{\overline{e' - c'}}{\overline{e' - b'}}.$$

This is further equivalent to

$$\frac{e^{\alpha i} - 1}{e^{\alpha i} - e^{(2\alpha+\beta)i}} = \frac{e^{-(\alpha+\beta)i} - e^{-(2\alpha+\beta)i}}{e^{-(\alpha+\beta)i} - 1}.$$

The right-hand side is transformed into the left-hand side by multiplying both the numerator and the denominator by $e^{(2\alpha+\beta)i}$, and we are done.

171 Invert with respect to the inscribed circle and let A', B', C', D' be the inverses of A, B, C, D, respectively. By Proposition 3.3, A', B', C', D' are the midpoints of MQ, MN, NP, and PQ, respectively (Fig. 11.9). In the triangles that are formed by three of the four vertices of the quadrilateral $MNPQ$, $A'B', B'C', C'D'$, and $D'A'$ are midlines, so they are parallel to the corresponding diagonals of this quadrilateral. Thus $A'B'C'D'$ is a parallelogram.

On the other hand, since A, B, C, D are concyclic, their images A', B', C', D' are concyclic as well. Thus $A'B'C'D'$ is both a cyclic quadrilateral and a parallelogram; hence it is a rectangle. It follows that $A'B'$ is perpendicular to $B'C'$. As NQ

Fig. 11.9 Inversion about the
inscribed circle

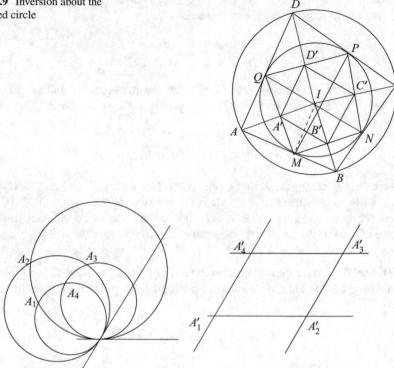

Fig. 11.10 Parallelogram obtained by inverting tangent circles

and MP are parallel to these two lines, NQ is perpendicular to MP, and we are
done.

172 *First solution.* Invert about a circle of center P and radius R. The images
of the four circles form a quadrilateral whose opposite sides are parallel since
their preimages are circles tangent at the center of inversion. This quadrilateral
is therefore a parallelogram (Fig. 11.10). If A_1', A_2', A_3', A_4' are the images of
A_1, A_2, A_3, A_4, respectively, then formula (3.1) yields

$$A_j' A_{j+1}' = \frac{R^2}{PA_j \cdot PA_{j+1}} \cdot A_j A_{j+1}, \quad j = 1, 2, 3, 4.$$

But because $A_1' A_2' A_3' A_4'$ is a parallelogram, $A_1' A_2' = A_3' A_4'$ and $A_2' A_3' = A_1' A_4'$, so

$$\frac{R^2}{PA_1 \cdot PA_2} \cdot A_1 A_2 = \frac{R^2}{PA_3 \cdot PA_4} \cdot A_3 A_4,$$

Fig. 11.11 Two pairs of circles intersecting at concyclic points

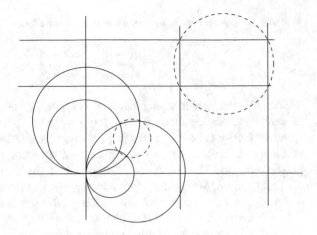

$$\frac{R^2}{PA_2 \cdot PA_3} \cdot A_2A_3 = \frac{R^2}{PA_1 \cdot PA_4} \cdot A_1A_4.$$

Multiply the two equalities, then divide the resulting identity by $R^4/PA_1 \cdot PA_3$, to obtain

$$\frac{A_1A_2 \cdot A_2A_3}{PA_4^2} = \frac{A_3A_4 \cdot A_1A_4}{PA_2^2}.$$

And this is equivalent to the identity from the statement.

(b) Using the same inversion, we note that A_1, A_2, A_3, A_4 are concyclic if and only if their images are. But the images of these four points are concyclic if the parallelogram they form is a rectangle (as in Fig. 11.11), namely, if we are in the presence of two pairs of orthogonal circles. But the orthogonality of circles translates into the orthogonality of ℓ_1 and ℓ_2, and we are done.

Second solution. We will prove that

$$\frac{A_1A_2}{A_1A_4} : \frac{PA_2}{PA_4} = \frac{PA_2}{PA_4} : \frac{A_3A_2}{A_3A_4}.$$

Switching to complex coordinates, we prove the stronger equality of cross-ratios

$$(a_1, p, a_2, a_4) = (p, a_3, a_2, a_4).$$

Use a Möbius transformation ϕ such that $\phi(p) = \infty$ and the invariance of the cross-ratio (Theorem 3.10) to reduce the problem to

$$(a_1', \infty, a_2', a_4') = (\infty, a_3', a_2', a_4'),$$

where $a'_i = \phi(a_i)$, $i = 1, 2, 3, 4$. This is equivalent to

$$\frac{a'_2 - a'_1}{a'_4 - a'_1} = \frac{a'_4 - a'_3}{a'_2 - a'_3}.$$

We can interpret this is saying that if a'_2 is the image of a'_4 under a spiral similarity of center a'_1, scaling factor k, and angle α, then a'_4 is the image of a'_2 under a spiral similarity of center a'_3, the same scaling factor k, and the same angle α. And this is true because a'_1, a'_2, a'_3, a'_4 form a parallelogram, as Möbius transformations preserve angles. This proves (a). The proof of (b) is analogous to that shown in the first solution: Möbius transformations preserve the property of four points to be concyclic, and the vertices of a parallelogram form a cyclic quadrilateral if and only if they form a rectangle. But this means that the images of ℓ_1 and ℓ_2 are perpendicular, and since Möbius transformations preserve angles, so are the lines themselves.

Third solution. A clumsier solution is possible if we use a Möbius transformation that maps A_2 to ∞. The new configuration is shown in Fig. 11.12, with the images of points denoted by dashes. We have seen in the second solution that the identity from (a) is invariant under Möbius transformations, so it suffices to prove it in this new setting. And in this setting, the terms that contain A'_2 cancel out because they are equal infinities, and so the identity reads

$$\frac{P'A'^2_4}{A'_1A'_4 \cdot A'_3A'_4} = 1 \iff \frac{P'A'_4}{A'_1A'_4} = \frac{A'_3A'_4}{P'A'_4}.$$

And this is a consequence of the similarity of the triangles $A'_4 P' A'_1$ and $A'_4 A'_3 P'$, as you can guess by examining Fig. 11.12. To prove that the triangles are similar, consider the spiral similarity of center A'_4 that maps ω'_3 to ω'_2. The tangent to ω'_1 at the image of P' forms with ω'_2 an angle equal to $\angle(\omega'_3, \omega'_2)$, so the image of P' must be A'_1. Consequently the image of A'_3 is P', and so the triangles are mapped into each other by the spiral similarity.

For (b), notice that the points A_1, A_2, A_3, A_4 are concyclic if and only if A'_1, A'_4, A'_3 are collinear, and because of the similarity of the two triangles, this happens if and only if $\angle P'A'_4A'_1 = \angle P'A'_4A'_3 = 90°$. But then $P'A'_1$ and $P'A'_3$ are diameters, so ω'_2 and ω'_3 are orthogonal, so the two pairs of circles in the original configuration are orthogonal. Consequently the points are concyclic if and only if ℓ_1 and ℓ_2 are orthogonal.

Remark The wisdom of this problem is to recognize the conclusion as being invariant under Möbius transformations, even under circular transformations, and then to map everything to a convenient configuration. And as the second and third solutions show, this convenient configuration may not be unique.

Note the similarity of the identity in (a) to Steiner's identity from Problem 170. Regarding the proof of (b), the original configuration contains five circles, the four circles passing through the origin, plus the circle formed by the four pairwise

Fig. 11.12 A clumsy
solution

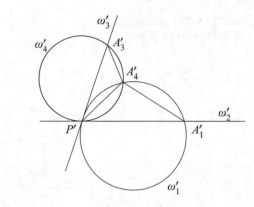

Fig. 11.13 Circle orthogonal
to another circle and to its
chord

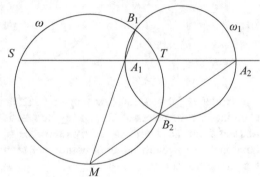

intersection points. A transformation that moves P at infinity significantly simplifies
the problem, transforming four of the five circles into lines. It is always easier
to prove that points obtained as intersections of lines are concyclic because, for
example, angle chasing is simpler.

Source Part (a) is a short listed problem of the 2003 International Mathematical
Olympiad, proposed by Armenia.

173 You can follow the argument on Fig. 11.13. We use the inversion of center M
and radius $MS = MT$. This inversion maps the circle ω to the line ST, and because
ω_1 is orthogonal to both ω and to ST, so is its image. But the image of ω_1 is a circle
that is homothetic to ω_1, under a homothety centered at M. These conditions can be
satisfied simultaneously only by ω_1 itself, so ω_1 is invariant under the inversion. As
the image of the intersection is the intersection of the images, the inversion maps
A_1 to B_1 and A_2 to B_2. So the lines $A_1 B_1$ and $A_2 B_2$ pass through the center M of
inversion, as desired.

174 *First solution.* Consider an inversion with center the tangency point O of Ω_1
and Ω_2. Then Ω_1 and Ω_2 transform into two parallel lines, Ω_1' and Ω_2', perpendicular
to the line of centers, as can be seen in Fig. 11.14. The circles ω_n, $n \geq 0$, transform

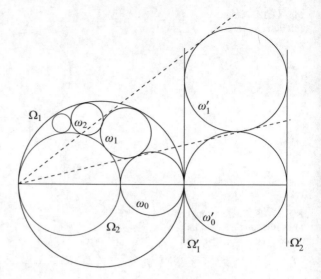

Fig. 11.14 The shoemaker's
knife problem

into circles ω_n', $n \geq 0$, that are tangent to Ω_1' and Ω_2' and so that the center of ω_0' is on the line of centers of Ω_1 and Ω_2, and ω_n' is tangent to Ω_1', Ω_2', and ω_{n-1}'. The relation from the statement clearly holds for ω_n', $n \geq 1$. Since the circles ω_n' are homothetic to ω_n with center of homothety O, the ratios are the same for ω_n's, and the problem is solved.

Second solution. Let us consider complex coordinates with the origin at O so that the ray connecting O to the centers of Ω_1 and Ω_2 is the positive semiaxis, and let us use the inversion from the first solution, with R being the radius of inversion. We have seen in the analytic proof of Theorem 3.27 that the image through the inversion of center 0 and radius R of the circle $|z - x| = r$, $x \in \mathbb{R}$ is the circle

$$\left| w - \frac{Rx}{x^2 - r^2} \right| = \left(\frac{Rr}{x^2 - r^2} \right)^2 ,$$

If we rotate the center to $z_0 = x e^{i\theta}$, then both the circle and its inverse are rotated by θ, turning the equation of the circle into $|z - z_0| = r$ and the equation of its image into

$$\left| w - \frac{Rz_0}{|z_0|^2 - r^2} \right| = \left(\frac{Rr}{|z_0|^2 - r^2} \right)^2 ,$$

Now if the lines Ω_1' and Ω_2' are $\Re z = c$ and $\Re z = c + 2r$, respectively, then the center of ω_n' is $c + r + 2nri$, $n \geq 1$. Consequently, the coordinate of the center of ω_n is

$$z_n = \frac{R(c + r + 2nri)}{c^2 + 2cr + r^2 + 4n^2r^2 - r^2} = \frac{Rc + r}{c^2 + 2cr + 4n^2r^2} + i\frac{2nRr}{c^2 + 2cr + rn^2r^2},$$

while its diameter is

$$d_n = \frac{2Rr}{c^2 + 2cr + 4n^2r^2}.$$

Taking the ratio of $h_n = \Im z_n$ to the diameter d_n we obtain n, and the problem is solved.

Remark It is worth mentioning that the centers of ω_n, $n \geq 1$, lie on the ellipse whose foci are the centers of Ω_1 and Ω_2 and which passes through O. Indeed, if O_1, O_2, R_1, R_2 are the centers and radii of Ω_1 and Ω_2, respectively, with Ω_2 inside Ω_1, and if C_n and ρ_n are the center and radius of ω_n, then $O_1C_n = R_1 + r_n$ and $O_2C_n = R_2 - r_n$, so

$$O_1C_n + O_2C_n = R_1 + r_n + R_2 - r_n = R_1 + R_2 = \text{constant}.$$

The name *shoemaker's knife* (*arbelos* in Greek) is motivated by the figure, more precisely, the region inside Ω_1 but outside of both Ω_2 and ω_0 looks like a shoemaker's knife.

Source Pappus of Alexandria.

175 The solution can be followed on Fig. 11.15. Invert with respect to the circumcircle of ABC. The points A, B, C are fixed while, as a consequence of Proposition 3.3, the points A_1, B_1, C_1 are mapped to the intersections of the tangents to the circumcircle at A, B, C, as shown in the figure. The circumcircles of AA_1O, BB_1O, and CC_1O are mapped into lines (Theorem 3.4), and these lines are AA_1', BB_1', CC_1'. The problem reduces to showing that AA_1', BB_1', CC_1' are concurrent.

Fig. 11.15 The points
A_1, B_1, C_1 and their inverses

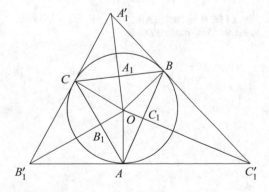

The equality of the tangents from a point to a circle implies that $A_1'B = A_1'C$, $B_1'A = B_1'C$, $C_1'A = C_1'B$, hence

$$\frac{A_1'B}{C_1'B} \cdot \frac{C_1'A}{B_1'A} \cdot \frac{B_1'C}{A_1'C} = 1,$$

which by the reciprocal of Ceva's Theorem implies that AA_1', BB_1', CC_1' are concurrent, and the problem is solved.

Remark If the triangle ABC is acute, then its circumcircle is the incircle of the triangle $A_1'B_1'C_1'$, and AA_1', BB_1', CC_1' meet at the Gergonne point of this triangle. If the triangle ABC is obtuse, then its circumcircle is an excircle of the triangle $A_1'B_1'C_1'$ and the lines AA_1', BB_1', CC_1' meet at an adjoint of the Gergonne point.

Using the same inversion, we can prove that, as a consequence of the fact that the medians of the triangle ABC intersect, the circumcircles of $AA_1'O$, $BB_1'O$, and $CC_1'O$ have a second common point. Said differently, if ABC is a triangle, A_1, B_1, C_1 are the points of tangency of the incircle with the sides, and I is the incenter, then the circumcircles of AA_1I, BB_1I, CC_1I have a second common point.

Source Romanian certification exam for high school teachers, 1984, solution by Ion D. Ion.

176 Let R and r be the radii of Ω and ω, respectively. Arguing on Fig. 11.16, we notice that $OD \cdot OA = OB \cdot OE = OC \cdot OF = r \cdot R$. It is natural to take an inversion of center O and radius $\sqrt{rR}$, so that A, B, C are mapped to D, E, F, respectively. The points A, B, C are on a circle passing through the center O of inversion, so their images are collinear. Done.

Remark For the same reason, the converse is also true: if Ω and ω are concentric, with ω inside Ω, and with common center O, and if the chord EF of Ω is tangent

Fig. 11.16 The circle (ABC)
is mapped into the line DEF

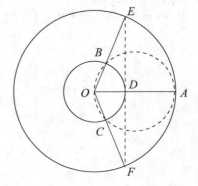

Fig. 11.17 The orthogonality
of AO and EF

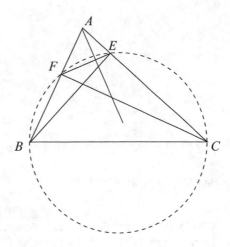

to ω, then the intersections of ω with the radii OE and OF are on the circle of
diameter OA.

177 Let ABC be the triangle, and let BE and CF be the altitudes from the vertices
B and C, respectively (Fig. 11.17). If we showed that AO is orthogonal to EF,
then, by symmetry, the corresponding orthogonalities would hold for the other two
vertices and the problem would be solved. Because the angles formed by BE and
AC, and by CF and AB are right, the quadrilateral $BFEC$ is cyclic, and so by
writing the power of the point A with respect to its circumcircle, we obtain

$$AE \cdot AB = AF \cdot AC.$$

This motivates us to consider an inversion of center A and radius $\sqrt{AF \cdot AB} =
\sqrt{AE \cdot AC}$. This inversion maps the line EF to the circumcircle of ABC, and
the line that connects the center of inversion A to the circumcenter O of ABC
is orthogonal to EF, as Theorem 3.4 states. This shows that AO is orthogonal to
EF, and we are done.

Remark The fact that the perpendiculars from the vertices onto the sides of
the orthic triangle intersect can be derived from the Theorem of Orthological
Triangles, which states that if the perpendiculars from the vertices of a triangle
onto the respective sides of a second triangle meet, then the same is true for the
perpendiculars from the vertices of the second triangle onto the sides of the first.
Applying this theorem to ABC and DEF solves the problem.

   The same inversion can be used to show that the altitudes of a triangle intersect.
Let H be the intersection of BE and CF. The inversion maps the side BC into the
circumcircle of $AEHF$, and so the diameter AH of this circle is perpendicular to
BC. In other words, the third altitude also passes through H.

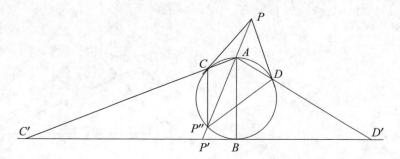

Fig. 11.18 Proof that $C'P' = P'D'$

178 Denote by P'' the second intersection point of AP with Ω (see Fig. 11.18). By the construction of the symmedian (Theorem 1.22), $ACP''D$ is harmonic, and so the complex coordinates of its vertices satisfy

$$\frac{a-c}{a-d} : \frac{p''-c}{p''-d} = -1.$$

Take the inversion of center A and radius AB. Then the circle Ω is mapped to the line t, and A, C, P'', D are mapped to ∞, C', P', D'. As inversion preserves real valued cross-ratios (Theorem 3.18), the points ∞, C', D', P' form a harmonic division on t, which means that P' is the midpoint of $C'D'$.

Remark It is important to know that the inversion from A that maps Ω to t is called the stereographic projection of the circle onto the line.

179 We invert about a circle centered at C and use our convention to mark the image of a geometric object with a dash. The images k' and k'_3 of k and k_3 are lines, in fact parallel lines because inversion preserves angles. Also, k'_1 and k'_2 are circles tangent to both k' and k'_3 and to each other, so they must be equal circles; the tangency points are A', B', X', Y', as shown in Fig. 11.19. The distance between the centers of k'_1 and k'_2 is equal to two radii, which is the same as the distance between k' and k'_3. Hence $A'B'Y'X'$ is a square.

On the other hand, because $A' \in CA$ and $B' \in CB$,

$$\angle A'CB' = \angle ACB = 180° - \angle ASB/2 = 135°.$$

Consequently

$$\angle A'CB' + \angle A'X'B' = 135° + 45° = 180°,$$

showing that the quadrilateral $A'CB'X'$ is cyclic. We deduce that C is on the circumcircle of the square $A'B'Y'X'$, and hence

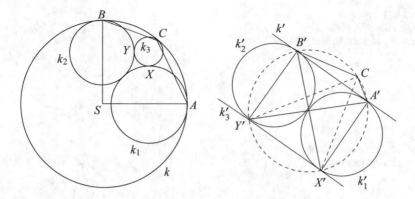

Fig. 11.19 The circles k, k_1, k_2, k_3 and their images through the inversion

Fig. 11.20 Inverted
configuration for the proof
that A, B, M are collinear

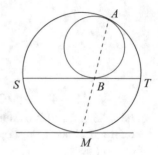

$$\angle XCY = \angle X'CY' = \angle X'A'Y' = 45°,$$

as desired.

Remark It is easier to chase angles with lines than with circles, and the inversion considerably simplifies the figure, from four circles to two circles and two lines.

Source Czech and Slovak Mathematical Olympiad, 1995, solution by Leandro Maia.

180 (a) *First solution.* The configuration is shown in Fig. 11.20. Consider an inversion χ_1 of center M and radius $MS = MT$. Then the circle ω and the line ST are inverses of each other, so the circle ω_1 is mapped to a circle $\chi_1(\omega_1)$ that is tangent to both ST and ω. But $\chi_1(\omega_1)$ must be homothetic to ω_1 under a homothety of center M, and this is only possible if $\chi_1(\omega_1) = \omega_1$. It follows that the point of tangency B of ω_1 and ST is mapped to the point of tangency A of ω_1 and ω. And the center of inversion M is collinear with B and $A = \chi_1(B)$, as desired.

Second solution. Consider the circular transformation ϕ that maps ω into itself obtained as the composition of an inversion of center B and the reflection over B (prove that it exists!). This transformation maps ω_1 into a line $\phi(\omega_1)$ that is tangent

Fig. 11.21 Inverted
configuration for the proof
that A, B, M are collinear

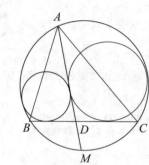

Fig. 11.22 The case where
the circles are tangent

to the arc $\overset{\frown}{SMT}$ and is parallel to ST. The point of tangency of $\phi(\omega_1)$ and the arc is therefore M. Since ϕ preserves tangencies, $\phi(M) = A$, and therefore B, A, $\phi(A) = M$ are collinear.

Third solution. Consider an inversion χ_2 of center A and arbitrary radius. Then $\omega' = \chi_2(\omega)$ and $\omega_1' = \chi_2(\omega_1)$ are parallel lines, and $\chi_2(ST)$ is a circle that passes through the center A of the inversion, crosses ω' at $S' = \chi_2(S)$, $T' = \chi_2(T)$, and is tangent to ω_1' at $B' = \chi_2(B)$, as shown in Fig. 11.21. Then $\overset{\frown}{S'B'} = \overset{\frown}{T'B'}$ so $\angle S'AB' = \angle B'AT'$. It follows that $\angle SAB = \angle BAT$, so SB is the angle bisector of $\angle SAT$. But AM is also the angle bisector of $\angle SAT$, so the two lines coincide, proving the collinearity of A, B, M.

(b) Consider again the inversion of center M and radius $MS = MT$. As seen in the first solution to (a), ω_1 and ω_2 are invariant under this inversion, hence so is $\omega_1 \cap \omega_2$. It cannot happen that both P and Q are invariant under this inversion, since $MP \neq MQ$, so they cannot both be equal to MS. The only possibility is that P is mapped to Q, and so P, Q, and the center M of the inversion are collinear, as desired.

(c) With changed notation, if the circles are tangent, then we are in the particular case of (b) where $P = Q$ (Fig. 11.22), and then the common tangent AD to the two circles passes through the midpoint M of the arc $\overset{\frown}{BC}$. Hence

$$\angle BAD = \angle BAM = \frac{1}{2}\ \overset{\frown}{BM} = \frac{1}{2}\ \overset{\frown}{CM} = \angle CAM = \angle CAD.$$

Fig. 11.23 AB is the
common tangent of ω and ω_1

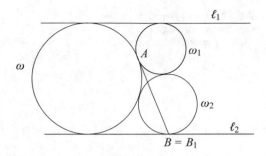

Conversely, if $\angle BAD = \angle DAC$, then the intersection M of AD with the circumcircle is the midpoint of the arc $\overset{\frown}{BC}$ that does not contain A. As seen in (a), the circles ω_1 and ω_2 are fixed by the inversion with center M and radius MB. The points where they are tangent to AD are also fixed by the inversion. But the inversion has only one fixed point on the ray $|OA$, so these points coincide. Done.

Remark The power of M with respect to any circle ω_1 interior tangent to ω, tangent to ST, and on the other side of ST than M is equal to MS^2; thus M is on the radical axis of any two such circles and is the radical center of any three such circles.

Homothety is the most natural approach to (a) consider the homothety h of center A that maps ω_1 to ω. This homothety maps the tangent to ω_1 at B to the tangent to ω at $h(B)$. The two tangents must therefore be parallel, so the tangent at $h(B)$ is parallel to the chord ST. This implies that $h(B) = M$. And A, B, $h(B) = M$ are collinear, as desired. But the proof by inversion allows progress toward (b) and (c).

Source Part (a) is a variation of Archimedes' Lemma. Part (c) was given at a Romanian Team Selection Test for the International Mathematical Olympiad in 1997.

181 The configuration is shown in Fig. 11.23.

Let B_1 be the intersection of the common interior tangent of ω and ω_1 with ℓ_2. Consider an inversion of center B_1 and radius $B_1 A$. Because the tangent is orthogonal to the radius at the point of contact, the inversion circle is orthogonal to both ω and ω_1. So ω and ω_1 are invariant under the inversion (Proposition 3.25). The line ℓ_1 is mapped to a circle that passes through B_1 and is tangent to both ω and ω_1. Moreover, because ℓ_1 and ℓ_2 are parallel, and ℓ_2 passes through the center of inversion, the image of ℓ_1 is also tangent to ℓ_2 (as ℓ_1 and ℓ_2 form a $0°$ angle). Thus the image of ℓ_1 is tangent to ω, ω_1, and ℓ_2. But there is only one such circle, namely, ω_2. But then ω_2 must pass through the center B_1 of inversion, because it is the image of a line. And this can only happen if $B_1 = B$.

Remark This is yet another example of a "position argument."

182 Rephrasing the question, we will prove that the line CO passes through M. To this end, consider an inversion of center C and radius $\sqrt{CA \cdot CD} = \sqrt{CB \cdot CF}$,

Fig. 11.24 The inversion
used for proving that
O, C, M are collinear

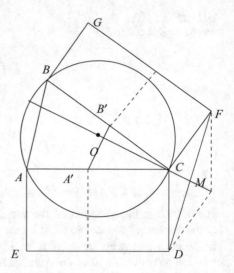

and let A', B', O' be the images of A, B, O, respectively (see Fig. 11.24). Note first
that $CA' = CD$ and $CB' = CF$, because $CA \cdot CA' = CA \cdot CD$ and $CB \cdot CB' = CB \cdot CF$.

By Theorem 3.4, CO is perpendicular to $A'B'$, as $A'B'$ is the image of the
circle under the inversion. We are now to show that CO passes through M,
which is equivalent to showing that CM is perpendicular to $A'B'$. This is a direct
consequence of Problem 15 from the first chapter, applied to the triangle $CA'B'$
with squares constructed on the sides CA' and CB', since CM passes through the
fourth vertex of the parallelogram determined by C, D, F.

Remark Like in the previous problem, the radius of inversion is determined by the
length of a geometric object present in the statement, in this case the square root of
the equal areas. With this choice, C becomes the natural center of inversion.

183 We invert about a circle of center E and radius $EB = EC$. The image of
N through this inversion is the center O of the circle (by Proposition 3.7), and the
inverses of K and M are the points K' and M' which are the other intersection
points of EK and EM with the circle, respectively (Fig. 11.25). It follows that line
through K, M, N is is mapped by the inversion to the circle through E, K', M', O,
so these four points are concyclic. Therefore

$$\angle OM'E = \angle OK'K = \angle OKK'.$$

But then

$$\angle M'MO = \angle MM'O = \angle OKE.$$

Fig. 11.25 The inversion of center E

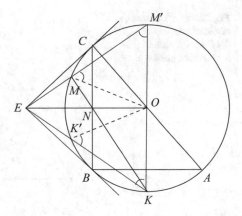

It follows that $EMOK$ is cyclic. From here we deduce that $\angle EMK = \angle EOK = 90°$, and the conclusion follows.

Remark We could have also concluded the solution by noting that $\angle EM'O = \angle EKO$ implies that M is the reflection of K' over EO, because it is on the arc traced by the vertex of an angle of measure $\angle EMO$ and this arc intersects the semicircle that lies above the line EO in only one point, the reflection of M.

Source Tournament of the Towns, 2002, solution by Leandro Maia.

184 The point H is the incenter of the orthic triangle $A_1B_1C_1$, and A_2, B_2, C_2 are the points of tangency of the incircle of this triangle to its sides. Thus (a) is a consequence of the following result:

Lemma *If ABC is a triangle that is not isosceles, I is its incenter, and D, E, F are the tangency points of the incircle to the sides BC, AC, and AB, respectively, then the circumcircles of AID, BIE, and CIF have a second intersection point besides I.*

Proof Consider the inversion with respect to the incircle of the triangle ABC (Fig. 11.26). Because AI, BI, and CI are angle bisectors, and hence medians, in the isosceles triangles EAF, FBD, and DCE, respectively, A', B', C' are the midpoints of EF, FD, and DE (by Proposition 3.3). It follows that the lines DA', EB', and FC', which are the images of the three circles, are medians in the triangle DEF. So the second point of intersection of the three circles is the inverse of the centroid of DEF. $\qquad\square$

For (b) consider again the inversion with respect to the incircle of the orthic triangle, in order to reduce the problem to showing that $A'A'_2$, $B'B'_2$, and $C'C'_2$ are concurrent. As seen above, A'_1, B'_1, C'_1 are the midpoints of the segments B_2C_2, C_2A_2, A_2B_2, and A', B', C' lie on the lines $B'_1C'_1$, $C'_1A'_1$, respectively $A'_1B'_1$ (because HB_1AC_1, HC_1BA_1, HA_1CB_1 are cyclic). In fact A', B', C' are the feet of the altitudes of the triangle $A'_1B'_1C'_1$, and in particular HA', HB', and HC' are concurrent. We are done with the solution once we prove the following result:

Fig. 11.26 Proof that the
circumcircles of AID, BIE,
CIF have a second
intersection point

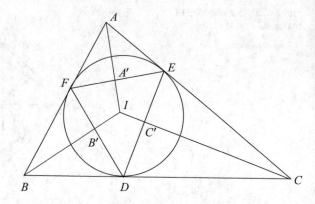

Fig. 11.27 Configuration
with three concurrent triples
of lines

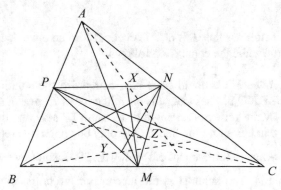

Lemma *If ABC is a triangle, and if M is on BC, N is on AC, and P is on AB such that AM, BN, and CP are concurrent, and if moreover X is on NP, Y is on MP, and Z is on MN such that MX, NY, and PZ are concurrent, then AX, BY, and PZ are also concurrent.*

Proof Such a configuration is shown in Fig. 11.27.

Note that if PQR is a triangle and S is on QR (Fig. 11.28), then by applying the Law of Sines in the triangles PQS and PSR, we obtain

$$\frac{PR}{SR} = \frac{\sin \angle PSR}{\sin \angle SPR} \text{ and } \frac{PQ}{QS} = \frac{\sin \angle PSQ}{\sin \angle QPS}.$$

Using the fact that supplementary angles have equal sines, we obtain

$$\frac{\sin \angle QPS}{\sin \angle SPR} = \frac{PR}{SR} : \frac{PQ}{QS} = \frac{QS}{SR} \cdot \frac{PR}{PQ}.$$

Applying this in the triangles ANP, BPM, and CMN, for the points X, Y, Z on the respective sides, and multiplying, we obtain

Fig. 11.28 Proof of
trigonometric formula for a
cevian in a triangle

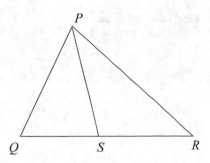

$$\frac{\sin \angle PAX}{\sin \angle XAN} \cdot \frac{\sin \angle NCY}{\sin \angle YCM} \cdot \frac{\sin \angle MBZ}{\sin \angle ZBP} = \frac{PX}{XN} \cdot \frac{AN}{AP} \cdot \frac{NY}{MY} \cdot \frac{CM}{CN} \cdot \frac{MZ}{PZ} \cdot \frac{BP}{BM}$$

$$= \left(\frac{PX}{XN} \cdot \frac{NY}{MY} \cdot \frac{MZ}{PZ}\right) \cdot \left(\frac{AN}{AP} \cdot \frac{CM}{CN} \cdot \frac{BP}{BM}\right).$$

Each of the products in the parentheses is equal to 1, by Ceva's Theorem in the
triangles ABC and MNP. Thus

$$\frac{\sin \angle PAX}{\sin \angle XAN} \cdot \frac{\sin \angle NCY}{\sin \angle YCM} \cdot \frac{\sin \angle MBZ}{\sin \angle ZBP} = 1.$$

Thus AX, BY, CZ are concurrent by the trigonometric form of Ceva's Theorem.

□

Remark Examining the proof of the second lemma, we realize that the following
stronger statement is true:

Let ABC be a triangle, let M, N, and P be on BC, AC, and AB, respectively, and
let X, Y, Z be on NP, MP, and MN, respectively. Then the concurrency of any two
of the triples of lines (AM, BN, CP), (MX, NY, PZ), and (AX, BY, PZ) implies
the concurrency of the third.

Source *Gazeta Matematică (Mathematics Gazette, Bucharest).*

185 *First solution.* A possible configuration is shown in Fig. 11.29. Let us consider
the inversion whose center is the tangency point of C_3 and ℓ and has the property
that maps C into itself. The circle C_3 is mapped to a line C_3' that is parallel to ℓ,
while the circles C_1 and C_2 are mapped to the circles C_1' and C_2' that maintain the
same tangency pattern, as illustrated in Fig. 11.30. Of course, C_1' and C_2' have equal
radii.

Let O, O_1, O_2 be the centers of C, C_1', C_2', respectively, and let T be the point of
contact of C_1' and C_2'; let also x be the radius of C_1' (and also of C_2'). The Pythagorean
Theorem in the right triangle OTO_1 yields

$$x^2 + (x - 1)^2 = (x + 1)^2,$$

Fig. 11.29 The circles
C, C_1, C_2, C_3 and the line ℓ

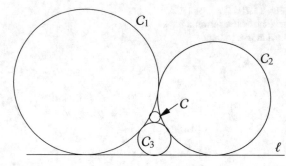

Fig. 11.30 The inverted
circles C, C_1, C_2, C_3 and the
line ℓ

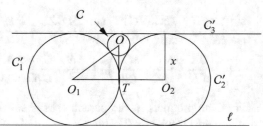

that is, $x^2 - 4x = 0$. It follows that $x = 4$, and hence the distance from O to ℓ is $2x - 1 = 7$.

Second solution. The proof of Theorem 3.27 works in the limiting case where one of the circles is a line, so there is an inversion χ that maps C and ℓ into two concentric circles. Now the circles C_1, C_2, and C_3 have equal radii and lie between the two concentric circles, being tangent to them. The center of the inversion χ lies on the image of ℓ, so there is a rotation that maps C_3 into a circle that passes through the center of χ. Compose χ with this rotation to map C and ℓ back to their original position, while C_3 is mapped to a line C_3' parallel to ℓ (the two lines are parallel because they form a $0°$ angle), and C_2 and C_3 have become two equal circles. And we are again in the configuration from Fig. 11.30, and the solution continues with the same computation.

Remark The reader can recognize that we are in a limiting case of Steiner's Porism (Theorem 3.28), with the two circles being C and ℓ, and the chain of circles being C_1, C_2, and C_3. This explains why the distance from the center of C to ℓ only depends on the radius of C and not on the radii of C_1, C_2, C_3. Both solutions reduce the problem to a particular chain of three circles in which one of the three circles is a line, where the computation is easy.

Source Short list of the International Mathematical Olympiad, 1982, proposed by Finland, second solution from D. Djukić, V. Janković, I. Matić, N. Petrović, *The IMO Compendium*, Springer, 2006.

Fig. 11.31 The configuration
for Problem 186 before and
after inversion

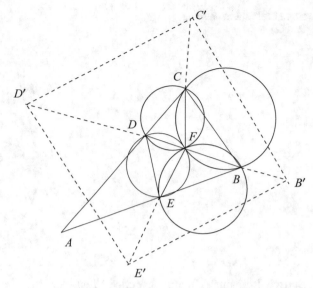

186 Examining Fig. 11.31 we realize that the problem reduces to showing that
$\angle FCD = \angle FEB$. Consider an inversion of center F and arbitrary power, and use
the standard convention that the image of a point is denoted by the same letter,
but with a dash. Theorem 3.2 allows us to reduce the problem to showing that
$\angle FD'C' = \angle FB'E'$.

Looking again at the figure, we notice that the configuration is simplified
considerably after inversion since the circumcircles of BCF and DEF have become
the lines $B'C'$ and $D'E'$, which form the same angle as the original circles,
and hence are parallel. Similarly $B'E'$ and $C'D'$ are parallel, and so $B'C'D'E'$
is a parallelogram. From here we obtain $\angle E'D'C' = \angle E'B'C'$, which can be
decomposed as

$$\angle FD'E' + \angle FD'C' = \angle FB'E' + \angle FB'C'.$$

On the other hand $\angle BED = \angle ACB$ yields $\angle FED + \angle FEB = \angle FCD + \angle FCB$,
and again from Theorem 3.2, we obtain

$$\angle FD'E' + \angle FB'E' = \angle FD'C' + \angle FB'C'.$$

Subtracting the two relations, we obtain the desired $\angle FD'C' = \angle FB'E'$.

Remark This solution relies on the behavior under inversion of both the angle
between two curves and the angle determined by two points and the center of
inversion.

Fig. 11.32 Inversion about
the excircle

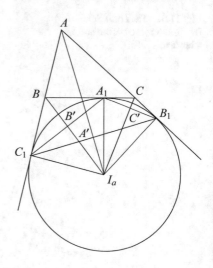

Fig. 11.32 Inversion about the excircle

Source Romanian Team Selection Test for the International Mathematical Olympiad, 2008.

187 We consider the inversion χ about the excircle. By Proposition 3.3 $\chi(A) = A'$, $\chi(B) = B'$, and $\chi(C) = C'$. Thus we are in the situation from Fig. 11.32. Since χ maps the circumcircle of ABC to the circumcircle of $A'B'C'$, the line ℓ_a, which passes through the center of inversion and the circumcenter of $A'B'C'$, also passes through the circumcenter of ABC. From here we infer that ℓ_a, ℓ_b, and ℓ_c pass through the circumcenter of ABC, and the problem is solved.

Source Petru Braica.

188 We consider an inversion of center A and examine the configuration before and after inversion on Fig. 11.33. With the convention of denoting the image of a point by the same letter but with a dash, the image of the circumcircle of ABC is the line $B'C'$, and the image of the line BC is the circumcircle of $AB'C'$.

Because AE is symmedian, $ABEC$ is a harmonic quadrilateral (Proposition 1.23), and so the cross-ratio of the complex coordinates of A, B, E, C is -1. As inversion preserves real valued cross-ratios, the cross-ratio of the complex coordinates of the images of these points is also -1. But the images of these points are ∞, B', E', C', which are collinear and therefore form a harmonic division. This means that E' is the midpoint of $B'C'$.

Because inversion preserves the angle between two curves, and because AD and BC are orthogonal, AD' and the circumcircle of $AB'C'$ are orthogonal, too, showing that A and D' are diametrically opposite in this circumcircle. Also, because AF and the circumcircle of ABC are orthogonal, so are AF' and $B'C'$, with F' on the circumcircle of $AB'C'$. But then $\angle AF'D' = \overset{\frown}{AD'}/2 = 90°$, so AF' is orthogonal to both $B'C'$ and $D'F'$. This implies that $D'F'$ is parallel to $B'C'$, and

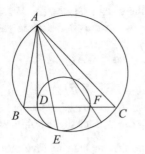
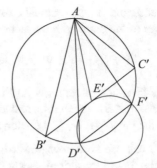

Fig. 11.33 The circumcircles of ABC and DEF are tangent

Fig. 11.34 Proof of the collinearity of M, N, P, Q

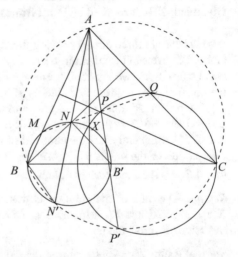

so $B'D'F'C'$ is an isosceles trapezoid. And since E' is the midpoint of $B'C'$, the circumcircle of $D'E'F'$ is tangent to $B'C'$. Because inversion preserves tangencies, the circumcircle of DEF is tangent to the circumcircle of ABC, as desired.

189 We argue on Fig. 11.34. Let B' be the point that is diametrically opposite to B in C_1. Then $\angle BNB' = 90°$, so $B'N$ is parallel to AC. Hence $\angle NB'B = \angle ACB$.

Now let N' be the second intersection point of intersection of AN and C_1. Then because $BN'B'N$ is cyclic it follows that

$$\angle BN'A = \angle BN'N = \angle BB'N = \angle ACB,$$

which implies that A, B, N', C lie on a circle, so N' is on the circumcircle of ABC. Similarly, if P' is the second intersection point of AP and C_2, then P' is on the circumcircle of ABC as well.

The altitude from A is the radical axis of the circles C_1 and C_2. Writing the fact that the power of the point with respect of the two circles is the same yields

$$AM \cdot AB = AN \cdot AN' = AP \cdot AP' = AQ \cdot AC.$$

Let χ be the inversion of center A and radius $\sqrt{AM \cdot AB}$. Then χ transforms the circumcircle ω of ABC into a line, and since $\chi(B) = M$, $\chi(N') = N$, $\chi(P') = P$, $\chi(C) = Q$, the points M, N, P, Q belong to $\chi(\omega)$. Hence these points are collinear.

Remark You can see here how power-of-a-point can lead to inversion. The case where X is the foot of the altitude is a well-known problem.

Source Romanian Team Selection Test for the International Mathematical Olympiad, 2014, proposed by Petru Braica.

190 There are three circles passing through A: the circumcircles of ABL, ACK, and AKL. We can turn them into lines by taking an inversion of center A. With the usual convention, we denote the image of a point X by X'. The original figure and the inverted one are presented side-by-side in Fig. 11.35.

The points B' and C' are the midpoints of the segments AK' and AL', respectively. So P' is the intersection of the medians $K'C'$ and $L'B'$ in the triangle $AK'L'$. Consequently it is the centroid of this triangle. On the other hand, Q' is the intersection of the median AP' with the side $K'L'$, thus $AP' = \frac{2}{3}AQ'$. It follows that $AP = \frac{3}{2}AQ$, the desired equality.

Remark The idea of the solution is that, for an inversion of center O that maps X and Y to X' and Y', respectively, if $OX/OY = m/n$ then $OX'/OY' = n/m$. Inversion "inverts" the ratio!

Source Baltic Way, 2006, communicated by Titu Andreescu.

191 Let ϕ_A be the $\sqrt{bc}$ inversion corresponding to the vertex A in the triangle ABC (see Fig. 11.36). Because ϕ_A maps ℓ into a line ℓ' that is parallel to BC and

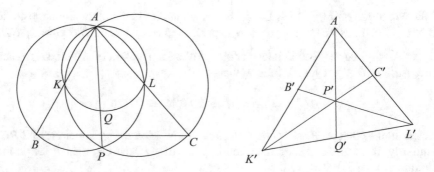

Fig. 11.35 $2AP = 3AQ$ is equivalent to $3AP' = 2AQ'$

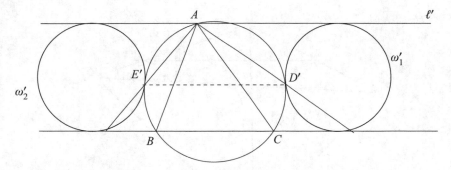

Fig. 11.36 Proof of tangency of circumcircles of ABC and ADE

the circumcircle of ABC and the line BC into each other, it maps ω_1 and ω_2 into two circles, ω'_1 and ω'_2, that are exterior tangent to ℓ', BC, and the circumcircle of ABC.

The circles ω'_1 and ω'_2 must be equal, since they are tangent to two parallel lines. Set $D' = \phi_A(D) \in \omega'_1$ and $E' = \phi_A(E)$ in ω'_2. Then D' and E' are tangency points of ω'_1 and ω'_2 with the circumcircle of ABC. Because the configuration formed by the three circles is symmetric with respect to the diameter of this latter circle that is perpendicular to BC, $D'E'$ is parallel to BC. But then the images of these lines through ϕ_A are tangent circles, that is, the circumcircles of ADE and ABC are tangent.

Remark This is a "position argument" that uses a Möbius transformation.

192 *First solution.* Consider an inversion of center B and radius 1. Let A', C', and D' be the images through the inversion of the points A, C, and D, respectively, as shown in Fig. 11.37. Using Theorem 3.2 we obtain $\angle BD'A' = \angle BAD$ and $\angle BD'C' = \angle BCD$, so

$$\angle C'D'A' = \angle BAD + \angle BCD = 360° - \angle ABC - \angle ADC = 360° - 2 \cdot 135° = 90°.$$

Applying formula (3.1) we obtain

$$C'D' = \frac{CD}{BD \cdot BC}, \quad D'A' = \frac{DA}{AB \cdot BD}, \quad A'C' = \frac{AC}{AB \cdot BC}.$$

The Pythagorean Theorem in the right triangle $A'C'D'$ yields

$$\left(\frac{AC}{AB \cdot BC}\right)^2 = \left(\frac{CD}{BD \cdot BC}\right)^2 + \left(\frac{DA}{AB \cdot BD}\right)^2,$$

hence

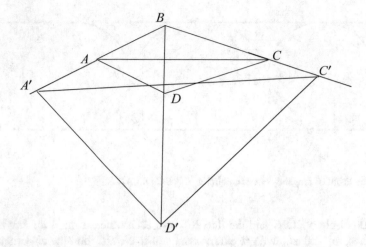

Fig. 11.37 The image of triangle ACD through the inversion

$$(AC \cdot BD)^2 = (AB \cdot CD)^2 + (BC \cdot DA)^2.$$

Combining this with the relation in the statement, we obtain

$$(AB \cdot CD)^2 + (BC \cdot DA)^2 = 2(AB \cdot CD) \cdot (BC \cdot DA).$$

Thus we have the equality case in the AM-GM inequality, which implies that $AB \cdot CD = BC \cdot DA$. In other words $AB/DA = BC/CD$. Since $\angle ABC = \angle ADC = 135°$, it follows that the triangles ABC and ADC are similar. These two triangles share the side AC, so they are congruent. So the figure is a kite, and consequently AC and BD are orthogonal.

Second solution. Write the equality from the statement as an equality of cross-ratios of segments

$$\frac{DB}{DC} : \frac{AB}{AC} = 2\frac{BC}{BD} : \frac{AC}{AD}.$$

Take an inversion of center A, and let B', C', D' be the images of the vertices B, C, D, respectively. Then $\angle AC'B' = \angle AC'D' = 135°$ (Theorem 3.2), so $\angle B'C'D' = 360° - 2 \times 135° = 90°$, i.e., the triangle $B'C'D'$ is right. Because inversion preserves the cross-ratio of segments, and because A is mapped to ∞,

$$\frac{D'B'}{D'C'} = 2\frac{B'C'}{B'D'}.$$

Let M be the midpoint of $B'D'$. Using trigonometric functions in the right triangle $B'C'D'$, we translate the above equality into

$$\frac{D'M'}{D'C'} = \sin \angle B'D'C'.$$

Then, using the Law of Sines in the isosceles triangle $MC'D'$, we transform this equality further into

$$\frac{\sin \angle B'D'C'}{\sin \angle C'MD'} = \sin \angle B'D'C'.$$

Hence $\sin \angle C'MD' = 1$. It follows that $\angle C'MD' = 90°$, showing that the right triangle $C'B'D'$ is isosceles. We thus have $\angle AC'B' = \angle AC'D'$ and $C'B' = C'D'$, which implies that D' is the reflection of B' over the angle bisector of AC. And so D is the reflection of B over the angle bisector of AC. From here we deduce that $ABCD$ is a kite, so its diagonals are orthogonal.

Remark The emergence of the right triangle becomes transparent when using complex coordinates. The angle condition from the statement is translated into

$$\frac{c-b}{a-b} = re^{\frac{3\pi i}{4}} \text{ and } \frac{a-d}{c-d} = se^{\frac{3\pi i}{4}}, \quad r, s \in \mathbb{R}.$$

Thinking in terms of cross-ratios, we can write

$$(a, c, b, d) = \frac{c-b}{a-b} : \frac{c-d}{a-d} = irs \in i\mathbb{R}.$$

If we consider a circular transformation that maps b to ∞, then (using dashes for the images) we have

$$\frac{c'-d'}{a'-d'} = \pm irs \in i\mathbb{R},$$

showing that a', c', d' form a right triangle (with the right angle at d').

Source United States of America Team Selection Test for the International Mathematical Olympiad, proposed by Răzvan Gelca.

193 Let AA', BB', CC' be the altitudes; let also A'', B'', C'' be the projections of the orthocenter H onto the lines MA, MB, MC, respectively (Fig. 11.38). Note that A'', B'', C'' lie on the circle of diameter MH because $\angle MA''H = \angle MB''H = \angle MC''H = 90°$.

Let A_1 be the intersection of HA'' and BC, B_1 the intersection of HB'' and AC, and C_1 the intersection of HC'' and AB. The quadrilateral $AA''A'A_1$ is cyclic, because $\angle AA''A_1 = \angle AA'A_1 = 90°$. Hence $HA_1 \cdot HA'' = HA \cdot HA'$. But $HA \cdot HA' = HB \cdot HB' = HC \cdot HC'$, and so $HA_1 \cdot HA'' = HB_1 \cdot HB'' = HC_1 \cdot HC''$. It follows that the points A_1, B_1, C_1 are the images of A'', B'', C'' with respect to the composition of an inversion of center H with a reflection over H, also known

Fig. 11.38 The construction
of the points A'', B'', C''

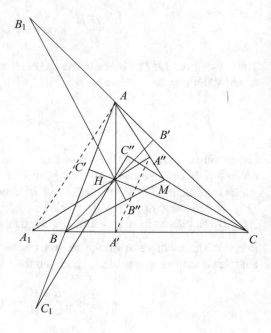

as an inversion of negative ratio (we have seen this inversion before). Consequently, they lie on the line that is the inverse of the circle of diameter MH, and this line is perpendicular to MH.

Remark Here we have used the circular transformation that maps the nine-point circle to the circumcircle, obtained as the composition of an inversion of center H and the reflection over H that was discussed in Sect. 3.1.7.

Source Gh. Țiţeica, *Probleme de Geometrie (Problems in Geometry)*, Ed. Tehnică, Bucharest, 1929.

194 *First solution.* Consider an inversion of center A and arbitrary radius. Because inversion preserves angles, the circumcircles of ABD and ACD are mapped to the lines $B'D'$ and $C'D'$ that are parallel to the lines AC' and AB', respectively. So $AB'D'C'$ is a parallelogram (Fig. 11.39 shows both the original figure and the inverted one).

As B is the midpoint of AE, E' is the midpoint of AB'. It follows that the triangles $AF'E'$ and $B'D'E'$ are congruent. Hence $AF' = B'D' = AC'$, that is, A is the midpoint of $C'F'$. But then A is the midpoint of CF, and the problem is solved.

Second solution. Let ϕ_A be the $\sqrt{bc}$ inversion corresponding to the vertex A in the triangle ABC. The image of the original configuration through this Möbius transformation is shown in Fig. 11.40. We obtain that $\phi_A(E) = E'$ is the midpoint of AC, because $C = \phi_A(B)$ and B is the midpoint of AE. Also, tangencies yield parallelism, so $ABD'C$ is a parallelogram, where $D' = \phi_A(D)$. And as the

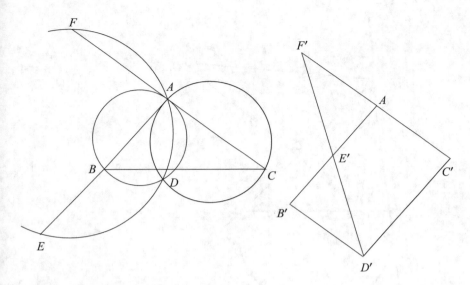

Fig. 11.39 The triangle ABC with the circumcircles of ABD, ACD, and ADE and their images through the inversion

Fig. 11.40 The image of the circumcircles of ABD, ACD, and ADE through ϕ_A

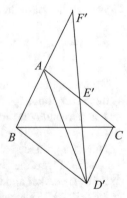

circumcircle of ADE is mapped to the line $D'E'$, $F' = \phi_A(F) \in D'E'$. Now since AF' is parallel to $D'C$ and E' is the midpoint of AD, $AF'CD'$ is a parallelogram, showing that $AF' = D'C = AB$. Hence $AF = AC$, as desired.

Remark Note that AD' is a median in the triangle ABC, and hence AD is a symmedian in this triangle, which gives a solution based on inversion to Problem 132.

Source Turkish Mathematical Olympiad, 1998.

195 Denote the circle that is tangent to AD at A and has the center on B_1C_1 by ω and its center by O_1. The configuration is shown in Fig. 11.41.

Writing the power of A with respect to the circumcircle of OBC, we obtain

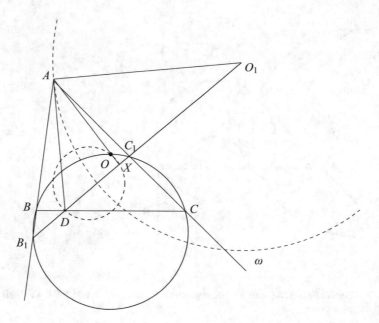

Fig. 11.41 The dotted circles are orthogonal

$$AB \cdot AB_1 = AC \cdot AC_1.$$

It is therefore natural to consider the inversion of center A and radius $\sqrt{AB \cdot AB_1}$, as this inversion maps the circumcircle of ABC to the line B_1C_1. By Theorem 3.4, AO is perpendicular to B_1C_1. Denote by X the intersection of the lines AO and B_1C_1.

Now we consider a different inversion; this is χ of center D and radius DA, which transforms the circle of diameter OD into a line. But note that the radii of the circle of inversion and of ω are orthogonal at A, so the two circles are orthogonal. Proposition 3.25 shows that $\chi(\omega) = \omega$. And the problem is simplified by using this inversion, because now we have to show that a line is orthogonal to a circle, which is a simpler task.

All we have to do is identify the line that is the image through χ of the circle of diameter OD. The Leg Theorem in the right triangle AO_1D in which AX is an altitude gives $DA^2 = DX \cdot DO_1$ and hence $\chi(X) = O_1$. But since $\angle AXD = 90°$, X is on the circle of diameter OD, so the image of this circle passes through the image of X, which is O_1. And any line that passes through the center O_1 of ω is orthogonal to ω. The problem is solved.

Remark The idea of the solution is that it is easier to compute the angle formed by a line and a circle than that the angle formed by two circles.

Source Mathematical Olympiad Summer Program, 1997, solution communicated by Leandro Maia.

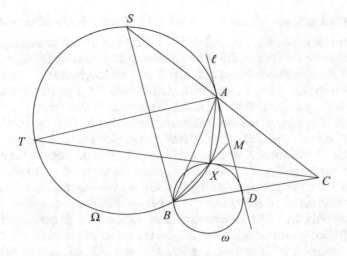

Fig. 11.42 Reformulation of the problem about proving that AX is orthogonal to XC

196 We will rephrase the problem and then solve it by inversion. Tracing the approach on Fig. 11.42, we start by observing that triangles MXA and MAB share the angle $\angle AMB$, which together with

$$\angle MXA = \angle DAC = \angle DAB = \angle MAB$$

implies that the two triangles are similar. This similarity implies that $\angle MAX = \angle ABX$, which then shows that the line AD is tangent to the circumcircle of ABX. The similarity also gives

$$\frac{AM}{XM} = \frac{BM}{AM},$$

so

$$MD^2 = AM^2 = MB \cdot MX.$$

This shows that MD^2 is equal to the power of M with respect to the circumcircle of BDX, which implies that AD is also tangent to this circle. We denote the line AD by ℓ, the circumcircle of ABX by Ω, and the circumcircle of BDX by ω.

Next, let S and T be the other intersections of the lines AC and CX with the circle Ω, respectively. Because

$$\angle BAD = \angle DAC = \frac{1}{2}\ \overset{\frown}{AS} = \angle ABS,$$

it follows that the lines SB and AD are parallel. Hence A is the midpoint of the arc $\overset{\frown}{BAS}$. The conclusion will then follow if we show that T is diametrically opposite to A in the circle Ω. Thus we are left with solving the following problem:

Two circles Ω and ω intersect at B and X and are tangent to the line ℓ at A and D, respectively. Let S and T be points on Ω such that BS is parallel to ℓ and AT is a diameter. Then the lines BD, SA, and TX intersect (at some point C).

To solve this problem, perform an inversion of center B. With the usual notation, the line ℓ becomes the circle $\ell' = A'BD'$, the circles Ω and ω become the lines $X'A'$ and $X'D'$ tangent at A' and D' to the circle ℓ', S' is the intersection of the line $X'A'$ with the tangent at B to the circle ℓ', and T' is the point on the line $X'A'$ for which $\angle A'BT'$ is a right angle (see Fig. 11.43). We have to show that the line BD' and the circumcircles of $BS'A'$ and $BT'X'$ have a second common point.

To prove this, let C' be the intersection of BD' and the perpendicular bisector of $A'D'$ (which passes through both X' and the center of the circle ℓ'). Using the isosceles triangle $C'A'D'$ and the fact that $S'A'$ and $S'B$ are tangents to the circle ℓ', we obtain

$$\angle A'S'B = 180° - 2\angle A'BS' = 180° - 2\angle A'D'B = 180° - 2\angle A'D'C' = \angle A'C'D'.$$

This shows that C' is on the circumcircle of $BA'S'$. On the other hand

$$\angle BT'X' = 90° - \angle BA'S' = \angle A'S'B/2 = \angle A'C'D'/2 = \angle D'C'X'.$$

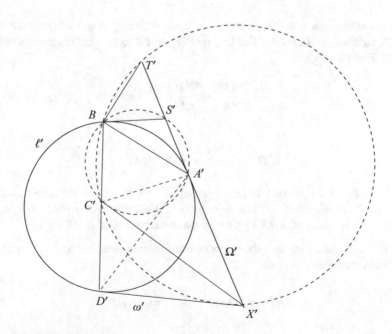

Fig. 11.43 Inverted figure for the problem about proving that AX is orthogonal to XC

This shows that C' is also on the circumcircle of $BT'X'$. So C' lies on BD' and also on the circumcircles of $BA'S'$ and $BT'X'$. The problem is solved.

Remark The official solution (which we sketch here) does not use geometric transformations. Once the similarity of the triangles MXA and MAB and its consequences are established, one considers the foot H of the perpendicular from A to BC. There are two analogous cases, that when H is between B and D and that when it is not. Working in the assumption that H is between B and D, in the right triangle AHD, $AM = HM$, and so $XM/HM = HM/BM$. It follows that the triangles XMH and HMB are similar. Hence $\angle HXM = \angle BHM$, and so

$$BHM = 180° - \angle MHD = 180° - \angle MDH = B - A/2.$$

Therefore

$$\angle AXH = \angle MXH + \angle AXM = (B + A/2) + A/2 = B + A.$$

So $\angle AXH + \angle ACH = B + A + C = 180°$, showing that $AXHC$ is cyclic. From here we deduce that $\angle AXC = \angle AHC = 90°$, as desired.

Source Italian Mathematical Olympiad, 2019.

197 Consider the line $z = t + i$, $t \in \mathbb{R}$. The complex inversion $z \mapsto \frac{1}{z}$ maps this line to the circle $|z + i/2| = 1/2$. Consider now two rational parameters $s = m/n$ and $t = p/q$, $m, n, p, q \in \mathbb{Z}$. The distance between the inverses of $s + i$ and $t + i$ is

$$\left| \frac{1}{t+i} - \frac{1}{s+i} \right| = \frac{|pn - qm|}{\sqrt{m^2 + n^2}\sqrt{p^2 + q^2}}.$$

So in order to find the desired points, it suffices to find a family of Pythagorean triples (a_k, b_k, c_k), $k = 1, 2, \ldots, 1975$ (with $a_k^2 + b_k^2 = c_k^2$) such that $a_k/b_k \neq a_j/b_j$ for $j \neq k$. For this let p_k be the kth prime, and set $a_k = p_{2k}^2 - p_{2k-1}^2$, $b_k = 2p_{2k}p_{2k-1}$ and $c_k = p_{2k}^2 + p_{2k-1}^2$, $k = 1, 2, \ldots, 1975$. The inverses of the points $z_k = a_k/b_k + i$ satisfy the desired condition on the circle $|z + i/2| = 1/2$. Now map these points by $z \mapsto 2z + i$ to the unit circle.

Remark If we multiply by i the Möbius transformation from the line to the unit circle that was used in the solution, and thus work with $z \mapsto \frac{2i}{z} - 1$, then the points $t + i$, $t \in \mathbb{R}$ are mapped to

$$\frac{1 - t^2}{1 + t^2} + i\frac{2t}{1 + t^2},$$

which yields the standard parametrization of the circle $t \mapsto \left(\frac{1-t^2}{1+t^2}, \frac{2t}{1+t^2} \right)$. This parametrization arises from the writing of $\cos x$ and $\sin x$ in terms of $\tan \frac{x}{2}$ and shows that the circle is what is called a rational algebraic curve.

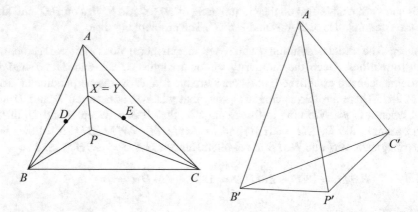

Fig. 11.44 The configuration of the triangle ABC with the point P inside and its inverted version

Source International Mathematical Olympiad, 1975, proposed by the Soviet Union.

198 Let X and Y be the points where the angle bisectors BD and CE of $\angle ABP$ and $\angle ACP$ intersect AP, respectively. We want to show that $X = Y$, which by the Bisector Theorem applied to the triangles ABP and ACP amounts to showing that

$$\frac{BA}{BP} = \frac{CA}{CP}.$$

Rewrite this equality as

$$\frac{BA}{CA} : \frac{BP}{CP} = 1 \iff \left| \frac{b-a}{c-a} : \frac{b-p}{c-p} \right| = 1.$$

Consider an inversion χ of center A (and use the convention $X' = \chi(X)$). Reasoning on Fig. 11.44 and using Theorem 3.2, we have

$$\angle B'C'P' = \angle AC'P' - \angle AC'B' = \angle APC - \angle B$$
$$= \angle APB - \angle C = \angle AB'P' - \angle AC'B' = \angle C'B'P'.$$

It follows that the triangle $P'B'C'$ is isosceles, so $P'B' = P'C'$. In other words

$$\left| \frac{b' - \infty}{c' - \infty} : \frac{b' - p'}{c' - p'} \right| = 1,$$

which is the same as

$$\left| \frac{\chi(b) - \chi(a)}{\chi(c) - \chi(a)} : \frac{\chi(b) - \chi(p)}{\chi(c) - \chi(p)} \right| = 1,$$

and since inversion preserves the absolute value of the cross-ratio, we are done.

Remark The argument implies that the locus of the points P with the given property is the circle of Apollonius defined by the points B and C and the ratio AB/AC.

To summarize the idea of the solution: transform the conditions about angles in the hypothesis and the conclusion into metric relations (using the Bisector Theorem) and isosceles triangles, and then use the invariance of the cross-ratio of segments under inversion.

Source International Mathematical Olympiad, 1996, proposed by Canada.

199 We reason on Fig. 11.45. We are supposed to show that BF is the reflection of the median from B over the angle bisector of $\angle B$. For this we consider ϕ_B, the $\sqrt{bc}$ inversion determined by the triangle ABC and the vertex B. Note that ϕ_B maps the circle Ω to the line AC and the angle bisector of $\angle B$ to itself. Consequently it maps D and E into each other.

As DE is the diameter of ω, ϕ_B maps ω into itself (ω is invariant under both the inversion and the reflection that define ϕ_B). It follows that $M = \phi_B(F)$ is the point where ω intersects the side AC. Because DE is diameter, $\angle EMD = 90°$, showing that M is the projection of E onto AC. But E is the midpoint of the arc $\overset{\frown}{AEC}$ so the projection E onto AC is the midpoint of AC. This shows that BM is the median and consequently $\phi_B(BM) = BF$ is the symmedian, as desired.

Remark This problem exhibits another procedure for constructing the symmedian.

Source Russian Mathematical Olympiad, 2009.

Fig. 11.45 Proof that BF is symmedian

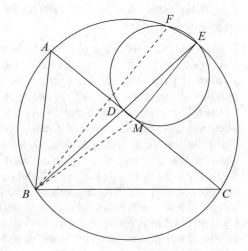

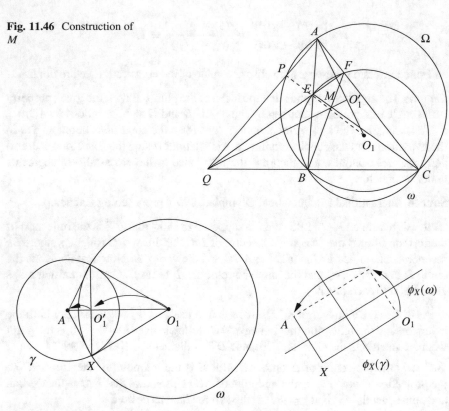

Fig. 11.47 O'_1 is both the reflection of O_1 over γ and of A over the circle ω

200 Let M be the intersection of CE and BF (Fig. 11.46). Then, by Theorem 3.33, QM is the polar of A with respect to the circle ω, and, consequently, QM is perpendicular to AO_1.

Consider the inversion χ over the circle γ of center A and radius $\sqrt{AE \cdot AB} = \sqrt{AF \cdot AC}$, which is the square root of the power of A with respect to ω. Then $\chi(\omega) = \omega$, so ω and γ are orthogonal, also $\chi(\Omega) = EF$, and $\chi(EF) = \Omega$. Since P is the intersection of AQ with Ω and Q is on EF, it follows that $\chi(Q) = P$.

Let $O'_1 = \chi(O_1)$. Then A and O'_1 are reflections of each other over the circle ω. There are many possible ways of proving this fact; here is one of them (Fig. 11.47). Let X be one of the two intersection points of the orthogonal circles γ and ω. Consider the $\sqrt{bc}$ inversion ϕ_X determined by the triangle XAO_1 and the vertex X. Then $\phi_X(A) = O_1$, $\phi_X(O_1) = A$, while $\phi_X(\gamma)$ and $\phi_X(\omega)$ are the perpendicular bisectors of XO_1 and XA, respectively. Then the composition of the reflection over $\phi_X(\gamma)$ with the reflection over $\phi_X(\omega)$ maps A to O_1. By the Symmetry Principle, O_1 is mapped to A by the composition of the reflection over γ with the reflection over ω. Hence the reflection of A over ω and the reflection of O_1 over γ coincide.

As QM is perpendicular to AO_1, we obtain $\angle AO'_1Q = 90°$. By Theorem 3.2,

Fig. 11.48 The image of Ω
through the inversion

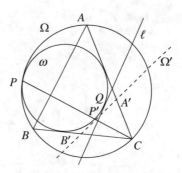

$$\angle APO_1 = \angle A\chi(O_1)\chi(P) = \angle AO_1'Q = 90°,$$

as desired.

Remark A fact worth remembering is that if ω_1 and ω_2 are two orthogonal circles, of centers O_1 and O_2, respectively, and if O_1' is the reflection of O_1 over ω_2, then O_1' is the reflection of O_2 over ω_1. This fact can also be proved in an elementary way by writing the Leg Theorem for both legs of the right triangle whose vertices are O_1, O_2, and one of the intersection points of the two circles.

201 We are supposed to prove yet another property of mixtilinear circles. Mixtilinear circles go hand in hand with $\sqrt{bc}$ inversions, but for this problem we will change slightly the $\sqrt{bc}$ inversion associated with the vertex C so that ω is invariant under the resulting Möbius transformation.

Consider first the inversion of center C that maps the circle ω into itself. If $A' \in CA$ and $B' \in CB$ are the images of A and B, respectively, under this inversion, then the circle Ω is mapped by the inversion to the line $A'B'$ (the dotted line in Fig. 11.48). This line is tangent to the circle ω at the point $P' \in CP$ that is the inverse of P, and it is also antiparallel to the line AB since the quadrilateral $AA'B'B$ is cyclic.

Next, the reflection over the angle bisector of $\angle ACB$ maps line $A'B'$ into ℓ (since the image of $A'B'$ should become parallel to AB and should remain tangent to circle ω).

Hence under the composition of the inversion and the reflection, the point P is mapped to the point Q, and so

$$\angle ACP = \angle BCQ,$$

as desired.

Source European Girls' Mathematical Olympiad, 2013.

202 Let the triangle be ABC; let the feet of the altitudes from A, B, C be A', B', C', respectively; and let M, N, P be the midpoints of BC, AC, AB,

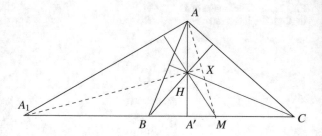

respectively. Denote by H the orthocenter, by A_1, B_1, C_1 the intersections of the perpendiculars from H to AM, BN, CP with BC, AC, and AB, respectively, and by X, Y, Z the intersections of these perpendiculars with AM, BN, CP, respectively.

Let us look at Fig. 11.49. The segments A_1X and AA' are altitudes in the triangle AA_1M, so H is the orthocenter of this triangle. Then

$$A_1H \cdot HX = AH \cdot HA'.$$

As $AH \cdot HA' = HB \cdot HB' = HC \cdot HC'$, we have

$$A_1H \cdot HX = B_1H \cdot HY = C_1H \cdot HZ.$$

If we show that X, Y, Z are on a circle that passes through H and its diameter lies on the Euler line of the triangle ABC, then the circular transformation that is the composition of the inversion of center H and power $\sqrt{AH \cdot HA'}$ and the reflection over H will transform this circle into the desired line (see Sect. 3.1.7), which will then be perpendicular to Euler's line. But notice that

$$\angle HXG = \angle HYG = \angle HZG = 90°,$$

where G is the centroid, and we are done.

Remark We have used the same circular transformation that maps the nine-point circle to the circumcircle, obtained as the composition of an inversion of center H and a reflection over H, that was used in Problem 193.

Source Gh. Țițeica, *Probleme de Geometrie (Problems in Geometry)*, Ed. Tehnică, Bucharest, 1929.

203 *First solution.* Consider an inversion of center A and denote, as it is the convention, by X' the image of the point X. Then on the one hand, C' is the midpoint of the segment AM' (inversion "inverts" ratios), and on the other hand, it is collinear with the points B' and D' because the circumcircle of BCD passes through the center of inversion (Fig. 11.50). The line AC is the bisector of $\angle BMD$ if and only if $\angle AMB = \angle AMD$, which is equivalent to $\angle AB'M' = \angle AD'M'$. And

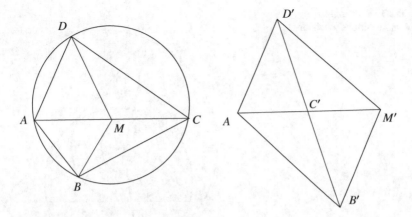

Fig. 11.50 Quadrilateral in which a diagonal bisects the angle formed by the segments connecting its midpoint with the other two vertices, and its inverse

this is equivalent to the fact that $AB'M'D'$ is a parallelogram, which happens if and only if $C'B' = C'D'$. Rewrite this equality in terms of cross-ratios of complex coordinates:

$$\frac{b' - c'}{d' - c'} = -1 \iff \frac{b' - c'}{d' - c'} : \frac{b' - \infty}{d' - \infty} = -1.$$

Here ∞ is the image of a, and so, as real valued cross-ratios are invariant under inversion,

$$\frac{b - c}{d - c} : \frac{b - a}{d - a} = -1.$$

This is the condition that the quadrilateral is harmonic. And this condition is symmetrical, so, by symmetry, it is equivalent to BD bisecting $\angle ANC$.

Second solution. Examining carefully the solution given to Problem 3.7 from the introduction, we notice that if M is the midpoint of the diagonal AC of the harmonic quadrilateral $ABCD$, then the triangle DAB is similar to both triangles DMC and CMB. And this proves that in a harmonic quadrilateral, the diagonal AC is the angle bisector of $\angle BMD$. Similarly the diagonal BD is the angle bisector of $\angle ANC$. Now it is important to notice that for every configuration DAC, there is a unique point B on the arc $\overset{\frown}{AC}$ that does not contain D such that $ABCD$ is harmonic, the point being at the intersection of the circle of Apollonius defined by D with respect to the points A and C. For that point B, AC bisects $\angle BMD$, and for any other point B' on that arc, $\angle B'MA \neq \angle BMA = \angle DMA$. Hence for no other point B' is AC the bisector of $\angle B'MD$. We conclude that any of the two properties from the statement holds only in a harmonic quadrilateral and in such a quadrilateral both properties hold.

Fig. 11.51 The tangency
point and a curious inversion

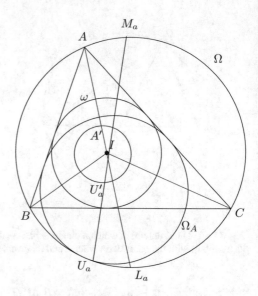

Remark We have seen this idea before: transform a condition about angles into a
metric relation involving cross-ratios, and then use the invariance of cross-ratios
under transformations. It is worth remembering this property that characterizes
harmonic quadrilaterals.

204 We work on Fig. 11.51. Let X' denote the inverse of X with respect to the
incircle ω, and, as usual, let r be the inradius. Since ω is inside Ω, the inversion
with center I maps Ω to a circle Ω' homothetic to Ω, where the homothety has
center I and negative ratio.

Consider the midpoints L_a and M_a of the arcs $\overset{\frown}{BC}$. From the inversion we have
$IA' \cdot IA = IU_a' \cdot IU_a = r^2$, while the power of I with respect to Ω is $IA \cdot IL_a = IU_a \cdot IM_a$, and so

$$\frac{IU_a'}{IM_a} = \frac{IA'}{IL_a}.$$

This means the $M_a L_a$ is mapped by the homothety to $U_a' A'$. Because $M_a L_a$ is a
diameter in Ω, $U_a' A'$ a diameter in Ω'.

Remark It is worth pointing out that when inverting about the incircle, the sides
become three circles of equal radii that pass through the center of inversion. The
circumcircle is mapped to the circle that passes through the pairwise intersection
points of these three circles, and we arrive at the configuration from Tzitzeica's
Five-Lei Coin Problem (Theorem 3.38).

205 We will find the exact locations of the points X, Y, I in a three-step process.
In Problem 180 (a) we have proved that Q is the midpoint of the arc $\overset{\frown}{AB}$ that does

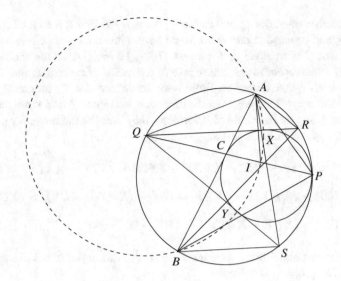

Fig. 11.52 Finding the location of the points I, X, Y

not contain P. From here it follows that RQ, PQ, and SQ are the respective angle bisectors of $\angle ARB$, $\angle APB$, and $\angle ACB$, so $X \in RQ$, $I \in PQ$ and $Y \in SQ$.

Next, using angle chasing we will prove that A, B, X, Y, I lie on a circle of center Q, that is

$$QA = QB = QX = QY = QI.$$

The reasoning can be followed on Fig. 11.52. We will only check that $QA = QI$, because checking $QA = QX$ and $QB = QY$ involves exactly the same argument. In the triangle APB,

$$\angle AIC = 180° - \angle AIP = \angle PAB/2 + \angle APB/2.$$

On the other hand,

$$\angle IAQ = \angle IAB + \angle BAQ = \angle PAB/2 + \overset{\frown}{QB}/2$$

$$= \angle PAB/2 + \overset{\frown}{AQB}/4 = \angle PAB/2 + \angle APB/2.$$

Hence $\angle AIQ = \angle IAQ$, showing that the triangle QAI is isosceles. We deduce that $QA = QI$ and this completes the second step.

It is at this moment that we employ inversion, and we do it in order to show that X and Y are the tangency points of QR and QS to ω (a fact that the reader might have guessed by looking at the figure).

The inversion of center Q and radius QA maps the line AB to a circle passing through Q, and because A and B are fixed by the inversion, this circle must be Ω. It follows that C is mapped to P and so $QC \cdot QP = QA^2$. This shows that the power of Q with respect to ω is the square of the radius of inversion, and hence ω is mapped to itself by the inversion. From here we deduce that the tangency points of QR and QS to ω are fixed points of the inversion, and since X and Y are themselves fixed points (being on the circle of inversion), they must be the tangency points.

The rest of the solution is just angle chasing:

$$\angle PXI + \angle PYI = 360° - \angle XPY - \angle XIP - \angle YIP = \angle XIY - \angle XPY$$

$$= 180° - \angle XQY/2 - \angle XPY = 180° - (\overset{\frown}{XPY} - \overset{\frown}{XCY})/4 - \overset{\frown}{XCY}/2$$

$$= 180° - (\overset{\frown}{XPY} + \overset{\frown}{XCY})/4 = 180° - 90° = 90°,$$

where for the first equality we have used the sum of the angles of the triangles PXI and PYI. The problem is solved.

Remark The geometric transformation need not be the only trick of the solution, but it can produce an essential breakthrough.

Source Romanian Team Selection Test for the International Mathematical Olympiad, 2013.

206 Invert with respect to a circle centered at B, and use the convention that the inverse of a point is denoted by the same letter, but with a dash. The original configuration and the inverted one are shown in Fig. 11.53.

By Theorems 3.4 and 3.5, the points A', C', M' are collinear; the points K', M', N' are collinear; and the points A', C', N', K' are concyclic. Let the circumcircle of $A'C'N'K'$ be ω. By Theorem 3.33 the polar p_B of B with respect to ω passes through M'.

But p_B is passes through the points B_1 and B_2 where the tangents from B to ω touch the circle. And by Proposition 3.7, the image O' of O is the midpoint of $B_1 B_2$. Note that p_B (which is the line $B_1 B_2$) makes with BO' an angle of $90°$. Using Theorem 3.2 we deduce that $\angle BMO = \angle BO'M' = 90°$, and the problem is solved.

Source International Mathematical Olympiad, 1985, proposed by the Soviet Union.

207 The solution can be followed on Fig. 11.54. Let ϕ_B be the $\sqrt{bc}$ inversion defined by the triangle ABC and the vertex B. By Theorem 3.4, $E' = \phi_B(E)$ is the point that is diametrically opposite to B in the circumcircle of ABC (said differently $\angle BEA = \angle BCE'$). Using Proposition 3.7 (i) we obtain that $O' = \phi_B(O)$ is the reflection of B over AC. Because $\phi_B(O') = O$ and BE' is diameter in ω, again by Theorem 3.4, ω is mapped to the line ℓ through O that is perpendicular to OB at O.

The circumcircle of AOC is mapped to the circumcircle of $AO'C$, which is the nine-point circle of the triangle A_1BC_1, where A_1 and C_1 are the reflections of B

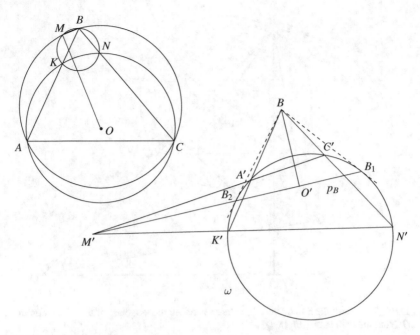

Fig. 11.53 $\angle BMO = 90°$

over A and C, respectively. So $P' = \phi_B(P)$ must be on this nine-point circle, it is also on the line AC, and it cannot be the midpoint C of the segment BC_1, because C is the image of A. The only possibility is that P' is the foot of the altitude from A_1 in the triangle A_1BC_1. Similarly $Q' = \phi_B(Q)$ is the foot of the altitude from C_1. Consequently, the line PQ is mapped to the circle of diameter BH_1, where H_1 is the orthocenter of triangle A_1BC_1.

The tangency of ω and the circumcircle of AOC is equivalent to the fact that ℓ is tangent to the nine-point circle of A_1BC_1, which is the same as saying that the distance from the center of the nine-point circle of A_1BC_1 to ℓ is equal to the circumradius of ABC. The fact that P, Q, E are collinear is equivalent to the fact that E' is on the circle of diameter H_1E', so it is equivalent to the condition $\angle H_1E'B = 90°$. But since E' and H_1 are the images of O and of the orthocenter H of the triangle ABC through a homothety of center B, the last statement is equivalent to saying that the Euler line OH of the triangle ABC is perpendicular to the radius OB. We are left to prove the equivalence of these two facts, summarized in the following simpler question:

In the triangle ABC, let O be the circumcenter. Then the perpendicular ℓ to the radius OB through O passes through the orthocenter H of ABC if and only if the distance from the reflection of O over BC to this perpendicular is equal to the circumradius R of ABC.

To prove this, let O_1 be the reflection of O over BC, and let M be the projection of O_1 on ℓ (Fig. 11.55). Then $O_1M = R$ if and only if BOO_1M is a parallelogram.

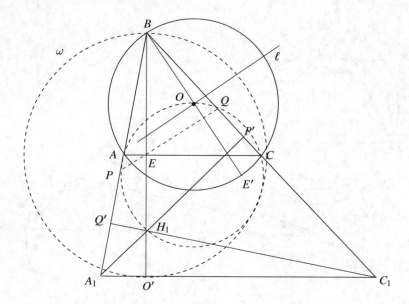

Fig. 11.54 The simplification of the equivalence between the tangency of ω and the circumcircle of AOC and the collinearity of P, Q, E

Fig. 11.55 Solution to the simplified problem

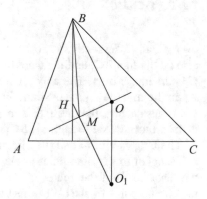

But BOO_1H is a parallelogram, so the latter happens if and only if $M = H$, that is, ℓ passes through H, and we are done.

Remark The condition that two circles are tangent is simplified via an inversion to the condition that a circle is tangent to a line. In our case it is natural to try an inversion of center B. But which inversion? We would like to have as many elements of the original configuration preserved, and it should be easy to locate the images of the elements that are not preserved. The $\sqrt{bc}$ inversion keeps the triangle ABC in place, while the image of E is on the circumcircle of ABC, and, also very important, the image of O is the point diametrically opposite to B in ω.

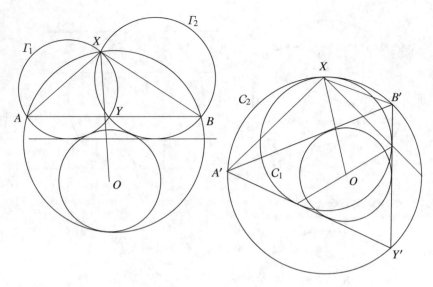

Fig. 11.56 Inverting the diagram

208 Consider the inversion of center X that maps the circle of center O into itself. The original configuration and the inverted one can be seen in Fig. 11.56. Here it is how things are transformed:

- X, Y', A', B' are concyclic;
- the common tangent of Γ_1 and Γ_2 and the line AB become two circles C_1 and C_2 that are interior tangent at X;
- the circles Γ_1 and Γ_2 become the lines $A'Y'$ and $B'Y'$, tangent to C_1;
- the circle of center O, invariant under the inversion, is now the incircle of the triangle $A'B'Y'$;
- the lines XA, XB, XO are the lines XA', XB', XO.

We must show that XO bisects $\angle A'XB'$.

We recognize C_1 to be a mixtilinear incircle of $A'B'Y'$, with X being the point where the mixtilinear circle touches the circumcircle. And we have seen in Proposition 3.43 that XO passes through the midpoint of the arc $\overset{\frown}{A'Y'B'}$. Hence XO is the angle bisector of $\angle A'XB'$, and we are done.

Source The Competition of the Institute of Mathematics of the Romanian Academy (IMAR), 2008, proposed by Radu Gologan.

209 The first observation is that $ADBC$ is harmonic. Indeed, if we invert with respect to a circle centered at P, then AB, γ, and δ are mapped to three parallel lines, AB, γ', and δ', while ω is mapped to a circle ω' that is tangent to γ' at the image C' of C and to δ' at the image D' of D (see Fig. 11.57). Hence $A'D'B'C'$

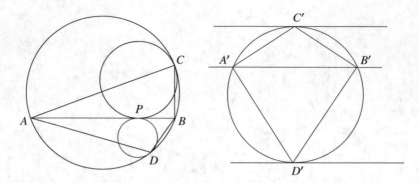

Fig. 11.57 $ADBC$ is harmonic

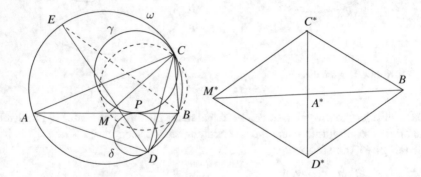

Fig. 11.58 DB is tangent to the circumcircle BMC

is a cyclic kite, and so it is harmonic. Because real valued cross-ratios are invariant under inversion (Theorem 3.18), $ADBC$ is harmonic.

Now we switch to Fig. 11.58. From the fact that DE is a diameter, we obtain that $\angle DBE = 90°$, and the desired conclusion that the circumcenter of triangle BMC lies on BE is equivalent to DB being tangent to the circumcircle of BMC. Take another inversion, this time centered at B, and mark the images by a star, to avoid a conflict with the notation from the previous inversion. Then the harmonic quadrilateral $ADBC$ is mapped to a harmonic division on a line, and since B is mapped to ∞, A^* is the midpoint of C^*D^*. Moreover, because M is the midpoint of AB, A^* is the midpoint of M^*B. So $M^*D^*BC^*$ is a parallelogram, showing that C^*M^* is parallel to BD^*; they form a $0°$ angle. But then the preimage of C^*M^*, which is the circumcircle of BMD, is tangent to the preimage of BD^*, which is the line BD. The problem is solved.

Remark Here we have learned another way to construct the symmedian in the triangle ABC: take the circle through A that is tangent to both the line BC and the circumcircle, consider the other circle that is tangent to BC at the same point

Fig. 11.59 The inversion that maps the circumcircle to the nine-point circle

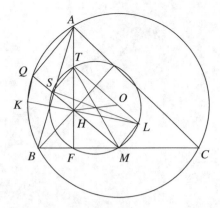

and is also tangent to the circumcircle, and let D be this latter tangency point. Then AD is the symmedian from A.

Source Stars of Mathematics Competition, Bucharest, 2009.

210 We argue on Fig. 11.59. Consider the inversion χ of center H and negative ratio that maps the circumcircle and the nine-point circle into each other (the ratio of this inversion is $-AH \cdot HF$). Then $F = \chi(A)$ and $M = \chi(Q)$, the second because the similar right triangles HQA and HFM yield $QH \cdot HM = AH \cdot HF$. Also, if $L = \chi(K)$ then $\angle HML = 90°$ (similar triangles HKQ and HML). Consequently the inversion maps the circumcircle of KHQ to the line LM and the circumcircle of FKM to the circumcircle of ALQ. So we are left to show that line LM is tangent to the circumcircle of ALQ.

We prove this as follows. If S and T are the midpoints of the segments HQ and HA (and therefore are on the nine-point circle, which is a consequence of the existence of the homothety that maps the circumcircle and the nine-point circle into each other), then the quadrilateral $LMST$ is a rectangle, since LS and MT are diameters in this circle, and the center of this rectangle is the center of the nine-point circle. The orthocenter H and the circumcenter O are reflections of each other over the center of the nine-point circle, and since H is on the side MS of the rectangle, O must lie on side LT. But then, since LT is perpendicular to ST and the latter is parallel to AQ, the line LT is the perpendicular bisector of the segment AQ. This perpendicular bisector contains the circumcenter of ALQ. And since LM is perpendicular to LT, it is perpendicular to the diameter at L, so it is tangent to the circumcircle of ALQ, and we are done.

Remark There are two pieces of information that hint to the inversion to be used: the condition that $\angle HQA = 90°$ tells that Q is mapped to M, and the more subtle condition that $\angle HQK = 90°$, which allows a nice description of the image of K in terms of H and M.

Source International Mathematical Olympiad, 2015, proposed by Ukraine.

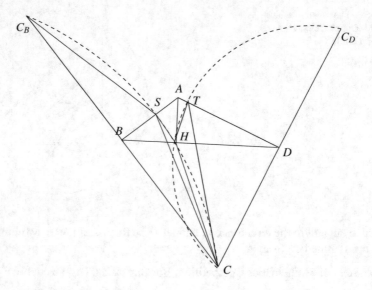

Fig. 11.60 How to locate S and T

211 We start with some constructions and observations that can be followed on Fig. 11.60. Let C_B and C_D be the reflections of C over B and D, respectively. Then in the right triangle BSC_B,

$$\angle SC_B B = 90° - \angle C_B SB = 90° - \angle BSC.$$

Using the condition from the statement, we obtain

$$\angle SC_B B + \angle SHC = 90° - \angle BSC + \angle SHC = 90° + 90° = 180°,$$

so S, C_B, C, H are concyclic. Similarly T, C_D, C, H are concyclic. Thus we have replaced the unfriendly angle conditions from the statement with an intuitive description of the locations of S and T. We should also note that, because AH is perpendicular to BD, AB is a diameter in the circumcircle of the triangle AHB. And since BC is perpendicular to AB, BC is tangent to this circumcircle. By a similar argument, BD is tangent to the circumcircle of the triangle AHD.

It is now time for a geometric transformation, and we consider an inversion of center H. As always, we denote the image under the inversion of a point P by P'. The problem is solved if we show that $S'T'$ is parallel to $B'D'$, as these two lines are the images of the circle through S, T, H and of the line BD: the angle formed by the circumcircle of SHT with line BD at the point of contact is zero if and only if the angle between $S'T'$ and $B'D'$ is zero, that is, if these lines are parallel.

Because the original quadrilateral is cyclic ($\angle B + \angle D = 90° + 90° = 180°$), the quadrilateral $A'B'C'D'$ is cyclic as well. Also, because B, C, C_B are collinear,

Fig. 11.61 The configuration after inversion

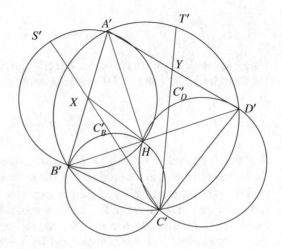

B', C', C'_B, H are concyclic, and because D, C, C_D are collinear, D', C', C'_D, H are concyclic. On the other hand, because S, C_B, C, H are concyclic, it follows that S', C'_B, C' are collinear, and because T, C_D, C, H are concyclic, T', C'_D, C' are collinear. Note also that S' and T' are on the circumcircles of $A'B'H$ and $A'D'H$, respectively, as these circles are the images of the lines AB and AD. These properties are illustrated in Fig. 11.61.

Let X and Y be the midpoints of $A'B'$ and $A'D'$, respectively. We claim that $X \in C'C'_B$ and $Y \in C'C'_D$. To prove this claim, we note that because B is the midpoint of CC_B, ∞, C, B, C_B form a harmonic division, and since inversion preserves real-valued cross-ratios, $HC'B'C'_B$ is a harmonic quadrilateral. By Proposition 1.23, $C'C'_B$ is a symmedian in the triangle $C'HB'$. On the other hand, the fact that BC is tangent to the circumcircle of AHB implies that $A'B'$ is tangent to the circumcircle of $HB'C'$. Consequently XB' is tangent to this circumcircle. But because X is the midpoint of the hypotenuse $A'B'$, $\angle XHB' = \angle XB'H$, so XH is also tangent to the circumcircle of $HC'B'C'_B$. So X is the intersection point of the tangents at B and H, which, by Theorem 1.22, implies that X is on the symmedian $C'C'_B$ of the triangle $HB'C'$, so $X \in C'C'_B$. We conclude that C', C'_B, X, S' are collinear. Similarly, C', C'_D, Y, T' are collinear. Now we know the precise locations of all elements in the inverted figure.

Proving that $S'T'$ is parallel to $B'D'$ is the same as proving that $S'T'$ is parallel to XY. And this is equivalent to

$$\frac{C'X}{XS'} = \frac{C'Y}{YT'}.$$

But now notice that A', S', B', H are concyclic (because A, S, B are collinear), and since X is the circumcenter of triangle $A'B'H$, it follows that $XS' = XH = XA' = XB'$. Similarly $YT' = YH = YA' = YD'$. Using this we transform the relation we have to prove into

$$\frac{XH}{C'X} = \frac{YH}{C'Y}.$$

Using the similarity of the triangles XHC'_B and $XC'H$ as well as the similarity of the triangles YHC'_D and $YC'H$, we transform the last relation into

$$\frac{HC'_B}{C'H} = \frac{HC'_D}{C'H}.$$

And this is equivalent to $HC'_B = HC'_D$, which before inversion is $HC_B = HC_D$. So we have to return to the original picture and prove $HC_B = HC_D$.

The point C_B maps to C_D by the composition of the reflection over AB and the reflection over AD, which is the rotation about A by twice $\angle BAD$. But then C_B maps to C_D by the reflection over the angle bisector of $\angle C_B A C_D$. On the other hand, the quadrilateral $ABCD$ being cyclic, we have $\angle ABD = \angle ACD$, so their complements $\angle BAH$ and $\angle CAD$ are equal as well. Similarly, $\angle HAD = \angle CAB$. Hence

$$\angle C_B AH = \angle C_B AB + \angle BAH = \angle DAH + \angle C_D AD = \angle HAC_D.$$

So H is on the angle bisector of $\angle C_B A C_D$ and consequently $HC_B = HC_D$, which completes the solution.

Source International Mathematical Olympiad, 2014, proposed by Iran, solution by contestant Yang Liu, USA.

212 The "complicated" circle Ω is tangent to the sides AB and BC and to the circumcircle ω of AOC. The sides AB and BC are good references, but ω makes locating the circle Ω harder. So we should try an inversion that keeps AB and BC in place, meaning that the center of inversion should be B. In fact, one should try the $\sqrt{bc}$ inversion determined by the triangle ABC and the vertex B.

By Proposition 3.40 (v), O is mapped to the reflection O' of B across AC. Then the circle ω is mapped to the circumcircle ω' of $AO'C$; note that if we reflect B across A and C, obtaining A_1 and C_1, respectively, then ω' is the nine-point circle of A_1BC_1. Use Fig. 11.62 to understand the transformed configuration.

Now, if the reference triangle is A_1BC_1, Ω' is tangent externally to the nine-point circle ω' and to the extensions of the sides A_1B and BC_1. Feuerbach's Theorem, which is proved in the next chapter (Theorem 4.2), states that in every triangle, the nine-point circle is tangent to the incircle and the excircles. It follows that Ω' must be the B-excircle of A_1BC_1. Moreover, the line KL is mapped to circumcircle of $BK'L'$, in which K' and L' are tangency points of Ω' to BC and AB, respectively.

The symmetry of the figure with respect to the bisector of $\angle B$ leads us to conclude that the midpoint M of KL lies on BI. Because of that, since $BM \perp KL$, M' is the intersection of the perpendicular lines through K' and L' to the sides; that is, M' is the B-excenter I_b. Reversing the inversion yields $BI_b \cdot BM = AB \cdot BC$. But

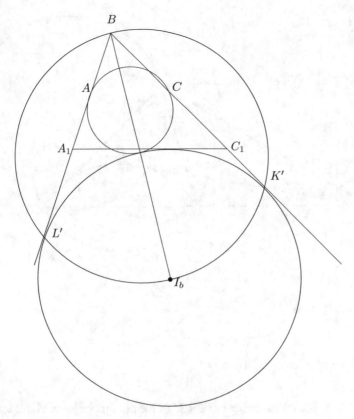

Fig. 11.62 Configuration after applying the $\sqrt{bc}$ inversion

$BI_b = 2BJ_b$, in which J_b is the B-excenter of ABC, and one quick computation of angles show that ABI and J_bBC are similar: in fact, $\angle BAI = \angle A/2$, and $\angle BJ_bC = (90° - \angle C/2) - \angle B/2 = \angle A/2$. Then

$$\frac{AB}{J_bB} = \frac{BI}{BC},$$

which is equivalent to $BI \cdot BJ_b = AB \cdot BC$. And this means that $I' = J_b$.

For the final step, $BI_b \cdot BM = AB \cdot BC$, which implies that $2BJ_b \cdot BM = BI \cdot BJ_b$, and this proves that $BM = BI/2$. So the point M, which is the midpoint of KL, is also the midpoint of BI, as desired.

213 In order to decide how to proceed, let us look at Fig. 11.63. From the experience of Problem 212 we have learned that the $\sqrt{bc}$ inversion determined by the triangle ABC and the vertex A transforms nicely the circumcircle of the triangle BOC. But after realizing that all three circles pass through C, we might be tempted to try an inversion centered at C. Yet another possibility is to take advantage of

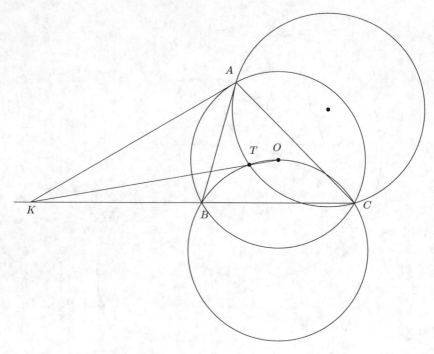

Fig. 11.63 Which inversion should we use?

the fact that K lies on the radical axis BC of the circumcircles of ABC and BOC and invert with respect to a circle centered at K, thus fixing both circles. All three choices lead to solutions, which we present here, leaving it to the readers to choose their favorite.

First solution, by inversion centered at A. We reason on Fig. 11.64. The $\sqrt{bc}$ inversion determined by the triangle ABC and the vertex A swaps B and C, and transforms O into the reflection of A across BC (Proposition 3.40 (v)). If B_1 and C_1 are the reflections of A across the points B and C, respectively, then the circumcircle of BOC is mapped to the nine-point circle of AB_1C_1. The circle ω through A and C that is tangent to AB (i.e., makes an angle of $0°$ with AB) is mapped to a line through $C' = B$ that is parallel to $AB' = AC$. Taking A_1BC_1 as reference, this line is the B_1-midline of this triangle.

Let us move forward to examining the point T. This point is the intersection of the circumcircle of BOC and ω, and therefore it is mapped to the intersection between the nine-point circle of AB_1C_1 and the B_1-midline, that is, T' is the midpoint of B_1C_1.

Next, let us find the image of K under the transformation. Since O' is the reflection of A across BC, O' lies on B_1C_1 and on the nine-point circle of AB_1C_1. The line OT is mapped to the circumcircle of $AO'T'$. The line BC is mapped to the circumcircle of ABC, so K' is the other intersection of the circumcircles of ABC

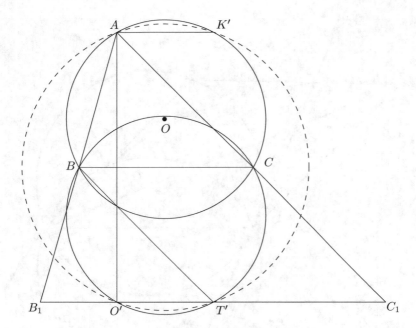

Fig. 11.64 Configuration after the $\sqrt{bc}$ inversion determined by ABC and A

and $AO'T'$. The angle $\angle AO'T'$ is right; thus AT' is a diameter of the circumcircle of $AO'T'$, and given that AT' is the median of AB_1C_1, the circumcenter of $AO'T'$ is the midpoint of BC.

Notice that the nine-point circle and the circumcircle of ABC are symmetric with respect to BC and that BC is an axis of symmetry of the circumcircle of $AO'T'$. Since O' and T' are the intersection points of the nine-point circle and the circumcircle of $AO'T'$, the reflection across BC maps O' and T' to the intersection points of the circumcircle of ABC and the circumcircle of $AO'T'$, which are A and K'. Then AK' is parallel to BC, and, reversing the transformation, this means that the angle between line AK and the circumcircle of ABC is $0°$, that is, AK is tangent to the circumcircle of ABC.

Second solution, by inversion centered at C. Recall that the motivation behind using an inversion centered at C is to turn all three circles into lines. In fact, an inversion about C maps the circles to the lines $A'B'$, $O'B'$, and one more line through A'. Since we will have to work with the image of O, using the $\sqrt{bc}$ inversion determined by ABC and the vertex C seems to be the right idea. This transformation, pictured in Fig. 11.65, swaps A and B and maps O to the reflection of C across AB. The circle through A and C that is tangent to AB is mapped to the line ℓ through $A' = B$ that is tangent to the image of AB, which is the circumcircle of ABC. The point T' is, then, the intersection of ℓ and AO', the latter being the reflection of AC across AB.

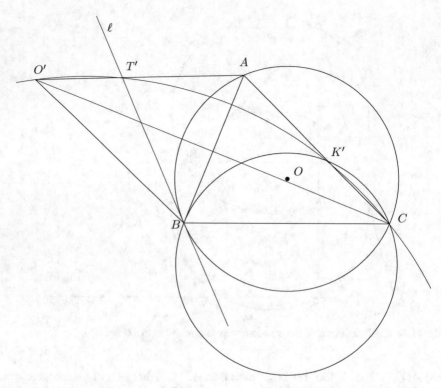

Fig. 11.65 Configuration after the $\sqrt{bc}$ inversion determined by ABC in C

Now let us find the image K' of K. The line OT is mapped to the circumcircle ω of $O'T'C$, and so K' is the intersection of this circumcircle and $CB' = CA$. Since the lines AC and $O'T'$ are symmetric with respect to AB, ω has AB as axis of symmetry, and hence K' is the reflection of T' across AB.

We are supposed to prove that AK is tangent to the circumcircle of ABC. The line AK is mapped by the $\sqrt{bc}$ inversion to the circumcircle of $A'K'C = BK'C$. And so the problem translates to proving that AB is tangent to the circumcircle of $BK'C$. But BT' is tangent to the circumcircle of ABC, and the symmetry of T' and K' with respect to AB yields

$$\angle ABK' = \angle ABT' = \angle ACB,$$

which proves the desired tangency and finishes the problem.

Third solution, by inversion centered at K. Recall that the main idea here is to keep the circumcircles of BOC and ABC in place by inverting about a circle centered at K and of radius the square root of the power of K with respect to the two circles. This inversion swaps B and C and, therefore, T and O. We want to prove that AK is tangent to the circumcircle of ABC, which is equivalent to the fact

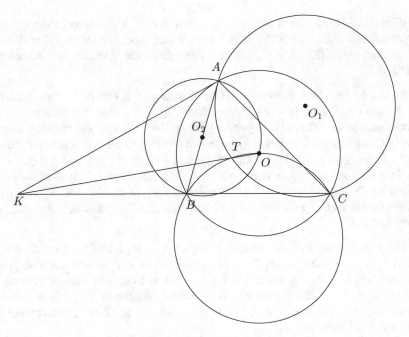

Fig. 11.66 Configuration after inverting about K

that $KA^2 = KB \cdot KC$, which, in turn, is the same as proving that A is fixed by the inversion. We work on Fig. 11.66.

It remains to investigate the image of the circle that is tangent to AB and passes through C, which is also the circumcircle of ACT. It is mapped to the circumcircle of $A'C'T' = A'BO$. So we are done if we prove that the circumcircle of ABO is this inverse, because then $A = A'$. One way of doing this is by using the invariance of the angle made with the line BC, which is fixed by the inversion. More specifically, if O_1 is the circumcenter of ACT and O_2 is the circumcenter of ABO, it suffices to prove that $\angle O_1 CB = \angle O_2 BC$. An angle chase shows that this is the case:

$$\angle O_1 CB = \angle O_1 CA + \angle ACB = \angle CAO_1 + \angle ACB = 90° - \angle BAC + \angle ACB,$$

and

$$\angle O_2 BC = \angle O_2 BO + \angle OBC = 90° - \angle BAO + 90° - \angle BAC$$

$$= 90° - (90° - \angle ACB) + 90° - \angle BAC = 90° - \angle BAC + \angle ACB.$$

And we are done.

Remark The first solution also proves that AT is the A-symmedian of ABC. It also proves that $(C', T', B', K') = -1$, which is the same as $(C, T, B, K) = -1$.

With a little extra effort, one can prove that $AT \perp KT$. Indeed, from the third inversion, $KT \cdot KO = KA^2$, which proves that the triangles KAT and KOA are similar; since $OA \perp AK$, which follows from the tangency, we deduce that $KT \perp AT$.

214 A $\sqrt{bc}$ inversion involves isogonal cevians, so it can be useful. Therefore we perform the $\sqrt{bc}$ inversion ϕ_A defined by ABC and the vertex A. The transformation ϕ_A takes P to a point Q on the line BC with the property that the lines AP and AQ are isogonal conjugates with respect to the vertex A in the triangle ABC. Then $A_1' = \phi_A(A_1)$ is the intersection of $\phi_A(PA) = QA$ and $\phi_A(BC) = \omega$.

Since A_1 and A_2 are symmetric with respect to the circumcircle ω of ABC, the Symmetry Principle (Theorem 3.15) implies that $A_1' = \phi_A(A_1)$ and $A_2' = \phi_A(A_2)$ are symmetric with respect to $\phi_A(\omega)$, which is the line BC. The line AA_2 is mapped to the line AA_2'.

The situation up to this moment is depicted in Fig. 11.67. Since $\angle BAP = \angle A_1'AC$ we obtain $BP = A_1'C = A_2'C$ and $PC = BA_1' = BA_2'$, which proves that $BPCA_2'$ is a parallelogram. Now we switch to complex coordinates, with the convention that the lowercase letter is the complex coordinate of the uppercase letter point. We have $a_2' + p = b + c$, hence $a_2' = b + c - p$. So the points on the line AA_2' are of the form

$$ta + (1 - t)a_2' = ta + (1 - t)(b + c - p), \quad t \in \mathbb{R}.$$

Taking $t = 1/2$ gives us the point T whose complex coordinate is

$$t = \frac{a + b + c - p}{2} = \frac{3g - p}{2},$$

Fig. 11.67 Configuration after the $\sqrt{bc}$ inversion determined by ABC and A

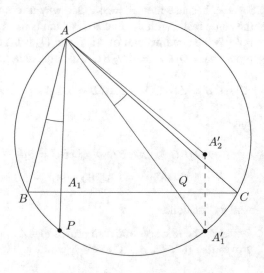

Fig. 11.68 Which of circles will be transformed into a line?

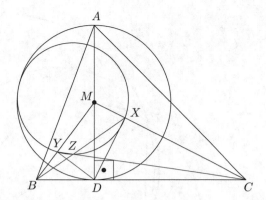

in which g is the coordinate of the centroid G of ABC. The expression for t is symmetric with respect to a, b, and c, so it belongs to all lines AA_2', BB_2', and CC_2', which are the isogonal lines of AA_2, BB_2, and CC_2, respectively. Therefore these three lines have a common point, namely, the isogonal conjugate of T.

It remains to prove that T lies in the nine-point circle of ABC. This can be done by rearranging the expression for t as $t - g = -\frac{1}{2}(p - g)$, that is, $\overrightarrow{GT} = -\frac{1}{2}\overrightarrow{GP}$. This equation means that P is mapped to T by a homothety with center G and ratio $-\frac{1}{2}$. This homothety also maps the circumcircle of ABC to the nine-point circle of ABC, and since P belongs to the circumcircle of ABC, T belongs to the nine-point circle of ABC. The proof is complete.

215 Proving that two circles are tangent to each other is usually harder than proving that a line is tangent to a circle, so we should try to invert so as to turn one of the circles into a line. Looking at Fig. 11.68 it becomes natural to pick the circle of diameter AD as the one that is transformed into a line.

Hence we invert about a circle of center D and arbitrary radius, and denote, as usual, the image of a point P by P'. The circle ω of diameter AD is mapped to a line ω'; since ω is tangent to BC, which is fixed by the inversion, ω' is parallel to BC. We also have $\angle DB'Y' = \angle DYB = 90°$ and, similarly, $\angle DC'X' = \angle DXC = 90°$ (Theorem 3.2). The quadrilateral $DYMX$ has two opposite right angles at vertices X and Y, so it is inscribed in the circle with diameter DM. Since this circle is also tangent to BC, its image is the line $X'Y'$, which is parallel to BC. Putting everything together we obtain that $B'Y'X'C'$ is a rectangle.

The point M' is the projection of D onto $X'Y'$. Since $DA = 2DM$, it follows that $DM' = 2DA'$ (inversion "inverts" ratios), and from here we deduce that ω' is the perpendicular bisector of DM'. We are required to prove that this perpendicular bisector is tangent to the circumcircle of $X'Y'Z'$. For a better intuition, let us examine Fig. 11.69.

Let O be the center of rectangle $B'Y'X'C'$. Then the power of O with respect to the circumcircles of the triangles $B'X'Z'$ and $C'Y'Z'$ is $OB' \cdot OX' = OC' \cdot OY'$,

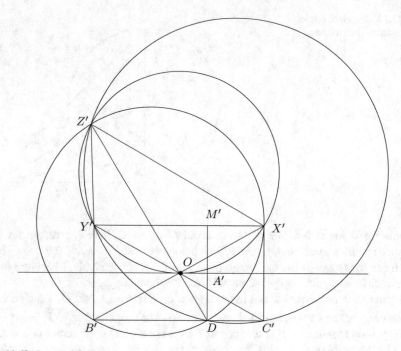

Fig. 11.69 Inverted figure

since all four segments in this equation are equal in length. So O lies on the radical axis of these circumcircles, that is, O lies on the line $Z'D$. Now

$$\angle Y'Z'O = \angle Y'Z'D = \angle Y'C'D = \angle DB'X' = \angle DZ'X' = \angle OZ'X',$$

meaning that $Z'O$ bisects $\angle X'Z'Y'$. But O also lies on the perpendicular bisector of $X'Y'$, and these two conditions can only happen if O lies on the circumcircle of $X'Y'Z'$. As the line $A'O$ being parallel to the chord $X'Y'$ in this circumcircle and as O is the midpoint of the arc $\overset{\frown}{X'Y'}$, it follows that the line $A'O$ is tangent to this circumcircle. And this completes the solution, because the line $A'O = \omega'$ is the image of the circle with diameter AD and the circumcircle of $X'Y'Z'$ is the image of the circumcircle of XYZ.

216 We argue on Fig. 11.70. Since $AP = AQ$, $\angle BPT = \angle CQT$; so we want to prove that the triangles BPT and CQT are similar. Now, looking at the triangle PQS, we notice that, since PA and QA are tangent to the circumcircle γ of PQS, ST is a symmedian, so, using the formula derived at the end of Sect. 2.4.3,

$$\frac{PT}{QT} = \frac{SP^2}{SQ^2},$$

Fig. 11.70 Where is the symmedian?

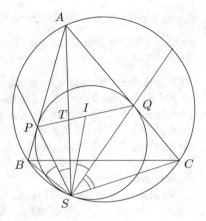

and this is a clue that we should try to prove that

$$\frac{PB}{QC} = \frac{SP^2}{SQ^2}.$$

Now we use Propositions 3.42 and 3.44 to deduce that I is the midpoint of DE and that SP and SQ meet the circumcircle of ABC at midpoints of the arcs $\overset{\frown}{AB}$ and $\overset{\frown}{AC}$ that do not contain the other vertex. From these observations we obtain that

$$\angle PSB = \angle ASP = \angle QSI,$$

where the latter equality follows from the fact that SI and ST are isogonal conjugates in the triangle SPQ or from Proposition 3.43. Since AB is tangent to γ at P, we have that $\angle SPB = \angle SQI$.

The two angle equalities show that the triangles SPB and SQI are similar, so

$$\frac{PB}{SP} = \frac{QI}{SQ}.$$

Analogously, the triangles SQC and SPI are similar (interchange B and C, as well as P and Q), and

$$\frac{QC}{SQ} = \frac{QI}{SP}.$$

Dividing these two equations yields

$$\frac{PB}{QC} = \frac{SP^2}{SQ^2},$$

as desired.

Remark But where is the inversion? In the proofs of Propositions 3.42 and 3.44.

Source Chinese Team Selection Test for the International Mathematical Olympiad, 2005.

217 Consider all three $\sqrt{bc}$ inversions ϕ_A, ϕ_B, ϕ_C determined by ABC, and let ω_b and ω_c be the B- and C-mixtilinear excircles of ABC, respectively. If ω is the incircle of ABC, by Proposition 3.41, $\phi_A(\omega_b) = \phi_C(\phi_B(\omega_b)) = \phi_C(\omega) = \omega_c$, and similarly $\phi_A(\omega_b) = \omega_c$. A nice corollary of this fact is that $\phi_A(AO_b) = AO_c$. Let ω_b touch AB at F, and let ω_c touch AC at G. Then, by the inversion, $\phi_A(F) = G$, and if $O'_b = \phi_A(O_b)$, $\angle AO'_bG = \angle AO'_bF' = \angle AFO_c = 90°$, so O'_b is the orthogonal projection of G onto AO_c.

We also have that $\phi_A(D)$ is the tangency point of $\phi_A(\omega_b) = \omega_c$ and $\phi_A(BC)$, which is the circumcircle of ABC, therefore $\phi_A(D) = S$.

Now focus on Fig. 11.71, paying particular attention to ω_c. The triangle O_cGA is right-angled at G, so, by the Leg Theorem, $O_cO'_b \cdot O_cA = O_cG^2$. This means that O'_b is also the image of A under an inversion χ_c about ω_c. Let $D_c = \chi_c(D)$, the image of D under this inversion. Recall that $O_cE \perp BC$. Then

$$\angle ED_cO_c = \angle DEO_c = 90°,$$

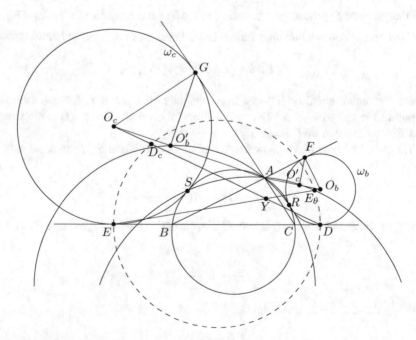

Fig. 11.71 Inversion, inversion, inversion! Point X not shown for the sake of clarity

that is, $ED_c \perp DD_c$. Moreover, since $\chi_c(A) = O_b'$ and $\chi_c(D) = D_c$, $AO_b'D_cD$ is cyclic. So D_c lies on the circumcircle of $AO_b'D = AO_b'S' = \phi_A(O_bS)$. Also notice that Y lies on the line D_cD.

Analogously, if χ_b is the inversion about ω_b and $E_b = \chi_b(E)$, then E_b lies on the circle $\phi_A(O_cR)$. We also have $DE_b \perp EE_c$, so DE_bED_c is cyclic, and Y is the intersection of the diagonals D_cD and EE_b. By power-of-a-point, $E_bY \cdot EY = D_bY \cdot DY$, which means that Y has the same power with respect to the circumcircles of $ADO_b'D_c = \phi_A(O_bS)$ and $AEO_c'E_b = \phi_A(O_cR)$. The other intersection point of these two circles is $\phi_A(O_bS \cap O_cR) = \phi_A(X) = X'$. So Y lies on the line AX', which completes our solution because we know that AY and AY' are isogonal conjugates.

Remark A piece of knowledge acquired from this solution is that the $\sqrt{bc}$ inversion corresponding to the vertex A in the triangle ABC swaps the B- and C-mixtilinear excircles.

Source WenWuGuangHua Mathematics Workshop, China.

 218 We argue on Fig. 11.72. First we prove that the point X with this property is unique. Indeed, suppose that X' is another point with this property, and suppose without loss of generality that X' is closer to B than X. Then, if E' and F' are the orthogonal projections of X' onto BI and CI, respectively, and if M' is the midpoint of $E'F'$, a first observation is that $BE' < BE$ and $CF' > CF$. If we regard BC as the x-axis of a coordinate system, and denote by x_T the x-coordinate of point T, and assume that $x_B < x_C$, we find that $BE' < BE$ implies $x_{E'} < x_E$, and $CF' > CF$ implies $x_{F'} < x_F$. Hence

$$x_{M'} = \frac{x_{E'} + x_{F'}}{2} < \frac{x_E + x_F}{2} = x_M.$$

But $PB = PC$ if and only if $x_P = \frac{x_B + x_C}{2}$, so the inequality $x_{M'} < x_M$ implies that M and M' cannot be at the same distance from B and C simultaneously. This proves that X is unique.

Now we work backward and prove that if X is constructed as above and is chosen such that $\angle BAD = \angle CAX$, namely, such that it is the point of tangency of the A-mixtilinear excircle and the circumcircle, then the point M used in the construction of X satisfies $MB = MC$. We will prove for later use the following result:

Lemma *Let I_a be the A-excenter of ABC, let N be the midpoint of the arc $\overparen{BC}$ that contains A, and let X be the tangency point of the A-mixtilinear excircle and the circumcircle of ABC. Then X, I_a, and N are collinear.*

Proof The proof is very similar to that of Proposition 3.43. Let ϕ_A be the $\sqrt{bc}$ inversion determined by ABC and the vertex A. If D is the tangency point of the incircle with the side BC then $\phi_A(D) = X$, and also $\phi_A(I_a) = I$. So by Proposition 3.40, part (iii), $\angle(AX, XI_a) = \angle(AI, DI)$. Now, if L is the midpoint of

Fig. 11.72 Point X asks:
should I stay or should I go?

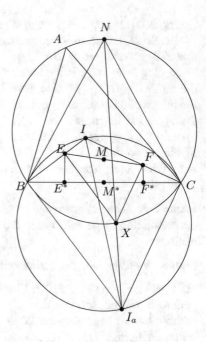

the arc $\overset{\frown}{BC}$ that does not contain A, then $LN \perp BC$ and $ID \perp BC$, so

$$\angle(AI, ID) = \angle(AI_a, LN) = \angle ALN = \angle(AX, XN),$$

and the lemma is proved.   □

With this lemma in tow, we can approach the actual problem. Consider triangle BI_aC and its circumcircle Γ. The incenter I belongs to this circle. Indeed, because $I = \phi_A(I_a)$ and $B = \phi_A(C)$, by using Theorem 3.2, we can infer that the triangles AIC and ABI_a are similar, and from here we obtain that $\angle AI_aB = \angle ACI = \angle C/2$, which combined with $\angle AIB = 180° - \angle A/2 - \angle B/2$ implies $\angle IBI_a = 90°$. Similarly $\angle ICI_a = 90°$, and so not only I is on Γ, but II_a is a diameter of Γ.

We also have $\angle CBI_a = 90° - \frac{1}{2}\angle B$, $\angle BCI_a = 90° - \frac{1}{2}\angle C$, and so

$$\angle BI_aC = 180° - \angle CBI_a - \angle BCI_a = 90° - \frac{1}{2}\angle A.$$

However, $\angle NBC = \angle NCB = 90° - \frac{1}{2}\angle A = \angle BI_aC$. So I_aN is a symmedian in the triangle BI_aC, and X is a point on it. Thus, since $BI_a \perp BI$ and $XE \perp BI$, BI_a is parallel to XE and

$$\frac{BE}{CF} = \frac{d(X, I_aB)}{d(X, I_aC)} = \frac{I_aB}{I_aC},$$

where for the second equality we used a property of symmedians proved in the solution to Problem 16.

Finally, if E^* and F^* are the orthogonal projections of E and F onto BC, respectively, then since $\angle EBE^* = \angle IBC = \angle II_aC$, it follows that the triangles BEE^* and I_aIC are similar, so

$$\frac{BE}{I_aI} = \frac{BE^*}{I_aC}.$$

Analogously,

$$\frac{CF}{I_aI} = \frac{CF^*}{I_aB}.$$

Dividing these two equations yields

$$\frac{BE}{CF} = \frac{BE^*}{I_aC} \cdot \frac{I_aB}{CF^*}.$$

This together with

$$\frac{BE}{CF} = \frac{I_aB}{I_aC}$$

implies that $BE^* = CF^*$. The orthogonal projection M^* of M onto BC is the midpoint of E^*F^*, so from $BE^* = CF^*$, we deduce that M^* is the midpoint of BC, so the line MM^* is the perpendicular bisector of BC. Consequently $MB = MC$, and the problem is solved.

Source Iranian Team Selection Test for the International Mathematical Olympiad, 2014.

219 The argument should be followed on Fig. 11.73. We perform the $\sqrt{bc}$ inversion determined by ABC and the vertex A, and we denote by a dash the image under this $\sqrt{bc}$ inversion. We also denote by a star the image under the reflection over the angle bisector of $\angle A$, which reflection is part of the definition of this $\sqrt{bc}$ inversion.

First notice that E and F are swapped by this transformation. Let γ be the circumcircle of DEF; we want to understand γ', the image of γ. If $X \in \gamma$, then X' satisfies $AX \cdot AX' = AB \cdot AC = AE \cdot AF$. On the other hand $X^* \in AX'$, and $AX^* \cdot AX' = AX \cdot AX'$. Hence $AX^* \cdot AX = AE \cdot AF$, so $X^* \in \gamma'$. Consequently γ' is the circumcircle of DEX^*, and so $\gamma' = \gamma^*$. From here we deduce that the point T' is the other intersection of $\gamma' = \gamma^*$, with BC. The circles γ and γ^* have the same radius, which we denote by R, and from the extended Law of Sines (which relates the circumradius to the pair angle–opposite side), we deduce that

Fig. 11.73 A perfect day for
a $\sqrt{bc}$ inversion

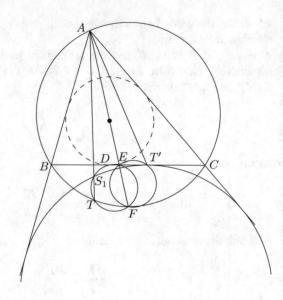

$$2R = \frac{DF}{\sin \angle DEF} = \frac{FT'}{\sin \angle FET'}.$$

Hence $DF = FT'$, which shows that the triangle DFT' is isosceles, and so the
projection M of F onto BC is the midpoint of DT'. However, since F is the
midpoint of $\overset{\frown}{BC}$, M is the midpoint of BC. Then D, the tangency point of the
incircle in BC, and T' are symmetric about M. This means that T' is the tangency
point of the A-excircle and BC (see Sect. 2.2.1).

Now we can complete the argument. The A-excircle is fixed by the reflection
across AE, while the circumcircle $\gamma^* = \gamma'$ of FET' is mapped to the circumcircle
γ of DEF. And T' is the intersection of γ^* and the A-excircle, so its reflection is
the intersection of the circumcircle γ of DEF and the A-excircle, that is, one of S_1
or S_2, say S_1. However, the reflection of the line AT' is the line AT, so AT passes
through S_1, and we are done.

Remark Reversing the transformation yields that T is the tangency point between
the A-mixtilinear incircle and Ω.

Source United States of America Team Selection Team for the International
Mathematical Olympiad, 2016, proposed by Evan Chen.

220 Before starting the solution, we introduce the notation $\kappa_i = 1/r_i$, $i =
1, 2, 3, 4$. These numbers are called the *curvatures* of the circles, and the Descartes
relation is written in terms of curvatures as

$$2(\kappa_1^2 + \kappa_2^2 + \kappa_3^2 + \kappa_4^2) = (\kappa_1 + \kappa_2 + \kappa_3 + \kappa_4)^2,$$

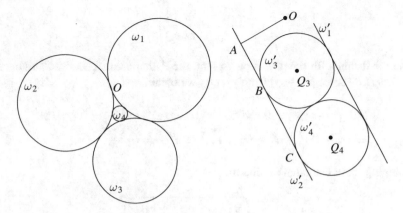

Fig. 11.74 Proof of Descartes' formula

or

$$\sum_{j=1}^{4} \kappa_j^2 = 2 \sum_{j<k} \kappa_j \kappa_k.$$

The proof of Descartes' formula can be followed on Fig. 11.74. The starting point is clear, too many circles, so we invert about a tangency point. Let therefore O be the tangency point of ω_1 and ω_2, and let us invert about the circle of center O and radius $2\sqrt{r_1 r_2}$. The lines ω_1' and ω_2' are parallel, with O lying between them, and ω_1' is at distance $2r_2$ from O, while ω_2' is at distance $2r_1$ from O. Let A be the projection of O onto ω_2'.

The circles ω_3' and ω_4' are tangent to the lines ω_1' and ω_2', are tangent to each other, and their radii are equal to $r_1 + r_2$. Let Q_3 and Q_4 be the centers of ω_3' and ω_4', and let B and C be their tangency points with ω_2', respectively, and assume that Q_3 is closer to O (O lies outside of both circles). Note that $BC = 2(r_1 + r_2)$. For convenience we will set $AB = a$, and for a circle ω we will denote by $\rho(\omega)$ the power of the point O with respect to ω.

At this moment we recall that the inverse about a circle of center O and radius R of a circle ω that does not pass through O is also the image of this circle through the homothety of center O and radius $R^2/\rho(\omega)$, and so the radius r of the circle and the radius r' of the inverse satisfy

$$\frac{r'}{r} = \frac{R^2}{\rho(\omega)}.$$

Using this formula we obtain

$$\frac{r_3}{r_1 + r_2} = \frac{4r_1r_2}{\rho(\omega_3')}.$$

Using the formula for the power-of-a-point, the Pythagorean Theorem, and the fact that O is at distance $|r_1 - r_2|$ from Q_1Q_2, we obtain

$$\rho(\omega_3') = OQ_3^2 - (r_1 + r_2)^2 = (r_1 - r_2)^2 + AB^2 - (r_1 + r_2)^2$$
$$= (r_1 - r_2)^2 + a^2 - (r_1 + r_2)^2 = a^2 - 4r_1r_2.$$

Combining these formulas we obtain

$$a = \sqrt{\frac{4r_1r_2(r_1 + r_2 + r_3)}{r_3}}.$$

Similar computations for $\rho(\omega_4')$ yield

$$\frac{r_4}{r_1 + r_2} = \frac{4r_1r_2}{\rho(\omega_4')}$$

and

$$\rho(\omega_4') = OQ_4^2 - (r_1 + r_2)^2 = (r_1 - r_2)^2 + AC^2 - (r_1 + r_2)^2$$
$$= (r_1 - r_2)^2 + (a + 2(r_1 + r_2))^2 - (r_1 + r_2)^2 = a^2 + 4a(r_1 + r_2) - 4r_1r_2.$$

Substituting $\rho(\omega_4')$ from one equation into the other, we obtain

$$\frac{r_1r_2(r_1 + r_2)}{r_4} = a^2 + 4a(r_1 + r_2) - 4r_1r_2.$$

Replace a by the formula that we have obtained above and switch to curvatures to obtain

$$\kappa_4 = \kappa_1 + \kappa_2 + \kappa_3 + 2\sqrt{\kappa_1\kappa_2 + \kappa_1\kappa_3 + \kappa_2\kappa_3}.$$

Write this as

$$\kappa_4 - \kappa_1 - \kappa_2 - \kappa_3 = 2\sqrt{\kappa_1\kappa_2 + \kappa_1\kappa_3 + \kappa_2\kappa_3},$$

then square both sides to arrive at Descartes' formula.

Remark Descartes' relation can be viewed as a quadratic equation in κ_4, which has two solutions. One solution is the curvature of ω_4; the other solution (or rather its absolute value) is the curvature of the other circle that is tangent to $\omega_1, \omega_2, \omega_3$ (this circle is either exterior tangent or interior tangent to all three). Descartes' relation

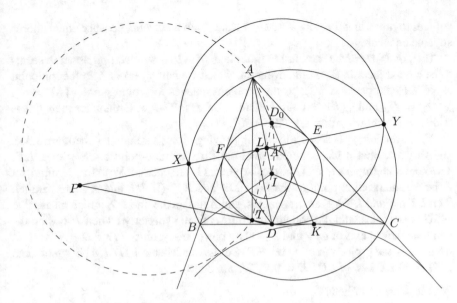

Fig. 11.75 How does T relate with the other points?

is an essential tool in studying Apollonian gaskets, which are obtained by starting with three circles and then filling in with new circles, each tangent to other three.

In differential geometry the concept of curvature is extended to any smooth curve; it is the reciprocal of the radius of the circle that best approximates the curve at a given point (also known as the oscullating circle, from the Latin *osculor* = to kiss). The reciprocal of the radius is the right choice because the more "curved" the curve is, the bigger its curvature is.

Source Descartes' relation was known to François Viète, as well. Frederick Soddy has extended it to spheres.

221 We work on Fig. 11.75. The most peculiar point in this problem is T, so we try to work as much as we can around it. Keep in mind that we are to prove that the line AT goes through the tangency point of the A-mixtilinear incircle and the circumcircle Γ. We have learned from Proposition 3.42, which itself is a consequence of the $\sqrt{bc}$-inversion centered at A determined by ABC, that the line AT is the isogonal of the line AK, where K is the tangency point of the A-excircle with BC. This information makes us a bit more confident, but the A-excircle does not play a big of a role in this problem; the incircle is more important. Is there anything relating the incircle and the A-excircle?

Yes, there is! Recall from Chap. 2 that the homothety with center A that takes the A-excircle to the incircle also takes K to the point D_0, diametrically opposite to D in the incircle. So now we have a more concrete goal: to prove that if I is the incenter of ABC, then AI bisects $\angle TAD_0$. Furthermore, since $IT = ID_0$, it suffices to prove that the points A, D_0, I, and T are concyclic, because equal chords subtend equal arcs.

The points D_0 and T are on the incircle ω of ABC, so inverting about ω seems to be a good idea. In fact, this inversion, which we call χ, takes A to the midpoint A' of EF (Proposition 3.3), so the problem reduces to proving that $\chi(A) = A'$, $\chi(D_0) = D_0$, and $\chi(T) = T$ are collinear. Because DD_0 is a diameter in ω, this is equivalent to showing that $\angle A'TD = 90°$.

The inversion χ has been used for proving Euler's relation (Theorem 3.36), and we know that it maps the circumcircle Γ to the nine-point circle γ of DEF. Theorem 3.26 implies that the circles ω, γ, and Γ are coaxial. With this in mind, let P be the intersection point of the lines DT and XY. (If DT and XY are parallel, then $DTFE$ and $DTXY$ are trapezoids, and A lies on the line IO, which means that ABC is isosceles, and the conclusion follows from symmetry.) Then P lies on the common radical axis of ω, γ and Γ, and by power-of-a-point, $PT \cdot PD = PA' \cdot PL$, where L is the projection of D on EF. So the quadrilateral $DTLA'$ is cyclic, and $\angle DTA' = \angle DLA' = \angle DLE = 90°$. We are done.

Source Cosmin Pohoață.

Chapter 12
A Synthesis

222 (a) *First solution.* In all solutions the triangle ABC is oriented counterclockwise. The first solution uses *spiral similarities* and can be followed on Fig. 12.1. Let A', B', and C' be the centers of the equilateral triangles A_1BC, B_1CA, and C_1AB, respectively. The point A' maps to B' by the spiral similarity of center B, angle $30°$, and ratio BC/BA', which maps A' to C, followed by the spiral similarity of center A, angle $30°$, and ratio AB'/AC, which maps C to B'. Because $AB'/AC = BA'/BC = \sqrt{3}/3$, the composition of the two spiral similarities is an isometry which rotates figures by $60°$; it is a *rotation*. The center of the rotation is the fixed point of the composition of the two spiral similarities. We show that this center is C'.

Because $\angle C'BC_1 = 30°$ and $BC'/BC_1 = \sqrt{3}/3$, the first spiral similarity maps C' to C_1. Also, because $\angle C'AC_1 = 30°$ and $C'A/C'C_1 = \sqrt{3}$, the second spiral similarity maps C_1 back to C'. So C' is indeed the fixed point of the rotation; it is the center of the rotation. Therefore $\angle A'C'B' = 60°$ and $C'A' = C'B'$. We conclude that the triangle $A'B'C'$ is equilateral, as desired.

Second solution. This solution and the next use isometries. Looking at Fig. 12.1, we notice that the problem can be rephrased as:

Problem On the sides of a triangle ABC construct in the exterior the triangles $A'BC$, $B'CA$, and $C'AB$ such that

$$\angle A'BC = \angle AB'C = \angle B'CA = \angle B'AC = \angle C'AB = \angle C'BA = 30°.$$

Prove that the triangle $A'B'C'$ is equilateral.

Consider the *rotations* ρ_a, ρ_b, and ρ_c by $120°$ about A', B', C', respectively. Their composition $\rho_a \circ \rho_b \circ \rho_c$ is an isometry that rotates figures by $120° + 120° + 120° = 360°$. So it is either a translation or the identity map. But $\rho_a \circ \rho_b \circ \rho_c(B) = B$, so it is the identity map, that is

© The Author(s), under exclusive license to Springer Nature Switzerland AG 2022
R. Gelca et al., *Geometric Transformations*, Problem Books in Mathematics,
https://doi.org/10.1007/978-3-030-89117-6_12

Fig. 12.1 Proof of
Napoleon's theorem

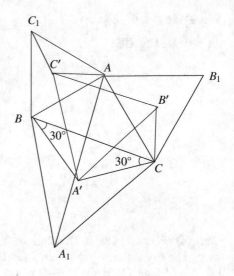

$$\rho_a \circ \rho_b \circ \rho_c = 1.$$

Then

$$\rho_b \circ \rho_c = \rho_a^{-1}.$$

This implies that the fixed point of $\rho_b \circ \rho_c$ is A', the center of the rotation ρ_a^{-1}. But it is not hard to see that the fixed point of $\rho_b \circ \rho_c$ forms with two centers of rotation B' and C' of ρ_b and ρ_c an equilateral triangle (oriented counterclockwise). This is because ρ_c maps this point to its reflection over $B'C'$, while ρ_b will bring it back to the initial position. Hence A', the fixed point of $\rho_b \circ \rho_c = \rho_a^{-1}$, forms with B' and C' an equilateral triangle.

Third solution. The complex number solution is also quite simple. Let the complex coordinates of the points A, B, C, A', B', C, be a, b, c, a', b', c', respectively, and assume that the triangle ABC is oriented counterclockwise. We work with angles measured in radians. Because the triangle $A'BC$ is isosceles with $\angle BA'C = \frac{2\pi}{3}$, C *rotates* to B about A' by $\frac{2\pi}{3}$. So

$$\frac{b - a'}{c - a'} = e^{\frac{2\pi i}{3}}.$$

Solving for a' we obtain

$$a' = \frac{b - ce^{\frac{2\pi i}{3}}}{1 - e^{\frac{2\pi i}{3}}}.$$

Similarly

$$b' = \frac{c - ae^{\frac{2\pi i}{3}}}{1 - e^{\frac{2\pi i}{3}}}, \quad c' = \frac{a - be^{\frac{2\pi i}{3}}}{1 - e^{\frac{2\pi i}{3}}}.$$

To prove that $A'B'C'$ is an equilateral triangle, it suffices to check that a rotation by $\frac{\pi}{3}$ around A' maps B' to C'. In complex coordinates, we have to check that $c' - a' = e^{\frac{\pi i}{3}}(b' - a')$. Ignoring the denominator, which is the same for a', b', and c', we have to check that

$$a - be^{\frac{2\pi i}{3}} - b + ce^{\frac{2\pi i}{3}} = e^{\frac{\pi i}{3}}(c - ae^{\frac{2\pi i}{3}} - b + ce^{\frac{2\pi i}{3}}).$$

And this is straightforward by using the identities

$$e^{\frac{\pi i}{3}} = -e^{\frac{4\pi i}{3}} \text{ and } 1 + e^{\frac{2\pi i}{3}} + e^{\frac{4\pi i}{3}} = 0.$$

Fourth solution. The triangles ABC_1, B_1CA, CA_1B are directly similar, so they are mapped into one another by *spiral similarities*. The general form of the Averaging Principle (Theorem 2.19) implies that the centroids of the triangles AB_1C, BCA_1, and C_1AB form a triangle that is similar to each of the three triangles. Hence this triangle is equilateral.

(b) *First solution.* The *spiral similarity* of center C, ratio $2\cos 30°$, and angle $30°$ maps B' to A and A' to A_1. It thus maps $B'A'$ to AA_1. This spiral similarity is depicted in Fig. 12.2. The spiral similarities of the same angles and ratios but of centers A and B map $C'B'$ to BB_1 and $A'C'$ to CC_1, respectively. This means that AA_1, BB_1, CC_1 are of length $2\cos 30°$ times the side-length of the equilateral triangle $A'B'C'$. This proves the equality of the three segments.

But moreover, the lines of support of these three segments form the same angles as $A'B', B'C', C'A'$, and these are $60°$ angles. Let P be the intersection of AA_1 and BB_1. Because $\angle(AA_1, BB_1) = 60°$, P is on the circumcircles of ABC_1, BCA_1, ACB_1. The same must be true, by symmetry, about the intersection of AA_1 and CC_1. But the three circles have at most one point of intersection (because their centers are not collinear). The intersection point must therefore be the common point of the three lines.

Second solution. Using the fact that A_1, B_1, C_1 are obtained by the $60°$ *rotations* about C, A, B of the points B, C, A, respectively, we can write

$$a_1 = e^{\frac{\pi i}{3}}(b - c) + c, \quad b_1 = e^{\frac{\pi i}{3}}(c - a) + a, \quad c_1 = e^{\frac{\pi i}{3}}(a - b) + b,$$

and consequently

$$a - a_1 = -a + e^{\frac{\pi i}{3}}b + (1 - e^{\frac{\pi i}{3}})c, \quad b - b_1 = -b + e^{\frac{\pi i}{3}}c + (1 - e^{\frac{\pi i}{3}})a,$$

$$c - c_1 = -c + e^{\frac{\pi i}{3}}a + (1 - e^{\frac{\pi i}{3}})b.$$

Fig. 12.2 The spiral
similarity that maps BA' to
AA_1

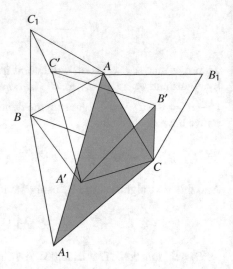

It is easy to check that $b - b_1 = e^{\frac{2\pi i}{3}}(a - a_1)$, $c - c_1 = e^{\frac{2\pi i}{3}}(b - b_1)$; hence the
three segments have equal lengths.

The equations of the lines AA_1, BB_1, and CC_1 are

$$(\overline{a} - \overline{a_1})z + (a_1 - a)\overline{z} + (a\overline{a_1} - a_1\overline{a}) = 0$$
$$(\overline{b} - \overline{b_1})z + (b_1 - b)\overline{z} + (b\overline{b_1} - b_1\overline{b}) = 0$$
$$(\overline{c} - \overline{c_1})z + (c_1 - c)\overline{z} + (c\overline{c_1} - c_1\overline{c}) = 0.$$

The fact that the three lines intersect at one point is reformulated by saying that this
system of equations in the unknowns z and $\overline{z}$ has solution. This is equivalent to the
fact that

$$\begin{vmatrix} \overline{a} - \overline{a_1} & a_1 - a & a\overline{a_1} - a_1\overline{a} \\ \overline{b} - \overline{b_1} & b_1 - b & b\overline{b_1} - b_1\overline{b} \\ \overline{c} - \overline{c_1} & c_1 - c & c\overline{c_1} - c_1\overline{c} \end{vmatrix} = 0,$$

and checking this is a long but routine computation.

Remark If the angles of the triangle ABC are all less than $120°$, then P is in the
interior of the triangle and is called the Fermat-Torricelli point of the triangle. It
is the point in the plane of the triangle with the sum of distances to the vertices
minimal.

Source While the result stated in (a) is usually attributed to Napoleon Bonaparte,
scholars were able to trace it back to W. Rutherford (1825). The fourth solution
of (a) is from T. Andreescu, M. Rolínek, J. Tkadlec, *Geometry Problems from the
AwesomeMath Year-Round Program*, XYZ Press, 2013.

223 *First solution.* We will employ the system of coordinates motivated by *homothety* and *reflection* that was introduced before Problem 4.1, with the coordinates of the vertices A, B, C, D being $1, a\omega, -ka, -k\omega$, respectively, where $k, a > 0$, $\omega\overline{\omega} = 1$. Then

$$
\begin{aligned}
AB \cdot CD + AD \cdot BC &= |1 - a\omega| \cdot |ka - k\omega| + |1 + k\omega| \cdot |ak + a\omega| \\
&= k|\omega| \cdot |1 - a\omega| \cdot |1 - a\overline{\omega}| + a|\omega| \cdot |1 + k\omega| \cdot |1 + k\overline{\omega}| \\
&= k|(1 - a\omega)\overline{(1 - a\omega)}| + a|(1 + k\omega)\overline{(1 + k\omega)}| \\
&= k(1 - a\omega)(1 - a\overline{\omega}) + a(1 + k\omega)(1 + k\overline{\omega}) \\
&= k + ka^2 - ka(\omega + \overline{\omega}) + a + ak^2 + ka(\omega + \overline{\omega}) \\
&= k + ka^2 + a + ak^2 = (k + a)(1 + k) = AC \cdot BD.
\end{aligned}
$$

Second solution. We can write Ptolemy's identity as

$$
\frac{AB}{AC} : \frac{DB}{DC} + \frac{AD}{AC} : \frac{BD}{BC} = 1,
$$

or, in complex coordinates, $|(a, d, b, c)| + |(a, b, d, c)| = 1$. We can remove the absolute value bars because the points lie on the circle in the order A, B, C, D, so the cross-ratios are positive (the two fractions in each cross-ratio have the same argument, by inscribed angles). We are supposed to prove that

$$
(a, d, b, c) + (a, b, d, c) = 1.
$$

As the cross-ratio is mapped by a circular transformation to either itself or its complex conjugate, if we consider a circular transformation that sends A to ∞, then Ptolemy's identity is equivalent to

$$
(\infty, d', b', c') + (\infty, b', d', c') = 1,
$$

where dashes denote the images of points. This is the same as

$$
\frac{d' - c'}{d' - b'} + \frac{c' - b'}{d' - b'} = 1,
$$

and this is obviously true.

Third solution. We have seen in the second solution that the property that a quadrilateral satisfies Ptolemy's identity is invariant under circular transformations. We can map A, B, C, D by a circular transformation to points on the real axis that have the coordinates $0, x, x + y, x + y + z$, $x, y, z > 0$. Here we use the fact that circular transformations preserve the relative position of points on a line (that contains ∞). Ptolemy's identity for these four points reads

Fig. 12.3 A slightly forced
configuration of
A, B, C, D, S to show what
happens when S is not the
midpoint of PQ

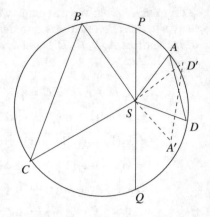

$$(x + y)(y + z) = x(x + y + z) + yz,$$

and this is obviously true.

Remark A lesson to be learned from this problem as well as from many others: If
you have to prove a property, find the largest group of transformations that preserve
it. Then, for each configuration find a configuration that is equivalent to it modulo
transformations and for which the property is easy to verify.

Working with absolute values in the last two approaches, and using the triangle
inequality, we can prove that, in general, if A, B, C, D are four points in the plane,
then $AC \cdot BD \leq AB \cdot CD + AD \cdot BC$, with equality if and only if A, B, C, D are
on a circle in this order.

224 *First solution.* Consider the *reflection* over the line through S that is perpendicular to PQ, and let A' and D' be the images of A and D, respectively (Fig. 12.3).
Then

$$\angle A'SQ = \angle ASP = \angle BSP$$

shows that A', S, B are collinear. Similarly D', S, C are collinear, so S is the
intersection of the diagonals of the quadrilateral $BCA'D'$. Then

$$\angle CD'A' = \angle SD'A' = \angle SDA = \angle CBS = \angle CBA',$$

which implies that $BCA'D'$ is a cyclic quadrilateral. Thus both $BCDA$ and
$BCA'D'$ are cyclic.

If the circumcircles of these two quadrilaterals are distinct, then one of the points
A' and D' lies outside of ω, and one is inside. But then ω intersects the circumcircle
of $BCA'D'$ in three points: B, C, and some point on the arc $\overset{\frown}{AD}$. This is impossible.
So the two circumcircles coincide. But then A', D' are on ω, showing that ω is

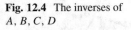

Fig. 12.4 The inverses of
A, B, C, D

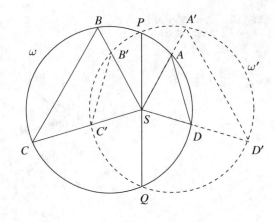

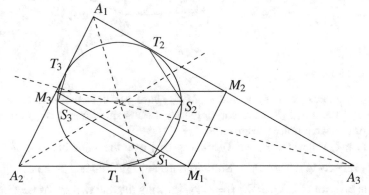

Fig. 12.5 $M_1 M_2 M_3$ and $S_1 S_2 S_3$ have parallel sides

invariant under the reflection. This implies that the line of reflection passes through the midpoint of the chord PQ, so S is the midpoint of this chord.

Second solution. The triangles SAD and SCB are similar, hence $SA \cdot SB = SC \cdot SD$. Consider the *inversion* of center S and radius $R = \sqrt{SA \cdot SB}$, let ω' be the circle that is the inverse of ω, and let A', B', C', D' be the inverses of A, B, C, D, respectively (Fig. 12.4). Then $SA' = SB, SB' = SA, SC' = SD, SD' = SA$, showing that A', B', C', D' are the *reflections* over PQ of B, A, D, C, respectively. Hence ω' is the reflection of ω over PQ, from where we infer that P and Q on both circles. But then P and Q are invariant under the inversion, so they are on the circle of inversion. Therefore $SP = R = SQ$, and the problem is solved.

Source Polish Mathematical Olympiad, 2005.

225 We will show that the triangles $M_1 M_2 M_3$ and $S_1 S_2 S_3$ have parallel sides, and then by Problem 84, they are *homothetic*. And then $M_1 S_1$, $M_2 S_2$, and $M_3 S_3$ intersect at the center of the homothety. The configuration can be seen in Fig. 12.5.

Fig. 12.6 Spiral similarity
keeps XYZ equilateral

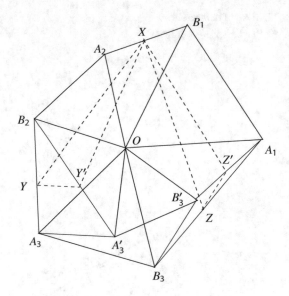

Of course, because $A_1A_2A_3$ and $M_1M_2M_3$ have parallel sides, it suffices to show that $A_1A_2A_3$ and $S_1S_2S_3$ have parallel sides. Let us show that A_1A_2 is parallel to S_1S_2. For this we employ... *reflections*!

Because T_2 reflects to T_3 and T_1 reflects to S_1 over the angle bisector of $\angle A_1$, the arc $\overparen{T_2T_1}$ reflects to the arc $\overparen{T_3S_1}$. Similarly, the arc $\overparen{T_2T_1}$ reflects to the arc $\overparen{T_3S_2}$ over the angle bisector of $\angle A_2$. So $\overparen{T_3S_1}=\overparen{T_3S_2}$, showing that the two arcs reflect into each other over the diameter from T_3, and in particular, S_1 reflects over this diameter to S_2. But A_2 also reflects over this diameter to A_1, since the tangent is perpendicular to the diameter. Therefore S_1S_2 is parallel to A_1A_2. The same argument can be repeated for the other sides, and the problem is solved.

Source International Mathematical Olympiad, 1982, proposed by the Netherlands.

226 *First solution.* The property is true if $A_1B_1A_2B_2A_3B_3$ is a regular hexagon. To arrive at an arbitrary configuration, you can transform the triangles one at a time by a *spiral similarity* centered at O. Let us show that if the triangle formed by the three specified midpoints is equilateral before such a spiral similarity is applied, then it stays equilateral afterward. Let therefore the configuration before the spiral similarity consist of the triangles OA_1B_1, OA_2B_2, and OA_3B_3 with the midpoints of B_1A_2, B_2A_3, and B_3A_1 being X, Y, Z, respectively, such that XYZ is equilateral. Let A_3 and B_3 map to A_3' and B_3', respectively, under a spiral similarity of center O (Fig. 12.6). Let also Y' and Z' be the midpoints of B_2A_3' and $B_3'A_1$. Then YY' and ZZ' are midlines in the triangles $B_2A_3A_3'$ and $A_1B_3B_3'$, and so YY' is parallel to A_3A_3' and half its length, and ZZ' is parallel to B_3B_3' and half its length.

Theorem 2.25 shows that spiral similarities come in pairs, and so the spiral similarity that maps OA_3B_3 to $OA_3'B_3'$ comes with a spiral similarity, also of center

Fig. 12.7 Proof by
isometries that the triangle
formed by midpoints is
equilateral

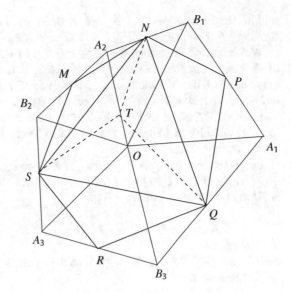

O, that maps $OA_3A'_3$ to $OB_3B'_3$. At a closer look we see that the latter is in fact a
$60°$ *rotation*. But then $A_3A'_3$ and $B_3B'_3$ are of equal lengths and form an angle of
$60°$. The same will be true about YY' and ZZ'. And the $60°$ rotation about X that
maps Y to Z will map Y' to Z'. Therefore $XY'Z'$ is equilateral and the problem is
solved.

Second solution. Place the origin of the coordinate system at O. If a_1, a_2, a_3 are
the complex coordinates of A_1, A_2, A_3, respectively, then since B_1, B_2, B_3 are the
$60°$ *rotations* of these points about O, their coordinates are $a_1e^{\frac{i\pi}{3}}, a_2e^{\frac{i\pi}{3}}, a_3e^{\frac{i\pi}{3}}$.
Let x, y, z be the coordinates of the midpoints X, Y, Z of B_1A_2, B_2A_3, and B_3A_1,
respectively. Then

$$x = \frac{1}{2}(a_1e^{\frac{\pi i}{3}} + a_2), \quad y = \frac{1}{2}(a_2e^{\frac{\pi i}{3}} + a_3), \quad z = \frac{1}{2}(a_3e^{\frac{\pi i}{3}} + a_1).$$

We want to show that y *rotates* to z about x by $60°$, which amounts to checking that
$2(z - x) = 2(y - z)e^{\frac{\pi i}{3}}$. This equality reads

$$a_1(1 - e^{\frac{\pi i}{3}}) - a_2 + a_3e^{\frac{\pi i}{3}} = [-a_1e^{\frac{\pi i}{3}} + a_2(e^{\frac{\pi i}{3}} - 1) + a_3]e^{\frac{\pi i}{3}},$$

and this is an easy consequence of the equality $e^{\frac{\pi i}{3}} - e^{\frac{2\pi i}{3}} = 1$.

Third solution. We explain this solution only for the case where $A_1B_1A_2B_2A_3B_3$
is not a skew hexagon. Let M, N, P, Q, R, S be the midpoints of A_1B_1, B_1A_2,
$A_2B_2, B_2A_3, A_3B_3, B_3A_1$, respectively. We have to show that the triangle NQS is
equilateral. We argue on Fig. 12.7.

The $60°$ *rotation* that maps A_1 to B_1 and A_2 to B_2 maps A_1A_2 to B_1B_2, so these
two segments have the same length and form an angle of $60°$. But MN and NP are

midlines in the triangles $B_1A_1A_2$ and $A_2B_1B_2$, so $MN = NP$ and $\angle MNP = 120°$. Similarly $PQ = QR$ and $\angle PQR = 120°$ and $RS = SM$ and $\angle RSM = 120°$.

We claim that the triangles MNS, PQN, and RSQ can be reflected over sides as to form a dissection of the triangle NQS. If this were indeed true, then each angle of this latter triangle is half of $120°$, namely, $60°$, and we are done. To prove our claim, *reflect M over NS to a point T*. By the reflection, $TN = MN = NP$, and $TS = MS = RS$. It is sufficient to show that $TQ = PQ = RQ$, because then, by equality of triangles, T would also be the reflection of P and R over the corresponding sides.

Let us assume that this is not true. If $TQ < PQ = RQ$, then in the triangles PQT and QRT, $\angle TPQ < \angle QTP$, and $\angle QRT < \angle RTQ$. Since the triangles NTP and SRT are isosceles,

$$\angle NPQ = \angle NPT + \angle TPQ < \angle PTN + \angle QTP = \angle NTQ,$$

and similarly $\angle QRS < \angle QTS$. Then on the one hand by using the sum of the angles of hexagon, we have

$$\angle SMN + \angle NPQ + \angle QRS = 4 \cdot 180° - \angle MNP - \angle PQR - \angle RSM$$
$$= 720° - 3 \cdot 120° = 360°.$$

and on the other hand

$$\angle SMN + \angle NPQ + \angle QRS < \angle NTS + \angle PTN + \angle STQ = 360°,$$

a contradiction. A similar reasoning rules out the situation $TQ > PQ = RQ$. The problem is solved.

Remark The third solution can be adapted to the general configuration where the hexagon can be skewed by using directed angles and taking into account the situation where T falls outside the triangle NQS.

The idea of the solution is similar to those used for solving Problem 1.5 and Problem 49.

The Averaging Principle implies that the midpoints of A_1B_1, A_2B_2, and A_3B_3 form a triangle that is similar to both $A_1A_2A_3$ and $B_1B_2B_3$.

Source The first solution was posted online by Alan Cooper.

227 An *inversion* of center A maps Γ_1 to a line Γ_1' that is parallel to t_1, and t_2 and Γ_2 are mapped to the circles t_2' and Γ_2' that are tangent to t_1 (at A and D') and to Γ_1', while t_3 becomes a circle t_3' that intersects t_1 at A and D' and intersects t_2' and Γ_2' a second time at B' and C', respectively, as shown in Fig. 12.8.

The new figure is invariant under *reflection* over the line $E'F'$, where E' and F' are the images of E and F and hence are the intersections of the circles t_2' and Γ_2'. The line $B'C'$ is therefore invariant under this reflection, so it is parallel to t_1.

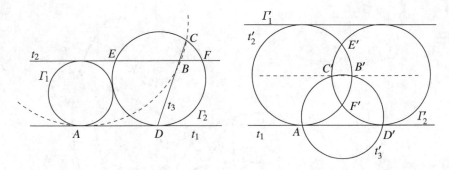

Fig. 12.8 Configuration of circles and lines and its image under inversion

Consequently, its preimage under the inversion, which is the circumcircle of ABC, is tangent to t_1.

Source Short list of the Balkan Mathematical Olympiad, 2003.

228 Examining Fig. 12.9, we notice that the Simson line of D with respect to the triangle ABC is mapped by a *homothety* of center D and ratio 2 to a line through the points E and F. Hence P, which is the intersection of BD and EF, must be the image of the intersection of the diagonals through the homothety, so it is the *reflection* of D over AC. It follows that A is on the perpendicular bisectors of ED and DP, showing that it is the circumcenter of the triangle DEP. From here we deduce that $AP = AQ$. Working with directed angles modulo $180°$ and using the facts that APD is isosceles and $AQBD$ is cyclic, we have

$$\angle AQB = \angle ADB = \angle DPA = \angle BPA.$$

This combined with $AP = AQ$ implies that Q is the reflection of P over AB. Thus EQ is the reflection of DP over AB. Similarly RF is the reflection of DP over BC. Therefore $EQ = DP = FR$, and we are done.

Source United States of America Junior Mathematical Olympiad, 2018, proposed by Ray Li.

229 *First solution.* We argue on Fig. 12.10. The *spiral similarity* s_B of center B, clockwise oriented angle of $30°$, and ratio $BA/BR = 2/\sqrt{3}$ maps R to A. Then, the spiral similarity s_C of center C, clockwise oriented angle of $30°$, and ratio $CQ/CA = \sqrt{3}/2$ maps A further to Q. The composition $s_C \circ s_B$ is a *rotation* of angle $60°$ which maps R to Q. Let us show that the fixed point of this rotation is P.

Let P' be the point in the exterior of the triangle ABC with the property that $\angle P'BC = \angle P'CB = 30°$. Then $PP'B$ and $PP'C$ are $90° - 60° - 30°$ triangles,

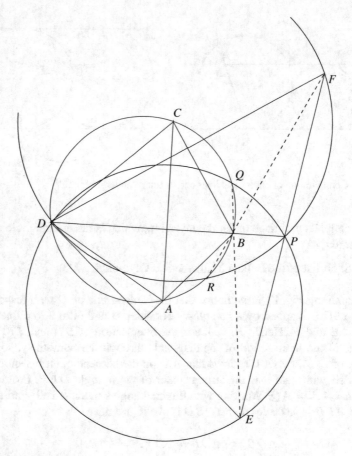

Fig. 12.9 *EQ* is equal to *RF*

so $s_B(P) = P'$ and $s_C(P') = P$. Hence $s_C \circ s_B(P) = P$, as claimed. We conclude that a 60° rotation about P maps R to Q, so PQR is equilateral.

Second solution. The *homothety* h_B of center B and ratio 2 maps P to C. The *spiral similarity* s_A of center A, angle 60°, and ratio $1/2$ maps C to Q. Then $s_A \circ h_B(P) = Q$. The composition $s_A \circ h_B$ is a spiral similarity of ratio $2 \times 1/2 = 1$ and angle 60°, so it is a 60° *rotation*. Let us find its fixed point, which is also the center of rotation.

We note that $h_B(R)$ is the reflection D of B over the line AN. The triangle ADN is equilateral, so $s_A(D) = R$. It follows that R is the fixed point of $s_A \circ h_B$. We conclude that $PR = QR$ and $\angle QRP = 60°$, which shows that triangle PQR is equilateral.

Third solution. We use complex numbers. Let the triangle ABC be oriented counterclockwise and denote by a, b, c the coordinates of A, B, C, respectively. Then P has the coordinate $p = \frac{b+c}{2}$.

Fig. 12.10 Proof that PQR is equilateral

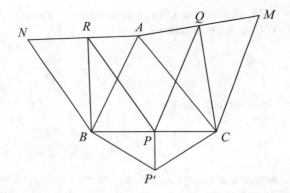

R is the image of A through a *spiral similarity* of center B, angle $\frac{\pi}{6}$, and ratio $\sqrt{3}/2$, so its complex coordinate r satisfies

$$\frac{r-b}{a-b} = \frac{\sqrt{3}}{2} e^{\frac{\pi i}{6}}.$$

We compute

$$r = \frac{\sqrt{3}}{2} e^{\frac{\pi i}{6}} a + \left(1 - \frac{\sqrt{3}}{2} e^{\frac{\pi i}{6}}\right) b.$$

Similarly, Q is the image of A through a spiral similarity of center C, angle $-\frac{\pi}{6}$, and ratio $\sqrt{3}/2$. So its complex coordinate is

$$q = \frac{\sqrt{3}}{2} e^{-\frac{\pi i}{6}} a + \left(1 - \frac{\sqrt{3}}{2} e^{-\frac{\pi i}{6}}\right) c.$$

The condition that PQR is equilateral

$$\frac{r-p}{q-p} = e^{\frac{\pi i}{3}}$$

translates to

$$\frac{\sqrt{3}}{2} e^{\frac{\pi i}{6}} a + \left(\frac{1}{2} - \frac{\sqrt{3}}{2} e^{\frac{\pi i}{6}}\right) b - \frac{c}{2} = e^{\frac{\pi i}{3}} \left[\frac{\sqrt{3}}{2} e^{-\frac{\pi i}{6}} a + \left(\frac{1}{2} - \frac{\sqrt{3}}{2} e^{-\frac{\pi i}{6}}\right) c - \frac{b}{2}\right].$$

And this is a straightforward computation using $e^{\pm\frac{\pi i}{6}} = \frac{\sqrt{3}}{2} \pm \frac{1}{2} i$ and $e^{\frac{\pi i}{3}} = \frac{1}{2} + \frac{\sqrt{3}}{2} i$.

Source Romanian Mathematics Competition, proposed by C. Ottescu, 1981.

230 *First solution.* We recognize two cross-ratios in the statement. The first cross-ratio is in the hypothesis

$$\frac{AD}{BD} : \frac{AC}{BC} = 1,$$

the other in the question:

$$\frac{AB}{DB} : \frac{AC}{DC}.$$

Looking at the hypothesis, we should realize that much more is true. The fact that C and D are on the same side of AB and $\angle ADB = 90° + \angle ACB$ means that, if we choose the appropriate orientation of the plane when putting on it complex coordinates, then

$$\arg \frac{a-d}{b-d} - \arg \frac{a-c}{b-c} = \frac{\pi}{2}.$$

Consequently

$$\frac{a-d}{b-d} : \frac{a-c}{b-c} = 1 \cdot e^{\frac{\pi}{2}i} = i.$$

Knowing this, let us apply a *Möbius transformation* that maps C to ∞. The fact that Möbius transformations preserve the cross-ratio implies that the complex coordinates of the images satisfy

$$\frac{a'-d'}{b'-d'} = i.$$

This means that the triangle $D'A'B'$ formed by the images is right isosceles, with the right angle at D'. So the lines $A'D'$ and $B'D'$ form a right angle, which means that their preimages, which are the circumcircles of ACD and BCD, are orthogonal. This solves half of the problem. In the right isosceles triangle $D'A'B'$, we have

$$\frac{a'-b'}{d'-b'} = \sqrt{2}e^{-\frac{\pi}{4}i},$$

or, equivalently,

$$\frac{a'-b'}{d'-b'} : \frac{a'-\infty}{d'-\infty} = \sqrt{2}e^{-\frac{\pi}{4}i}.$$

Again by the invariance of the cross-ratio,

Fig. 12.11 Construction of B'

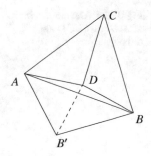

$$\frac{a-b}{d-b} : \frac{a-c}{d-c} = \sqrt{2}e^{-\frac{\pi}{4}i}.$$

Hence

$$\frac{AB}{DB} : \frac{AC}{DC} = \sqrt{2},$$

and this is the expression (written as a cross-ratio) that we had to compute.

Second solution. After this short and elegant solution, it seems pointless to look for more. But it might be worth to observe that this problem can be solved not only within the context of complex projective geometry but also in the complex affine setting. Consider a *spiral similarity* of center A that maps C to D and let B' be the image of B (Fig. 12.11). Then the triangle ACB is mapped to ADB', so on the one hand

$$\angle BDB' = \angle ADB - \angle ADB' = \angle ADB - \angle ACB = 90°$$

and on the other hand

$$\frac{B'D}{BC} = \frac{AD}{AC} = \frac{BD}{BC},$$

where for the last equality we used the hypothesis. Hence $B'D = BD$. It follows that the triangle DBB' is right isosceles.

As spiral similarities come in pairs (Theorem 2.25), the triangles ACD and ABB' are mapped into each other by a spiral similarity; hence they are similar. Therefore

$$\frac{AB}{AC} = \frac{BB'}{CD} = \frac{BD\sqrt{2}}{CD},$$

and hence

Fig. 12.12 The circumcircles of ACD and BCD are orthogonal

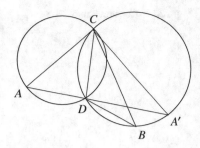

$$\frac{AB \cdot CD}{AC \cdot BD} = \sqrt{2}.$$

Let us also check the orthogonality of the two circles. In the concave quadrilateral $ADBC$,

$$\angle CAD + \angle CBD = 360° - \angle ACB - (360° - \angle ADB) = 90°.$$

Next, consider the *spiral similarity* of center C that maps the circumcircle of ACD to the circumcircle of CDB. By Problem 131, the image A' of A is the second intersection point of AD with the circumcircle of CDB (Fig. 12.12). The angle of the spiral similarity is

$$\angle ACA' = 180° - \angle CAD - \angle CA'D = 180° - \angle CAD - \angle CBD$$
$$= 180° - 90° = 90°.$$

And as the tangents at C to the two circles are mapped into each other by the spiral similarity, they form a 90° angle, showing that the circles are orthogonal.

Remark So the right isosceles triangle that is at the core of the solution can be produced not just by a projective transformation (the Möbius transformation) but also by an affine transformation (the spiral similarity).

Source International Mathematical Olympiad, 1993, proposed by the Great Britain.

231 We assume that the radius of ω' is smaller than that of ω, so that we are in the situation from Fig. 12.13. Let T be the intersection of PP' and QQ'. Consider the *homothety* of center T that maps ω to ω', and let A' be the image of A. Because O is mapped to O' and M to M', $\angle M'A'O' = \angle MAO$.

Now consider the *inversion* of center T and radius OP'. By power-of-a-point, this inversion maps ω into itself. By Theorem 3.2, the inversion maps O' to M' (because $\angle TP'O' = \angle TM'P = 90°$), and it also maps A to A'. It follows that $AA'M'O'$ is cyclic, so $\angle M'A'O' = \angle M'AO'$.

Combining the two equalities, we obtain $\angle MAO = \angle M'AO'$, and consequently $\angle MAM' = \angle OAO'$, as desired.

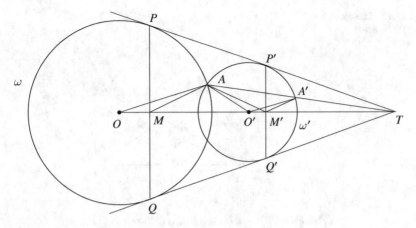

Fig. 12.13 Proof that $\angle MAM' = \angle OAO'$

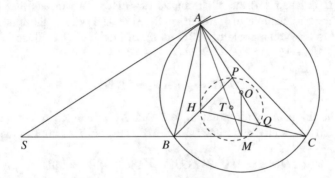

Fig. 12.14 Proof that AT is a median in triangle ABC

Source International Mathematical Olympiad, 1983, proposed by the Soviet Union.

232 We will prove that the circumcenter T of the triangle HPQ lies on the median from A. Without loss of generality we assume that $AB < AC$. The reasoning can be followed on Fig. 12.14.

Note that since $CH \perp AB$,

$$\angle HQP = 90° - \angle QAB = 90° - \angle OAB = \frac{1}{2}\angle AOB = \angle ACB,$$

and similarly $\angle HPQ = \angle ABC$. This shows that the triangles HPQ and ABC are similar. The triangles have opposite orientation, so there is a map λ that is the composition of a *spiral similarity* and a *reflection* over a line such that $\lambda(HPQ) = ABC$.

Because

Fig. 12.15 The proof that ℓ is angle bisector

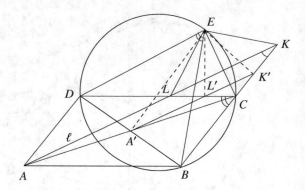

$$\angle AHP = 90° - \angle HAC = \angle ACB = \angle HQP,$$

the line AH is tangent to the circumcircle of HPQ. This means that $\lambda(AH)$ is tangent to the circumcircle of ABC at A. So $S = \lambda(A)$ lies at the intersection of the tangent to the circumcircle of ABC at A with the line BC. Since $\lambda(T) = O$, $\lambda(A) = S, \lambda(P) = B$,

$$\angle OAT = \angle PAT = \angle BSO.$$

Let M be the intersection point of AT and BC. Then $\angle MSO = \angle BSO = \angle TAO = \angle MAO$, so the quadrilateral $AOMS$ is cyclic. This implies that

$$\angle OMS = 180° - \angle OAS = 180° - 90° = 90°,$$

so $OM \perp BC$. Thus M is the midpoint of the side BC, so T belongs to the median AM.

Remark If two figures are similar but with opposite orientation, then by reflecting one of them over a line, we obtain a figure directly similar to the other, which can now be mapped into the other by a spiral similarity.

Source Short list of the International Mathematical Olympiad, 2017, proposed by Ukraine.

233 (a) Consider the *homothety* of center C and ratio $1/2$, and let A', K', and L' be the images of A, K, L, respectively, as shown in Fig. 12.15. Then A' is the midpoint of AC, so it is the intersection of the diagonals of the parallelogram, while K' and L' are the midpoints of the sides CK and CL of the triangles ECK and ECL. Because these triangles are isosceles, K' and L' are the projections of E onto the corresponding sides, so $K'L'$, which is the image of ℓ, is the Simson line of E with respect to the triangle BCD. It follows that EA' is perpendicular to BD, since A' belongs to the Simson line. We conclude that E is on the perpendicular bisector of BD.

In the right triangles $EA'B$ and $EL'C$,

Fig. 12.16 The proof that
$BCDE$ is cyclic

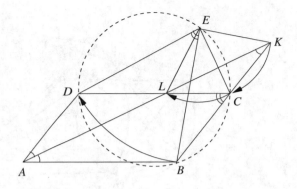

$$\angle A'BE = \frac{1}{2}\ \widehat{DE} = \angle L'CE,$$

so the triangles are directly similar. It follows that there is a *spiral similarity* of
center E that maps the triangle $EA'B$ to $EL'C$. And as spiral similarities come
in pairs (Theorem 2.25), there is a spiral similarity that maps the triangle $EA'L'$
to EBC. The angle of this spiral similarity is $\angle A'EB = \angle DEB/2 = \angle DCB/2$
(EDB is isosceles and $BCED$ is cyclic). So $A'L'$, as well as its preimage ℓ through
the homothety, make with BC an angle of $\angle DCB/2$, and consequently they make
with AB an angle of $\angle DCB/2 = \angle DAB/2$. It follows that ℓ is the angle bisector
of $\angle DAB$, and we are done.

(b) The converse is proved using a *rotation*. Let E be the circumcenter of CKL.
To prove that E is on the circumcircle of BCD amounts to showing that $\angle BED =
\angle BCD$. Let us prove this equality, following the argument on Fig. 12.16. The fact
that AD is parallel to BK and AB is parallel to CL implies that

$$\angle BKA = \angle KAD = \angle BAK = \angle CLK.$$

From here we obtain on the one hand that $AB = CD = BK$ and on the other that
$KC = LC$. These two chords are equal in the circle of center E, and hence they are
mapped into each other by rotation ρ of center E. Then ρ maps the line BK to the
line CD, so the angle of rotation is $\angle BCD$. On the other hand because the segments
KB and CD are equal and $\rho(K) = C$, it follows that $\rho(B) = D$. But then $\angle BED$
is equal to the angle of rotation, which is $\angle BCD$, and we are done.

Source The direct implication was given at the International Mathematical
Olympiad in 2007, being proposed by Luxembourg, the converse was published in
1987 in *Kvant (Quantum)* being proposed by Igor Fedorovich Sharygin.

234 *First solution.* This argument can be followed on Fig. 12.17. Let Q_1 be the
midpoint of the arc $\widehat{BAC}$ and let P_1 be the midpoint of the arc $\widehat{BAD}$; let also K and
L be the midpoints of PP_1 and QQ_1, respectively.

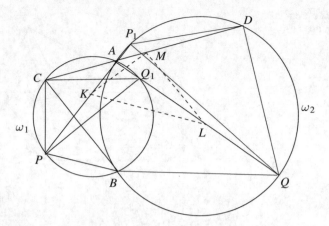

Fig. 12.17 Proof by spiral similarity that $\angle PMQ = 90°$

Consider the *spiral similarity* s_1 of center B that maps ω_1 to ω_2. By Problem 131 (b), $s_1(C) = D$. And since spiral similarities preserve midpoints of arcs, $s_1(P) = P_1$ and $s_1(Q_1) = Q$. So on the one hand, Problem 131 (b) implies that PP_1 and QQ_1 pass through A. On the other hand, s_1 maps the triangle PCQ_1 to P_1DQ. Applying the Averaging Principle (Theorem 2.18), we obtain that the triangle KML is similar to any of the triangles PCQ_1 and P_1DQ. In particular $\angle KML = 90°$ and $KM/ML = PC/CQ_1$.

Let us compute $\angle(PP_1, Q_1Q)$. For this we use the fact that spiral similarities come in pairs (Theorem 2.25), so s_1 comes with a second spiral similarity s_2, also of center B, satisfying $s_2(P) = Q_1$ and $s_2(P_1) = Q$. The angle of s_2 is $\angle PBQ_1 = 90°$, because PQ_1 is a diameter. Hence $\angle(PP_1, QQ_1) = 90°$. But from this second spiral similarity, we extract more information:

$$\frac{KP}{LQ} = \frac{PP_1}{Q_1Q} = \frac{BP}{BQ_1} = \frac{CP}{CQ_1} = \frac{KM}{ML}.$$

And from the fact that

$$\angle(KM, ML) = 90° = \angle(PP_1, QQ_1) = \angle(KP, LQ)$$

we obtain that $\angle MKP = \angle MLQ$. It follows that the triangles MKP and MLQ are similar, in fact directly similar, so they are mapped into each other by yet another spiral similarity. Spiral similarities come in pairs, so KML is mapped to PMQ by a fourth spiral similarity, showing that $\angle PMQ = \angle KML = 90°$.

Second solution. From the fact that P and Q are the midpoints of the arcs $\overset{\frown}{BC}$ and $\overset{\frown}{BD}$ it follows that AP and AQ are the angle bisectors of the supplementary angles $\angle BAC$ and $\angle BAD$, so $\angle PAQ = 90°$. The problem is solved if we show that A, M, P, Q lie on a circle.

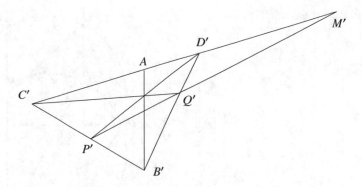

Fig. 12.18 Proof by inversion that $\angle PMQ = 90°$

To prove this, consider an *inversion* about a circle centered at A, and use the convention to denote the image of a point by adding a dash to its letter name (you can see the useful part of the inverted configuration in Fig. 12.18). The problem reduces to showing that M', P', Q' are collinear. As M is the midpoint of the segment CD, the points ∞, C, M, D form a harmonic division. As inversion preserves cross-ratios that are real (Theorem 3.18) A, C', M', D' also form a harmonic division (A is the image of ∞ and you should always think of the points as appearing in this circular order on the projective line $\mathbb{R} \cup \{\infty\}$).

On the other hand, if in the triangle $B'C'D'$ the cevians $B'A$, $C'Q'$, and $D'P'$ intersected at one point, then the intersection of $P'Q'$ with $C'D'$ would form with A, C', and D' a harmonic division (see the proof of Theorem 3.33) so this intersection would be M', showing that M', P', Q' are collinear. Thus we are left with showing that $B'A$, $C'Q'$, and $C'P'$ are concurrent.

The lines AP' and AQ' are the angle bisectors of $\angle B'AC'$ and $\angle B'AD'$, so by the Bisector Theorem,

$$\frac{P'C'}{P'B'} = \frac{AC'}{AB'} \text{ and } \frac{Q'B'}{Q'D'} = \frac{AB'}{AD'}.$$

It follows that

$$\frac{AD'}{AC'} \cdot \frac{P'C'}{P'B'} \cdot \frac{Q'B'}{Q'D'} = \frac{AD'}{AC'} \cdot \frac{AC'}{AB'} \cdot \frac{AB'}{AD'} = 1.$$

So by Ceva's Theorem (Problem 116) the lines $B'A$, $C'Q'$, $D'P'$ are indeed concurrent, and we are done.

Source Second solution by Flavian Georgescu.

235 *First solution.* (a) Let the point be A, let the triangle be BCD, and let its orthocenter be H (Fig. 12.19). Construct the *reflections* A_1 and A_2 of A over BC and DC, respectively. The composition of these reflections is a *rotation* about C by

Fig. 12.19 Proof of Steiner's
theorem

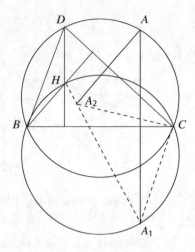

$2\angle BCD$, which maps A_1 to A_2. Notice that A_1 belongs to the circumcircle of BHC and A_2 belongs to the circumcircle of CHD. Taking into account the *homothety* of center A and ratio 2, we realize that the problem is solved if we show that A_1, A_2, and H are collinear. And this is true because the rotation maps the circumcircle of HBC to the circumcircle of HAC. Interpreting the rotation as a particular case of *spiral similarity*, we apply Problem 131 (b) to conclude that A_1, A_2 and the point of intersection of the two circles that is not the center of rotation are collinear. This point of intersection is H, and we are done.

Second solution. Like in the first solution, we start by considering the *reflections* σ_{CB} and σ_{CD} over lines CB and CD, respectively. Using the same *homothety*, we reduce the problem to showing that $A_1 = \sigma_{CB}(A)$, $A_2 = \sigma_{CD}(A)$, and H are collinear.

We switch to complex coordinates, place the origin of the coordinate system at the circumcenter of BCD, and let the circumcircle of BCD be the unit circle. The coordinates of A, B, C, D, H, A_1, A_2 are a, b, c, d, h, a_1, a_2. As proved in the solution to Problem 23, $h = b + c + d$.

In view of the formula (1.1) derived in the proof of Theorem 1.5 we choose the parametric equations of lines

$$CB: \quad z = c + \frac{b - c}{|b - c|}t, \quad t \in \mathbb{R},$$

$$CD: \quad z = c + \frac{d - c}{|d - c|}t, \quad t \in \mathbb{R},$$

then apply (1.1) to obtain

$$a_1 = \sigma_{CB}(a) = \frac{(b - c)^2}{(b - c)(\overline{b} - \overline{c})}\overline{a} + c - \frac{(b - c)^2}{(b - c)(\overline{b} - \overline{c})}\overline{c} = \frac{b - c}{\overline{b} - \overline{c}}\overline{a} + c - \frac{b - c}{\overline{b} - \overline{c}}\overline{c}$$

$$= \frac{b-c}{\underset{bc}{c-b}}\overline{a} + c - \frac{b-c}{\underset{bc}{c-b}}\overline{c} = -\overline{a}bc + c + b.$$

We have used the fact that $b\overline{b} = c\overline{c} = d\overline{d} = 1$, as they lie on the unit circle. Similarly

$$a_2 = -\overline{a}dc + c + d.$$

To show that a_1, a_2, h are on a line, we have to show that $(h - a_1)/(h - a_2) \in \mathbb{R}$. We compute

$$\frac{h - a_1}{h - a_2} = \frac{d + \overline{a}bc}{b + \overline{a}dc} = \frac{(d + \overline{a}bc)\overline{(b + \overline{a}dc)}}{(b + \overline{a}dc)\overline{(b + \overline{a}dc)}} = \frac{(d + \overline{a}bc)(\overline{b} + a\overline{dc})}{(b + \overline{a}dc)\overline{(b + \overline{a}dc)}}$$

$$= \frac{d\overline{b} + \overline{a}c + b\overline{d} + a\overline{c}}{(b + \overline{a}dc)\overline{(b + \overline{a}dc)}} = \frac{2\Re(d\overline{b} + \overline{a}c)}{|b + \overline{a}dc|^2}.$$

The latter is a real number and we are done.

(b) Let the quadrilateral be $ABCD$, and let H_1, H_2, H_3, H_4 be the orthocenters of the triangles BCD, CDA, DAB, and ABC. As we have seen in Problem 23, the quadrilaterals $ABCD$ and $H_1 H_2 H_3 H_4$ are mapped into each other by a *reflection* over a point M that is simultaneously the midpoint of the segments $AH_1, BH_2,$ CH_3, and DH_4. The conclusion now follows from Steiner's Theorem.

Remark The second solution to (a) combines a synthetic observation (the reduction of the problem to the collinearity of A_1, A_2, H) with a complex number computation.

236 (a) The key observation is that the triangles FAB and FDC are similar because they have equal angles, which is a consequence of the fact that the quadrilateral is cyclic. But they are not directly similar, so we cannot use homothety or spiral similarity. This can be fixed, by composing with the reflection over a line. The ratios from the statement suggest transformations that map AB and CD into each other, and we will use two transformations: f that is the composition of the *homothety* of center E and ratio AB/CD with the *reflection* over the angle bisector of $\angle AEB$ and g that is the composition of *homothety* of center E and ratio CD/AB with the same *reflection*. Let $G = f(F)$ and $H = g(F)$ (Fig. 12.20). Then G, H are on the reflection of EF over the angle bisector of $\angle AEB$, and

$$GH = |EG - EH| = EF \cdot \left| \frac{AB}{CD} - \frac{CD}{AB} \right|.$$

The problem reduces to showing that $MN = GH/2$. Examining the figure, we notice that f maps the triangle FCD to the triangle GAB, so the triangle GAB is similar to the triangle FBA. But these two triangles share the side AB, so they are

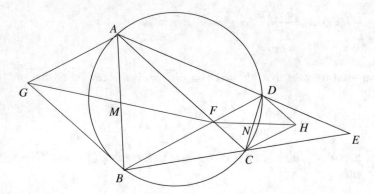

Fig. 12.20 Construction of F and G

Fig. 12.21 $F'AD$ and FBC
are directly similar

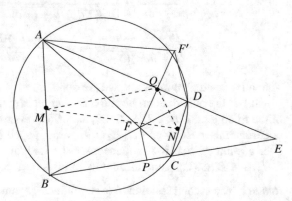

congruent. This implies that $GAFB$ is a parallelogram, so M is the midpoint of the diagonal FG. For a similar reason N is the midpoint of the segment FH, so MN is midline in the triangle FGH. Hence $MN = GH/2$, which proves (a).

(b) *First solution.* We use again *reflection* over a line, this time composed with a *spiral similarity*, and now we focus on the similar triangles FAD and FBC. The presence of midpoints hints to the use of the Averaging Principle (Theorem 2.18).

Let F' be the reflection of F over AD (Fig. 12.21). Now $F'AD$ and FBC are directly similar, so there is a spiral similarity that maps them into each other. Note that Q is the midpoint of FF', while M, N are the midpoints of BA and CD, respectively. By the Averaging Principle, the triangle QMN is similar to both $F'AD$ and FBC.

By using instead the reflection over BD, we infer that the triangle PMN is similar to both $F'AD$ and FBC. Consequently, the triangles QMN and PMN are similar, and since they share the side MN, they are congruent. They form a kite, so PQ is perpendicular to MN, which proves (b).

Second solution. We consider the system of coordinates inspired by a *homothety* and a *reflection* used in Problem 4.1 from the introduction to this chapter (with F at the origin). The equation of the line AD is

$$\begin{vmatrix} 1 & z & \overline{z} \\ 1 & 1 & 1 \\ 1 & -k\omega & -k\overline{\omega} \end{vmatrix} = 0,$$

that is

$$(1 + k\overline{\omega})z - (1 + k\omega)\overline{z} + k(\omega - \overline{\omega}) = 0.$$

The projection of the origin onto this line is obtained by intersecting this line with the one orthogonal to it passing through the origin, whose equation is $(1 + k\overline{\omega})z + (1 + k\omega)\overline{z} = 0$. Requiring both equations to be satisfied simultaneously yields a system of two equations in the unknowns z and $\overline{z}$, with the solution the coordinate of Q:

$$q = \frac{k(\overline{\omega} - \omega)}{2(1 + k\overline{\omega})}.$$

Similarly, the equation of the line BC is $(k + \overline{\omega})z - (k + \omega)\overline{z} + ka(\overline{\omega} - \omega) = 0$, and so the projection P of F onto the line BC has the coordinate

$$p = \frac{ka(\omega - \overline{\omega})}{2(k + \overline{\omega})} = \frac{ka\omega(\omega - \overline{\omega})}{2(1 + k\omega)}.$$

The coordinates of M and N are, respectively,

$$m = \frac{1 + a\omega}{2}, \quad n = \frac{-ka - k\omega}{2}.$$

Then

$$p - q = \frac{k(\omega - \overline{\omega})}{4} \cdot \frac{a\omega + ak + 1 + k\omega}{(1 + k\omega)(1 + k\overline{\omega})} = \frac{k(\omega - \overline{\omega})}{2(1 + k\omega)(1 + k\overline{\omega})}(m - n).$$

In the fraction the numerator is imaginary, while the denominator is real, so the fraction itself is imaginary. That is $(p - q)/(m - n)$ is imaginary, proving that PQ is perpendicular to MN.

Remark This solution presents three ways in which, in a cyclic quadrilateral, the two similar triangles formed by diagonals and sides can be mapped into each other:

(i) by a direct homothety with center at the intersection of two sides followed by a reflection over the bisector of the angle formed by those two sides (solution to (a));

(ii) by the reflection over a side followed by a spiral similarity (first solution to (b));

(iii) by an inverse homothety centered at the intersection of the diagonals followed
       by a reflection over the bisector of the angle formed by the diagonals (second
       solution to (b)).

Source Part (a) was given at the Bulgarian Mathematical Olympiad in 1997. Part
(b) was given at Team Selection Test in the United States of America in 2000. First
solution to (b) from T. Andreescu, M. Rolínek, J. Tkadlec, *Geometry Problems from
the AwesomeMath Year-Round Program*, XYZ Press, 2013.

237 The solution can be followed on Fig. 12.22. The triangle DAI *rotates* by 90°
to the triangle DLC, so $AI \perp LC$, and for similar reasons $BK \perp FD$, $CE \perp HA$, and
$DG \perp JB$. The quadrilaterals $P_1 Q_1 R_1 S_1$ and $P_2 Q_2 R_2 S_2$ have orthogonal sides; if

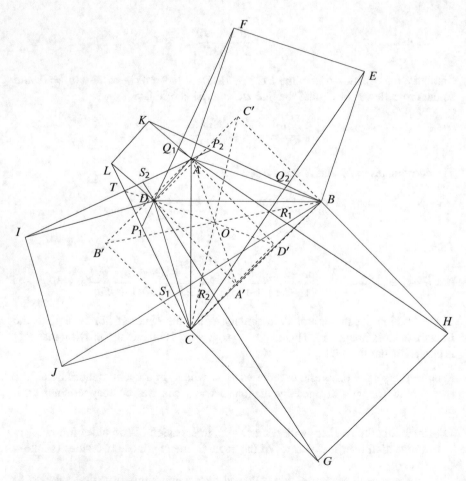

Fig. 12.22 Finding the center of the rotation that maps $P_1 Q_1 R_1 S_1$ to $P_2 Q_2 R_2 S_2$

they are to be equal, they must map into each other by some 90° rotation ρ. We want to identify the center O of ρ.

We expect that $\rho(AI) = LC$, $\rho(BK) = FD$, $\rho(CE) = HA$, and $\rho(DG) = JB$. So the center of rotation must be at the intersection of the bisectors of the angles formed by AI and LC, BK and FD, CE and HA, and DG and JB. If T is the intersection of AI with CL, then TD is the angle bisector of $\angle ATC$ (as DAI rotates to DLC), and so T, D, O have to be collinear. A similar situation happens at each vertex of $ABCD$, the four bisectors are the lines that connect the intersection points to the corresponding vertices of $ABCD$, and they should pass through O.

But are the four angle bisectors concurrent? This is where the orthogonality of the diagonals of $ABCD$ comes into play. Construct two squares $AB'CD'$ and $A'BC'D$ whose orientations are opposite to that of $ABCD$ (Fig. 12.22). Because AC is perpendicular to BD, the squares have parallel sides, and as a consequence of Problem 84, they are *homothetic*. This implies that AA', BB', CC', DD' intersect at a certain point O. As the figure suggests, O is the center of the rotation, and we will prove it.

The quadrilateral $AD'CT$ is cyclic, having two opposite right angles, which implies that $\angle CTD' = \angle CAD' = 45°$, and consequently the points T, D, D' are collinear. It follows that O lies on the angle bisector of $\angle ATC$. The same is true for the other angles, and so there is a rotation ρ about O by 90° that takes those pairs of lines into each other, as desired. This rotation ρ maps $P_2 Q_2 R_2 S_2$ into $P_1 Q_1 R_1 S_1$, and we are done.

Source Short list of the 36th International Mathematical Olympiad, 1995, proposed by Germany, solution from D. Djukić, V. Janković, I. Matić, N. Petrović, *The IMO Compendium, A Collection of Problems Suggested for the International Mathematical Olympiads: 1959–2004*, Springer, 2006.

238 *First solution.* Consider a *spiral similarity* that takes the triangle ABO to the triangle AOQ'. We will prove that $Q = Q'$, with the help of Fig. 12.23.

We have $\angle Q'AO = \angle OAB$ and $\angle Q'OA = \angle OBA$. But $\angle DOA = \angle OAB + \angle OBA$ (by the Exterior Angle Theorem), hence $\angle Q'AO = \angle Q'OD$. Also, because of the spiral similarity, $AQ'/AO = Q'O/OB$. But the later equals $Q'O/OD$, and we conclude that the triangles $AQ'O$ and $OQ'D$ are similar, so $\angle Q'DO = \angle Q'OA = \angle ODC$. It follows that $Q = Q'$ and hence Q is the center of the spiral similarity that maps triangle QAO to triangle QOD. In particular it is the center of the spiral similarity that maps AO to OD.

Second solution. Let B' be the *reflection* of B over the line AC (Fig. 12.24). Note first that AC is midline in the triangle BDB', so AC is parallel to DB'. Also, because AB' is the reflection of AB over AC, it follows that $Q \in AB'$. And because reflection preserves angles, we have that

$$\angle AB'O = \angle ABO = \angle BDC = \angle BDQ.$$

Consequently, the quadrilateral $DQOB'$ is cyclic, hence

Fig. 12.23 Q is the center of
the spiral similarity, proved
by... spiral similarity

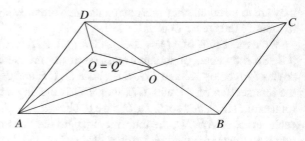

Fig. 12.24 Q is the center of
the spiral similarity, proved
by reflection

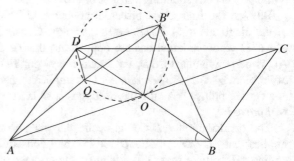

$$\angle DOQ = \angle DB'Q = \angle B'AC = \angle QAO.$$

Applying the Exterior Angle Theorem in the triangle AOB gives $\angle AOD = \angle ABO + \angle BAO$, hence

$$\angle AOQ = \angle AOD - \angle DOQ = \angle ABO = \angle BDQ.$$

It follows that the triangles QAO and QOD are similar (having two pairs of equal angles), and hence they are mapped into each other by a spiral similarity of center Q. This spiral similarity maps AO to OD, and we are done.

   Third solution. This approach uses symmedians; thus it is logically related to *isometries*. Let Q' be the intersection of the circle through A tangent to BD at O with the circle through D tangent to AC at O (Fig. 12.25). Problem 132 shows that OQ' is symmedian in the triangle OAD. Note that if M is the midpoint of AD, then OM is parallel to AB. Thus

$$\angle BAO = \angle MOA = \angle Q'OD = \angle Q'AO,$$

where the last equality follows from inscribed angles. By a similar argument

$$\angle CDO = \angle MOD = \angle Q'OA = \angle Q'DO.$$

The equalities $\angle BAO = \angle Q'AO$ and $\angle CDO = \angle Q'DO$ imply $Q = Q'$. The equalities $\angle QOD = \angle QAO$ and $\angle QOA = \angle QDO$ imply that the triangles QAO

Fig. 12.25 Proof that Q is the center of the spiral similarity using symmedians

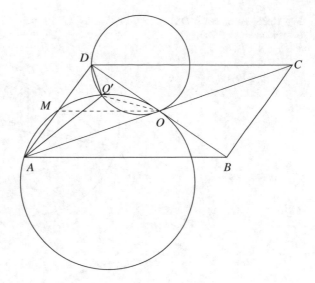

and QOD are similar, so they are mapped into each other by a spiral similarity centered at Q. The conclusion follows.

Remark The third solution is probably the most natural given the discussion in Sect. 2.4.3.

Source The second and third solutions were found by the students of the Brazilian Mathematical Olympiad Program.

239 *First solution.* Let N be the midpoint of AD. The problem asks us to show that MN is perpendicular to BC. We argue on Fig. 12.26.

Let A' be the point that is diametrically opposite to A in the circle of center M and radius MA. Then, in the triangle $AA'D$, the segment MN is midline, so it is parallel to $A'D$. It remains to prove that $A'D$ is perpendicular to BC.

We begin by showing that the triangles $C'CA'$ and $C'MD$ are similar. We have $\angle CC'A' = \angle MC'D = 90°$, the first because AA' is diameter, the second because MC' is radius and $C'D$ is tangent. Then $\angle C'MD$ is half of $\angle C'MB'$ and so it measures half of the arc $\overset{\frown}{B'C'}$. Thus $\angle C'MD = \angle BAC$. Also, $ABA'C$ is a parallelogram because M is the midpoint of both BC and AA', so $\angle C'CA' = \angle BAC$. This shows that $\angle C'CA' = \angle C'MD$. It follows that the triangles $C'CA$ and $C'MD$ have two pairs of equal angles and so they are similar, as claimed. From here we deduce that there is a *spiral similarity* of center C' mapping the triangle $C'CA'$ to the triangle $C'MD$. By Theorem 2.25 (spiral similarities come in pairs), there is a spiral similarity of center C' that maps the triangle $C'CM$ to the triangle $C'A'D$. Because $\angle(MC', C'D) = \angle MC'D = 90°$, the angle of the spiral similarity is $90°$. It follows that $A'D$ is perpendicular to CM, and we are done.

Fig. 12.26 Proof that $A'D$ is
perpendicular to BC

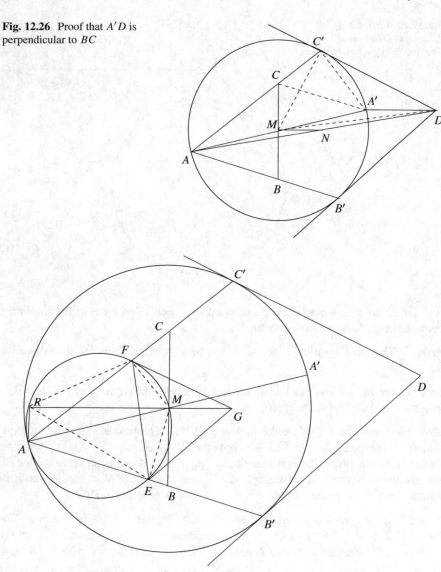

Fig. 12.27 Proof that G is on the perpendicular to BC

Second solution. Consider the *homothety* of center A and ratio $1/2$. Let E, F, G
be the images of B', C', D, respectively. Note that G is the intersection of the
tangents at E and F to the circumcircle of triangle AEF, while M, the image of A',
is on this circumcircle. It suffices to prove that G is on the perpendicular bisector of
the side BC. We argue on Fig. 12.27.

Let R be on the circumcircle of AEF such that AR is parallel to BC. Then the
pencil AB, AC, AM, AR is harmonic, because AR is parallel to BC and AM runs

through the midpoint of the segment BC, so using the line BC as the secant to this pencil, we compute $MB/MC : \infty B/\infty C = -1$. From Lemma 4.3 it follows that the cyclic quadrilateral $EMFR$ is harmonic. So the tangents at E and F to the circumcircle intersect on MR, that is, $G \in MR$ (see Sect. 1.2.1). But since AM is a diameter in the circumcircle of AEF, $\angle ARM = 90°$. We deduce that $MR = MG$ is orthogonal to BC, and we are done.

Source The Competition of the "Simion Stoilow" Institute of Mathematics of the Romanian Academy, 2014, second solution by Ştefan Rareş Tudose.

240 The centers of three circles passing through the same point I and not tangent to one another are collinear if and only if the circles have a second common point. It is therefore enough to show that the circumcircles of $A_i B_i I$ have a second point of intersection.

To simplify the problem, apply an *inversion* of center I. With the convention of denoting the image of an object by adding a dash to its name, C_i' is the line $B_{i+1}' B_{i+2}'$ while the image of the line $A_{i+1} A_{i+2}$ is the circumcircle of $I A_{i+1}' A_{i+2}'$, which is tangent to $B_i' B_{i+1}'$ and $B_i' B_{i+2}'$ (the configuration before and after inversion is shown in Fig. 12.28). As I is at equal distance from the three sides of $A_1 A_2 A_3$, these three circles have equal radii. So their centers O_1, O_2, O_3 form a triangle that has sides parallel to, and therefore is *homothetic* to, $B_1' B_2' B_3'$.

The point I is the circumcenter of $O_1 O_2 O_3$, so its *reflections* over the sides of this triangle, which are A_1', A_2', A_3', form a triangle that is homothetic to the median triangle of $O_1 O_2 O_3$ and hence to $O_1 O_2 O_3$ itself. Composing homotheties we conclude that the triangles $A_1' A_2' A_3'$ and $B_1' B_2' B_3'$ are homothetic. So the lines $A_1' B_1'$, $A_2' B_2'$, $A_3' B_3'$ intersect at one point, which is the image through the inversion of the second intersection point of the circumcircles of $A_1 B_1 I$, $A_2 B_2 I$, $A_3 B_3 I$. The problem is solved.

Remark The inverted figure is the configuration from Tzitzeica's Five-Lei Coin Problem (Theorem 3.38). The observation that $A_1' A_2' A_3'$ is homothetic to the median

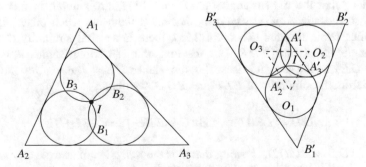

Fig. 12.28 The image of $A_1 A_2 A_3$ under the inversion centered at I

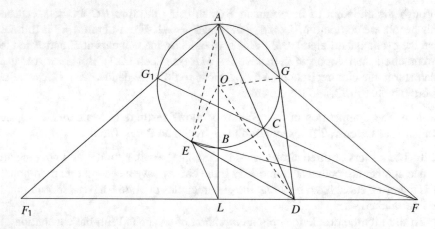

Fig. 12.29 Proof that CF intersects Γ at the reflection of G over AB

triangle of $O_1 O_2 O_3$ and so it is mapped into $O_1 O_2 O_3$ by a homothety of ratio -1 gives another solution to Tzitzeica's problem.

Source Short list of the International Mathematical Olympiad, 1997, proposed by the United States of America.

241 Let O be the center of Γ and let L be the intersection point of the lines AB and ℓ (Fig. 12.29). Then $\angle AEF = 90°$ because AB is diameter, so $\angle AEF = \angle ALF$. This shows that the quadrilateral $AELF$ is cyclic.

Also $\angle OED = 90°$ because ED is tangent to Γ, and so $\angle OED = \angle OLD = 90°$, showing that $OELD$ is cyclic. Using the cyclic quadrilaterals $OELD$ and $AELF$, we can write

$$\angle ODE = \angle OLE = \angle ALE = \angle AFE.$$

We deduce that the right triangles EAF and EOD are similar, in fact directly similar, so there is a *spiral similarity* that maps them into each other. As spiral similarities come in pairs (Theorem 2.25), there is a spiral similarity that maps the triangle EOA into EDF. But the first of these two triangles is isosceles ($OA = OE$), so the second is isosceles, too. Hence $DE = DF$.

Also from the similarity of EAF and EOD, we obtain

$$\angle EOD = \angle EAF = \angle EAG = \overset{\frown}{EBG}/2 = \angle EOG/2.$$

Hence $\angle EOD = \angle GOD$, showing that G is the *reflection* of E over OD (because Γ is invariant under this reflection, and reflection preserves intersections). So DG is the other tangent from D to the circle, $DG = DE = DF$, and hence $\angle DFG = \angle DGF$.

Let G_1 be the other intersection point of the line CF with Γ, and let F_1 be the intersection of the line AG_1 with ℓ. Consider the *inversion* χ of center A that maps Γ to ℓ. Then $\chi(C) = D$, $\chi(F) = G$, $\chi(G_1) = F_1$, so the image of the line CF is a circle passing through A and also through the points D, G, F_1. It follows that AF_1DG is cyclic, so $\angle AF_1D = \angle DGF$. But the latter is equal to $\angle DFG$. Hence F_1 is the *reflection* of F over the line through A perpendicular to ℓ, and this line is AB. As Γ is invariant under this reflection, the intersections G and G_1 of AF and AF_1 with Γ are images of each other under the reflection over AB, and the problem is solved.

Remark The fact that the diameter AB is perpendicular to ℓ tells that there is an inversion of center A that maps Γ to ℓ. The collinearity of C, F, G_1 is equivalent to the concyclicity of their images and the center of inversion.

Source Moldovan Team Selection Test for the International Mathematical Olympiad, 2005, solution by Liubomir Chiriac.

242 The situation is described in Fig. 12.30. We have three circles passing through Q that we are supposed to show intersect a second time. Certainly it is easier to check that three lines intersect, so it reasonable to use an inversion of center Q. But that inversion would alter too much of the original figure. We would be better off with an "inversion" that preserves most of the figure, especially one that keeps P and R in place. For this reason we use the $\sqrt{bc}$ *inversion* ϕ_Q determined by the triangle

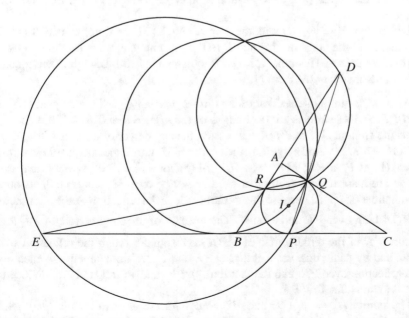

Fig. 12.30 The circumcircles of QEP, QDR, and QBI

Fig. 12.31 The images of
B, D, E through the $\sqrt{bc}$
inversion

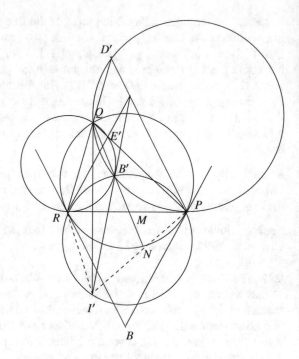

QRP and the vertex Q. By Proposition 3.40 (v), $I' = \phi_Q(I)$ is the reflection of Q
over PR.

Let us find $B' = \phi_Q(B)$ with the help of Fig. 12.31. First notice that B is at the
intersection of the tangents at P and R to the circumcircle of the triangle PQR, so,
as a consequence of Theorem 1.22, QB is symmedian in this triangle. Consequently,
B' is on the median QM from Q.

Also, B' is at the intersection of the circles through Q that are tangent at P and
R to PR, as these circles are the images of the sides AB and BC. If the line QB'
meets the circumcircle of PQR again at N, then by power-of-a-point $MN \cdot MQ =
MP \cdot MR = MP^2$. And by writing the power of M with respect to the circle through
Q tangent at P to PR, $MP^2 = MB' \cdot MQ$ shows that $MB' = MN$, that is, B'
and N are mapped into each other by a *reflection* over M. But then B' is on the
circumcircle of $I'PR$, because this circumcircle is the reflection of the circumcircle
of PQR both over PR and over M. The arc $\overset{\frown}{NP}$ of the circumcircle of PQR and
the arc $\overset{\frown}{B'R}$ of the circumcircle of $I'PR$ correspond through the reflection so are
equal, and by using this, the fact that QPR and $I'PR$ are mapped into each other
by a reflection over PR, and the fact that QN is median in QPR, we deduce that
$I'B'$ is symmedian in $I'PR$.

The points $D' = \phi_Q(D)$ and $E' = \phi_Q(E)$ are easy to locate; they are the
intersections of the circles through Q tangent to PR at P and R with the lines RQ
and PQ, respectively. If we examine carefully Fig. 12.31, we notice that the lines

PD' and RE' seem tangent to the circumcircle of $PI'R$. This would be the case if in the original figure the circumcircle of PIR were tangent to the circumcircles of both PEQ and RDQ. And this can be checked easily by angle chasing on Fig. 12.30: the angle formed by the tangent to the circumcircle of PIR at P with PR is $\angle PIR$, so the angle it makes with PE is $\angle PIR - \angle BPR = \angle PIR/2$. But the angle made by the tangent to the circumcircle of PQE at P with PE is equal to $\angle PQR$, which is $\angle PIR/2$. So the two tangents make the same angle with PE; thus they coincide, and the circles are tangent. The same argument works for checking the tangency of the circumcircles of PIR and QRD.

By Theorem 1.22, the symmedian $I'B'$ passes through the intersection point of the tangents PE' and RD'. The preimages of the three lines, which are the three circles in question, have therefore a second intersection point.

Remark The property remains true if we replace the incircle with the excircle corresponding to the vertex B and I with the center I_b of this excircle.

243 *First solution.* Add to the original configuration the point B where Γ and t intersect the second time. Then take an *inversion* of center R, which solves the problem through a "position argument." To see how this works, with the usual convention that X' is the image of X, let us examine the inverted configuration as depicted in Fig. 12.32.

First, since in the original configuration t is tangent to Ω, and so they form an angle of $0°$, in the inverted figure the line RA' and the line through J', S', K' also form an angle of $0°$, they are parallel. This means that $A'B'S'J'$ is a trapezoid

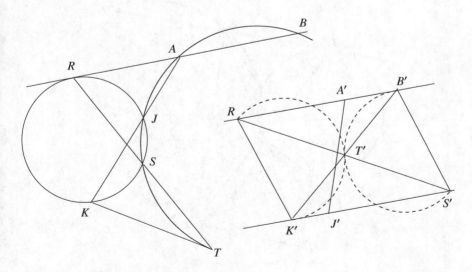

Fig. 12.32 Proof by inversion that KT is tangent to Γ

$(A'B'||S'J')$, but this trapezoid is inscribed in the circle that is the image of Γ, and so it must be an isosceles trapezoid.

Similarly $RA'J'K'$ is a trapezoid ($RA'||J'K'$), and since line the line through A, J, K is mapped by the inversion to a circle passing through R, the trapezoid $RA'J'K'$ is also cyclic, hence isosceles.

Joining the two isosceles trapezoids, we obtain the parallelogram $RB'S'K'$. Because S is the midpoint of RT, T' is the midpoint of RS' and thus also the midpoint of the other diagonal, $B'K'$, of the parallelogram. This means that a *reflection* over T' maps the triangle $T'B'S'$ to the triangle $T'K'R$, and so it maps the circumcircle of the first to the circumcircle of the second. In particular the two circumcircles are tangent. In the original figure, the preimage of the circumcircle of $T'B'S'$ is the circumcircle of TBS, which is Γ, while the preimage of the circumcircle of $T'K'R$ is the line TK. So TK and Γ are tangent, and we are done.

Second solution. An approach based entirely on a reflection is also possible and can be followed on Fig. 12.33. Chasing angles in the cyclic quadrilaterals $RJSK$ and $AJST$, we obtain

$$\angle KRS = \angle KJS = \angle ATS,$$

which implies that RK is parallel to AT. Now *reflect* A over S to obtain the point A'. Then $ARA'T$ is a parallelogram with center S, RA' is parallel to AT, so K is on the line RA'.

From

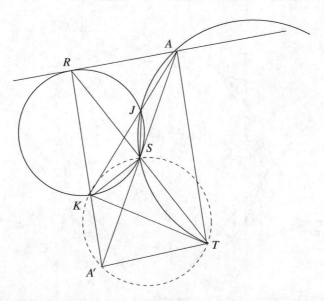

Fig. 12.33 Proof by reflection that KT is tangent to Γ

$$\angle STA' = \angle SRA = \widehat{SJR}\,/2 = \angle SKR$$

we obtain that the points S, K, A', T are concyclic. This implies that

$$\angle STK = \angle SA'K = \angle SAT,$$

which shows that KT is tangent to Γ at T.

Remark If we invert with respect to a circle centered at S instead, we obtain an analogous version of the problem. Another version of the second solution can be obtained by reflecting K over S. The second approach shows how a geometric transformation can contribute to a solution without necessarily dominating it.

Source International Mathematical Olympiad, 2017, proposed by Charles Leytem, Luxembourg.

244 First, the fact that the lines AA', BB', and CC' intersect at one point is a classical fact; they are the radical axes of the circles ω_A, ω_B, ω_C, so they intersect at the radical center. The difficulty lies in identifying this radical center and showing that it lies on IO.

The solution can be followed on Fig. 12.34. Let γ be the incircle of the triangle ABC and let A_1, B_1, C_1 be its contact points with the sides BC, CA, AB, respectively. Let X_a be the contact point of the circles γ and ω_A. Let M_A be the point where $X_A A_1$ intersects the circumcircle. Then, as seen in Problem 180 (a) and its solution, M_A is the midpoint of the arc $\widehat{BC}$ not containing X_A, and A_1

Fig. 12.34 Proof that AA', BB', CC' meet on OI

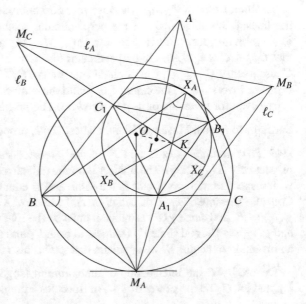

is mapped to M_A by the *inversion* of center M_A and radius $M_A B$. The equality $M_A B^2 = M_A A_1 \cdot M_A X_A$ shows that M_A is on the radical axis ℓ_B of (the degenerate circle) B and γ. A similar argument shows that M_A is on the radical axis ℓ_C of C and γ.

Define the points X_B, X_C, M_B, M_C and the line ℓ_A similarly. Note that $\ell_A = M_B M_C$, $\ell_B = M_A M_C$, and $\ell_C = M_A M_B$. Also, the fact that ℓ_A is the radical axis of A and γ implies that ℓ_A is perpendicular to AI and hence parallel to $B_1 C_1$. Similarly ℓ_B is parallel to $A_1 C_1$ and ℓ_C is parallel to $A_1 B_1$. Consequently, there is a *homothety* h that maps the triangle $A_1 B_1 C_1$ to $M_A M_B M_C$; let K be the center of this homothety. Then $M_A A_1$, $M_B B_1$, and $M_C C_1$ meet at K.

Since the points A_1, B_1, X_A, X_B are concyclic, $A_1 B_1$ and $X_A X_B$ are antiparallel. And because $A_1 B_1$ and $M_A M_B$ are parallel, $M_A M_B$ and $X_A X_B$ are antiparallel; thus X_A, X_B, M_A, M_B are concyclic. Consequently $M_A K \cdot K X_A = M_B K \cdot K X_B$ (which is the power of K with respect to the circle passing through these points). But, by symmetry, these products are also equal to $M_C K \cdot K X_C$. The equality

$$M_A K \cdot K X_A = M_B K \cdot K X_B = M_C K \cdot K X_C$$

shows that K is the radical center of $\omega_A, \omega_B, \omega_C$. We have identified the intersection of AA', BB', CC'!

To prove that $K \in IO$, consider the *circular transformation* χ that is the composition of the inversion with center K and radius $\sqrt{M_A K \cdot K X_A}$ with the reflection over K (the inversion of negative power). The circles $\omega_A, \omega_B, \omega_C$ are invariant under this transformation, from which it follows that $\chi(\gamma)$ is the circumcircle of $M_A M_B M_C$. This circle must be tangent to ω_A, ω_B, and ω_C because circular transformations preserve tangencies (Theorem 3.21). Let O' be the circumcenter of $M_A M_B M_C$. Because this circumcircle is tangent to $\omega_A, \omega_B, \omega_C$, the lines OM_A, OM_B, OM_C are perpendicular to the tangents at M_A, M_B, M_C to $\omega_A, \omega_B, \omega_C$, respectively, and since M_A, M_B, M_C are the midpoints, it follows that OM_A, OM_B, OM_C are perpendicular bisectors of BC, AC, AB, respectively. Consequently $O' = O$, the circumcenter of ABC. Finally, by Theorem 3.5, the center of inversion, K, the center I of γ and the center O of $\chi(\gamma)$ are collinear (this is true even for inversion of negative power). The problem is solved.

Source Romanian Master of Mathematics, 2012, proposed by Fedor Ivlev.

245 *First solution.* Let E and F be the intersections of Γ_2 with AN and BN, respectively (Fig. 12.35). The point A is on the radical axis of Γ_1 and Γ_2, so its power with respect to the two circles is the same and is equal to $AC \cdot AM = AE \cdot AN$. Consider the *inversion* χ of center A and radius $\sqrt{AE \cdot AN}$. Then $\chi(M) = C$, $\chi(N) = E$, and since $\chi(\Gamma)$ is a line, it must be the line CE. Moreover, $\chi(\Gamma_1) = \Gamma_1$ and $\chi(\Gamma_2) = \Gamma_2$, and since Γ is tangent to both Γ_1 and Γ_2, so is $\chi(\Gamma) = CE$. Using an inversion of center B, we conclude that DF is also tangent to both Γ_1 and Γ_2.

Denote by P the intersection of the common tangents CE and DF of Γ_1 and Γ_2, and let O_2 be the center of Γ_2. An isosceles triangle arises in the configuration

Fig. 12.35 Proof by
inversion that CD is tangent
to Γ_2

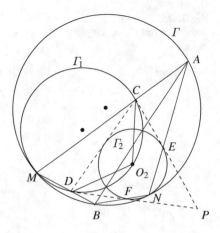

PCD, and we claim that O_2 is its incenter. Indeed, the circle Γ_1 is tangent to PC
and PD; thus PC reflects to PD over PO_2, and so $\overset{\frown}{CO_2}{=}\overset{\frown}{DO_2}$. Using inscribed
angles in this circle, we have

$$\angle PCO_2 = \overset{\frown}{CO_2}/2 = \overset{\frown}{DO_2}/2 = \angle DCO_2,$$

showing that CO_2 is the angle bisector of $\angle PCD$. Similarly, DO_2 is angle bisector
of $\angle PDC$. The claim is proved. Because Γ_2 is tangent to PC and PD and it is
centered at the incenter O_2, it follows that Γ_2 is the incircle of the triangle PCD;
hence CD is tangent to Γ_2, as desired.

Second solution. The second solution can be followed on Fig. 12.36. We denote
by O, O_1, O_2 the centers of Γ, Γ_1, Γ_2, respectively, by X and Y the intersections
of AB with the two circles, with X closer to A, and by R the intersection of O_1O_2
with XY. We also let r, r_1, r_2 be the radii of Γ, Γ_1, Γ_2, respectively. Also, given a
point T and a line ℓ, we denote by $d(T, \ell)$ the distance from T to ℓ. It suffices to
show that the $d(O_2, CD) = r_2$.

Consider the *homothety* of center M and ratio r/r_2 takes Γ_1 to Γ. This homothety
maps O_1 to O, and so M, O, O_1 are collinear. It also maps C to A and D to B, so
CD is parallel to AB and $d(M, CD)/d(M, AB) = r_1/r$. Consequently

$$d(C, AB) = d(M, AB) - d(M, CD) = d(M, AB) - \frac{r_1}{r}d(M, AB)$$

$$= \frac{r - r_1}{r}d(M, AB).$$

But the distance from M to AB is $O_1R = O_1O_2 - O_2R$ plus the distance from M to
the parallel through O_1 to AB. The latter is $OM \cos \angle OO_1O_2 = r_1 \cos \angle OO_1O_2$
(since M, O_1, O are collinear). It follows that

Fig. 12.36 Proof by
homothety that CD is tangent
to Γ_2

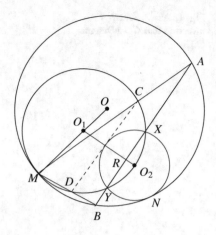

$$d(O_2, CD) = d(O_2, AB) + d(C, AB) = O_2R + \frac{r - r_1}{r} d(M, AB)$$

$$= O_2R + \frac{r - r_1}{r} |O_1O_2 - O_2R + r_1 \cos \angle OO_1O_2|.$$

We can compute the value of the cosine that appears in this expression using the
Law of Cosines in the triangle OO_1O_2. Because $OO_1 = r - r_1$, $OO_2 = r - r_2$,
and $O_1O_2 = r_1$, we have

$$\cos \angle OO_1O_2 = \frac{r_1^2 + (r - r_1)^2 - (r - r_2)^2}{2r_1(r - r_1)} = \frac{2r_1^2 - 2rr_1 + 2rr_2 - r_2^2}{2r_1(r - r_1)}.$$

Also, in the triangle XO_1O_2, we have $O_1X = O_1O_2 = r_1$, and $O_2X = r_2$, so,
using the Law of Cosines, we can compute

$$O_2R = O_2X \cos \angle O_1O_2X = r_2 \frac{r_1^2 + r_2^2 - r_1^2}{2r_1r_2} = \frac{r_2^2}{2r_1}.$$

We therefore have

$$d(O_2, CD) = \frac{r_2^2}{2r_1} + \frac{r - r_1}{r} \left| r_1 - \frac{r_2^2}{2r_1} + \frac{2r_1^3 - 2rr_1^2 + 2rr_1r_2 - r_1r_2^2}{2r_1(r - r_1)} \right|$$

$$= \frac{r_2^2}{2r_1} + \frac{r - r_1}{r} \cdot \frac{2r_1r_2r - rr_2^2}{2r_1(r - r_1)} = \frac{r_2^2}{2r_1} + \frac{2r_2r_1 - r_2^2}{2r_1} = r_2.$$

The problem is solved.

Remark The first solution relies on the following fact, which can prove useful in
other situations: if ABC is an isosceles triangle, and ω is a circle tangent to AB

and AC at B and C, respectively, then ω passes through the incenter of the triangle ABC.

Source International Mathematical Olympiad, 1999, proposed by Russia, first solution communicated by Leandro Maia, second solution from D. Djukić, V. Janković, I. Matić, N. Petrović, *The IMO Compendium*, Springer, 2006.

246 We should first locate O_1 and O_2 in a more practical way. To this end, let L_a and L_b be the midpoints of the arcs $\overset{\frown}{BC}$ and $\overset{\frown}{AC}$ that do not contain the other vertex. Let γ_a be the circle with center L_a that passes through B and C; define γ_b in the same fashion. Then O_1 is the other intersection point of XL_b and γ_b, and O_2 is the other intersection point of XL_a and γ_a (Fig. 12.37). Also, notice that γ_a and γ_b meet at incenter I of ABC, and, by Theorem 2.22, the intersection T of the circumcircles of XO_1O_2 and ABC is the center of the *spiral similarity* that takes O_1 to O_2 and L_b to L_a. As spiral similarities come in pairs, this is also the center of the spiral similarity that maps O_1 to L_b and O_2 to L_a.

There are far too many circles in the diagram; let us turn them into lines: perform a $\sqrt{bc}$ *inversion* determined by the triangle ABC and the vertex C. Then I is taken to the center I_c of the C-excircle, and γ_a, which passes through C, I, and B, is taken to the line AI_c. Therefore the center L_a of γ_a is taken to the reflection of C across the line γ_a' (this is a consequence of Proposition 3.7). Since L_a lies on the circumcircle Ω of ABC, L_a' is on its image Ω', which is the line BC (see Fig. 12.38). Similar properties hold for L_b'.

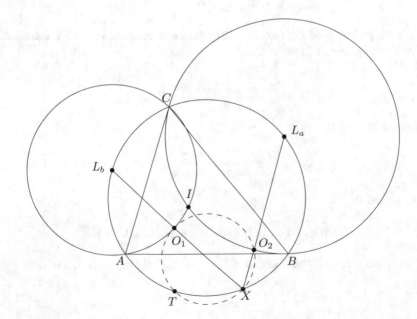

Fig. 12.37 Spiral similarity?

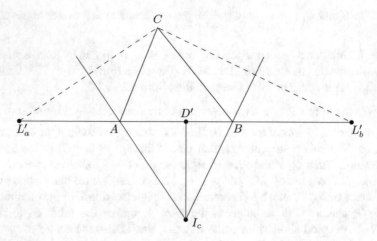

Fig. 12.38 Inverting the diagram

We infer that AI_c is the perpendicular bisector of CL'_a and BI_c is the perpendicular bisector of CL'_b. This means that I_c is the circumcenter of the triangle $CL'_aL'_b$, which means that its projection D' onto BC satisfies $D'L'_a = D'L'_b$, in other words, in complex coordinates,

$$\frac{l'_a - d'}{l'_b - d'} : \frac{l'_a - \infty}{l'_b - \infty} = -1.$$

If D is the preimage of D', then by using the invariance of real valued cross-ratio under inversion, we find that

$$\frac{l_a - d}{l_b - d} : \frac{l_a - c}{l_b - c} = -1,$$

which can be rewritten as

$$\frac{l_a - d}{l_b - d} = -\frac{l_a - c}{l_b - c}.$$

By inscribed angles

$$\angle CL_bO_1 + \angle CL_aO_2 = \angle CL_bX + \angle CL_aX = 180°,$$

and combining this with the fact that $CL_a = O_2L_a$ and $CL_b = O_2L_b$, we can write

$$\frac{l_a - c}{l_b - c} = -\frac{l_a - o_2}{l_b - o_1}.$$

Consequently

$$\frac{l_a - d}{l_b - d} = \frac{l_a - o_2}{l_b - o_1} \iff \frac{d - l_a}{o_2 - l_a} = \frac{d - l_b}{o_1 - l_b}.$$

We interpret this equality as saying that the spiral similarity of center L_a that maps O_2 to D and the spiral similarity of center L_b that maps O_1 to D have the same ratio and angle. But this means that the triangles DL_aO_2 and DL_bO_1 are directly similar. This implies that D is the center of the *spiral similarity* that takes O_2 to L_a and O_1 to L_b. This means that $D = T$, and we are done, because D is fixed when X varies.

Remark The point $D = T$ is the tangency point between Ω and the C-mixtilinear incircle. The solution seems to be a bit tricky because it looks like we have guessed the point D; however, midpoints in inversion usually lead to equal ratios, which interact well with spiral similarities. So in this sense, the point D' is quite natural. You can try to rewrite in geometric terms our complex number manipulations, in order to see that they are less mysterious than they look.

Source Short list of the International Mathematical Olympiad, 1999, proposed by Russia.

247 We *invert* about the *incircle* ω, because this circle is the focus of the problem, and use dashes for the inverses. Let ω touch the sides BC, CA, and AB at the points D, E, and F, respectively. The proof to Euler's Theorem (Theorem 3.36) shows that the inverses of the vertices of the triangle ABC with respect to the incircle are the midpoints of the sides of the triangle DEF and that the inverse of the circumcircle of ABC is the nine-point circle of DEF (Fig. 12.39).

Notice that ω is the M-excircle of the triangle MK_1K_2. Let ω touch MK_1 and MK_2 at T_1 and T_2, respectively. Then, by Proposition 3.3, K_1' is the midpoint of DT_1, K_2' is the midpoint of DT_2, and M' is the midpoint of T_1T_2. That is, the

Fig. 12.39 Nine-point circles

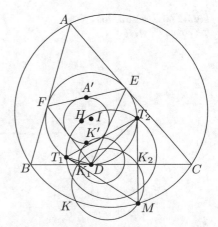

circumcircle of MK_1K_2 is mapped to the nine-point circle γ of DT_1T_2. Notice that the circumcircle of DT_1T_2 is ω. It is tempting to consider the *homothety* between the two circles, but we will divide this homothety into steps.

First, consider the homothety with center D and ratio 2, which maps γ to the circle γ^* that passes through T_1 and T_2 and has the same radius as ω. This circle does not coincide with ω because D is not on γ. So γ^* is the *reflection* of ω across T_1T_2. This means that γ^* passes through the orthocenter H of DEF. So γ passes through the midpoint K' of DH, which lies on the nine-point circle of DEF, that is, on the inverse of the circumcircle of ABC. Now, K' is diametrically opposite to A' on this nine-point circle, and by Problem 204, K is the tangency point of the circumcircle of ABC with the A-mixtilinear incircle. Done.

Source Taiwanese Team Selection Test for the International Mathematical Olympiad, 2014.

248 Let P and Q be the intersection points of the line FG with the sides AB and AC, respectively, and let I and J be the second intersection points of FG with the circumcircles of the triangles BDF and CEG, respectively, as shown in Fig. 12.40. Because $AF = AG$, AO is the perpendicular bisector of the segment FG. Note also that in the triangle ABC, the line AO and the altitude from A are isogonal, and so FG is antiparallel to BC, meaning that $\angle APQ = \angle ACB$.

Let Y be the intersection of BC and FG. Writing the power of the point Y with respect to the circles Ω and Γ, we obtain $YB \cdot YC = YF \cdot YG = YD \cdot YE$. This leads us to considering the *inversion* of center Y that maps Ω and Γ into themselves. This inversion maps F, B, and D to G, C, and E, respectively, so it maps the circumcircle of FBD to the circumcircle of GCE.

Since Y is the center of the inversion that maps the circumcircle of FBD to the circumcircle of GCE, it is also the center of a *homothety* that maps these circles into each other. Consequently the arcs $\overset{\frown}{BF}$ and $\overset{\frown}{EJ}$ have the same measure because they are mapped by the homothety into each other. We obtain

Fig. 12.40 Construction of
the points P, Q, I, J

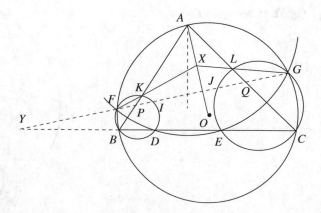

$$\angle XFG = \widehat{KI}\,/2 = \angle APQ - \widehat{BF}\,/2 = \angle ACB - \widehat{EJ}\,/2 = \widehat{LJ}\,/2 = \angle XGF.$$

Hence X is on the perpendicular bisector of segment FG, which is the line AO. The problem is solved.

Remark The key step of the solution was the interplay between inversion and homothety.

Source International Mathematical Olympiad, 2015, proposed by Greece.

249 *Invert* with respect to Γ. Since no reference will be made to the original configuration, the images will be denoted by the same letters (with no dash), except for G, the center of Γ, which will remain the center of inversion (and not the point at infinity). The inversion maps lines tangent to Γ to circles with diameters equal to the radius of Γ that are internally tangent to Γ and pass through G. After inversion, the problem becomes as follows:

Fix a circle Γ of center G and radius r, a circle ℓ of radius $r/2$ which passes through G, and a circle Ω inside ℓ and disjoint from ℓ. The circles ω_1 and ω_2 of radii $r/2$ pass through G and through a variable point X on Ω and cross again ℓ at Y and Z. Prove that as X traces Ω, the circumcircle of XYZ is tangent to two fixed circles.

Let's solve it! Since ω_1 and ω_2 are the *reflections* of the circumcircle ℓ of the triangle GYZ over the sides GY and GZ, respectively, they pass through the orthocenter of this triangle. The second intersection point of these circles is X, so X is the orthocenter of the triangle GYZ (Fig. 12.41). Hence the circumcircle ω_3 of the triangle XYZ is the reflection of ℓ over YZ; in particular the radius of ω_3 is $r/2$.

We continue our reasoning on Fig. 12.42. Let O and L be the centers of Ω and ℓ, respectively, and let R be the (variable) center of ω_3. The circle ℓ is mapped into ω_3 by a *translation* of vector $\overrightarrow{LR}$, and this translation maps G to X. Hence $GLRX$ is a parallelogram, which implies that we have the equality of vectors $\overrightarrow{XR} = \overrightarrow{GL}$, showing that $\overrightarrow{XR}$ is constant, and so R is the is image of X under a *translation* by the vector $\overrightarrow{GL}$.

We use this translation of vector $\overrightarrow{GL}$ to construct the point N as the translate of O. Then $XRNO$ is a parallelogram, so the distance $RN = OX$ is constant. Consequently ω_3 is tangent to the fixed circles centered at N and of radii $r/2 - OX$ and $r/2 + OX$.

But we are not done yet! We have to check that the inversion maps these two circles into circles and not lines. The latter happens only if one of the circles contains G, the center of inversion. Let us verify that this cannot happen. Since Ω lies inside ℓ, $OL < r/2 - OX$, so

$$NG = \|\overrightarrow{NG}\| = \|\overrightarrow{GL} + \overrightarrow{LO} + \overrightarrow{ON}\| = \|2\overrightarrow{GL} + \overrightarrow{LO}\|$$

$$\geq 2\|\overrightarrow{GL}\| - \|\overrightarrow{LO}\| > r - (r/2 - OX) = 1/2 + OX.$$

Fig. 12.41 Inverted figure

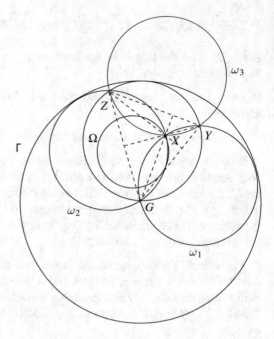

This shows that G is necessarily outside each of the two circles.

Remark At first glance, the inverted figure does not look simpler than the original. But the fact that the new X is the orthocenter of GYZ gives us a significant advantage.

We should point out that the required fixed circles are also tangent to Ω.

Source Romanian Master of Mathematics, 2018, proposed by Ivan Frolov, Russia.

250 We consider the system of coordinates inspired by a *homothety* and a *reflection* used in Problem 4.1 from the introduction to this chapter (with E at the origin). Let us show that the Euler lines of ABE, ADE, and BCE are concurrent (and then applying the argument to some other triple of triangles we obtain that all Euler lines are concurrent).

We use the formulas for the centroid and circumcenter of a triangles with vertices z_1, z_2, z_3:

$$g = \frac{z_1 + z_2 + z_3}{3}, \qquad o = \frac{\begin{vmatrix} 1 & 1 & 1 \\ z_1 & z_2 & z_3 \\ |z_1|^2 & |z_2|^2 & |z_3|^2 \end{vmatrix}}{\begin{vmatrix} 1 & 1 & 1 \\ z_1 & z_2 & z_3 \\ \overline{z_1} & \overline{z_2} & \overline{z_3} \end{vmatrix}}.$$

Fig. 12.42 Construction of
the two circles

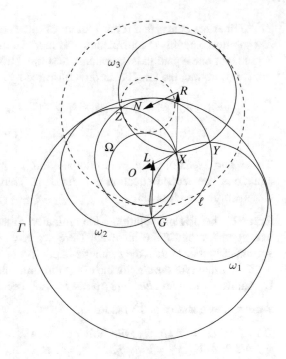

A short computation shows that the centroids and circumcenters of ABE, ADE, BCE are, respectively,

$$g_1 = \frac{1 + a\omega}{3}, \qquad o_1 = \frac{\omega - a}{\omega - \overline{\omega}}$$

$$g_2 = \frac{1 - k\omega}{3}, \qquad o_2 = \frac{k + \omega}{\omega - \overline{\omega}}$$

$$g_3 = \frac{a\omega - ak}{3}, \qquad o_3 = \frac{-ka\omega - a}{\omega - \overline{\omega}}.$$

Using the equation of a line through the points z_1, z_2:

$$\begin{vmatrix} 1 & 1 & 1 \\ z & z_1 & z_2 \\ \overline{z} & \overline{z_1} & \overline{z_2} \end{vmatrix} = 0,$$

after algebraic manipulations, we obtain the equations of the three Euler lines to be

$$[-a(\overline{\omega}^2 + 2) + \overline{\omega}(\omega^2 + 2)]z + [-a(\omega^2 + 2) + \omega(\overline{\omega}^2 + 2)]\overline{z} + (a^2 - 1)(\omega + \overline{\omega}) = 0$$

$$[k(\overline{\omega}^2 + 2) + \overline{\omega}(\omega^2 + 2)]z + [k(\omega^2 + 2) + \omega(\overline{\omega}^2 + 2)]\overline{z} + (k^2 - 1)(\omega + \overline{\omega}) = 0$$

$$[(\overline{\omega}^2 + 2) + k\overline{\omega}(\omega^2 + 2)]z + [(\omega^2 + 2) + k\omega(\overline{\omega}^2 + 2)]\overline{z} - a(k^2 - 1)(\omega + \overline{\omega}) = 0.$$

These lines are concurrent if this system of three equations in the unknowns z and $\bar{z}$ has solution, and this is equivalent to the fact that one of the equations is redundant. To see that one equation is redundant, add the third equation multiplied by a to the first, and subtract the third equation multiplied by k from the second to obtain

$$(ak + 1)[\bar{\omega}(\omega^2 + 2)z + \omega(\bar{\omega}^2 + 2)\bar{z}] - (ak + 1)(ak - 1)(\omega + \bar{\omega}) = 0$$

$$(1 - k^2)[\bar{\omega}(\omega^2 + 2)z + \omega(\bar{\omega}^2 + 2)\bar{z}] - (1 - k^2)(ak - 1)(\omega + \bar{\omega}) = 0.$$

These two equations are proportional, and so in the original system of three equations, the second equation is redundant. Thus the three lines are concurrent, and the problem is solved.

Remark The original solution communicated to us by the authors of the problem was synthetic and quite involved. Here we have an example where the analytic argument is shorter than the synthetic one.

The formula for the coordinate of the circumcenter can be found in T. Andreescu, D. Andrica, *Complex Numbers from A to ... Z*, second ed., Birkhäuser, 2014.

Source Petru Braica and Vlad Robu.

251 Let $T_n = A_n B_n C_n$ with altitudes $A_n A_{n+1}$, $B_n B_{n+1}$, and $C_n C_{n+1}$. Let also $\alpha_n = \angle B_n A_n C_n$, $\beta_n = \angle A_n B_n C_n$, $\gamma_n = \angle B_n C_n A_n$. Then we have the recursions

$$\alpha_{n+1} = \pi - 2\alpha_n, \quad \beta_{n+1} = \pi - 2\beta_n, \quad \gamma_{n+1} = \pi - 2\gamma_n \quad \text{if } A_n B_n C_n \text{ is acute,}$$

$$\alpha_{n+1} = 2\alpha_n - \pi, \quad \beta_{n+1} = 2\beta_n, \quad \gamma_{n+1} = 2\gamma_n \quad \text{if } \alpha_n > \frac{\pi}{2},$$

$$\alpha_{n+1} = 2\alpha_n, \quad \beta_{n+1} = 2\beta_n - \pi, \quad \gamma_{n+1} = 2\gamma_n \quad \text{if } \beta_n > \frac{\pi}{2},$$

$$\alpha_{n+1} = 2\alpha_n, \quad \beta_{n+1} = 2\beta_n, \quad \gamma_{n+1} = 2\gamma_n - \pi \quad \text{if } \gamma_n > \frac{\pi}{2}.$$

These recursions can be visualized using Fig. 12.43.

Because we only care about triangles up to a similarity, these recursions carry all the information of interest to us. So we may view H as the map

$$H : \Delta \to \Delta, \quad H(\alpha_n, \beta_n, \gamma_n) = (\alpha_{n+1}, \beta_{n+1}, \gamma_{n+1}),$$

where $\Delta \subset (0, \infty)^3$ consists of those triples (x, y, z) satisfying $x + y + z = \pi$. The problem asks us to find the triples $(\alpha_0, \beta_0, \gamma_0)$ for which $(\alpha_n, \beta_n, \gamma_n)$ is a permutation of $(\alpha_0, \beta_0, \gamma_0)$.

Let us draw an equilateral triangular region S with altitude of length π (Fig. 12.44). To every triple (α, β, γ) with $\alpha + \beta + \gamma = \pi$ and $\alpha > 0, \beta > 0, \gamma > 0$, we associate a unique point P inside S such that the distances from P to the sides are, respectively, α, β, and γ. Thus the interior of S parametrizes triangles up to a

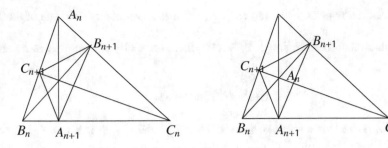

Fig. 12.43 The map H

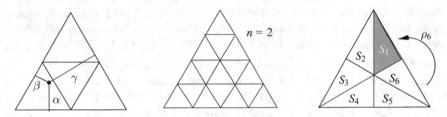

Fig. 12.44 The parametrization of triangles using an equilateral triangle

similarity. Moreover, the segments joining the midpoints of the sides parametrize right triangles.

Examining the geometric content of the above recursions, we deduce that

- H is not defined on the segments joining the midpoints,
- H maps the interior of each of the 4 triangles determined by segments that join the midpoints by a *homothety* into the original triangle (this homothety has ratio 2 for the triangles in the corners and -2 for the triangle in the middle).

Inductively we prove that H^n is not defined on the lines parallel to sides that divide the altitudes in 2^n equal parts, and on the interior of each of the 2^{2n} triangles determined by these lines, H^n is a homothety that maps that triangle to S (the case $n = 2$ is shown in the middle of Fig. 12.44).

Again, because we are only interested in triangles up to similarity, we restrict ourselves to the case where $\alpha_0 \geq \beta_0 \geq \gamma_0$. The region inside S parametrizing such triangles is one of the six congruent right triangles determined by sides and altitudes. Denote by S_1 this triangle and by S_2, S_3, S_4, S_5, S_6 the other five. Let also ρ_j, $j = 1, 2, \ldots, 6$, be the *isometries* of S that map S_j, Respectively, to S_1 (Fig. 12.44). The fact that T_n is similar to T_0 is equivalent to the fact that $(\rho_j \circ H^n)(x) = x$ for some $x \in S_1$.

If each of the 2^{2n} triangles in which S is divided is further divided in six by its altitudes, then in each of the equilateral triangles appears a triangle that corresponds to S_1 and is similar to S_1 with similarity ratio $1/2^n$. Let these triangles be $R_1, R_2, \ldots, R_{2^n}$. Now extend H to the boundaries of these triangles and view them as closed regions. Note that $H^n : R_k \to S$ is a homothety onto one of the S_j, say S_{j_k}.

We claim that for each $j = 1, 2, \ldots, 2^n$, there is exactly one point x_k in R_k such that

$$(\rho_{j_k} \circ H^n)(x_k) = x_k.$$

The map $\rho_{j_k} \circ H^n$ is a spiral similarity. Then $\rho_{j_k}^{-1} : S_1 \to S_1$ is a *spiral similarity* with ratio $1/2^n$. We will show that $\rho_{j_k}^{-1}$ has a unique fixed point.

Uniqueness is obvious, since $\rho_{j_k}^{-1}$ shrinks distances. On the other hand, $\rho_{j_k}^{-1}$, being a spiral similarity of the plane with ratio different from 1, does have a fixed point, and this fixed point necessarily lies in S_1, because by choosing m sufficiently large, we can make sure that $(\rho_{j_k}^{-1})^m(S_1)$ contains the fixed point. The unique fixed point of $\rho_{j_k}^{-1}$ is the unique fixed point of ρ_{j_k} so the claim is proved.

However, not all points x_k are admissible, since some might lie on the lines that parametrize right triangles or degenerate triangles. In fact it is not hard to see that this is the case precisely when the triangle R_k has the larger leg on the larger leg of S_1, for in this case the fixed point is on the boundary, and hence yields a degenerate triangle. There are 2^n such triangles, which have to be removed. So the answer to the problem is

$$2^{2n} - 2^n.$$

Source *Kvant (Quantum)*, proposed by Nikolai Borisovich Vassiliev.

252 (a) *First solution.* Suppose without loss of generality that A is the closest vertex to the line EF, as in Fig. 12.45. Then, since $ABCD$, $ADEM$, and $ABFM$ are cyclic, we have the following chain of equalities of directed angles modulo $180°$:

$$\angle AME = \angle ADE = \angle ADC = \angle ABC = \angle ABF = \angle AMF,$$

so M, E, and F are collinear.

Second solution. To be in tune with the rest of the book, we can work with complex coordinates and write

$$\arg \frac{e - m}{f - m} = \arg \left(\frac{e - m}{a - m} : \frac{f - m}{a - m} \right) = \arg \left(\frac{e - d}{a - d} : \frac{f - b}{a - b} \right)$$

$$= \arg \left(\frac{c - d}{a - d} : \frac{c - b}{a - b} \right) = 0 \pmod{\pi},$$

where we have used again the cyclicity of $ABCD$, $ADEM$, and $ABFM$, and the collinearities of A, B, C and C, D, E. Hence M, E, F are collinear.

Fig. 12.45 Cyclic complete
quadrilateral

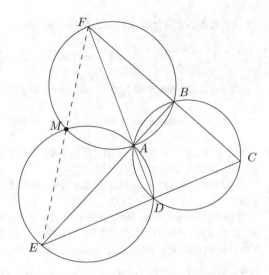

(b) Since the $\sqrt{bc}$ *inversion* swaps A with C, and B with D, the circumcircle of
$ABCD$ is taken to the circumcircle of $A'B'C'D' = CDAB$, which, of course, is
the same circle.

(c) From (b), since the circumcircle $ABCD$ is fixed by the $\sqrt{bc}$ *inversion*, we
deduce that its center O lies on the common bisector of $\angle AMC$, $\angle BMD$, and
$\angle EMF$.

We will now prove that O and P are swapped by the *inversion*, which will finish
this part of the problem. To see why this is true, it suffices to prove that O lies on
the circumcircles of both MAC and MBD, which are the images of AC and BD
under this inversion. If $MA \neq MC$, this is true because MO is the internal bisector
of $\angle AMC$, $OA = OC$, and the intersection of the perpendicular bisector of AC and
the internal bisector of $\angle AMC$ is the midpoint of the arc $\overset{\frown}{AC}$ of the circumcircle of
AMC that does not contain M. Such midpoint is O. If $MA = MC$, the inversion
circle goes through A and C, and the circumcircle of $ABCD$, being fixed by this
transformation, is orthogonal to the circle of inversion. This implies $OC \perp MC$
and $OA \perp MA$, which prove the claim in this case as well. We deduce that $MAOC$
and $MBOD$ are both cyclic, completing the proof.

(d) By (a) M lies on EF, so $\angle EMF = 180°$. Hence the bisector of $\angle EMF$
divides it into two angles of $90°$; since this common bisector is also the line OP, it
follows that $EF \perp OP$.

(e) Part (c) already establishes that $MAOC$ is cyclic. For the other part, let ω_{AC}
meet say, line AB again at X. Line AC divides the plane in two half-planes and will
be taken as reference: points are *opposites* if they are in opposite half-planes and
neighbors if they are in the same half-plane. For now, suppose that B and M are
opposites. Then either B and X are opposites (and X and M are neighbors), and

$$\angle BXC = \angle AXC = \angle AMC,$$

or B and X are neighbors (and X and M are opposites), so

$$\angle BXC = 180° - \angle AXC = \angle AMC$$

In any case, since M and O are always opposites,

$$\angle BXC = \angle AMC = 180° - \angle AOC = 180° - 2\angle ABC = 180° - 2\angle XBC,$$

which implies that triangle BXC is isosceles with $BX = XC$, that is, X lies in the perpendicular bisector of BC.

The case in which B and M are neighbors is handled similarly and is left to the reader.

Finally, notice how the pattern works: every side contains exactly one of points A, C and another point $P \in \{B, D\}$; then ω_{AC} meets the perpendicular bisector of the side at the side through P and the other point from A, C.

253 We work in the complex plane and assign the corresponding lowercase letters to uppercase points. We choose the coordinate system discussed in the introduction to this section on complete quadrilaterals (Sect. 4.2), with the Miquel point at the origin and such that $c = 1/a$ and $d = 1/b$. Rotating the figure appropriately, we can make the common bisector of $\angle AMC$ and $\angle BMD$ be the real axis. Then p is a positive real number, that is, $\bar{p} = p$. Since P lies on AC,

$$\begin{vmatrix} p & p & 1 \\ a & \bar{a} & 1 \\ \frac{1}{a} & \frac{1}{\bar{a}} & 1 \end{vmatrix} = 0 \iff p = \frac{a + \bar{a}}{1 + a\bar{a}} \iff \left(a - \frac{1}{p} \right)\left(\bar{a} - \frac{1}{p} \right) = \frac{1 - p^2}{p^2}$$

$$\iff \left| a - \frac{1}{p} \right| = \frac{\sqrt{1 - p^2}}{p}.$$

In the above formula for p, if we divide both the numerator and the denominator by $a\bar{a}$, we obtain

$$p = \frac{a + \bar{a}}{1 + a\bar{a}} = \frac{1/a + 1/\bar{a}}{1 + 1/(a\bar{a})},$$

so

$$\left| \frac{1}{a} - \frac{1}{p} \right| = \frac{\sqrt{1 - p^2}}{p}$$

as well.

Similarly,

$$\left| b - \frac{1}{p} \right| = \frac{\sqrt{1-p^2}}{p} \quad \text{and} \quad \left| \frac{1}{b} - \frac{1}{p} \right| = \frac{\sqrt{1-p^2}}{p}.$$

We deduce that A, B, C, and D lie on the circle with center $1/p$ and radius $\frac{\sqrt{1-p^2}}{p}$.

Remark We can read in this formula what has been proved in the solution to Problem 252, namely, that the center of the circle is the *inverse* of P. In fact this problem is the reciprocal of part (c) of Problem 252.

254 *First solution.* Let the circumcircles of AKN and CKN be ω_1 and ω_2, respectively. Let us first consider the case where ω_1 is tangent to ω and, arguing on Fig. 12.46, prove that ω_2 is tangent to ω.

Because ω and ω_1 are interior tangent, there is a direct *homothety* that maps the first into the second. This homothety maps BD to KN, so BD is parallel to KN. But then there is an inverse homothety that maps CBD to CNK, and this homothety maps ω to ω_2. This is possible only if the two circles are tangent. The reciprocal is similar.

Second solution. Perform the $\sqrt{bc}$ inversion that fixes the complete quadrilateral. Then A and C are swapped, B and D are swapped, K and N are swapped, and ω is fixed. Therefore the circumcircles ω_1 of AKN and ω_2 of CKN are mapped into each other, and ω and ω_1 are tangent if and only if $\omega' = \omega$ and $\omega'_1 = \omega_2$ are tangent.

Source Russian Mathematical Olympiad (regional level), 2001, proposed by L. Emelyanov.

255 We argue with the help of Fig. 12.47. This problem is a showcase for *spiral similarities* and Miquel points as discussed in Chapter 2. In fact, the ratios

$$\frac{AE}{ED} = \frac{BF}{FC}$$

suggest that we consider the center M of the spiral similarity that takes A to B, D to C, and, by proxy, E to F.

Fig. 12.46 ω_1 is tangent to ω

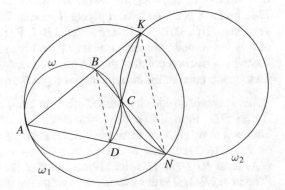

Fig. 12.47 Spiral similarities
and Miquel points all around

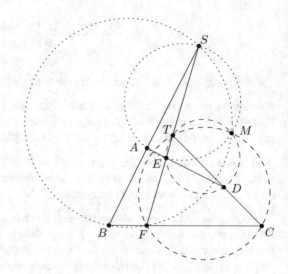

We already know that if a spiral similarity takes X to X' and Y to Y', then its center is the Miquel point of $XYY'X'$. Therefore M is the Miquel point of $AEFB$, so, by Miquel's Theorem (Theorem 2.24), M lies on the circumcircles of SAE and SBF, and also M is the Miquel point of $EDCF$, so M lies on the circumcircles of TCF and TDE. We are done.

Source United States of America Mathematical Olympiad, 2006, proposed by Zuming Feng and Zhonghao Ye.

256 We reason on Fig. 12.48 and solve the problem in a more general case, replacing the assumption that $BC = AD$ and $BE = DF$ with the only condition that

$$\frac{BE}{EC} = \frac{DF}{FA}.$$

We are now in a situation very similar to that of the previous problem, and it is natural to consider the *spiral similarity* s that maps the segment BC to DA, because $s(E) = F$. Applying Theorem 2.22 successively to the pairs of segments (BC, DA), (BE, DF), and (EC, FD), we deduce that the center O of s is the second intersection point of the circumcircles of PBC and PDA, the second intersection point of the circumcircles of RBE and RFD, and the second intersection point of the circumcircles of QAF and QEC.

Two complete quadrilaterals arise in the configuration, $BEQPRC$ and $APRFQD$. From the above considerations, we deduce that O is the second intersection point of the circumcircles of PBC and RBE, so it is the Miquel point of $BEQPRC$, and also O is the second intersection point of the circumcircles of QAF and RFD, so it is the Miquel point of $APRFQD$. By Miquel's Theorem (Theorem 2.24), O is the second intersection point of the circumcircles of PQR and

Fig. 12.48 Spiral similarities
and Miquel points again!

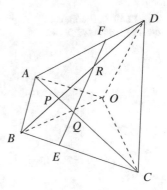

PBC and the second intersection point of the circumcircles of PQR and PAD. So
the circumcircles of all triangles PQR pass through the second intersection point
of the circumcircles of PAD and PBC, and this point is fixed because so are the
two triangles.

Source International Mathematical Olympiad, 2005, proposed by W. Pompe,
Poland.

257 *First solution.* Examining Fig. 12.49, we recognize Q as the Miquel point of
the complete quadrilateral $AMPNBC$. So in addition to the direct and inverse
homotheties that map AMN to ABC and MNP to CBP, which are manifest in
the statement of the problem, we also have the *spiral similarities* that map QAB to
QCP and QAC to QBP. Switching to complex coordinates and working with the
convention that the lowercase letter is the coordinate of the uppercase point, we can
use the abovementioned transformations to write

$$\frac{a-n}{p-n} = \frac{a-n}{m-n} \cdot \frac{m-n}{p-n} = \frac{a-c}{b-c} \cdot \frac{c-b}{p-b} = \frac{a-c}{b-p} = \frac{a-q}{b-q}.$$

So the triangles ANP and AQB are similar, and consequently $\angle PAN = \angle BAQ$,
as desired.

Second solution. Let ϕ_A be the $\sqrt{bc}$ *inversion* corresponding to the vertex A in
the triangle ABN. Recall that ϕ_A is obtained as a composition of the inversion χ
of center A and radius $\sqrt{AB \cdot AN}$ and the reflection σ with respect to the bisector
of $\angle BAN$. From the fact that AB and MN are parallel, we obtain that $AB \cdot AN =
AC \cdot AM$, so $\phi_A(C) = M$. The circumcircles of ABN and ACM are therefore
images of the lines CM and BN, respectively.

Examining once more the complete quadrilateral $ANPMBC$ from Fig. 12.49,
we notice that Q is its Miquel point, where the circumcircles of ACM, ABN, CNP,
and BMP meet, so Q is the second intersection point of the circumcircles of ACM
and ABN. Since this point is mapped by ϕ_A to the intersection point of CM and
BN, we have $\phi_A(Q) = P$. Consequently the lines AP and AQ are the images
of one another through ϕ_A, and since the lines through A are invariant under the

Fig. 12.49 The complete
quadrilateral $AMPNBC$

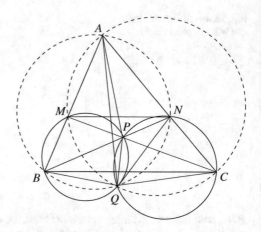

inversion χ, the lines AP and AQ are mapped into one another by the reflection σ (they are isogonals), and the equality of the two angles follows.

Remark So at the heart of the first solution lies the observation that all the information about the triangle ABC, up to a similarity, is encoded in the expression

$$\frac{c-a}{b-a}.$$

For any triangle that is similar to ABC, this expression is the same.

Source Balkan Mathematical Olympiad, 2007, proposed by Moldova.

258 We recognize the flavor of Problem 255, and inspired by that problem, we argue as follows. Because

$$\frac{AP}{AC} + \frac{BQ}{BD} = 1$$

is equivalent to

$$\frac{AP}{AC} = \frac{DQ}{BD},$$

we employ the *spiral similarity* that takes A to D and C to B and which therefore also takes P to Q. The center O of this spiral similarity is the Miquel point of $ACBD$ (it is the Miquel point of a skew quadrilateral).

Now we reason on Fig. 12.50. Recall that spiral similarities come in pairs. In fact, there is an (automatic) spiral similarity with center O that takes B to C and Q to P, so O is the Miquel point of $BQPC$ and belongs to the circumcircles of CNP and BNQ.

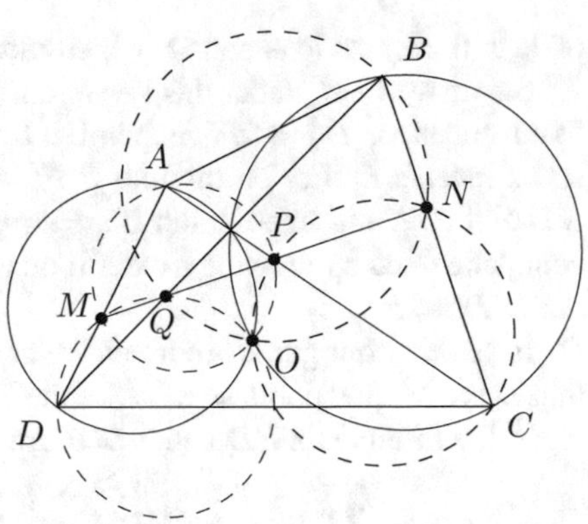

Fig. 12.50 More spiral
similarities and Miquel points
all around

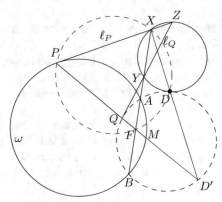

Fig. 12.51 Asian Pacific
Mathematical Olympiad
problem via inversion

Another automatic spiral similarity, that takes A to D and P to Q, finishes the
problem, as O is also the Miquel point of $APQD$ and belongs to the circumcircles
of AMP and DMQ.

Source Bulgarian Team Selection Test for the International Mathematical
Olympiad, 2004.

259 *First solution.* Our first solution uses *inversion*. Let D be the Miquel point of
the complete quadrilateral determined by ℓ_P, ℓ_Q, AB, and MP. A good diagram
(such as Fig. 12.51) suggests that the tangency point is D. We also define X as
$\ell_P \cap AB$, Y as $\ell_Q \cap AB$, and Z as $\ell_P \cap \ell_Q$. Also, let F be the intersection of PQ
and AB.

We start by noticing that, since AB is parallel to the tangent line to ω at M,

$$\angle XPF = \frac{1}{2}\ \widehat{PAM} = \frac{1}{2}(\widehat{PA} + \widehat{AM}) = \frac{1}{2}(\widehat{PA} + \widehat{MB}) = \angle XFP,$$

so $XP = XF$. Taking advantage of the fact that X lies on the radical axis of ω and
Ω, we invert with center X and radius equal to the square root of the common power

of X with respect to ω and Ω (which radius is $XP = XF$). As usual, we denote by K' the image of K under this inversion. Then $P' = P$, $F' = F$, ω, Ω, and AB are fixed, and since D is a Miquel point, it lies on the circumcircle of the triangle XPF, so its inverse D' lies on the line $P'F' = PF$. Therefore D' is the intersection of XD and PQ, and since A and B are swapped by the inversion, A, B, D, and D' are concyclic. And by drawing a careful diagram, we can guess that the circumcircle of $ABD'D$ is Ω!

To see why our guess is correct, we consider the circumcircle Γ of $DD'Q$. Recall that D is a Miquel point, so it lies on the circumcircle of YQF. Because $XD \cdot XD' = XF^2$, the triangles XFD and $XD'F$ are similar, so

$$\angle DD'Q = \angle XD'F = \angle DFX = \angle DFY = \angle DQY.$$

This implies that ℓ_Q is tangent to Γ at Q. Finally, notice that $XQ < XF$ implies that $XQ' > XF > XQ$ so $Q' \neq Q$. From $XD \cdot XD' = XQ \cdot XQ'$, using power-of-a-point, we deduce that Q' also lies on Γ. But Q' lies on Ω as well, because Ω is fixed by the inversion. And because two distinct points of a circle together with the tangent line at one of the points uniquely determine the circle, $\Gamma = \Omega$, and D lies on Ω.

Thus we have proved that D is on both Ω and the circumcircle γ of XYZ. It remains to show that D is a tangency point. The inversion helps here too: γ is mapped to the line $D'Y'$, so we need to show that this line is tangent to $\Omega' = \Omega$. Another angle chasing completes the solution: since $\angle XK'L' = \angle XLK$ for any points K, L,

$$\angle DD'Y' = \angle XD'Y' = \angle XYD = 180° - \angle DYF$$

$$= 180° - \angle DQF = 180° - \angle DQD' = \frac{1}{2}\,\overset{\frown}{D'D}.$$

Second solution. The official solution makes use of the idea that two tangent circles are *homothetic* with respect to an inverse homothety centered at the tangency point. We use the same notation as in the previous solution. The main idea is to construct a triangle inscribed in Ω that is (hopefully) homothetic to our main triangle XYZ. Let PQ meet Ω again at $R \neq Q$, and let us define the points S and T on Ω such that $RS \parallel AB$ and $RT \parallel \ell_P$ (see Fig. 12.52.) Then RST and XYZ are homothetic if and only if $ST \parallel YZ$, namely, if $ST \parallel \ell_Q$.

Fig. 12.52 Asian-Pacific
Mathematical Olympiad
problem via homothety

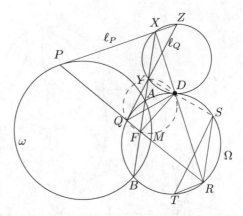

As in the previous solution $\angle FPX = \angle PFX$, so

$$\angle PRT = \angle RPX = \angle FPX = \angle PFX = \angle(AB, PR) = \angle(RS, PR) = \angle SRP.$$

We infer that Q is the midpoint of the arc $\overset{\frown}{SQT}$ of Ω. Therefore, $ST \parallel \ell_Q$, and RST and XYZ are homothetic. This homothety also takes the circumcircle of RST, Ω, to the circumcircle of XYZ, γ.

Now we only need to prove that D, the other intersection point of the circle Ω with the line XR, is the center of homothety. This happens if and only if D belongs to the line YS. Power-of-a-point and some angle chasing will do: since $XF^2 = XP^2 = XA \cdot XB = XD \cdot XR$, the triangles XFD and XRF are similar, so

$$\angle DFY = \angle DFX = \angle FRX = \angle QRD = \angle DQY,$$

that is, $DFQY$ is cyclic. Finally,

$$\angle YDQ = \angle YFQ = \angle(AB, QR) = \angle(SR, QR) = \angle SRQ = 180° - \angle SDQ,$$

so D lies on SY, and we are done.

Source Asian Pacific Mathematical Olympiad, 2014, proposed by Ilya Igorevich Bogdanov, Russia, and Medeubek Kungozhin, Kazakhstan.

260 Let the complete quadrilateral be obtained from the convex quadrilateral $BCB'C'$ by intersecting BC' with $B'C$ at A and BC with $B'C'$ at A', with B' on the segment AC and B on the segment $A'C$, as shown in Fig. 12.53. On this figure we have marked the respective circumcenters O, O_a, O_b, O_c of the triangles ABC, $AB'C'$, $BA'C'$, $CA'B'$, as well as the Miquel point M of the complete quadrilateral.

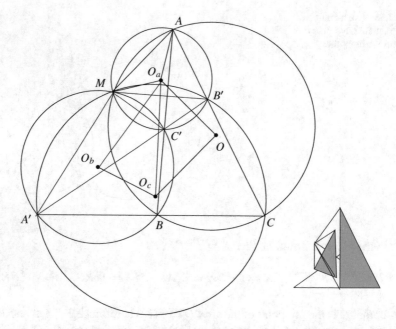

Fig. 12.53 The complete quadrilateral with circumcenters and the Miquel point

Theorem 4.7 proves the existence of the *spiral similarities* s, s_a, s_b, s_c centered at M that map the following pairs of triangles into each other:

$$ABC \mapsto O_aO_bO_c, \quad AB'C' \mapsto OO_bO_c, \quad A'BC' \mapsto O_aOO_c, \quad A'B'C \mapsto O_aO_bO.$$

Also, by Theorem 4.6 the quadrilateral $OO_aO_bO_c$ is cyclic, and its circumcircle passes through M.

Looking at the configuration from a different angle, we denote by H', H'_a, H'_b, H'_c the orthocenters of the triangles $O_aO_bO_c$, O_bO_cO, O_cOO_a, and OO_aO_b, respectively, and then, as shown in Problem 23, the quadrilateral $H'H'_aH'_bH'_c$ is the image of the quadrilateral $OO_aO_bO_c$ under a *reflection* over a point, and this reflection (which is a 180° rotation) we denote by σ. Let O_m and H_m be the circumcenters of $OO_aO_bO_c$ and $H'H'_aH'_bH'_c$, respectively, which correspond to each other through this reflection.

Now, the composition $s' = \sigma \circ s$ is a spiral similarity, being the composition of two spiral similarities, whose ratio is the same as that of s and whose angle is equal to the angle of s plus 180° (which is the angle of σ). We have

$$s'(O) = \sigma \circ s(O) = \sigma(O_m) = H_m,$$
$$s'(H) = \sigma \circ s(H) = \sigma(H') = O.$$

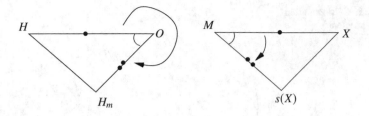

Fig. 12.54 Graphic aid for understanding the transformation $s' = \sigma \circ s$

It follows that $s'(HO) = OH_m$ and from here we can deduce that the triangle OHH_m is similar to all triangles of the form $MXs(X)$, which define the spiral similarity s (look at Fig. 12.54 to understand why this is true). In particular OHH_m is similar to MAO_a. And since the triangle MAO_a is isosceles because O_aM and O_aA are radii in the circle of center O_a, it follows that the triangle H_mOH is isosceles, with $H_mO = H_mH$. But then H_m is on the perpendicular bisector of OH, which, by Theorem 2.10, is the perpendicular to the Euler line of ABC at the center of its nine-point circle. A similar argument shows that the other three perpendiculars pass through the circumcenter H_m of $H'H'_aH'_bH'_c$, and we are done.

Source This result was proved by S. Kantor in *Bulletin des Sciences Mathématiques* in 1879 and by R.J. Hervey in *Educational Times*, 1891, the solution is taken from G. Mihalescu, *Geometria Elementelor Remarcabile (The Geometry of the Remarkable Elements)*, Ed. Tehnică, Bucharest, 1957.

261 We first argue on Fig. 12.55. By definition, D lies on the bisector of angle $\angle BAC$. The angles $\angle ADE$ and $\angle BCD$ feel a little "off" with each other. So we define D' as the isogonal conjugate of D in the triangle ABC. Then

$$\angle ECD' = \angle ACD' = \angle DCB = \angle EDA = \angle EDD',$$

and we discover the much friendlier cyclic quadrilateral $ECD'D$. And for a similar reason, the quadrilateral $FBD'D$ is cyclic as well.

Notice that A lies on the radical axis DD' of the circumcircles of $ECD'D$ and $FBD'D$, so by power-of-a-point $AF \cdot AB = AD' \cdot AD = AE \cdot AC$, and therefore the quadrilateral $BCEF$ is cyclic.

Now we can drop D' and complete the diagram, and we argue now on Fig. 12.56. Out of the three lines BC, EF, and O_1O_2, the "odd one out" is O_1O_2, so let the lines BC and EF meet at the point T. Our goal is to prove that T lies on O_1O_2.

Let M be the Miquel point of the cyclic quadrilateral $BCEF$. Notice that the line AT passes through M. As a consequence of Problem 252 (e), the point X, which is the intersection of the side EC of $BCEF$ and the perpendicular bisector of the side BC, lies on the circumcircle of BEM.

We take advantage of D being on both the circle with center O_1 and the circle with center O_2 and *invert* with center D and arbitrary radius. As usual, denote the

Fig. 12.55 Getting started
with the problem

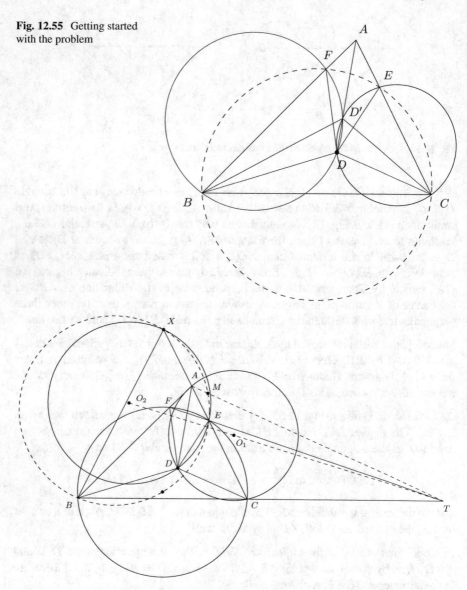

Fig. 12.56 Miquel point helps connecting X to the rest of the diagram

inverse of P by P' for any point $P \neq D$. Before we draw the inverted diagram, let
us locate the inverses by computing some angles. Recall that $\angle DP'Q' = \angle DQP$
for distinct points D, P, Q. Then

$$\angle F'B'C' = \angle F'B'D + \angle DB'C' = \angle BFD + \angle DCB.$$

Some additional angle chasing yields

$$\angle F'B'C' = \angle BFD + \angle DCB = \angle ADF + \angle BAD + \angle DCB$$

$$= \angle DBC + \angle DAC + \angle DCB,$$

which is symmetric with respect to B and C. So $\angle F'B'C' = \angle E'C'B'$. Since $BCEF$ is cyclic, so is $B'C'E'F'$, which means that $B'C'E'F'$ is an isosceles trapezoid. Or, speaking in terms of transformations, there is a *reflection* σ across a line ℓ that takes B' to $\sigma(B') = C'$ and E' to $\sigma(E') = F'$.

We argue now on Fig. 12.57. The reflection σ makes the inverse of almost all the other points fall into place: since P is the intersection of the lines EF and BC, P' is the second intersection of the circumcircles of $DE'F'$ and $DB'C'$, which by symmetry is $\sigma(D)$, and since the Miquel point M is the intersection of, say, the line AP and the circumcircle of AEF, M' is the intersection of the circumcircles of $DA'P'$ and $A'E'F'$; again by symmetry, $M' = \sigma(A')$.

Back to the problem, we need to prove that T and the circumcenters O_1 and O_2 of ADC and EDX, respectively, are collinear. Recall that the inverse of the center of a circle passing the center of inversion is the reflection of the center across the inverted circle, which is a line (Theorem 3.7). The inversion transforms the problem into proving that the four points D, T', and the reflections of D across $A'C'$ and $E'X'$ lie on a circle.

We know that T' is the reflection of D across ℓ, so ℓ, $A'C'$, and $E'X'$ are perpendicular bisectors of the line segments connecting D to the other three points. So it is sufficient to prove that ℓ, $A'C'$, and $E'X'$ meet at a single point.

Finally, since E, M, X, B lie on a circle, E', M', X', B' also lie on a circle α; since X, A, E, C lie on a line, X', A', E', C' lie on a circle β; and since M is a Miquel point, M, A, B, C lie on a circle, so M', A', B', C' also lie on a circle γ. Now, the radical axis of α and β is $E'X'$, the radical axis of α and γ is $M'B'$, and

Fig. 12.57 Inversion makes (almost) every point nicer

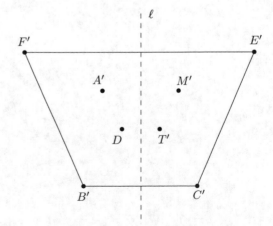

the radical axis of β and γ is $A'C'$. These three lines meet at the radical center of the three circles. This radical center is also the intersection point of $A'C'$ and $M'B'$, which by symmetry lies on ℓ. We are done.

Source International Mathematical Olympiad, 2021, proposed by Mykhailo Shtandenko, Ukraine.

Index

A

Affine transformation, 87, 101, 112, 207, 528
Apollonian circles, 36, 113, 158, 187, 190,
  415, 416, 443
Averaging Principle, 101, 102, 121, 124, 125,
  258, 265, 266, 418–421, 425, 515, 522,
  532, 536

C

Complete quadrilateral, 107, 108, 234–245,
  267, 405, 425, 428, 563–569, 571, 572
Complex affine transformations, 101, 102
Cross-ration, 12, 13, 35, 80, 131, 140, 141,
  143–145, 148, 173, 180, 183–187, 191,
  195, 196, 227–229, 240, 241, 261, 262,
  265, 441, 447, 449, 456, 466, 470, 471,
  479, 483, 484, 490, 493, 517, 526, 527,
  554

D

Directed angles, 11, 12, 114, 116, 120, 132,
  163, 188–191, 221, 223, 227, 240, 298,
  300, 301, 414, 430, 522, 523, 562

E

Euler's line, 79, 89, 90, 94–96, 123, 202, 233,
  245, 375, 379, 389, 394, 395, 482, 487,
  558, 559, 573

Excircle, 82, 83, 173–175, 177, 201, 204, 210,
  212, 255–257, 264, 374, 393, 400, 407,
  408, 454, 466, 494, 504, 505, 508, 511,
  547, 553, 555

G

Group of transformations, 7, 77, 112, 327, 518

H

Harmonic division, 79, 81, 131, 145, 159, 163,
  187, 229, 456, 466, 490, 493, 533
Harmonic pencil, 228, 229
Harmonic quadrilateral, 34–36, 145, 173, 196,
  227, 229, 466, 483, 484, 490, 493
Homothety, viii, 21, 71–128, 131, 136, 137,
  144, 146, 158, 160, 162, 163, 176, 177,
  207, 211–215, 222, 223, 236–238, 242,
  243, 255–267, 325, 369–437, 445, 451,
  452, 457, 459, 484, 487, 491, 501, 509,
  511, 517, 519, 523, 524, 528, 530, 531,
  534–538, 542–544, 550–552, 556–558,
  561, 562, 565, 567, 570, 571

I

International Mathematical Olympiad, ix, 89,
  123, 191, 193, 306, 310, 312, 332, 349,
  361, 388, 396, 408, 420, 427, 431–433,
  436, 451, 459, 464, 466, 468, 486, 507,
  530, 531, 539, 544, 545, 555–557, 567,
  569

Inversion, viii, 36, 82, 129–205, 207, 212,
    214, 215, 220, 222–224, 226, 227, 229,
    234–237, 239, 244, 261–267, 437–512,
    519, 522, 523, 528, 533, 543, 545–548,
    550, 551, 553–557, 563, 565, 567–570,
    575
Isogonal conjugate of a point, 36, 38, 39, 251
Isometries, viii, 3–69, 77, 109, 110, 112, 135,
    141–145, 149, 249–254, 266, 271–367,
    372, 410, 444, 513, 521, 540, 561

L
Linear fractional transformations, viii, 138,
    145–149, 183, 445

M
Mathematical Olympiads, ix, 46, 62, 89, 120,
    121, 196, 213, 239, 287, 292, 299, 303,
    308, 310, 315, 328, 330, 336, 349, 386,
    390, 396, 399, 409, 411, 423, 424, 435,
    445, 457, 473, 474, 477, 479, 481, 519,
    523, 538, 541, 565, 566, 568, 569, 571
Miquel's point, 107, 108, 234–237, 241, 243,
    244, 258, 267, 425, 428, 564–575
Mixtilinear circles, 176, 204, 481, 489
Möbius transformations, 138–145, 149, 151,
    152, 158–159, 171, 173, 180, 183–186,
    195, 196, 198–205, 207, 226, 227, 235,
    239, 240, 261, 437, 438, 441–444, 449,
    450, 469, 477, 481, 526, 528

N
Nine-point circle, 79–82, 94, 95, 153, 166,
    204, 210, 212, 226, 245, 256, 264, 379,
    391, 392, 394, 395, 472, 482, 486, 487,
    491, 494, 496, 497, 501, 512, 555, 556,
    573

O
Orientation of polygons, 15, 16

P
Polar of a point with respect to a circle, 191
Pole of a line with respect to a circle, 159, 160

R
Reflection, 3, 8–21, 24–28, 30–34, 36–39, 42,
    43, 49, 50, 54–56, 77, 80, 81, 94, 95,
    110–112, 121–123, 127, 129–131, 136,
    141–144, 151, 153, 154, 158, 166–168,
    170–172, 179–181, 187, 198, 207, 212,
    213, 220, 222, 227, 228, 230, 232, 233,
    236, 249–252, 254, 262–266, 271–278,
    284, 287–292, 294, 298–307, 315, 320,
    330, 332–335, 339, 340, 350, 353, 355,
    357, 359, 361, 362, 369, 374, 379, 380,
    383, 387, 388, 390, 392, 393, 425, 437,
    441, 442, 457, 461, 471, 472, 479–482,
    486, 487, 491, 492, 494, 496–498,
    507, 508, 514, 517–520, 522–524, 529,
    533–540, 543–546, 548, 550, 553,
    556–558, 567, 568, 572, 575
Rotation, 8–19, 21–30, 32, 39, 40, 42, 43, 45,
    46, 48, 49, 58, 59, 62, 64, 65, 69, 71,
    100, 101, 104, 109, 110., 112, 113,
    118, 124, 126, 144, 148, 157, 207–209,
    250–254, 265, 266, 271, 273–276,
    279–285, 294–297, 302–305, 308,
    310–312, 322–327, 333, 338, 339, 346,
    347, 350, 353–355, 358–365, 420, 426,
    435, 464, 494, 513–515, 521, 523, 524,
    531, 533, 534, 538, 539, 572

S
Simson line, 164, 231, 236, 265, 523, 530
Spiral similarity, 71–128, 135, 141–144, 149,
    171, 172, 176, 186, 189, 207, 210, 216,
    218–220, 226, 232, 234, 235, 237,
    243, 255–259, 265–267, 369–436, 450,
    513, 515, 516, 520, 523–525, 527–532,
    534–537, 539–541, 544, 553, 555, 562,
    566, 567, 569, 572
Symmedian, 32–36, 53, 81, 105–107, 113,
    119, 173, 201, 202, 220, 225–227, 250,
    258, 264, 283, 285, 415, 429, 430, 456,
    466, 473, 479, 490, 491, 493, 499, 502,
    503, 506, 507, 540, 541, 546, 547

T
Translation, 8–30, 39, 43, 49, 57, 63, 67, 75,
    77, 78, 84, 86, 88, 101–104, 109, 110,
    112, 113, 131, 141–144, 146, 186, 192,
    207, 209, 249–253, 255, 271–275, 277,
    282, 286, 293, 297, 298, 309, 311–313,
    315, 321, 322, 324, 326–330, 333, 334,
    336–338, 341, 343–345, 348, 350, 355,
    356, 361, 362, 365, 369, 371, 372, 393,
    398, 418, 445, 513, 557

Printed in the United States
by Baker & Taylor Publisher Services